Applied Mathematics

Applied Mathematics

Third Edition

J. David Logan

Willa Cather Professor
Department of Mathematics
University of Nebraska at Lincoln

WILEY-
INTERSCIENCE

A JOHN WILEY & SONS, INC., PUBLICATION

Library of Congress Cataloging-in-Publication Data:

Logan, J. David (John David)
   Applied mathematics / J. David Logan.—3rd ed.
      p. cm.
   Includes bibliographical references and index.
   ISBN-13 978-0-471-74662-1 (acid-free paper)
   ISBN-10 0-471-74662-2 (acid-free paper)
   1. Mathematics—Textbooks.   I. Title.

   QA37.3.L64 2006
   510—dc22   2005058223

Dedicated to

my mother, Dorothy Elizabeth Logan, and my father, George E. Logan (1919–1996)

Contents

Preface

This third edition of *Applied Mathematics* shares the same goals, philosophy, and writing style as the original edition, which appeared in 1987—namely, to introduce key ideas about mathematical methods and modeling to seniors and graduate students in mathematics, engineering, and science. The emphasis is upon the interdependence of mathematics and the applied and natural sciences. The prerequisites include elementary, sophomore-level courses in differential equations and matrices.

There are major changes in this new third edition. Summarizing, the material has been rearranged and basically divided into two parts. Chapters 1 through 4 involve models leading to ordinary differential equations and integral equations, while Chapters 6 through 8 focus on partial differential equations and their applications. Motivated by problems in the biological sciences, where quantitative methods are becoming central, there is a new chapter (Chapter 5) on discrete-time models, which includes some material on stochastic processes. Two sections reviewing elementary methods for solving systems of ordinary differential equations have been added at the end of Chapter 1. Many new exercises appear throughout the book. The Table of Contents details the specific topics covered.

My colleagues in Lincoln, who have often used the text in our core sequence in applied mathematics, deserve special thanks. Tom Shores provided me with an extensive *errata*, and he, Steve Cohn, and Glenn Ledder supplied several new exercises from graduate qualifying examinations, homework, and course exams. Bill Wolesensky read some of the manuscript and was often a sounding board for suggestions. Also, I wish to thank those who used earlier editions of the book and helped establish it as one of the basic textbooks in the area; many have given me corrections and suggestions. Any comments, or otherwise, will be greatly appreciated, and they may be forwarded to me at *dlogan@math.unl.edu*. Finally, my wife Tess has been a constant source of support for my writing, and I take this opportunity to publicly express my appreciation for her encouragement and affection.

Suggestions for use of the text. Chapters 1 through 4 can form the basis of a one-semester course involving differential and integral equations and the basic core of applied mathematics. Chapter 3, on calculus of variations, is independent from the others, so it may be deleted or covered in part. Portions of Chapter 5, on discrete models could be added to this course. A second semester, involving partial differential equation models, would cover Chapters 6, 7, and 8. Students may take the second semester, as is often done at the University of Nebraska, without having the first, provided a small portion of Chapter 1 on scaling is briefly reviewed.

There is much independence among the chapters, enabling an instructor to design his or her special one-semester course in applied mathematics.

David Logan
Lincoln, Nebraska

1
Dimensional Analysis, Scaling, and Differential Equations

The techniques of dimensional analysis and scaling are basic in the theory and practice of mathematical modeling. In every physical setting a good grasp of the possible relationships and comparative magnitudes among the various dimensioned parameters nearly always leads to a better understanding of the problem and sometimes points the way toward approximations and solutions. In this chapter we introduce some of the basic concepts from these two topics. A statement and proof of the fundamental result in dimensional analysis, the Pi theorem, is presented, and scaling is discussed in the context of reducing problems to dimensionless form. The notion of scaling points the way toward a proper treatment of perturbation methods, especially boundary layer phenomena in singular perturbation theory.

The second half of this chapter, Sections 1.3 and 1.4, contains a review of differential equations. This material may be perused or skipped if readers are familiar with the basic concepts and elementary solution methods.

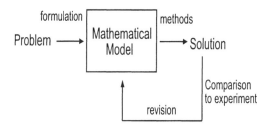

Figure 1.1 Schematic of the modeling process.

1.1 Dimensional Analysis

1.1.1 The Program of Applied Mathematics

Applied mathematics is a broad subject area in the mathematical sciences dealing with those topics, problems, and techniques that have been useful in analyzing real-world phenomena. In a very limited sense it is a set of methods that are used to solve the equations that come out of science, engineering, and other areas. Traditionally, these methods were techniques used to examine and solve ordinary and partial differential equations, and integral equations. At the other end of the spectrum, applied mathematics is applied analysis, or the theory that underlies the methods. But, in a broader sense, applied mathematics is about mathematical modeling and an entire process that intertwines with the physical reality that underpins its origins.

By a **mathematical model** we mean an equation, or set of equations, that describes some physical problem or phenomenon that has its origin in science, engineering, economics, or some other area. By **mathematical modeling** we mean the process by which we formulate and analyze the model. This process includes introducing the important and relevant quantities or variables involved in the model, making model specific assumptions about those quantities, solving the model equations by some method, and then comparing the solutions to real data and interpreting the results. Often the solution method involves computer simulation or approximation. This comparison may lead to revision and refinement until we are satisfied that the model accurately describes the physical situation and is predictive of other similar observations. This process is depicted schematically in Fig. 1.1. Therefore the subject of mathematical modeling involves physical intuition, formulation of equations, solution methods, and analysis. A good mathematical model is simple, applies to many situations, and in predictive.

Overall, in mathematical modeling the overarching objective is to make sense of the world as we observe it, often by inventing caricatures of reality. Scientific exactness is sometimes sacrificed for mathematical tractability. Model predictions depend strongly on the assumptions, and changing the assumptions changes the model. If some assumptions are less critical than others, we say the model is robust to those assumptions. They help us clarify verbal descriptions of nature and the mechanisms that make up natural law, and they help us determine which parameters and processes are important, and which are unimportant.

Another issue is the level of complexity of a model. With modern computer technology it is tempting to build complicated models that include every possible effect we can think of, with large numbers of parameters and variables. Simulation models like these have their place, but computer runs do not always allow us to discern which are the important processes, and which are not. Of course, the complexity of the model depends upon the data and the purpose, but it is usually a good idea to err on the side of simplicity, and then build in complexity as it is needed or desired.

Finally, authors have tried to classify models in several ways—stochastic vs. deterministic, continuous vs. discrete, static vs. dynamic, quantitative vs. qualitative, descriptive vs. explanatory, and so on. In our work we are interested in modeling the underlying reasons for the phenomena we observe (explanatory) rather than fitting the data with formulas (descriptive) as is often done in statistics. For example, fitting measurements of the size of an animal over its lifetime by a regression curve is descriptive, and it gives some information. But describing the dynamics of growth by a differential equation relating growth rates, food assimilation rates, and energy maintenance requirements tells more about the underlying processes involved.

The reader is already familiar with many models. In an elementary science or calculus course we learn that Newton's second law, force equals mass time acceleration, governs mechanical systems like falling bodies; Newton's inverse-square law of gravitation describes the motion of the planets; Ohm's law in circuit theory dictates the voltage drop across a resistor in terms of the current; the law of mass action in chemistry describes how fast chemical reactions occur; or the logistics equation models growth and competition in population.

The first step in modeling is to select the relevant variables (independent and dependent) and parameters that describe the problem. Physical quantities have *dimensions* like time, distance, degrees, and so on, or corresponding *units* like seconds, meters, and degrees Celsius. The equations we write down as models must be dimensionally correct. Apples cannot equal oranges. Verifying that each term in our model has the same dimensions is the first task in obtaining a correct equation. Also, checking dimensions can often give us insight into what a term in the model might be. We always should be aware of the dimensions of the quantities, both variables and parameters, in a model, and we should always try to identify the physical meaning of the terms in the equations we obtain. A general rule is to always let the physical problem drive the mathematics, and not vice-versa.

It would be a limited view to believe that applied mathematics consists only of developing techniques and algorithms to solve problems that arise in a physical or applied context. Applied mathematics deals with all the stages of the modeling process, not merely the formal solution. It is true that an important aspect of applied mathematics is studying, investigating, and developing procedures that are useful in solving mathematical problems: these include analytic and approximation techniques, numerical analysis, and methods for solving differential and integral equations. It is more the case, however, that applied mathematics deals with all phases of the problem. Formulating the model and understanding its origin in empirics are crucial steps. Because there is a constant interplay between the various stages, the scientist, engineer, or mathematician must understand each phase. For example, the solution stage sometimes involves making approximations that lead to a simplification; the approximations often come from a careful examination of the physical reality, which in turn suggests what terms may be neglected, what quantities (if any) are small, and so on. The origins and analysis are equally important. Indeed, physical insight forces us toward the right questions and at times leads us to the theorems and their proofs. In fact, mathematical modeling has been one of the main driving forces for mathematics itself.

In the first part of this chapter our aim is to focus on the first phase of the modeling process. Our strategy is to formulate models for various physical systems while emphasizing the interdependence of mathematics and the physical world. Through study of the modeling process we gain insight into the equations themselves. In addition to presenting some concrete examples of modeling, we also discuss two techniques that are useful in developing and interpreting the model equations. One technique is *dimensional analysis*, and the other is *scaling*. The former permits us to understand the dimensional (meaning length, time, mass, etc.) relationships of the quantities in the equations and the resulting implications of dimensional homogeneity. Scaling is a technique that helps us understand the magnitude of the terms that appear in the model equations by comparing the quantities to intrinsic reference quantities that appear naturally in the physical situation.

1.1.2 Dimensional Methods

One of the basic techniques that is useful in the initial, modeling stage of a problem is the analysis of the relevant quantities and how they must relate to each other in a dimensional way. Simply put, apples cannot equal oranges; equations must have a consistency to them that precludes every possible relationship among the variables. Stated differently, equations must be dimensionally homogeneous. These simple observations form the basis of the subject known as **dimensional analysis**. The methods of dimensional analysis have led to important results in determining the nature of physical phenomena, even when the governing equations were not known. This has been especially true in continuum mechanics, out of which the general methods of dimensional analysis evolved.

The cornerstone result in dimensional analysis is known as the **Pi theorem**. The Pi theorem states that if there is a physical law that gives a relation among a certain number of dimensioned physical quantities, then there is an equivalent law that can be expressed as a relation among certain dimensionless quantities (often noted by $\pi_1, \pi_2, \ldots$, and hence the name). In the early 1900s, E. Buckingham gave a proof of the Pi theorem for special cases, and now the theorem often carries his name. Birkhoff (1950) can be consulted for a bibliography and history.

To communicate the flavor and power of this classic result, we consider a calculation made by the British applied mathematician G. I. Taylor in the late 1940s to compute the yield of the first atomic explosion after viewing photographs of the spread of the fireball. In such an explosion a large amount of energy E is released in a short time (essentially instantaneously) in a region small enough to be considered a point. From the center of the explosion a strong shock wave spreads outward; the pressure behind it is on the order of hundreds of thousands of atmospheres, far greater than the ambient air pressure whose magnitude can be accordingly neglected in the early stages of the explosion. It is plausible that there is a relation between the radius of the blast wave front r, time t, the initial air density ρ, and the energy released E. Hence, we *assume* there is a physical law

$$g(t, r, \rho, E) = 0, \tag{1.1}$$

which provides a relationship among these quantities. The Pi theorem states that there is an equivalent physical law between the independent dimensionless quantities that can be formed from t, r, E, and ρ. We note that t has dimensions of time, r has dimensions of length, E has dimensions of mass $\cdot$ length$^2 \cdot$ time^{-2}, and ρ has dimensions of mass $\cdot$ length^{-3}. Hence, the quantity $r^5 \rho / t^2 E$ is dimensionless, because all of the dimensions cancel out of this quantity (this is easy to check). It is not difficult to observe that no other independent dimensionless quantities can be formed from t, r, e, and ρ. The Pi theorem then guarantees that the physical law (1.1) is equivalent to a physical law involving only the dimensionless quantities; in this case

$$f\left(\frac{r^5 \rho}{t^2 E}\right) = 0, \tag{1.2}$$

because there is only one such quantity, where f is some function of a single variable. From (1.2) it follows that the physical law must take the form (a root of (1.2))

$$\frac{r^5 \rho}{t^2 E} = C,$$

or

$$r = C\left(\frac{E t^2}{\rho}\right)^{1/5}, \tag{1.3}$$

where C is a constant. Therefore, just from dimensional reasoning it has been shown that the radius of the wave front depends on the two-fifths power of time. Experiments and photographs of explosions confirm this dependence. The constant C depends on the dimensionless ratio of the specific heat at constant pressure to the specific heat at constant volume. By fitting the curve (1.3) to experimental data of r versus t, the initial energy yield E can be computed, since C and ρ are known quantities. (See Exercise 3.) Although this calculation is only a simple version of the original argument given by Taylor, we infer that dimensional reasoning can give crucial insights into the nature of a physical process and is an invaluable tool for the applied mathematician, scientist, or engineer.

As an aside, there is another aspect of dimensional analysis that is important in engineering, namely the design of small-scale models (say of an airplane or ship) that behave like their real counterparts. A discussions of this important topic is not treated here, but can be found in many engineering texts, especially books on fluid mechanics.

Example 1.1

Those who bike have certainly noticed that the force F (mass times length, per time-squared) of air resistance appears to be positively related to their speed v (length per time) and cross-sectional area A (length-squared). But force involves mass, so it cannot depend only upon v and A. Therefore, let us add fluid, or air, density ρ (mass per length-cubed), and assume a relation of the form

$$F = f(\rho, A, v),$$

for some function f to be determined. What are the possible forms of f? To be dimensionally correct, the right side must be a force. What powers of ρ, A, and v would combine to give a force, that is,

$$F = k\rho^x A^y v^z,$$

where k is some constant without any units (i.e., dimensionless). If we denote mass by M, length by L, and time by T, then the last equation requires

$$MLT^{-2} = (ML^{-3})^x (L^2)^y (LT^{-1})^z.$$

Equating exponents of M, L, and T gives

$$x = 1, \quad -3x + 2y + z = 1, \quad -z = -2.$$

Therefore $x = 1$, $y = 1$, and $z = 2$. So the only relation must be

$$F = k\rho A v^2.$$

Consequently, the force must depend upon the square of the velocity, a fact often used in problems involving fluid resistance, and it must be proportional to density and area. Again, substantial information can be obtained from dimensional arguments alone.

EXERCISES

1. A pendulum executing small vibrations has period P and length l, and m is the mass of the bob. Can P depend only on l and m? If we assume P depends on l and the acceleration g due to gravity, then show that $T = $ constant $\cdot \sqrt{L/g}$.

2. A detonation wave moves through a high explosive with velocity D, thereby initiating an explosion which releases an energy per unit mass, e. If there is a relationship between e and D, what can be concluded?

3. In the blast wave problem take $C = 1$ (a value used by Taylor) in (1.3), and use $\rho = 1.25$ kg/m³. Some of the radius (m) vs. time (milliseconds) data for the Trinity explosion is given in the following table:

t	0.10	0.52	1.08	1.5	1.93	4.07	15.0	34.0
r	11.1	28.2	38.9	44.4	48.7	64.3	106.5	145

Using these data, estimate the yield of the Trinity explosion in kilotons (1 kiloton equals $4.186(10)^{12}$ joules). Compare your answer to the actual yield of approximately 21 kilotons.

4. In the blast wave problem assume, instead of (1.1), a physical law of
   the form

$$g(t, r, \rho, e, P) = 0,$$

where P is the ambient pressure. By inspection, find two independent dimensionless parameters formed from t, r, ρ, e, and P. Naming the two
dimensionless parameters π_1 and π_2 and assuming the law is equivalent to

$$f(\pi_1, \pi_2) = 0,$$

does it still follow that r varies like the two-fifths power of t?

5. The law governing the distance x an object falls in time t in a field of
   constant gravitational acceleration g with no air resistance is $x = \frac{1}{2}gt^2$.
   How many independent dimensionless quantities can be formed from t, x,
   and g? Rewrite the physical law in terms of dimensionless quantities. Can
   the distance a body falls depend on the mass m as well? That is, can there
   be a physical law of the form $f(t, x, g, m) = 0$?

6. A ball tossed upward with initial velocity v reaches a height at time t given
   by $x = -\frac{1}{2}gt^2 + vt$. Show this law is dimensionally consistent. Find two
   independent dimensionless variables (call them y and s) and rewrite the
   law only in terms of the dimensionless quantities. Plot the dimensionless
   law in the sy plane, and argue that the single, dimensionless plot contains
   all the information in the graphs of all the plots of $x = -\frac{1}{2}gt^2 + vt$ (x vs
   t), for all possible parameter values g and v.

1.1.3 The Pi Theorem

It is generally true that a physical law

$$f(q_1, q_2, q_3, \dots, q_m) = 0 \qquad (1.4)$$

relating m quantities $q_1, q_2, \dots, q_m$ is equivalent to a physical law that relates
the dimensionless quantities that can be formed from $q_1, q_2, \dots, q_m$. This is
the content of the Pi theorem. Before making a formal statement, however, we
introduce some basic terminology.

First, the m quantities $q_1, q_2, \dots, q_m$, which are like the quantities t, r, ρ,
and e in the blast wave example, are dimensioned quantities. This means that
they can be expressed in a natural way in terms of certain selected fundamental
dimensions $L_1, L_2, \dots, L_n$ ($n < m$), appropriate to the problem being studied.
In the blast wave problem, time T, length L, and mass M can be taken to be the
fundamental dimensions, since each quantity t, r, ρ, and e can be expressed

in terms of T, L, and M. For example, the dimensions of the energy e are ML^2T^{-2}. In general, the dimensions of q_i, denoted by $[q_i]$, can be written in terms of the fundamental dimensions as

$$[q_i] = L_1^{a_{1i}} L_2^{a_{2i}} \cdots L_n^{a_{ni}} \tag{1.5}$$

for some choice of exponents $a_{1i}, \ldots, a_{ni}$. If $[q_i] = 1$, then q_i is said to be **dimensionless**. The $n \times m$ matrix

$$A = \begin{pmatrix} a_{11} & \cdots & a_{1m} \\ a_{21} & \cdots & a_{2m} \\ \vdots & & \vdots \\ a_{n1} & \cdots & a_{nm} \end{pmatrix}$$

containing the exponents is called the **dimension matrix**. The elements in the ith column give the exponents for q_i in terms of the powers of $L_1, \ldots, L_n$.

The fundamental assumption regarding the physical law (1.4) goes back to the simple statement that apples cannot equal oranges. We assume that (1.4) is **unit free** in the sense that it is independent of the particular set of units chosen to express the quantities $q_1, q_2, \ldots, q_m$. We are distinguishing the word **unit** from the word **dimension**. By units we mean specific physical units like seconds, hours, days, and years; all of these units have dimensions of time. Similarly, grams, kilograms, slugs, and so on are units of the dimension mass. Any fundamental dimension L_i has the property that its units can be changed upon multiplication by the appropriate conversion factor $\lambda_i > 0$ to obtain $\overline{L}_i$ in a new system of units. We write

$$\overline{L}_i = \lambda_i L_i, \quad i = 1, \ldots, n.$$

The units of derived quantities q can be changed in a similar fashion. If

$$[q] = L_1^{b_1} L_2^{b_2} \cdots L_n^{b_n},$$

then

$$\overline{q} = \lambda_1^{b_1} \lambda_2^{b_2} \cdots \lambda_n^{b_n} q$$

gives q in the new system of units. The physical law (1.4) is said to be independent of the units chosen to express the dimensional quantities $q_1, q_2, \ldots, q_m$, or *unit-free*, if $f(q_1, \ldots, q_m) = 0$ and $f(\overline{q}_1, \ldots, \overline{q}_m) = 0$ are equivalent physical laws. More formally:

Definition 1.2

The physical law (1.4) is unit-free if for all choices of real numbers $\lambda_1, \ldots, \lambda_n$, with $\lambda_i > 0$, $i = 1, \ldots, n$, we have $f(\overline{q}_1, \ldots, \overline{q}_m) = 0$, if, and only if, $f(q_1, \ldots, q_m) = 0$.

Example 1.3

The physical law

$$f(x, t, g) \equiv x - \tfrac{1}{2}gt^2 = 0 \tag{1.6}$$

relates the distance x a body falls in a constant gravitational field g to the time t. In the cgs system of units, x is given in centimeters (cm), t in seconds, and g in cm/sec^2. If we change units for the fundamental quantities x and t to inches and minutes, then in the new system of units

$$\bar{x} = \lambda_1 x, \quad \bar{t} = \lambda_2 t,$$

where $\lambda_1 = \frac{1}{2.54}$ (in/cm) and $\lambda_2 = \frac{1}{60}$ (min/sec). Because $[g] = LT^{-2}$, we have

$$\bar{g} = \lambda_1 \lambda_2^{-2} g.$$

Then

$$f(\bar{x}, \bar{t}, \bar{g}) = \bar{x} - \tfrac{1}{2}\bar{g}\bar{t}^2 = \lambda_1 x - \tfrac{1}{2}(\lambda_1 \lambda_2^{-2} g)(\lambda_2 t)^2 = \lambda_1 \left(x - \tfrac{1}{2}gt^2 \right).$$

Therefore (1.6) is unit-free.

Theorem 1.4

(Pi Theorem) Let

$$f(q_1, q_2, \ldots, q_m) = 0 \tag{1.7}$$

be a unit free physical law that relates the dimensional quantities $q_1, q_2, \ldots, q_m$. Let $L_1, \ldots, L_n (n < m)$ be fundamental dimensions with

$$[q_i] = L_1^{a_{1i}} L_2^{a_{2i}} \cdots L_n^{a_{ni}}, \quad i = 1, \ldots, m,$$

and let $r = \text{rank } A$, where A is the dimension matrix (1.5). Then there exists $m - r$ independent dimensionless quantities $\pi_1, \pi_2, \ldots, \pi_{m-r}$ that can be formed from $q_1, \ldots, q_m$, and the physical law (1.7) is equivalent to an equation

$$F(\pi_1, \pi_2, \ldots, \pi_{m-r}) = 0, \tag{1.8}$$

expressed only in terms of the dimensionless quantities.

The proof of the Pi theorem is presented later. For the present we analyze an example. Before continuing, however, we note that the existence of a physical law (1.7) is an assumption. In practice one must conjecture which are the relevant variables in a problem and then apply the machinery of the theorem. The resulting dimensionless physical law (1.8) must be checked by experiment, or whatever, in an effort to establish the validity of the original assumptions.

Example 1.5

(Heat transfer)At time $t = 0$ an amount of heat energy e, concentrated at a
point in space, is allowed to diffuse outward into a region with temperature
zero. If r denotes the radial distance from the source and t is time, the problem
is to determine the temperature u as a function of r and t . This problem can
be formulated as a boundary value problem for a partial differential equation
(the heat equation, as in Chapter 6), but here we see what can be learned
from a careful dimensional analysis of the problem. The first step is to make
an assumption regarding which quantities affect the temperature u. Clearly t,
r, and e are relevant quantities. It also is reasonable that the heat capacity c,
with dimensions of energy per degree per volume, of the region plays a role,
as will the rate at which heat diffuses outward. The latter is characterized by
the thermal diffusivity k of dimensions length-squared per time. Therefore, we
conjecture a physical law of the form

$$f(t, r, u, e, k, c) = 0,$$

which relates the six quantities, t, r, u, e, k, c. The next step is to determine a
minimal set of fundamental dimensions $L_1, \dots, L_n$ by which the six-dimensional
quantities can be expressed. A suitable selection would be the four quantities
T (time), L (length), Θ (temperature), and E (energy). Then

$$[t] = T, \quad [e] = E, \quad [r] = L,$$
$$[k] = L^2 T^{-1}, \quad [u] = \Theta, \quad [c] = E\Theta^{-1}L^{-3},$$

and the dimension matrix A is given by

$$
\begin{array}{c}
\quad\quad t \quad r \quad u \quad e \quad k \quad c \\
\begin{array}{c} T \\ L \\ \Theta \\ E \end{array}
\left(
\begin{array}{cccccc}
1 & 0 & 0 & 0 & -1 & 0 \\
0 & 1 & 0 & 0 & 2 & -3 \\
0 & 0 & 1 & 0 & 0 & -1 \\
0 & 0 & 0 & 1 & 0 & 1
\end{array}
\right)
\end{array}
$$

Here $m = 6$, $n = 4$, and the rank of the dimension matrix is $r = 4$. Consequently, there are $m - r = 2$ dimensionless quantities that can be formed from
t, r, u, e, k, and c. To find them we proceed as follows: If π is dimensionless,
then for some choice of $\alpha_1, \dots, \alpha_6$

$$
\begin{aligned}
1 = [\pi] &= [t^{\alpha_1} r^{\alpha_2} u^{\alpha_3} e^{\alpha_4} k^{\alpha_5} c^{\alpha_6}] \\
&= T^{\alpha_1} L^{\alpha_2} \Theta^{\alpha_3} E^{\alpha_4} (L^2 T^{-1})^{\alpha_5} (E\Theta^{-1}L^{-3})^{\alpha_6} \\
&= T^{\alpha_1 - \alpha_5} L^{\alpha_2 + 2\alpha_5 - 3\alpha_6} \Theta^{\alpha_3 - \alpha_6} E^{\alpha_4 + \alpha_6}.
\end{aligned}
$$

Therefore, the exponents must vanish and we obtain four homogeneous linear equations for $\alpha_1, \ldots, \alpha_6$, namely

$$\alpha_1 - \alpha_5 = 0,$$
$$\alpha_2 + 2\alpha_5 - 3\alpha_6 = 0,$$
$$\alpha_3 - \alpha_6 = 0,$$
$$\alpha_4 + \alpha_6 = 0.$$

The coefficient matrix of this homogeneous linear system is just the dimension matrix A. From elementary matrix theory the number of independent solutions equals the number of unknowns minus the rank of A. Each independent solution will give rise to a dimensionless variable. Now the method unfolds and we can see the origin of the rank condition in the statement of the Pi theorem. By standard methods for solving linear systems we find two linearly independent solutions

$$\alpha_1 = -\tfrac{1}{2}, \quad \alpha_2 = 1, \quad \alpha_3 = \alpha_4 = 0, \quad \alpha_5 = -\tfrac{1}{2}, \quad \alpha_6 = 0,$$

and

$$\alpha_1 = \frac{3}{2}, \quad \alpha_2 = 0, \quad \alpha_3 = 1, \quad \alpha_4 = -1, \quad \alpha_5 = \frac{3}{2}, \quad \alpha_6 = 1.$$

These give two dimensionless quantities

$$\pi_1 = \frac{r}{\sqrt{kt}}$$

and

$$\pi_2 = \frac{uc}{e}(kt)^{3/2}.$$

Therefore the Pi theorem guarantees that the original physical law $f(t, r, u, e, k, c) = 0$ is equivalent to a physical law of the form

$$F(\pi_1, \pi_2) = 0.$$

Solving for π_2 gives

$$\pi_2 = g(\pi_1)$$

for some function g, or

$$u = \frac{e}{c}(kt)^{-3/2} g\left(\frac{r}{\sqrt{kt}}\right). \tag{1.9}$$

Again, without solving an equation governing the diffusion process, we are able via dimensional analysis to argue that the temperature of the region varies according to (1.9). For example, we can conclude that the temperature near the source $r = 0$ falls off like $t^{-3/2}$.

1.1.4 Proof of the Pi Theorem

To prove the Pi theorem we must demonstrate two propositions.

(i) Among the quantities $q_1, \ldots, q_m$ there are $m - r$ independent dimensionless variables that can be formed, where r is the rank of the dimension matrix A.

(ii) If $\pi_1, \ldots, \pi_{m-r}$ are the $m - r$ dimensionless variables, then (1.7) is equivalent to a physical law of the form $F(\pi_1, \ldots, \pi_{m-r}) = 0$.

The proof of (i) is straightforward; the general argument proceeds exactly like the construction of the dimensionless variables in Example 1.5. It makes use of the familiar result in linear algebra that the number of linearly independent solutions of a set of n homogeneous equations in m unknowns is $m - r$, where r is the rank of the coefficient matrix. For, let π be a dimensionless quantity. Then

$$\pi = q_1^{\alpha_1} q_2^{\alpha_2} \cdots q_m^{\alpha_m} \tag{1.10}$$

for some $\alpha_1, \alpha_2, \ldots, \alpha_m$. In terms of the fundamental dimensions $L_1, \ldots, L_n$,

$$\pi = (L_1^{a_{11}} L_2^{a_{21}} \cdots L_n^{a_{n1}})^{\alpha_1} (L_1^{a_{12}} L_2^{a_{22}} \cdots L_n^{a_{n2}})^{\alpha_2} \cdots (L_1^{a_{1m}} L_2^{a_{2m}} \cdots L_n^{a_{nm}})^{\alpha_m}$$
$$= L^{a_{11}\alpha_1 + a_{12}\alpha_2 + \cdots + a_{1m}\alpha_m} \cdots L_n^{a_{n1}\alpha_1 + a_{n2}\alpha_2 + \cdots + a_{nm}\alpha_m}.$$

Because $[\pi] = 1$, the exponents vanish, or

$$a_{11}\alpha_1 + a_{12}\alpha_2 + \cdots + a_{1m}\alpha_m = 0$$

$$\vdots$$

$$a_{n1}\alpha_1 + a_{n2}\alpha_2 + \cdots + a_{nm}\alpha_m = 0 \tag{1.11}$$

By the aforementioned theorem in linear algebra, the system (1.11) has exactly $m - r$ independent solutions $[\alpha_1, \ldots, \alpha_m]$. Each solution gives rise to a dimensionless variable via (1.10), and this completes the proof of (i). The independence of the dimensionless variables is in the sense of linear algebraic independence.

The proof of (ii) makes strong use of the hypothesis that the law is unit-free. The argument is subtle, but it can be made almost transparent if we examine a particular example.

Example 1.6

Consider the unit-free law

$$f(x, t, g) = x - \tfrac{1}{2}gt^2 = 0 \tag{1.12}$$

for the distance a particle falls in a gravitational field. If length and time are chosen as fundamental dimensions, a straightforward calculation shows there is a single dimensionless variable given by

$$\pi_1 = t^2 \frac{g}{x}.$$

The remaining variable g can be expressed as $g = x\pi_1/t^2$, and we can define a function G by

$$G(x, t, \pi_1) \equiv f(x, t, x\pi_1/t^2).$$

Clearly the law $G(x, t, \pi_1) = 0$, or

$$x - \frac{1}{2}\left(\frac{x\pi_1}{t^2}\right)t^2 = 0, \tag{1.13}$$

is equivalent to (1.12) and is unit-free because f is. Then F is defined by

$$F(\pi_1) \equiv G(1, 1, \pi_1) = 1 - \frac{1}{2}\pi_1 = 0,$$

which is equivalent to (1.13) and (1.12).

We present the proof of (ii) in the special case when $m = 4$, $n = 2$, and $r = 2$. The notation will be easier, and the general argument for arbitrary m, n, and r proceeds in exactly the same manner. Therefore we consider a unit-free physical law

$$f(q_1, q_2, q_3, q_4) = 0, \tag{1.14}$$

with

$$[q_j] = L_1^{a_{1j}} L_2^{a_{2j}}, \quad j = 1, \dots, 4,$$

where L_1 and L_2 are fundamental dimensions. The dimension matrix

$$A = \begin{pmatrix} a_{11} & a_{12} & a_{13} & a_{14} \\ a_{21} & a_{22} & a_{23} & a_{24} \end{pmatrix}$$

is assumed to have rank $r = 2$. If π is a dimensionless quantity, then

$$\pi = q_1^{\alpha_1} q_2^{\alpha_2} q_3^{\alpha_3} q_4^{\alpha_4},$$

where the exponents α_1, α_2, α_3, α_4 satisfy the homogeneous system (1.11), which in this case is

$$\begin{pmatrix} a_{11} \\ a_{21} \end{pmatrix} \alpha_1 + \begin{pmatrix} a_{12} \\ a_{22} \end{pmatrix} \alpha_2 + \begin{pmatrix} a_{13} \\ a_{23} \end{pmatrix} \alpha_3 + \begin{pmatrix} a_{14} \\ a_{24} \end{pmatrix} \alpha_4 = \begin{pmatrix} 0 \\ 0 \end{pmatrix}. \tag{1.15}$$

We wish to determine α_1, α_2, α_3, α_4, and the form of the two dimensionless variables. Without loss of any generality, we can assume the first two columns of A are linearly independent. This is because we can rearrange the indices on the q_j so that the two independent columns appear as the first two. Then columns three and four can be written as linear combinations of the first two, or

$$\begin{pmatrix} a_{13} \\ a_{23} \end{pmatrix} = c_{31} \begin{pmatrix} a_{11} \\ a_{21} \end{pmatrix} + c_{32} \begin{pmatrix} a_{12} \\ a_{22} \end{pmatrix}, \qquad (1.16)$$

$$\begin{pmatrix} a_{14} \\ a_{24} \end{pmatrix} = c_{41} \begin{pmatrix} a_{11} \\ a_{21} \end{pmatrix} + c_{42} \begin{pmatrix} a_{12} \\ a_{22} \end{pmatrix} \qquad (1.17)$$

for some constants c_{31}, c_{32}, c_{41}, and c_{42}. Substituting into (1.15) gives

$$(\alpha_1 + c_{31}\alpha_3 + c_{41}\alpha_4) \begin{pmatrix} a_{11} \\ a_{21} \end{pmatrix} + (\alpha_2 + c_{32}\alpha_3 + c_{42}\alpha_4) \begin{pmatrix} a_{12} \\ a_{22} \end{pmatrix} = \begin{pmatrix} 0 \\ 0 \end{pmatrix}.$$

The left side is a combination of linearly independent vectors, and therefore the coefficients must vanish,

$$\alpha_1 + c_{31}\alpha_3 + c_{41}\alpha_4 = 0,$$
$$\alpha_2 + c_{32}\alpha_3 + c_{42}\alpha_4 = 0.$$

Therefore, we can solve for α_1 and α_2 in terms of α_3 and α_4 and write

$$\begin{pmatrix} \alpha_1 \\ \alpha_2 \\ \alpha_3 \\ \alpha_4 \end{pmatrix} = \alpha_3 \begin{pmatrix} -c_{31} \\ -c_{32} \\ 1 \\ 0 \end{pmatrix} + \alpha_4 \begin{pmatrix} -c_{41} \\ -c_{42} \\ 0 \\ 1 \end{pmatrix}.$$

The two vectors on the right represent two linearly independent solutions of (1.15); hence, the two dimensionless quantities are

$$\pi_1 = q_1^{-c_{31}} q_2^{-c_{32}} q_3,$$
$$\pi_2 = q_1^{-c_{41}} q_2^{-c_{42}} q_4.$$

Next define a function G by

$$G(q_1, q_2, \pi_1, \pi_2) \equiv f(q_1, q_2, \pi_1 q_1^{c_{31}} q_2^{c_{32}}, \pi_2 q_1^{c_{41}} q_2^{c_{42}}).$$

The physical law

$$G(q_1, q_2, \pi_1, \pi_2) = 0 \qquad (1.18)$$

holds if, and only if, (1.14) holds, and therefore (1.18) is an equivalent physical law. Since $f = 0$ is unit free, it easily follows that (1.18) is unit-free (note that $\bar{\pi}_1 = \pi_1$, $\bar{\pi}_2 = \pi_2$ under any change of units; that is, dimensionless variables have the same value in all systems of units). For the final stage of the argument we show that (1.18) is equivalent to the physical law

$$G(1, 1, \pi_1, \pi_2) = 0, \qquad (1.19)$$

which will give the result, for (1.19) implies $F(\pi_1, \pi_2) = 0$, where $F(\pi_1, \pi_2) \equiv G(1, 1, \pi_1, \pi_2)$. Because (1.18) is a unit-free, we must have

$$G(\bar{q}_1, \bar{q}_2, \pi_1, \pi_2) = 0, \tag{1.20}$$

where

$$\bar{q}_1 = \lambda_1^{a_{11}} \lambda_2^{a_{21}} q_1, \quad \bar{q}_2 = \lambda_1^{a_{12}} \lambda_2^{a_{22}} q_2$$

for *every* choice of the conversion factors $\lambda_1, \lambda_2 > 0$. We select λ_1 and λ_2 so that $\bar{q}_1 = \bar{q}_2 = 1$. We are able to make this choice because

$$\lambda_1^{a_{11}} \lambda_2^{a_{21}} q_1 = 1, \quad \lambda_1^{a_{12}} \lambda_2^{a_{22}} q_2 = 1 \tag{1.21}$$

implies

$$a_{11} \ln \lambda_1 + a_{21} \ln \lambda_2 = -\ln q_1,$$
$$a_{12} \ln \lambda_1 + a_{22} \ln \lambda_2 = -\ln q_2. \tag{1.22}$$

And, because the coefficient matrix

$$\begin{pmatrix} a_{11} & a_{21} \\ a_{12} & a_{22} \end{pmatrix}$$

in (1.22) is nonsingular (recall the assumption that the first two columns of the dimension matrix A are linearly independent), the system (1.22) has a unique solution $(\ln \lambda_1, \ln \lambda_2)$, from which λ_1 and λ_2 can be determined to satisfy (1.21). Thus (1.19) is an equivalent physical law and the argument is complete.

The general argument for arbitrary m, n, and r can be found in Birkhoff (1950), from which the preceding proof was adapted. This classic book also provides additional examples and historical comments.[1]

In performing a dimensional analysis on a problem, two judgments are required at the beginning.

1. The selection of the pertinent variables.

2. The choice of the fundamental dimensions.

The first is a matter of experience and may be based on intuition or experiments. There is, of course, no guarantee that the selection will lead to a useful formula after the procedure is applied. Second, the choice of fundamental dimensions may involve tacit assumptions that may not be valid in a given problem. For example, including mass, length, and time but not force in a given problem assumes there is some relation (Newton's second law) that plays

---

[1] A simple proof of the Pi theorem in a linear algebraic setting is contained in W. D. Curtis, J. D. Logan, & W. A. Parker, 1982. Dimensional analysis and the Pi theorem, *Linear Algebra and Its Applications* **47**, 117–126.

an important role and causes force not to be an independent dimension. As a specific example, a small sphere falling under gravity in a viscous fluid is observed to fall, after a short time, at constant velocity. Since the motion is un-accelerated we need not make use of the proportionality of force to acceleration, and so force can be treated as a separate, independent fundamental dimension. In summary, intuition and experience are important ingredients in applying the dimensional analysis formalism to a specific physical problem.

EXERCISES

1. The speed v of a wave in deep water is determined by its wavelength λ and the acceleration g due to gravity. What does dimensional analysis imply regarding the relationship between v, λ, and g?

2. An ecologist postulated that that there is a relationship among the mass m, density ρ, volume V, and surface area S of certain animals. Discuss this conjecture in terms of dimensionless analysis.

3. A small sphere of radius r and density ρ is falling at constant velocity v under the influence of gravity g in a liquid of density ρ_l and viscosity μ (given in mass per length per time). It is observed experimentally that

$$v = \frac{2}{9} r^2 \rho g \mu^{-1} \left(1 - \frac{\rho_l}{\rho} \right).$$

Show that this law unit-free.

4. A physical phenomenon is described by the quantities P, l, m, t , and ρ, representing pressure, length, mass, time, and density, respectively. If there is a physical law

$$f(P, l, m, t, \rho) = 0$$

relating these quantities, show that there is an equivalent physical law of the form $G(l^3 \rho/m, t^6 P^3/m^2 \rho) = 0$.

5. A physical system is described by a law $f(E, P, A) = 0$, where E, P, and A are energy, pressure, and area, respectively. Show that $PA^{3/2}/E = \text{const}$.

6. The length L of an organism depends upon time t, its density ρ, its resource assimilation rate a (mass per area per time), and its resource use rate b (mass per volume per time). Show that there is a physical law involving two dimensionless quantities.

7. A piece of shrapnel of density ρ is driven off an explosive device at velocity v. The density of the explosive is ρ_e, and E is its Gurney energy (joules/kilogram), or the specific energy available in the explosive to do work. Determine how the velocity of the shrapnel depends upon E.

8. In an indentation experiment, a slab of metal of thickness h is subjected to a constant pressure P on its upper surface by a cylinder of radius a. The technician then measures the vertical displacement of U of the indentation. The displacement also depends upon two material properties, Poisson's ratio v, which is dimensionless, and the Lamé constant μ, which has dimensions M/L^3T^2, where M is mass, L is length, and T is time. Determine a set of dimensionless variables and show that the functional form of U is

$$U = aG\left(\frac{P}{\mu a^2}, \frac{h}{a}, v\right).$$

9. In modeling the digestion process in insects, it is believed that digestion yield rate Y, in mass per time, is related to the concentration C of the limiting nutrient, the residence time T in the gut, the gut volume V, and the rate of nutrient nutrient breakdown r, given in mass per time per volume. Show that for fixed T, r, C, the yield is positively related to the gut volume.

10. A chemical **C** flows continuously into a reactor with concentration C_i and volumetric flow rate q (volume/time). While in the reactor, which has volume V, the substances are continuously stirred and a chemical reaction **C**$\to$ products, with rate constant k (1/time), consumes the chemical. The mixture exits the reactor at the same flow rate q. The concentration of **C** in the reactor at any time t is $C = C(t)$, and $C(0) = C_0$. Use dimensional analysis to deduce that

$$C = C_i F(tk, a, b),$$

where $a = C_0/C_i$ and $b = Vk/q$ are dimensionless constants and F is some function.

11. The problem is to determine the power P that must be applied to keep a ship of length l moving at a constant speed V. If it is the case, as seems reasonable, that P depends on the density of water ρ, the acceleration due to gravity g, and the viscosity of water ν (in length-squared per time), as well as l and V, then show that

$$\frac{P}{\rho l^2 V^3} = f(\text{Fr}, \text{Re}),$$

where Fr is the Froude number and Re is the Reynolds number defined by

$$\text{Fr} \equiv \frac{V}{\sqrt{lg}}, \quad \text{Re} \equiv \frac{Vl}{\nu}$$

12. A spherical gas bubble with ratio of specific heats γ is surrounded by an infinite sea of liquid of density ρ_l. The bubble oscillates with growth and contraction periodically with small amplitude at a well-defined frequency ω. Assuming a physical law

$$f(P, R, \rho_l, \omega, \gamma) = 0,$$

where P is the mean pressure inside the bubble and R is the mean radius, show that the frequency must vary inversely with the mean radius R. The constant γ is dimensionless.

13. A perfect gas in equilibrium has specific energy E (energy per mass), temperature T, and Boltzmann constant k (specific energy per degree). Derive a functional relationship of the form $E = f(k, T)$.

14. In tests for fuel economy, cars are driven at constant speed V on a level highway. With no acceleration, the force of propulsion F must be in equilibrium with other forces, such as the air resistance, and so on. Assume that the variables affecting F are the velocity V, the rate C that fuel is burned, in volume per time, and the amount of energy K in a gallon of fuel, in mass per length per time-squared. Determine F as a function of V, C, and K.

15. Across a cell membrane of thickness w the mass concentration of a biochemical molecule is C_i on the inside of the cell, and C_o on the outside. The diffusion constant in the membrane is d, measured in length-squared per time. Use dimensional analysis (not guessing) to obtain the most general relationship for the flux φ of the molecules (mass/(area·time)) through the membrane in terms of the other quantities.

1.2 Scaling

1.2.1 Characteristic Scales

Another procedure useful in formulating a mathematical model is *scaling*. Roughly, scaling means selecting new, usually dimensionless variables and reformulating the problem in terms of those variables. Not only is the procedure useful, but it often is a necessity, especially when comparisons of the magnitudes of various terms in an equation must be made in order to neglect small terms. This idea is particularly crucial in the application of perturbation methods to identify small and large parameters. Further, scaling usually reduces the number of parameters in a problem, thereby leading to great simplification, and it identifies what combinations of parameters are important.

For motivation let us suppose that time t is a variable in a given problem, measured in units of seconds. If the problem involved the motion of a glacier, clearly the unit of seconds is too fast because significant changes in the glacier could not be observed on the order of seconds. On the other hand, if the problem involved a nuclear reaction, then the unit of seconds is too slow; all of the important action would be over before the first second ticked. Evidently, every problem has an intrinsic **time scale**, or *characteristic time* t_c, which is appropriate to the given problem. This is the shortest time for discernible changes to be observed in the physical quantities. For example, the characteristic time for glacier motion would be of the order of years, whereas the characteristic time for a nuclear reaction would be of the order of microseconds. Some problems have multiple time scales. A chemical reaction, for example, may begin slowly and the concentration change little over a long time; then, the reaction may suddenly go to completion with a large change in concentration over a short time. There are two time scales involved in such a process. Other examples are in the life sciences, where multi-scale processes are the norm. Spatial scales vary over as much as 10^{15} orders of magnitude as we progress from processes involving genes, proteins, cells, organs, organisms, communities, and ecosystems; time scales vary from times that it takes for protein to fold to times for evolution to occur. Several scales can occur in the same problem. Yet another example occurs in fluid flow, where the processes of heat diffusion, advection, and possible chemical reaction all have different scales.

Once a characteristic time has been identified, at least for a part of a process, then a new dimensionless variable $\bar{t}$ can be defined by

$$\bar{t} = \frac{t}{t_c}$$

If t_c is chosen correctly, then the *dimensionless time* $\bar{t}$ is neither too large nor too small, but rather of order unity. The question remains to determine the time scale t_c for a particular problem, and the same question applies to other variables in the problem (e.g., length, concentration, and so on). The general rule is that the characteristic quantities are formed by taking combinations of the various dimensional constants in the problem and should be roughly the same order of magnitude of the quantity itself.

After characteristic scales, which are built up from the parameters in the model, are chosen for the independent and dependent variables, the model can then reformulated in terms of the new dimensionless variables. The result will be a model in dimensionless form, where all the variables and parameters in the problem are dimensionless. This process is called **non-dimensionalization**, or scaling a problem. By the Pi theorem, it is guaranteed that we can always non-dimensionlize a problem. The payoff is simpler model that is independent of units and dimensions and that has fewer parameters, which is an economy of savings.

Example 1.7

(Population growth) Let $p = p(t)$ denote the population of an animal species located in a fixed region at time t. The simplest model of population growth is the classic **Malthus model**,[2] which states that the per capita growth rate $\frac{1}{p}\frac{dp}{dt}$ is constant. This means that the growth rate $\frac{dp}{dt}$ is proportional to the population p, or $\frac{dp}{dt} = rp$, where r is the *growth rate*, given in dimensions of inverse-time. Easily, the Malthus model predicts that the population will grow exponentially for all time, that is, $p = p_0 e^{rt}$, where p_0 is the initial population. Many books on an impending world population explosion have been written with the Malthusian model as a premise. Clearly, however, as a population grows, intraspecific competition for food, living space, and natural resources limit the growth. We can modify the Malthus model to include a competition term. The simplest approach is to notice that if there are p individuals in the system, then the number of encounters, which is a measure of competition, is approximately p^2; therefore we subtract a term proportional to p^2 from the growth rate to obtain the model

$$\frac{dp}{dt} = rp\left(1 - \frac{p}{K}\right), \quad p(0) = p_0. \tag{1.23}$$

This is saying that the per capita growth rate is not constant, but decreases linearly with population. The parameter K is the *carrying capacity*, which is the number of individuals that the ecosystem can sustain. When $p = K$ the growth rate is zero. The model (1.23) is called the **logistics model**[3]; as the population grows, the negative p^2 term will kick in and limit the growth. To reduce (1.23) to dimensionless form we select new dimensionless independent and dependent variables. The time scale and population scale are formed from the constants in the problem, r, K, and p_0. Of these, only r contains the dimensions of time, and therefore we scale time by $1/r$ giving a new, dimensionless time τ defined by

$$\tau = rt.$$

There are two choices for the population scale, K or p_0. Either will work, and

---

[2] Thomas Malthus (1766–1834) was an English essayist who was one of the first individuals to address demographics and food supply.

[3] The logistics model was introduced by P. Verhulst (1804–1849).

so we select K to obtain a dimensionless population P given by

$$P = \frac{p}{K}$$

Thus, we measure population in the problem relative to the carrying capacity. Using these dimensionless variables, it is straightforward to obtain

$$\frac{dP}{d\tau} = P(1 - P), \quad P(0) = \alpha, \tag{1.24}$$

where $\alpha \equiv p_0/K$ is a dimensionless constant. The scaled model (1.24) has only one constant (α), a significant simplification over (1.23) where there are three constants (r, K, p_0). The constant α represents a scaled, initial population. There is a single combination of the parameters in the original that is relevant to the dynamics. The initial value problem (1.24) can be solved by separating variables to obtain

$$P(\tau) = \frac{\alpha}{\alpha + (1 - \alpha)e^{-\tau}}. \tag{1.25}$$

Clearly

$$\lim_{\tau \to \infty} P(\tau) = 1.$$

It follows that, confirming our earlier statement, the limiting population p is equal to the carrying capacity K. We observe that there are two *equilibrium populations*, or constant solutions of (1.23), $p = K$ and $p = 0$. The population $p = K$ is an *attractor*; that is, regardless of the initial population, the population $p(t)$ tends to the value K as time gets large.

1.2.2 A Chemical Reactor Problem

A basic problem in chemical engineering is to understand how the concentration of chemical species vary when undergoing a reaction in a chemical reactor. To illustrate the concept of non-dimensionalization and scaling we analyze a simple model of an isothermal, continuously stirred, tank reactor (see Fig. 1.2). The reactor has a fixed volume V, and a chemical $\mathbf{C}$ of fixed concentration c_i, given in mass per volume, enters the reactor through the feed at a constant flow rate q, given in volume per time; initially the concentration of the chemical in the reactor is c_0. When the chemical enters the reactor, the mixture is perfectly stirred while undergoing a chemical reaction, and then the mixture exits the reactor at the same flow rate q. At any time t, we denote the concentration of the chemical $\mathbf{C}$ in the reactor by $c = c(t)$. The reactant chemical is assumed to disappear, that is, it is consumed by reaction, with a rate $r = r(c)$, given in mass per unit volume, per unit time. We are thinking of a simple reaction of the form $\mathbf{C} \to$ products. Usually reaction rates depend on temperature, but

Figure 1.2 Continuously stirred tank reactor.

here we assume that r depends only on the concentration c; this is what makes our problem isothermal. The *perfectly stirred* assumption is an idealization and implies that there are no concentration gradients in the reactor; otherwise, c would also depend on a spatial variable.

To obtain a mathematical model of this problem we look for a physical law. A common principle that is fundamental to all flow problems is mass balance. That is, the time rate of change of the mass of the chemical inside the reactor must equal the rate mass flows in (qc_i), minus the rate that mass flows out (qc), plus the rate that mass is consumed by the reaction (Vr). At any given time the mass of the chemical in the reactor is Vc. In symbols, the mass balance equation is

$$\frac{d}{dt}(Vc(t)) = qc_i - qc(t) - Vr(c(t)).$$

Observe that the factor V is required on the reaction term to obtain consistent dimensions. To fix the idea, let us take the reaction rate r to be proportional to c, that is $r = kc$, where k is the rate constant having dimensions of inverse time. Recall that this is the law of mass action in elementary chemistry, which states that the rate of a reaction is proportional to the product of the concentrations of the reactants. Then, the mathematical model is given by the initial value problem

$$\frac{dc}{dt} = \frac{q}{V}(c_i - c) - kc, \quad t > 0, \tag{1.26}$$

$$c(0) = c_0. \tag{1.27}$$

To non-dimensionalize the problem we choose dimensionless independent and dependent variables. This means we must select a characteristic time and a characteristic concentration by which to measure the real time and concentration. These characteristic values are formed from the constants in the problem: c_i, c_0, V, q, k. Generally, we measure the dependent variable relative to some maximum value in the problem, or any other value that represents the order of magnitude of that quantity. There are two constant concentrations, c_i and c_0, and either one of them is a suitable concentration scale. Therefore, we define a dimensionless concentration C by

$$C = \frac{c}{c_i}.$$

Therefore, all concentrations in the problem are measured relative to c_i, the concentration of the feed. To select a time scale we observe that there are two quantities with dimensions of time that can be formed from the constants in the problem, V/q and k^{-1}. The former is based on the flow rate, and the latter is based on the reaction rate. So the choice of a time scale is not unique. Either choice leads to a correct dimensionless problem. Let us hold up on making a selection and define a dimensionless time τ by

$$\tau = \frac{t}{T},$$

where T is either V/q or k^{-1}. We recast the model in dimensionless form. By the chain rule

$$\frac{dc}{dt} = \frac{c_i}{T}\frac{dC}{d\tau},$$

and therefore the model becomes

$$\frac{dC}{d\tau} = \frac{qT}{V}(1 - C) - kTC, \quad \tau > 0,$$
$$C(0) = \gamma,$$

where γ is a dimensionless constant given by the ratio

$$\gamma = \frac{c_0}{c_i}.$$

If we choose $T = V/q$, then we obtain the dimensionless model

$$\frac{dC}{d\tau} = 1 - C - \beta C, \quad \tau > 0, \tag{1.28}$$
$$C(0) = \gamma, \tag{1.29}$$

where

$$\beta = \frac{kV}{q}$$

is a dimensionless constant representing a ratio of the two time scales. Therefore, the problem has been reduced to dimensionless form, and any results that are obtained are free of any specific set of units that we select. Moreover, the number of parameters has been decreased from five to two, and the problem is simpler. If we choose the other time scale, $T = k^{-1}$, then the problem reduces to the dimensionless form

$$\frac{dC}{d\tau} = \frac{1}{\beta}(1 - C) - C, \quad \tau > 0, \tag{1.30}$$
$$C(0) = \gamma. \tag{1.31}$$

Is there any advantage of one over the other? Yes. In some problems, where the terms differ in order of magnitude, it is important to choose the *correct* scale, so that each term reflects the correct magnitude. For example, suppose reaction occurs on a slow time scale compared to the flow through the reactor. Then k is small compared to the flow rate q/V. We write $k \ll q/V$. Therefore the dimensionless parameter β is small, or $\beta \ll 1$. Of the two dimensionless models (1.28)–(1.29) or (1.30)–(1.31), which best reflects this assumption? We expect the reaction term to be small compared to the flow rate term, so the best choice is (1.28)–(1.29), which means we should scale time on the more dominant flow rate. If reaction is fast compared to the flow rate, then β is large ($\beta \gg 1$), and we should choose a reaction time scale, giving (1.30)–(1.31); this choice puts the small term $1/\beta$ correctly on the flow rate term.

Making the correct choice when terms have different orders of magnitude is especially essential when we want to make an approximation by deleting small terms in the equation. This is a very common strategy in equations that we cannot solve, for example, most nonlinear equations. For example, if $\beta \ll 1$ and we scale by by flow rate, the model is (1.28)–(1.29). Ignoring the small term gives the approximation

$$\frac{dC}{d\tau} = 1 - C, \quad C(0) = \gamma,$$

which gives approximation $C(\tau) = 1 + (\gamma - 1)e^{-\tau}$. This approximation is believable; the reaction is slow and the concentration approaches the concentration of the feed. On the other hand, the model (1.30)–(1.31) gives $\beta\frac{dC}{d\tau} = (1 - C) - \beta C$, which, upon neglecting the small terms, provides the approximation $C = 1$. This approximation does not even satisfy the initial condition.

In summary, if approximations are to be made by deleting small terms, it is important how we non-dimensionalize the problem. The common scaling strategy is to non-dimensionalize the problem using a generic time scale T that is chosen later to make the coefficients of the terms in the equation reflect their size or reflect what terms balance in the process. Ultimately, proper scaling is learned through experience and careful analysis.

1.2.3 The Projectile Problem

The projectile problem, as first pointed out by Lin and Segel (1974), is a good illustration of the importance of choosing correct scales, particularly when it is desired to make a simplification by neglecting small quantities. Terms in an equation that appear small are not always as they seem, and proper scaling is essential in determining the orders of magnitude of the terms.

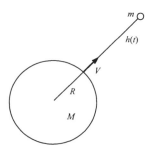

Figure 1.3 The projectile problem.

We analyze the motion of a projectile thrust vertically upward from the surface of the earth. At time $t = 0$ on the surface of the earth, with radius R and mass M, an object of mass m is given a vertical upward velocity of magnitude V. To be determined is the height h above the earth's surface that the mass reaches at time t (see Fig. 1.3). Generally, forces on the object are the gravitational force and the force due to air resistance. But we assume the force due to air resistance is negligible in the particular problem we are considering. In general, as a first approximation it is common to neglect what are believed to be small effects, since in that case the equations are more tractable for analysis. Should the analytic results compare unfavorably with experiment or should a more detailed description be required, then additional effects can be included.

The governing equation, or mathematical model, comes from a physical law. Newton's universal gravitational law, which states that the force between the two objects is proportional to the product of the masses and inversely proportional to the square of the distance between them, where the mass of each object can be regarded as concentrated at its center. By Newton's second law the force equals the mass times acceleration, or

$$m\frac{d^2h}{dt^2} = -G\frac{Mm}{(h+R)^2},$$

where G is the proportionality constant in the universal gravitational law. When $h = 0$, that is, at the earth's surface, the gravitational force equals $-mg$, and therefore

$$\frac{GM}{R^2} = g.$$

Thus

$$\frac{d^2h}{dt^2} = -\frac{R^2g}{(h+R)^2}, \tag{1.32}$$

with initial conditions

$$h(0) = 0, \quad \frac{dh}{dt}(0) = V. \tag{1.33}$$

The initial value problem (1.32) and (1.33) is the mathematical model for the problem.

At this point we undertake a dimensional analysis of the problem and gain considerable insight without actually attempting a solution. From our model the relevant dimensional quantities are t, h, R, V, and g having dimensions

$$[t] = \text{time } (T), \qquad [V] = \text{velocity } (LT^{-1}), \tag{1.34}$$

$$[h] = \text{length } (L), \qquad [g] = \text{acceleration } (LT^{-2}), \tag{1.35}$$

$$[R] = \text{length } (L). \tag{1.36}$$

We can use T (time) and L (length) as fundamental dimensions. Following the procedure described earlier, if π is a dimensionless combination of t, h, R, V, and g, then

$$[\pi] = [t^{\alpha_1} h^{\alpha_2} R^{\alpha_3} V^{\alpha_4} g^{\alpha_5}] = T^{\alpha_1 - \alpha_4 - 2\alpha_5} L^{\alpha_2 + \alpha_3 + \alpha_4 + \alpha_5} = 1.$$

Therefore,

$$\alpha_1 - \alpha_4 - 2\alpha_5 = 0,$$

$$\alpha_2 + \alpha_3 + \alpha_4 + \alpha_5 = 0. \tag{1.37}$$

This system has rank two and so there are three independent dimensionless variables. Either by inspection or solving (1.37), we find dimensionless quantities

$$\pi_1 = \frac{h}{R}, \quad \pi_2 = \frac{t}{R/V}, \quad \pi_3 = V/\sqrt{gR}. \tag{1.38}$$

By the Pi theorem, if there is a physical law relating t, h, R, V, and g (and we assume there must be, since in theory we could solve (1.32) and (1.33) to obtain that law), then there is an equivalent law that can be expressed as

$$\frac{h}{R} = f\left(\frac{t}{R/V}, \frac{V}{\sqrt{gR}}\right) \tag{1.39}$$

for some function $f(\pi_2, \pi_3)$.

Actually there is considerable information in (1.39). For example, suppose we are interested in finding the time $t_{\max}$ that is required for the object to reach its maximum height for a given velocity V. Then differentiating (1.39) with respect to t and setting $h'(t)$ equal to zero gives

$$\frac{\partial f}{\partial \pi_2}\left(\frac{t_{\max}}{R/V}, \frac{V}{\sqrt{gR}}\right) = 0,$$

or

$$\frac{t_{\max}}{R/V} = F\left(\frac{V}{\sqrt{gR}}\right) \tag{1.40}$$

for some function F. Remarkably, with little analysis beyond dimensional reasoning, we have found that the time to maximum height depends on the dimensionless combination $V/\sqrt{gR}$. The value in knowing this kind of information lies in the efficiency of (1.40); a *single* graph of $t_{max}/(R/V)$ vs. $V/\sqrt{Rg}$ contains all of the data of the graphs of t_{max} versus V for *all* choices of g and R. For example, an experimenter making measurements on different planets of t_{max} versus V would not need a separate plot of data for each planet. An entire atlas can be replaced by a single map when we plot dimensionless quantities.

Now we select characteristic time and length scales and recast the problem represented by (1.32) and (1.33) into dimensionless form. The problem is more subtle than it originally appears. The general method requires us to choose a new dimensionless dependent variable $\bar{h}$ and independent variable $\bar{t}$ by

$$\bar{t} = \frac{t}{t_c}, \quad \bar{h} = \frac{h}{h_c}, \tag{1.41}$$

where t_c is an intrinsic time scale and h_c is an intrinsic length scale; the values of t_c and h_c should be chosen by taking combinations of the constants in the problem, which in this case are R, V, and g. This problem presents several choices. For a length scale h_c we could take either R or V^2/g. Possible time scales are R/V, $\sqrt{R/g}$, and V/g. Which choice is the most appropriate? Actually, equations (1.41) represent a legitimate transformation of variables for any choice of t_c and h_c; after the change of variables an equivalent problem results. From a scaling viewpoint, however, one particular choice is advantageous. The three choices

$$\bar{t} = \frac{t}{R/V}, \quad \bar{h} = \frac{h}{R}, \tag{1.42}$$

$$\bar{t} = \frac{t}{\sqrt{R/g}}, \quad \bar{h} = \frac{h}{R}, \tag{1.43}$$

and

$$\bar{t} = \frac{t}{Vg^{-1}}, \quad \bar{h} = \frac{h}{V^2g^{-1}} \tag{1.44}$$

lead to the following three dimensionless problems, which are equivalent to (1.32) and (1.33):

$$\epsilon\frac{d^2\bar{h}}{d\bar{t}^2} = -\frac{1}{(1+\bar{h})^2}, \quad \bar{h}(0) = 0, \quad \frac{d\bar{h}}{d\bar{t}}(0) = 1, \tag{1.45}$$

$$\frac{d^2\bar{h}}{d\bar{t}^2} = -\frac{1}{(1+\bar{h})^2}, \quad \bar{h}(0) = 0, \quad \frac{d\bar{h}}{d\bar{t}}(0) = \sqrt{\epsilon}, \tag{1.46}$$

and

$$\frac{d^2\bar{h}}{d\bar{t}^2} = -\frac{1}{(1+\epsilon\bar{h})^2}, \quad \bar{h}(0) = 0, \quad \frac{d\bar{h}}{d\bar{t}}(0) = 1, \tag{1.47}$$

respectively, where ϵ is a dimensionless parameter defined by

$$\epsilon = \frac{V^2}{gR}.$$

To illustrate how difficulties arise in selecting an incorrect scaling, let us modify our original problem by examining the situation when ϵ is known to be a small quantity; that is, V^2 is much smaller than gR. Then one may be tempted, in order to make an approximation, to delete the terms involving ϵ in the scaled problem. Problem (1.45) then becomes

$$(1+\overline{h})^{-2} = 0, \qquad \overline{h}(0) = 0, \qquad \frac{d\overline{h}}{d\overline{t}}(0) = 1,$$

which has no solution, and problem (1.46) becomes

$$\frac{d^2\overline{h}}{d\overline{t}^2} = -\frac{1}{(1+\overline{h})^2}, \qquad \overline{h}(0) = 0, \qquad \frac{d\overline{h}}{d\overline{t}}(0) = 0,$$

which has no physically valid solution (the graph of $\overline{h}(\overline{t})$ passes though the origin with zero slope and is concave downward, thereby making $\overline{h}$ negative). Therefore it appears that terms involving small parameters cannot be neglected. This is indeed unfortunate because this kind of technique is common practice in making approximations in applied problems. What went wrong was that (1.42) and (1.43) represent *incorrect* scalings; in these cases, terms that appear small may not in fact be small. For example, in the term $\epsilon d^2\overline{h}/d\overline{t}^2$, the parameter ϵ is small but $d^2\overline{h}/d\overline{t}^2$ may be large, and hence the term may not be negligible compared to other terms in the equation.

If, on the other hand, the term $\epsilon\overline{h}$ is neglected in (1.47), then $d^2\overline{h}/d\overline{t}^2 = -1$, or $\overline{h} = \overline{t} - \overline{t}^2/2$, after applying the initial conditions. Therefore

$$h = -\tfrac{1}{2}gt^2 + Vt,$$

and we have obtained an approximate solution that is consistent with our experience with falling bodies close to the earth. In this case we are able to neglect the small term and obtain a valid approximation because the scaling is correct. That (1.44) gives the correct time and length scales can be argued physically. If V is small, then the body will be acted on by a constant gravitational field; hence, launched with speed V, it will uniformly decelerate and reach its maximum height in V/g units of time, which is the characteristic time. It will travel a distance of about (V/g) times its average velocity $\tfrac{1}{2}(V + 0)$, or $V^2/2g$. Hence V^2/g is a good selection for the length scale. Measuring the height relative to the radius of the earth is not a good choice for small initial velocities.

In general, if a correct scaling is chosen, then terms in the equations that appear small are indeed small and may be safely neglected. In fact, one goal of scaling is to select intrinsic, characteristic reference quantities so that each term in the dimensional equation transforms into a term that the dimensionless coefficient in the transformed term represents the order of magnitude or approximate size of that term. Pictorially,

$$
\left[\begin{pmatrix} \text{Dimensional} \\ \text{term} \end{pmatrix} \right] \rightarrow \begin{bmatrix} \text{coefficient} \\ \text{representing} \\ \text{the order of} \\ \text{magnitude} \\ \text{of the term} \end{bmatrix} \cdot \left[\begin{pmatrix} \text{dimensionless} \\ \text{factor} \\ \text{of order unity} \end{pmatrix} \right].
$$

By *order of unity* we mean a term that is neither extremely large nor small.

EXERCISES

1. Let $u = u(t)$, $0 \le t \le b$, be a given smooth function. If $M = \max |u(t)|$, then u can be scaled by M to obtain the dimensionless dependent variable $U = u/M$. A time scale can be taken as $t_c = M/\max |u'(t)|$, the ratio of the maximum value of the function to the maximum slope. Find M and t_c for the following functions:

   a) $u(t) = A \sin \omega t$, $\quad t > 0$.

   b) $u(t) = Ae^{-\lambda t}$, $\quad t > 0$.

   c) $u(t) = Ate^{-\lambda t}$, $\quad 0 \le t \le 2/\lambda$.

2. Consider a process described by the function $u(t) = 1 + e^{-t/\epsilon}$ on the interval $0 \le t \le 1$, where ϵ is a small number. Use Exercise 1 to determine a time scale. Is this time scale appropriate for the entire interval $[0, 1]$? (Sketch a graph of $u(t)$ when $\epsilon = 0.05$.) Explain why two time scales might be required for a process described by $u(t)$.

3. The growth rate of an organism is often measured using carbon biomass as the "currency." The von Bertalanffly growth model is

$$
m' = ax^2 - bx^3,
$$

   where m is its biomass, x is some characteristic length of the organism, a is its biomass assimilation rate, and b is its biomass use rate. Thus, it assimilates nutrients proportional to its area, and it uses nutrients proportional to its volume. Assume $m = \rho x^3$ and rewrite the model in terms of the length x. Determine the dimensions of the constants a, b, and ρ. Select time and length scales ρ/b and a/b, respectively, and reduce the problem to dimensionless form. If $x(0) = 0$, find the length x at time t. Does this seem like a reasonable model?

4. A mass hanging on a spring is given a positive initial velocity V from equilibrium. The ensuing displacement $x = x(t)$ is governed by

$$mx'' = -ax|x'| - kx,$$
$$x(0) = 0, \quad x'(0) = V,$$

where x is the displacement, $-ax|x'|$ is a nonlinear damping force, and $-kx$ is a linear restoring force of the spring. What are the possible time scales, and on what physical processes are they based. If the restoring force is small compared to the damping force, choose appropriate time and spatial scales and non-dimensionalize the model.

5. The dynamics of a nonlinear spring–mass system is described by

$$mx'' = -ax' - kx^3,$$
$$x(0) = 0, \quad mx'(0) = I,$$

where x is the displacement, $-ax'$ is a linear damping term, and $-kx^3$ is a nonlinear restoring force. Initially, the displacement is zero and the mass m is given an impulse I that starts the motion.

a) Determine the dimensions of the constants I, a, and k.

b) Recast the problem into dimensionless form by selecting dimensionless variables $\tau = t/T$, $u = ax/I$, where the time scale T is yet to be determined.

c) In the special case that the mass is very small, choose an appropriate time scale T and find the correct dimensionless model. (A small dimensionless parameter should occur on the terms involving the mass in the original model.)

6. In a classic work modeling the outbreak of the spruce bud worm in Canada's balsam fir forests, researchers proposed that the bud worm population $n = n(t)$ was governed by the law

$$\frac{dn}{dt} = rn\left(1 - \frac{n}{K}\right) - P(n),$$

where r and K are the growth rate and carrying capacity, respectively, and $P(n)$ is a bird *predation* term given by

$$P(n) = \frac{bn^2}{a^2 + n^2},$$

where a and b are positive constants.

a) Determine the dimensions of the constants a and b.

b) Graph the predation rate $P(n)/b$ for $a = 1, 5, 10$ and make a qualitative statement about the effect that the parameter a has on the model.

c) Select dimensionless variables $N = n/a$ and $\tau = t/(a/b)$ and reduce the differential equation to dimensionless form (introduce constants $q \equiv K/a$ and $s \equiv ar/b$).

d) Researchers are often interested in equilibrium populations, or constant solutions of the differential equation. Working with the dimensionless model, and ignoring the zero population, show that there is always at least one equilibrium population, and show that there may be two or three, depending on the values of the parameters. (Hint: Plot the growth rate and the predation rate on the same set of axes.) Find the equilibrium populations when $q = 12$ and $s = 0.25$ and when $q = 35$ and $s = 0.4$.

e) In the case $q = 35$ and $s = 0.4$ use a numerical differential equations solver to graph the population curve when the initial population is given by $N(0) = 40, 25, 2$, and 0.05, respectively.

7. A rocket blasts off from the earth's surface. During the initial phase of flight, fuel is burned at the maximum possible rate α, and the exhaust gas is expelled downward with velocity β relative to the velocity of the rocket. The motion is governed by the following set of equations:

$$m'(t) = -\alpha, \quad m(0) = M,$$

$$v'(t) = \frac{\alpha\beta}{m(t)} - \frac{g}{(1 + x(t)/R)^2}, \quad v(0) = 0,$$

$$x'(t) = v(t), \quad x(0) = 0,$$

where $m(t)$ is the mass of the rocket, $v(t)$ is the upward velocity, $x(t)$ is the height above the earth's surface, M is the initial mass, g is the gravitational constant, and R is the radius of the earth. Reformulate the problem in terms of dimensionless variables using appropriate scales for m, x, v, t. (Hint: Scale m and x by obvious choices; then choose the time scale and velocity scale to ensure that the terms in the v equation and x equation are of the same order. Assume that the acceleration is due primarily to fuel burning and that the gravitational force is relatively small.)

8. In the chemical reactor problem assume that the reaction is $\mathbf{C} + \mathbf{C} \rightarrow$ products, and the chemical reaction rate is $r = kc^2$, where k is the rate constant. What is the dimension of k? Define dimensionless variables and reformulate the problem in dimensionless form. Solve the dimensionless problem to determine the concentration.

9. The temperature $T = T(t)$ of a chemical sample in a furnace at time t is governed by the initial value problem

$$\frac{dT}{dt} = qe^{-A/T} - k(T - T_f), \quad T(0) = T_0,$$

where T_0 is the initial temperature of the sample, T_f is the temperature in the furnace, and q, k, and A are positive constants. The first term on the right side is the heat generation term, and the second is the heat loss term given by Newton's law of cooling.

a) What are the dimensions of the constants q, k, A?

b) Reduce the problem to dimensionless form using T_f as the temperature scale and choosing a time scale to be one appropriate to the case when the heat loss term is large compared to the heat generated by the reaction.

10. A ball of mass m is tossed upward with initial velocity V. Assuming the force caused by air resistance is proportional to the square of the velocity of the ball and the gravitational field is constant, formulate an initial value problem for the height of the ball at any time t. Choose characteristic length and time scales and recast the problem in dimensionless form.

11. A particle of mass m moves in one dimension on the x-axis under the influence of a force

$$F(x, t) = -\frac{k}{x^2} e^{-t/a},$$

where k and a are positive constants. Initially the particle is located at $x = L$ and has zero velocity.

a) Set up an initial value problem for the location $x = x(t)$ of the particle at time t.

b) Identify the length and time scales in the problem, and introduce dimensionless variables in two different ways, reformulating the problem in dimensionless form in both cases.

12. An aging spring is sometimes modeled by a force $F(x, t) = -kxe^{-t/a}$, where k is the stiffness of the spring and a is constant. Set up and non-dimensionalize a model for oscillation of a mass m on an aging spring if $x(0) = L$ and $x'(0) = V$.

13. (Fishery management) A fishing industry on the east coast has fished an area to near depletion and has stopped its activity (assume this occurs at time zero with fish population x_0). The management assumes that the fish will recover and grow logistically with growth rate r and carrying

capacity K. The question is: when should fishing resume? Let $x(t)$ be the fish population, $b(t)$ the number of fishing boats, and q be the catchability, given in units of per boat per time, so that the harvesting rate is $h(t) = qb(t)x(t)$.

a) If t_f is the time fishing is resumed, find differential equations that govern the fish population for $t > t_f$ and $t < t_f$.

b) Assume that the fleet is constant, $b(t) = B$, and non-dimensionalize the model in Part (a), choosing the growth rate as the time scale and the carrying capacity as the population scale.

c) Determine the (dimensionless) time τ that fishing begins as a function of the dimensionless fleet size $\beta = qB/r$, subject to the condition that the fish population should remain at constant for times greater than τ. Translate this relation into dimensioned form to find $t_f = t_f(B)$.

d) Under the conditions of Part (c), what is the optimum number of boats and the maximum sustainable catch, i.e., the maximum harvesting rate?

e) Suppose p is the selling price per fish, c_b is the cost of maintenance per boat per time, w is a fisherman's wage per time, and n is the number of fishermen per boat. Argue that the rate money is earned from the fishing is $ph(t) - b(t)(c_b + nw)$. If the value of money is discounted at rate R over time, conclude that the net profit can be written as a functional of the fleet size as

$$J(b(t)) = \int_0^\infty e^{-Rt}[ph(t) - b(t)(c_b + nw)]\, dt.$$

f) Under the condition of a constant fleet size, $b(t) = B$, at what time t_f should the fleet resume fishing to maximize the profit $J(B)$? What is the optimal fleet size?

g) Use parameter values $R = 0.7$, $p = 2$, $q = 0.5$, $K = 50$, $r = 0.55$, $x_0 = 2$, $c_b + nw = 2$, to plot $J(B)$ in Part (f), and find the maximum starting time, the optimal fleet size, and the maximum profit. (This exercise was adapted from Illner et al (2005).)

14. A pendulum of length l with a bob of mass m executes a (dimensionless) angular displacement $\theta = \theta(t)$ from its attachment point, with $\theta = 0$ when the pendulum is vertically downward. See Fig. 1.4. Use Newton's law to derive the equation of motion for the pendulum by noting that the acceleration is d^2s/dt^2, where $s = l\theta$ is the length of the circular arc traveled

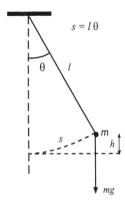

Figure 1.4 Pendulum.

by the bob; the force is $-mg$ vertically downward, and only the component tangential to the arc of oscillation affects the motion. The model is

$$\frac{d^2\theta}{dt^2} + \frac{g}{l}\sin\theta = 0.$$

If the bob is released from a small angle θ_0 at time $t = 0$, formulate a dimensionless initial value problem describing the motion. (Observe that the radian measurement θ is dimensionless, but one can still scale θ by θ_0.)

15. The initial value problem for the damped pendulum equation is

$$\frac{d^2\theta}{dt^2} + k\frac{d\theta}{dt} + \frac{g}{l}\sin\theta = 0,$$
$$\theta(0) = \theta_0, \quad \theta'(0) = \omega_0.$$

a) Find three time scales and comment upon what process each involves. Which involves the period of undamped oscillations?

b) Non-dimensionalize the model with a time scale appropriate to expecting a small-amplitude, oscillatory behavior.

1.3 Differential Equations

The first subsection below contains a very brief review of elementary solution techniques for first- and second-order differential equations. The Exercises

review standard applications with which the reader should be familiar. This review may be safely omitted for those who have fresh knowledge of elementary
methods.

1.3.1 Review of Elementary Methods

In calculus, one focus is on finding derivatives of given functions. In differential
equations the focus in on the reverse problem, namely, given a relationship
between an unknown function $u = u(t)$ and some of its derivatives, find the
function itself. The given relationship is called a differential equation. The
simplest differential equation is the **pure time** equation

$$u' = g(t), \tag{1.48}$$

where g is a given continuous function. We recognize, by definition, that a
solution $u = u(t)$ must be an antiderivative of g. By the fundamental theorem
of calculus, all antiderivatives are given by

$$u(t) = \int_a^t g(s) \, ds + C, \tag{1.49}$$

where C is an arbitrary constant of integration and a is any lower limit of
integration. Thus (1.49) gives all solutions of the pure time equation (1.48).
Antiderivatives are often denoted by the usual indefinite integral sign $u(t) =
\int g(t) dt + C$, but this representation is useless if we cannot find an expression
for $\int g(t) \, dt$. For example, if $g(t) = \exp(-t^2)$, then there is no simple formula
for the antiderivative, and thus we cannot find, for example, the antiderivative
evaluated at a specific value of t. Therefore we prefer to write the antiderivative
as $\int_0^t \exp(-s^2) \, ds + C$. A multiple of this function is the common special function
$\mathrm{erf}(t) = \frac{2}{\sqrt{\pi}} \int_0^t \exp(-s^2) \, ds$, which is the error or "erf" function.
    In general, a first order differential has the form

$$u' = f(t, u), \tag{1.50}$$

where t ranges over some specified interval I of time. A solution is a differentiable function $u = u(t)$ that, when substituted into (1.50), satisfies it identically for all t in the interval I. Often the differential equation is accompanied
by an initial condition of the form

$$u(t_0) = u_0, \tag{1.51}$$

which specifies that the solution curve pass through the point (t_0, u_0), or that
the system is in state u_0 at time t_0. The problem of solving (1.50) subject to

(1.51) is called the **initial value problem** associated with the equation. If f is a continuous function in some open region of the tu-plane containing the point (t_0, u_0), then one can show that there is a solution to the initial value problem in some interval containing t_0. If, in addition, the partial derivative f_u is continuous in the domain, then the solution is unique.

Some differential equations of first order can be solved, at least in principle. A differential equation is **separable** if it can be written in the form

$$u' = h(u)g(t).$$

In this case we can divide both sides by $h(u)$ and integrate over t to obtain

$$\int \frac{u'(t)dt}{h(u(t))} = \int g(t)\, dt + C.$$

Here we have written explicit dependence on t. Also, as noted above, sometimes the antiderivatives must be written as definite integrals with a variable upper limit of integration. Next we can change variables in the integral on the left: $u = u(t)$, $du = u'(t)\, dt$. Then

$$\int \frac{du}{h(u)} = \int g(t)\, dt + C,$$

which, when the integrals are resolved, gives the solution u implicitly as a function of t. One may, or may not, be able to solve for u and find an explicit form for the solutions.

An example of a simple, separable equation is the **growth-decay** equation, $u' = \lambda u$, from which one easily obtains the general solution $u = Ce^{\lambda t}$. The solution models exponential growth if $\lambda > 0$ and exponential decay if $\lambda < 0$.

A **first-order linear equation** has the form

$$u' + p(t)u = q(t). \tag{1.52}$$

Multiplying this equation by an integrating factor $e^{\int p(t)dt}$ turns the left side into a total derivative and (1.52) becomes

$$\frac{d}{dt}\left(ue^{\int p(t)dt}\right) = q(t)e^{\int p(t)dt}.$$

Now, both sides can be integrated to determine u. We illustrate this procedure with an example.

Example 1.8

Find an expression for the solution to the initial value problem

$$u' + 2tu = 2\sqrt{t}, \quad u(0) = 3.$$

The integrating factor is $\exp(\int 2t\, dt) = \exp(t^2)$. Multiplying both sides of the equation by the integrating factor makes the left side a total derivative, or

$$(ue^{t^2})' = 2\sqrt{t}e^{t^2}.$$

Integrating from 0 to t (while changing the dummy variable of integration to s) gives

$$u(t)e^{t^2} - u(0) = 2\int_0^t \sqrt{s}e^{s^2}\, ds.$$

Solving for u gives

$$u(x) = e^{-t^2}(3 + 2\int_0^t \sqrt{s}e^{s^2}\, ds) = 3e^{-t^2} + 2\int_0^t \sqrt{s}e^{s^2-t^2}\, ds.$$

As is often the case, the integrals in this example cannot be performed easily, if at all, and we must write the solution in terms of integrals with variable limits.

Some first-order equations can be solved by substitution. For example, **Bernoulli equations** are differential equations having the form

$$u' + p(t)u = q(t)u^n.$$

The transformation of dependent variables $w = u^{1-n}$ turns a Bernoulli equation into a linear equation for $w = w(t)$.

Some second-order equations can be immediately reduced to a first-order equation. For example, if the equation has the form

$$G(t, u', u'') = 0,$$

where the dependent variable u is missing, then the substitution $v = u'$ immediately reduces the equation to the first order equation

$$G(t, v, v') = 0.$$

If the second order equation does not depend explicitly on the independent variable t, that is, it has the form

$$G(u, u', u'') = 0,$$

then we again define $v = u'$. Then, using the chain rule, $u'' = \frac{d}{dt}u' = \frac{dv}{dt} = \frac{dv}{du}\frac{du}{dt} = \frac{dv}{du}v$, and the equation becomes

$$G(u, v, \frac{dv}{du}v) = 0,$$

which is a first order equation for $v = v(u)$.

The *linear equation with constant coefficients,*

$$au'' + bu' + cu = 0, \tag{1.53}$$

where a, b and c are constants, occurs frequently in modeling RCL circuits and spring–mass–damping systems. We recall that if $u_1(t)$ and $u_2(t)$ are two independent solutions to a second order, linear, homogeneous equation, then the general solution is $u = c_1 u_1(t) + c_2 u_2(t)$, where c_1 and c_2 are arbitrary constants. Equation (1.53) has solutions of the form $u = e^{mt}$, where m is to be determined. Substitution of this exponential form into the equation leads to the **characteristic equation**

$$am^2 + bm + c = 0.$$

This quadratic will have two roots m_1 and m_2. Three possibilities can occur: unequal real roots, equal real roots, and complex roots (which must be complex conjugates).

- m_1, m_2 real and unequal. Two independent solutions of (1.53) are $e^{m_1 t}$ and $e^{m_2 t}$.

- m_1, m_2 real and equal, i.e., $m_1 = m_2 \equiv m$. Two independent solutions of (1.53) are e^{mt} and te^{mt}.

- $m = \alpha \pm i\beta$ are complex conjugate roots. Two real, independent solutions of (1.53) are $e^{\alpha t} \sin \beta t$ and $e^{\alpha t} \cos \beta t$.

Of particular importance are the two equations $u'' + a^2 u = 0$, which has oscillatory solutions $u = c_1 \cos at + c_2 \sin at$, and $u'' - a^2 u = 0$, which has exponential solutions $u = c_1 e^{-at} + c_2 e^{at}$; we can also write this latter solution in terms of hyperbolic functions as $u = c_1 \cosh at + c_2 \sinh at$. These two equations occur so frequently that it is best to memorize them.

It is difficult to find analytic solution formulas for second order linear equations with variable coefficients. Often power series methods are applied, and they lead to series representations that define special functions like Bessel functions, and so on. However, there is a special equation that can be solved with simple formulas, namely a **Cauchy–Euler equation** of the form

$$at^2 u'' + btu' + cu = 0. \tag{1.54}$$

This equation admits power functions as solutions of the form $u = t^m$, where m is to be determined. Upon substitution, we obtain the characteristic equation

$$am(m - 1) + bm + c = 0.$$

This quadratic equation has two roots m_1 and m_2, and there are three cases:

- m_1, m_2 are real and unequal. Two independent solutions of (1.54) are t^{m_1} and t^{m_2}.

- m_1, m_2 are real and equal, i.e., $m_1 = m_2 \equiv m$. Two independent solutions of (1.54) are t^m and $t^m \ln t$.

- $m = \alpha \pm i\beta$ are complex conjugate roots. Two real, independent solutions of (1.54) are $t^\alpha \sin(\ln \beta t)$ and $t^\alpha \cos(\ln \beta t)$.

The general solution of the **linear nonhomogeneous equation**

$$u'' + p(t)u' + q(t)u = f(t) \tag{1.55}$$

is

$$u = c_1 u_1(t) + c_2 u_2(t) + u_P(t),$$

where u_1 and u_2 are independent solutions of the homogeneous equation (when $f(t) \equiv 0$), and u_P is any particular solution to (1.55). This structure result is valid for all linear equations of any order: The general solution is the sum of the general solution to the homogeneous equation and a particular solution to the nonhomogeneous equation. For constant coefficient equations a particular solution can sometimes be inferred from the form of the forcing term $f(t)$, especially when it is a polynomial, sine or cosine function, exponential function, or the sum and product of these. This is the *method of undetermined coefficients*. In any case, however, there is a general formula, called the **variation of parameters** formula, that gives a particular solution in terms of the linear independent solution of the homogeneous equation. In the case of the second-order equation (1.55) with linearly independent solutions u_1 and u_2, the formula is

$$u_P(t) = \int_a^t \frac{u_1(s)u_2(t) - u_2(s)u_1(t)}{u_1(s)u_2'(s) - u_2(s)u_1'(s)} f(s)\, ds.$$

EXERCISES

1. Find the general solution of the following differential equations.

   a. $u' + 2u = e^{-t}$.
   b. $u'' + 4u = t \sin 2t$.
   c. $u' - ty = t^2 u^2$.
   d. $t^2 u'' - 3tu' + 4u = 0$.
   e. $u'' + 9u = 3 \sec 3t$.
   f. $u'' + tu'^2 = 0$
   g. $u'' - 3u' - 4u = 2 \sin t$.

   h. $uu'' - u'^3 = 0$.
   i. $2t^2 u'' + 3tu' - u = 0$.
   j. $u' - 2tu = 1$.
   k. $u'' + 5u' + 6u = 0$.
   l. $u'' + u' + u = te^t$.
   m. $u'' + \omega^2 u = \sin \beta t$, $\quad \omega \neq \beta$.
   n. $u'' + \omega^2 u = \cos \omega t$.

2. A differential equation of the form $u' = f\left(\frac{u}{t}\right)$ is called a homogeneous equation (the word homogeneous in this case is used differently from that of an equation whose right side is zero.) Show that the change of variables $y = u/t$ transforms the equation into a separable equation for y.

3. (Particle dynamics) An object of mass m is shot vertically upward with initial velocity v_0. The force of gravity, of magnitude mg, acts on the object, as well as a force due to air resistance that is proportional to the square of the velocity.

   a) Use Newton's second law to write down the governing initial value problem.

   b) Non-dimensionalize the model and find the object's velocity as a function of time.

   c) At what time will the object reach it maximum height? How high will it go?

4. (Reaction dynamics) Consider a chemical reactor of constant volume where a chemical $\mathbf{C}$ is pumped into the reactor at constant concentration and constant flow rate. While in the reactor it reacts according to $\mathbf{C} + \mathbf{C} \rightarrow$ products. The law of mass action dictates that the rate of the reaction is $r = kC^2$, where k is the rate constant. If the concentration of $\mathbf{C}$ in the reactor is given by $C(t)$, show that mass balance leads the governing equation $(VC)' = qC_{in} - qC - kVC^2$. Assume $C(0) = c_0$. Non-dimensionalize the model by selecting the concentration scale to be concentration scale C_{in} and a time scale based on the flow-through rate. Determine the equilibria and analyze their stability. Finally, find a formula for the concentration as a function of time.

5. (Heat transfer) Consider an object of uniform temperature T_0 that is placed in an oven of temperature T_e. Over time the object heats up and eventually its temperature becomes that of the oven environment T_e. A fundamental assumption is that the heat is uniformly distributed throughout the body and there are no temperature gradients, or spatial variations, in the body itself. *Newton's law of cooling* (heating), a constitutive model inferred from experiment, requires that the rate of change of the temperature of the object is proportional to the difference between the temperature of the object and the environmental temperature, or

$$T' = -h(T - T_e).$$

The positive proportionality constant h is the *heat loss coefficient*.

a) Determine the temperature $T = T(t)$ of the object and sketch a generic plot of temperature vs. time.

b) Find an analytic formula for the temperature if the environmental temperature varies according to $T_e = \theta(t)$.

6. (Sales Response to Advertising) This exercise develops a simple model to assess the effectiveness of an advertising campaign. Let $S = S(t)$ be the monthly sales of an item. In the absence of advertising it is observed from sales history data that sales decrease at a constant per capita rate a. If there is advertising, then sales rebound and tend to saturate at some maximum value $S = M$, because there are only finitely many consumers. The rate of increase in sales due to advertising is jointly proportional to the advertising rate $A(t)$ and to the degree the market is not saturated, that is,

$$rA(t)\left(\frac{M-S}{M}\right).$$

The constant r measures the effectiveness of the advertising campaign, and $\frac{M-S}{M}$ is a measure of the market share that has still not purchased the product. Combining natural sales decay and advertising gives the model

$$S' = -aS + rA(t)\left(\frac{M-S}{M}\right).$$

If $S(0) = S_0$ and the advertising is constant over a fixed time period T and is then removed, that is,

$$A(t) = \begin{cases} a, & 0 \le t \le T, \\ 0, & t > T, \end{cases}$$

find a formula for the sales $S(t)$.

7. (Biogeography) The *MacArthur–Wilson model* of the dynamics of species (e.g., bird species) that inhabit an island located near a mainland was developed in the 1960s. Let P be the number of species in the source pool on the mainland, and let $S = S(t)$ be the number of species on the island. Assume that the rate of change of the number of species is

$$S' = \chi - \mu,$$

where χ is colonization rate and μ is the extinction rate. In the MacArthur–Wilson model,

$$\chi = I(1 - \frac{S}{P}) \quad \text{and} \quad \mu = \frac{E}{P}S,$$

where I and E are the maximum colonization and extinction rates, respectively.

Figure 1.5 RCL circuit with electromotive force $E(t)$.

a) Over a long time, what is the expected equilibrium for the number of species inhabiting the island? Is this equilibrium stable?

b) Given $S(0) = S_0$, find an analytic formula for $S(t)$.

c) Suppose there are two islands, one large and one small, with the larger island having the smaller maximum extinction rate. Both have the same colonization rate. Show that the smaller island will eventually have fewer species.

8. (Chemical reactors) A large industrial retention pond of volume V, initially free of pollutants, was subject to the inflow of a contaminant produced in the factory's processing plant. Over a period of b days the EPA found that the inflow concentration of the contaminant decreased linearly (in time) to zero from its initial initial value of a (grams per volume), its flow rate q (volume per day) being constant. During the b days the spillage rate to the local stream was also q. What is the concentration in the pond after b days? Take $V = 6000$ cubic meters, $b = 20$ days, $a = 0.03$ grams per cubic meter, and $q = 50$ cubic meters per day. With these data, how long would it take for the concentration in the pond to get below a required EPA level of 0.00001 grams per cubic meter if fresh water is pumped into the pond at the same flow rate, with the same spill over?

9. (Circuits) Consider the RCL circuit (Fig. 1.5) equation

$$Lq'' + Rq' + \frac{1}{C}q = V_0,$$

$$q(0) = q_0, \quad q'(0) = I_0,$$

where q is the charge on the capacitor (coulombs), L is the inductance (henrys), R is the resistance (ohms), C is the capacitance (farads), I_0 is the initial current (amperes per time), and $E(t)$ is the electromotive force (the voltage generated by a battery or generator).

a) Non-dimensionalize the model in the case that the electromotive force is a battery with $E(t) = V_0$ volts.

b) From part (a), discuss the dynamics and long time behavior of the
circuit in the three cases: $R^2C > 4L$, $R^2C = 4L$, and $R^2C < 4L$.

10. The charge $q = q(t)$ on a capacitor in an RC circuit is modeled by the first
order equation

$$Rq' + \frac{1}{C}q = E(t),$$

where C is the capacitance (farads), R is the resistance (ohms), and $E(t)$
the electromotive force (volts). If $q(0) = q_0$, find the charge as a function
of time. If the electromotive force is constant, what is the long-time charge
on the capacitor?

1.3.2 Stability and Bifurcation

First-order **autonomous** equations, which arise frequently in applications,
have the form

$$u' = f(u), \tag{1.56}$$

where there is no explicit dependence on time. Because (1.56) is separable, we
can immediately obtain an implicit, general solution of the form $\int \frac{1}{f(u)} du =$
$t + C$. In only the simplest equations can the integral be resolved, and so we
almost always resort to a geometric, or qualitative, analysis of (1.56), which
gives the essential behavior of the solutions. Usually, the qualitative behavior is
what we want to know; accurate approximations can be obtained by numerical
methods (e.g., Runge–Kutta methods).

Key to the qualitative analysis is to determine the equilibria of an equation
and the dynamics near the equilibria. An *equilibrium solution* of (1.56) is a
constant solution, which must be clearly a root of the algebraic equation $f(u) =$
0. Thus, $u(t) = u^*$ is an equilibrium solution if, and only if, $f(u^*) = 0$. The
values u^* are the values where the graph of $f(u)$ versus u intersects the u
axis. We always assume the equilibria are *isolated*, i.e., if u^* is an equilibrium,
then there is an open interval containing u^* that contains no other equilibria.
Fig. 1.6 shows a generic plot of $f(u)$, where the equilibria are $u^* = a$, b, c. In
between the equilibria we can observe the values of u for which u is increasing
($f(u) > 0$) or decreasing ($f(u) < 0$). We may place arrows on the u axis,
or **phase line**, in between the equilibria showing direction of the *flow* of u
(increasing or decreasing) as time increases. If desired, the information from
the phase line can be translated into time series plots of $u(t)$ versus t (Fig. 1.7).

On the phase line, if arrows on both sides of an equilibrium point toward
that equilibrium point, then we say the equilibrium point is an **attractor**. If

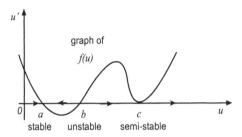

Figure 1.6 Generic phase line plot.

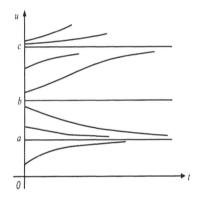

Figure 1.7 Time series plots for the equation whose phase line plot is shown in Fig. 1.6.

both of the arrows points away, the equilibrium is a **repeller**. If one points toward the equilibrium and one points away, the equilibrium is called **hyperbolic**. Attractors are **locally asymptotically stable** because if the system is in that constant equilibrium state and is given a small *perturbation* (i.e., a change or "bump") to a nearby state, then it returns to that state as $t \to +\infty$. In one-dimensional systems we often us the word **stable** in place of locally asymptotic stable; in higher-dimensional systems we must distinguish between these two terms. It seems plausible that real systems will seek out stable states.

Repellers and hyperbolic points are **unstable** because a small perturbation can cause the system to go to a different equilibrium or even go off to infinity.

When we state that an equilibrium u^* is locally asymptotically stable, the understanding is that it is with respect to *small* perturbations. To fix the idea, consider a population of fish in a lake that is in a locally asymptotically stable state u^*. A small death event, say caused by some toxic chemical that is dumped into the lake, will cause the population to drop. Local asymptotic stability means that the system will return the original state u^* over time. If many fish are killed by the pollution event, then the perturbation is not small and there is no guarantee that the fish population will return to the original state u^*; for example, it may go to some other equilibrium state or become extinct. If the population returns to the state u^* for all perturbations, no matter how large, then the state u^* is called **globally asymptotically stable**. A more precise definition of local asymptotic stability can be given as follows. An isolated equilibrium state u^* of (1.56) is **locally asymptotically stable** if there is an open interval I containing u^* with $\lim_{t \to +\infty} u(t) = u^*$ for any solution $u = u(t)$ of (1.56) with $u(0)$ in I. That is, each solution starting in I converges to u^*.

Graphically, stability can be determined from the slope of the tangent line $f'(u^*)$ to the graph of $f(u)$ vs. u at the equilibrium state. If $f'(u^*) < 0$, the graph of $f(u)$ through the equilibrium falls from positive to negative values, giving an arrow pattern of an attractor. Similarly, if $f'(u^*) > 0$, we obtain the arrow pattern of a repeller. If $f'(u^*) = 0$, then u^* may be an attractor, repeller, or hyperbolic.

Using linearization, we can confirm these graphical criteria analytically. If (1.56) is in an equilibrium state u^* and $U = U(t)$ represents a small perturbation from that state, then $u(t) = u^* + U(t)$ and

$$u' = U' = f(u^* + U) = f(u^*) + f'(u^*)U + \cdots,$$

where the dots denote higher-order terms in U. We get

$$U' = f'(u^*)U,$$

which is called the **linearization** about the equilibrium u^*. This is a linear growth-decay equation and its solution is $U(t) = Ce^{\lambda t}$, where $\lambda = f'(u^*)$, which is a constant. If $\lambda < 0$ the perturbation decay, and if $\lambda > 0$ the perturbation grows, implying local asymptotic stability and instability, respectively.

Example 1.9

(Logistics model) The logistics model of population growth is

$$\frac{dp}{dt} = rp\left(1 - \frac{p}{K}\right),$$

where r is the growth rate and K is the carrying capacity. The equilibria are $p^* = 0$, K, found by setting the right side $f(p) = rp\left(1 - \frac{p}{K}\right) = 0$. The equilibrium $p^* = K$ is asymptotically stable because $f'(K) = -r < 0$, and the zero population is unstable because $f'(0) = r > 0$.

Example 1.10

(**Harvesting a population**) More often than not, a differential equation contains parameters. Let us modify the logistics model in the last paragraph by adding a harvesting term. That is, consider a population $p(t)$ that is governed by logistics growth, while at the same time it is harvested at a constant rate $H > 0$. Then the dynamics is

$$\frac{dp}{dt} = rp\left(1 - \frac{p}{K}\right) - H. \tag{1.57}$$

There are three constants in this problem, r, K, and H, and we want to examine how the solutions change as the harvesting rate H varies, with r and K held fixed. To aid in the analysis we non-dimensionalize the problem by introducing a scaled population u and a scaled time τ defined by

$$u = \frac{p}{K}, \quad t = r\tau.$$

In other words, we measure population relative to the carrying capacity and time relative to the reciprocal of the growth rate. Then equation (1.57) becomes

$$\frac{du}{d\tau} = u(1 - u) - h \tag{1.58}$$

where $h \equiv H/rK$ is the dimensionless harvesting parameter. Treating h as a control parameter, we inquire if equilibrium populations exist. These are easily found by setting $du/dt = 0$, or

$$u(1 - u) - h = 0$$

Solving for u gives

$$u^* = \frac{1 \pm \sqrt{1 - 4h}}{2} \tag{1.59}$$

Therefore, if the harvesting parameter exceeds $\frac{1}{4}$, then no equilibrium populations exist. If $0 < h < \frac{1}{4}$, then there are two possible equilibrium states. Usually we sketch the locus (1.59) on a hu^*-coordinate system to graphically indicate the dependence of the equilibrium solutions on the parameter h. Such a graph is shown in Fig. 1.8 and it is called a **bifurcation diagram**; the harvesting parameter h is called a **bifurcation parameter**. We observe that if h is too large (i.e., there is too much harvesting of the population), then no equilibrium

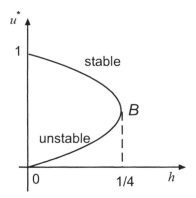

Figure 1.8 Bifurcation diagram.

populations exist and (1.58) shows that $du/d\tau < 0$, so that the population dies out. As harvesting is decreased, there is a critical value of h ($h = 1/4$) below which two equilibrium populations can exist; in these two populations, there is a balance between growth and harvesting. The point B on the graph where the two solutions appear is called a bifurcation point. Now let us ask if nature prefers one of these equilibrium populations over the other. The resolution of this question lies in the stability properties of those states. We would expect nature to select out a stable state over an unstable one; for, if a small change is made in an unstable state (and small deviations are always present), then the system would leave that state permanently; if the state were stable, then small perturbations will decay away. We could perform a linearized stability analysis as in the preceding subsection to determine which, if any, of the two equilibrium populations given by (1.59) is unstable. But let us proceed a different way and phase line of the differential equation (1.58) when $0 < h < \frac{1}{4}$. Let u^+ and u^- denote the two equilibrium populations in (1.59) with the plus and minus signs, respectively. A plot of the right side of (1.58) for a fixed $0 < h < \frac{1}{4}$ is shown in Fig. 1.9. The phase line clearly indicates that the larger equilibrium u^+ is stable, and the smaller u^- is unstable. Consequently, nature would select out the larger equilibrium population for a given harvesting rate $0 < h < \frac{1}{4}$. In terms of the bifurcation diagram in Fig. 1.8, the upper branch is stable and the lower branch is unstable. A bifurcation is said to occur at the point B.

Example 1.11

(**Non-isothermal tank reactor**) In this section we set up a model for deter-

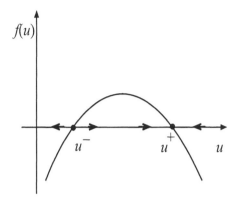

Figure 1.9 Plot of $f(u)$ vs. u and the phase line.

mining the temperature $\bar{\theta}$ and concentration $\bar{c}$ of a heat-releasing, chemically reacting substance in a continuously stirred tank reactor. (We are using over-bars in the variables in anticipation of using unbarred dimensionless variables later.) The reaction takes place in a tank of volume V (see Fig. 1.2) that is stirred continuously in order to maintain uniform temperature and concentration. It is fed by a stream of constant flow velocity q, constant reactant concentration c_i, and constant temperature θ_i. After mixing and reacting the products are removed at the same volume rate q. We assume that the exothermic reaction $\mathbf{C} \to$ products is first-order and irreversible, and the reactant disappears at the rate

$$-k\bar{c}e^{-A/\bar{\theta}},$$

where A and k are positive constants. The temperature-dependent factor $e^{-A/\bar{\theta}}$ in the rate is the Arrhenius factor, where A is proportional to the activation energy, or the amount of kinetic energy needed for the reaction to occur. The amount of heat released is assumed to be given by

$$hk\bar{c}e^{-A/\bar{\theta}},$$

where h is a positive constant, the specific heat of reaction, measured in energy per mass.

We are able to write a system of two differential equations that govern the concentration $\bar{c}(\bar{t})$ and the temperature $\bar{\theta}(\bar{t})$ of the reactant in the tank. One equation comes from balancing the mass of the reactant, and the other arises from conservation of (heat) energy. First, mass balances gives (see Section 2.2)

$$V \frac{d\bar{c}}{d\bar{t}} = qc_i - q\bar{c} - Vk\bar{c}e^{-A/\bar{\theta}}. \tag{1.60}$$

This equation states that the rate of change of mass of reactant in the tank equals the mass in, less the mass out, plus the rate at which the mass of the reactant disappears in the chemical reaction. Next, heat balance gives

$$VC \frac{d\bar{\theta}}{d\bar{t}} = qC\theta_i - qC\bar{\theta} + hVk\bar{c}e^{-A/\bar{\theta}}, \tag{1.61}$$

where C is the heat capacity of the mixture, measured in energy per volume per degree. These equations can be reduced to dimensionless form by introducing dimensionless variables

$$t = \frac{\bar{t}}{V/q}, \quad \theta = \frac{\bar{\theta}}{\theta_i}, \quad c = \frac{\bar{c}}{c_i}$$

and dimensionless constants

$$\mu = \frac{q}{kV}, \quad b = \frac{hc_i}{C\theta_i}, \quad \gamma = \frac{A}{\theta_i}.$$

In this case the mass and energy balance equations (1.60) and (1.61) become

$$\frac{dc}{dt} = 1 - c - \frac{c}{\mu} e^{-\gamma/\theta}, \tag{1.62}$$

$$\frac{d\theta}{dt} = 1 - \theta + \frac{bc}{\mu} e^{-\gamma/\theta}. \tag{1.63}$$

This pair of equations can be reduced to a single equation using the following argument. If we multiply (1.62) by b and add it to (1.63), we get

$$\frac{d}{dt}(\theta + bc) = 1 + b - (\theta + bc),$$

which integrates to

$$\theta + bc = 1 + b + De^{-t},$$

where D is a constant. Assuming $\theta + bc$ at $t = 0$ is $1 + b$ gives $D = 0$; then $\theta + bc = 1 + b$. Consequently the heat balance equation becomes

$$\frac{d\theta}{dt} = 1 - \theta + \frac{1 + b - \theta}{\mu} e^{-\gamma/\theta},$$

which is a single differential equation for θ. Introducing $u = \theta - 1$, this equation becomes

$$\frac{du}{dt} = -u + \frac{b - u}{\mu} e^{-\gamma/(u+1)} = \frac{1}{\mu} \left(-\mu u + (b - u) e^{-\gamma/(u+1)} \right), \tag{1.64}$$

which is the final form that we shall study.

By definition, the parameter μ in (1.64) can be identified as the ratio of the flow rate to the reaction rate, and it acts as the bifurcation parameter in our

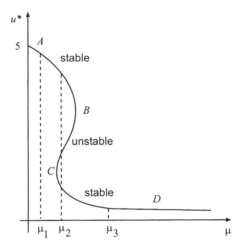

Figure 1.10 Bifurcation diagram.

analysis because we wish to study equilibrium solutions as a function of flow rate. The equilibrium solutions are given by solutions to the equation

$$\mu u^* = (b - u^*)e^{-\gamma/(u^*+1)}. \tag{1.65}$$

Because (1.65) cannot be solved in analytic form, we resort to graphical techniques. We can plot μ vs. u^* easily and then rotate the plot to obtain a bifurcation diagram with abscissa μ. Figure 1.10 shows one case ($b = 5$, $\gamma = 8$, not drawn to scale). For a small value $\mu = \mu_1$ there is one large temperature equilibrium. As μ increases slowly, to maintain equilibrium, there is a bifurcation at point C to two, then three, equilibria. As μ increases still further, say up to $\mu = \mu_3$, there is a bifurcation at point C back to a single, lower, equilibrium temperature. To determine the stability of the branches AB, CB, and CD let us write $u' = \frac{1}{\mu}(-\mu u + g(u))$, where $g(u) = (b - u) e^{-\gamma/(u+1)}$. The equilibria occur when the graphs of $g(u)$ and μu intersect. Figure 1.11 shows three cases: $\mu = \mu_1, \mu = \mu_2, \mu = \mu_3$. For μ values where $g(u) > \mu u$ we have $u' > 0$, and or μ values where $g(u) < \mu u$ we have $u' > 0$. This means the upper temperature branch AB must be asymptotically stable, the middle branch BC must be unstable, and the lower temperature branch must be asymptotically stable.

Physically, if the flow rate μ is fast, say $\mu = \mu_3$, the reactor is operating at a low temperature. If it is slowly decreased, the temperature changes along the lower, stable branch CD until it reaches the point C. At that time a bifurcation must occur, and further decrease in μ will cause the temperature to jump suddenly (and dynamically) to the upper, hotter branch. We say *ignition* has

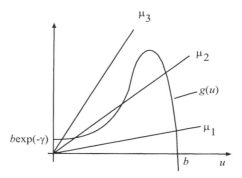

Figure 1.11 Plot of the left and right sides of equation (1.65).

occurred. Conversely, if the flow rate is increased from μ_1, on reaching state B there is a sudden jump to the lower temperature state along the stable branch CD. This bifurcation is called *quenching* the reaction. It is not difficult to understand why bifurcation phenomena is important in nuclear reactors.

EXERCISES

1. In the usual Malthus growth law $p' = rp$ for a population of size p, assume the growth rate is a linear function of food availability F; that is, $r = bF$, where b is the conversion factor of food into newborns. Assume that F_T is the total, constant food in the system with $F_T = F + cp$, where cp is amount of food already consumed. Write down a differential equation for the population p. What is the population as t gets large? Find the population as a function of time if $p(0) = p_0$.

2. A model of tumor growth is the Gompertz equation

$$R' = -aR\ln\left(\frac{R}{k}\right),$$

where $R = R(t)$ is the tumor radius, and a and k are positive constants. Find the equilibria and analyze their stability. Find the solution $R = R(t)$ and plot several solution curves.

3. A fish population in a lake is harvested at a constant rate H, and it grows logistically. The growth rate is 0.2 per month, the carrying capacity is 40 (thousand), and the harvesting rate is 1.5 (thousand per month). Write down the model equation, find the equilibria, and classify them. Will the fish population ever become extinct? What is the most likely long-term fish population?

4. For the following equations, find the equilibria and sketch the phase line. Determine the type and stability of all the equilibria.

   a) $u' = u^2(3 - u)$.

   b) $u' = 2u(1 - u) - \frac{1}{2}u$.

   c) $u' = (4 - u)(2 - u)^3$.

5. For the following models, which contain a parameter h, find the equilibria in terms of h and determine their stability. Construct a bifurcation diagram showing how the equilibria depend upon h, and label the branches of the curves as unstable or stable.

   a) $u' = hu - u^2$.

   b) $u' = (1 - u)(u^2 - h)$.

6. Consider the model $u' = (\lambda - b)u - au^3$, where a and b are fixed positive constants and λ is a parameter that may vary.

   a) If $\lambda < b$ show that there is a single equilibrium and that it is asymptotically stable.

   b) If $\lambda > b$ find all the equilibrium and determine their stability.

   c) Sketch the bifurcation diagram showing how the equilibria vary with λ. Label each branch of the curves shown in the bifurcation diagram as asymptotically stable or unstable.

7. An animal species of population u grows exponentially with growth rate r. At the same time it is subjected to predation at a rate $\frac{au}{b+u}$, which depends upon its population. The constants a and b are positive, and the dynamics is
$$u' = ru - \frac{au}{b + u}.$$

   a) Non-dimensionalize the problem with the simplest choice of time and population scales, and introduce a single dimensionless parameter h.

   b) Sketch a bifurcation diagram.

   c) Determine the stability of the equilibria as a function of h and indicate the results on the diagram.

8. The biomass P of a plant grows logistically with intrinsic growth rate r and carrying capacity K. At the same time it is consumed by herbivores at a rate
$$\frac{aP}{b + P},$$

per herbivore, where a and b are positive constants. The model is

$$P' = rP(1 - \frac{P}{K}) - \frac{aPH}{b + P},$$

where H is the density of herbivores, and r, K, a, and b are fixed constants.

a) Plot, as a function of P, the growth rate and the consumption rate for several values of H on the same set of axes, and identify the values of P which give equilibria.

b) Draw the bifurcation diagram (equilibria vs. H) and label each branch as stable or unstable.

c) Explain what happens to steady states if herbivory H is slowly increased from a small value to a large value? Next explain how steady states evolve if herbivory is slowly decreased back to a small value.

9. A deer population grows logistically, but it is harvested at a rate proportional its population size. Conclude that the dynamics of population growth is given by

$$P' = rP(1 - \frac{P}{K}) - \mu P,$$

where μ is the *per capita* harvesting rate. Non-dimensionalize the problem and use a bifurcation diagram to explain the effects on the equilibrium deer population when μ is slowly increased from a small value.

10. Draw a bifurcation diagram for the model $u' = u^3 - u + h$, where h is the bifurcation parameter, using h as the abscissa. Label branches of the curves as stable or unstable.

11. A one-dimensional system is governed by the dynamical equation

$$u' = 4u(a - u) - \eta e^{-u},$$

where a and η are positive constants. Holding η constant, draw a bifurcation diagram with respect to the parameter a. Indicate the stable and unstable branches.

1.4 Two-Dimensional Problems

1.4.1 Phase Plane Phenomena

We now extend the analysis to a system of two simultaneous equations

$$x' = P(x, y), \quad y' = Q(x, y), \tag{1.66}$$

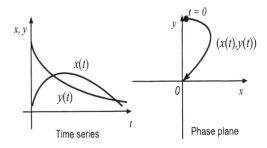

Figure 1.12 Time series and orbits in phase space.

in two unknowns x and y. The functions P and Q are assumed to possess continuous partial derivatives of all orders in a domain D of the xy plane. A system of the type (1.67) in which the independent variable t does not appear in P and Q is said to be **autonomous**. Under these assumptions on P and Q, it can be shown that there is a unique solution $x = x(t)$, $y = y(t)$ to the **initial value problem**

$$x' = P(x, y), \quad y' = Q(x, y), \qquad (1.67)$$
$$x(t_0) = x_0, \quad y(t_0) = y_0,$$

where t_0 is an instant of time and $(x_0, y_0) \in D$. The solution is defined in some interval $\alpha < t < \beta$ containing t_0. There are two ways to plot the solution. It can be plotted as time series in state space (x and y vs. t). Or, if $x(t)$ and $y(t)$ are not both constant functions, then $x = x(t)$ and $y = y(t)$ define the parametric equations of a curve that can be plotted in the xy plane, called the **phase plane**. This curve is called an **orbit** of the system (1.67). Both the time series and a corresponding orbit are shown in Fig. 1.12. The orbit is directed in the sense that it is traced out in a certain direction as t increases; this direction is often indicate by an arrow. Because the solution to the initial value problem is unique, it follows that at most one orbit passes through each point of the phase plane, and all of the orbits cover the entire phase plane without intersecting each other. A constant solution $x(t) = x_0$, $y(t) = y_0$ of (1.67) is called an **equilibrium** or steady-state solution. Such a solution does not define an orbit in the phase plane, but rather a point, often called a **critical point**. Clearly, critical points occur where both P and Q vanish, that is,

$$P(x_0, y_0) = Q(x_0, y_0) = 0.$$

It is evident that no orbit can pass through a critical point; otherwise, uniqueness would be violated. The totality of all the orbits, their directions, and critical points graphed in the phase plane is called the **phase diagram** of the system (1.67). Indeed, the qualitative behavior of all the orbits in the phase plane is determined to a large extent by the location of the critical points and the local behavior of the orbits near those points. One can prove the following results, which essentially form the basis of the Poincaré[4]–Bendixson theorem (Section 4.3).

(i) An orbit cannot approach a critical point in finite time; that is, if an orbit approaches a critical point, then necessarily $t \to \pm\infty$.

(ii) As $t \to \pm\infty$, an orbit is either a critical point, approaches a critical point, moves along a closed orbit, approaches a closed orbit, or leaves every bounded set.

A closed orbit corresponds to a periodic solution of (1.67). These matters are developed further in the sequel.

Example 1.12

Consider the system

$$x' = y, \quad y' = -x. \tag{1.68}$$

Here $P = y$ and $Q = -x$ and $(0,0)$ is the only critical point corresponding to the equilibrium solution $x(t) = 0$, $y(t) = 0$. To find the orbits we note that

$$x'' + x = 0,$$

and so

$$x = x(t) = c_1 \cos t + c_2 \sin t, \quad y = y(t) = -c_1 \sin t + c_2 \cos t$$

give the solution, where c_1 and c_2 are constants. We may eliminate the parameter t by squaring and adding to obtain

$$x^2 + y^2 = c_1^2 + c_2^2.$$

Therefore the orbits are circles centered at the origin and the phase diagram is shown in Fig. 1.13. The critical point is called a *center*. Alternatively, we may divide the two equations in the system (1.68) to obtain

$$\frac{dy}{dx} = -\frac{x}{y}.$$

---

[4] Henri Poincaré (1860–1934) was called the *last universalist*. His contributions to nearly all areas of mathematics made him one of the most influential mathematicians of all time. His work in dynamical systems paved the way to modern topological methods.

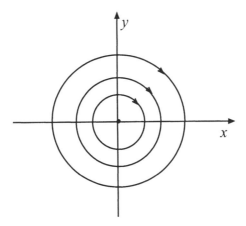

Figure 1.13 Orbits representing a center.

This can be integrated directly to obtain the orbits

$$x^2 + y^2 = C.$$

In this example all the paths are closed and each corresponds to a periodic solution of period 2π because

$$x(t + 2\pi) = x(t), \quad y(t + 2\pi) = y(t)$$

Example 1.13

The autonomous system

$$x' = 2x, \quad y' = 3y,$$

is *uncoupled* and each equation can be solved separately to obtain the solution

$$x = c_1 e^{2t}, \quad y = c_2 e^{3t}.$$

The origin $(0, 0)$ is the only critical point. The orbits can be found by eliminating t, or by integrating

$$\frac{dy}{dx} = \frac{3y}{2x}.$$

We obtain $y^2 = cx^3$, where c is a constant and the orbits are plotted in Fig. 1.14. A critical point having this behavior is called a node. Because the orbits are leaving the origin, the origin is called an unstable critical point, or repeller.

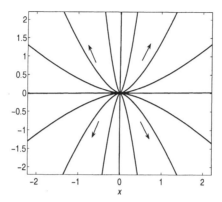

Figure 1.14 Phase diagram showing an unstable node where all the orbits are emitted from the origin.

Example 1.14

Consider a particle of mass m moving in one dimension x subject to a force $f(x, x')$. By Newton's second law the governing equation of motion is

$$mx'' = f(x, x').\tag{1.69}$$

Equation (1.69) may be recast as a system of two first-order equations by introducing as a new variable $y = x'$. Then

$$x' = y,$$
$$y' = \frac{1}{m}f(x, y).\tag{1.70}$$

Critical points occur at $(x^*, 0)$, where $f(x^*, 0) = 0$; these correspond to points where the velocity and the acceleration vanish. Thus the particle is in equilibrium and no forces are acting on it. If f depends only on position, $f = f(x)$, then the force is conservative and there is a potential function $V = V(x)$ defined by $\frac{dV}{dx}(x) = -f(x)$, or $V(x) = -\int f(x)\,dx$. In this special case the total energy (kinetic plus potential energy) is conserved, or

$$\frac{1}{2}my^2 + V(x) = E,\tag{1.71}$$

where E is constant. This follows from

$$\frac{d}{dt}\left(\frac{1}{2}my^2 + V(x)\right) = myy' + \frac{dV}{dx}x' = myy' - f(x)y = 0.$$

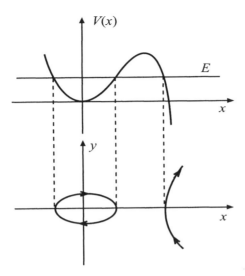

Figure 1.15 Orbit showing a constant energy curve.

For conservative systems we can draw the phase diagram by solving (1.71) for y (velocity) as a function of x (position) to get

$$y = \pm\sqrt{\frac{2}{m}(E - V(x))}. \qquad (1.72)$$

Each energy level E, determined by the particle's initial position and velocity, defines an orbit in regions where $E > V(x)$. The orbits can be obtained quickly by plotting the potential $V(x)$, selecting different values of E, and sketching the curves (1.72), with symmetry about the x axis. Figure 1.15 shows a generic potential energy function and one such orbit. In regions where $E < V(x)$ there are no orbits; values of x for which $E = V(x)$ are called turning points.

The functions P and Q on the right side of (1.67) can be regarded as the components of a velocity vector field $\mathbf{v}(x,y) = \langle P(x,y), Q(x,y)\rangle$ in the plane representing the velocity of a two-dimensional fluid motion or flow. The orbits $x = x(t)$, $y = y(t)$ of the fluid particles are tangent to $\mathbf{v}$ and therefore satisfy (1.67). The critical points are points where $\mathbf{v} = \mathbf{0}$, or where the fluid particles are at rest. In fluid mechanics these points are called stagnation points. In plotting the orbital structure in the phase plane, we sometimes plot separately the locus $P(x,y) = 0$ (where the vector field is vertical) and the locus $Q(x,y) = 0$ (where the vector field is horizontal); these loci are called the x- and y-**nullclines**, respectively. Sketching nullclines and critical points, and plotting

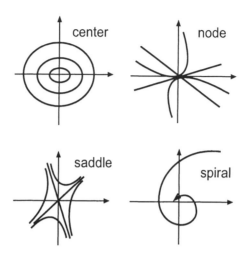

Figure 1.16 Types of critical points for linear systems.

the directions of the vector field often facilitate drawing a generic phase plane. Of course, software can plot the orbits for us if the parameters have specific values.

In principle the critical points of (1.67) can be found by solving the simultaneous equations $P(x, y) = 0$ and $Q(x, y) = 0$. The orbits in the phase plane can be found by integrating the differential relationship

$$\frac{dy}{dx} = \frac{Q(x, y)}{P(x, y)}.$$

A critical point of (1.67) is said to be **isolated** if there is a neighborhood of the critical point that contains no other critical points. There are four types of isolated critical points that occur in linear problems, where both P and Q are linear functions of x and y; these are called *centers, nodes, saddles*, and *spirals*. In Fig. 1.16 these critical points are drawn schematically with the local orbit structure. In nonlinear systems critical points can occur where, for example, a node structure is on one side and a saddle structure is on the other, as shown in Fig. 1.17. It is an important exercise to determine the location and orbital structure of the critical points for a given problem. Critical points represent equilibrium solutions of (1.67), and another important issue is that of stability. That is, does the equilibrium solution have some degree of permanence to it when it is subjected to small perturbations? Roughly speaking, a critical point is stable if all paths that start sufficiently close to the point remain close to the point. To formulate this in mathematical terms, let us suppose that the origin $(0, 0)$ is an isolated critical point of (1.67). We say that $(0, 0)$ is **stable** if for

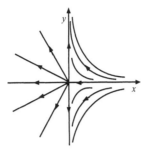

Figure 1.17 Example of the orbital structure near an critical point in a nonlinear system.

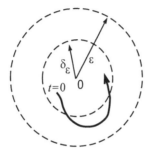

Figure 1.18 Definition of stability.

each $\varepsilon > 0$ there exists a positive number δ_ε such that every path that is inside the circle of radius δ_ε at some time t_0 remains inside the circle of radius ε for all $t > t_0$ (see Fig. 1.18). The critical point is **locally asymptotically stable** if it is stable and there exists a circle of radius δ_ε such that every path that is inside this circle at $t = t_0$ approaches $(0,0)$ as $t \to \infty$. If $(0,0)$ is not stable, then it is **unstable** . We note that a center is stable, but not asymptotically stable; a saddle is unstable. A spiral is either asymptotically stable or unstable, depending on the direction of the paths, and similarly for nodes. To determine analytic criteria for stability we can proceed as we did for a single differential equation, namely to linearize near the critical point. Suppose (x_0, y_0) is an isolated critical point of (1.67).[5] This means $x_0(t) = x_0$, $y_0(t) = y_0$, is an equilibrium solution. Let

$$x(t) = x_0 + u(t),$$
$$y(t) = y_0 + v(t),$$

---

[5] If (x_0, y_0) is a critical point, then the transformation $\overline{x} = x - x_0, \overline{y} = y - y_0$ translates the critical point to the origin.

where $u(t)$ and $v(t)$ represent small perturbations from the equilibrium state. Substituting into (1.67) gives the perturbation equations

$$\frac{du}{dt} = P(x_0 + u, y_0 + v), \qquad \frac{d\tilde{y}}{dt} = Q(x_0 + u, y_0 + v).$$

By Taylor's theorem,

$$P(x_0 + u, y_0 + v) = P(x_0, y_0) + P_x(x_0, y_0)u + P_y(x_0, y_0)v + O(u^2 + v^2)$$
$$= au + bv + O(u^2 + v^2),$$

where $a = P_x(x_0, y_0)$ and $b = P_y(x_0, y_0)$. Similarly,

$$Q(x_0 + u, y_0 + v) = cu + dv + O(u^2 + v^2),$$

where $c = Q_x(x_0, y_0)$ and $d = Q_y(x_0, y_0)$. Neglecting the higher-order terms, we obtain the linearized perturbation equations, or the **linearization** near the critical point:

$$\frac{du}{dt} = au + bv, \qquad (1.73)$$

$$\frac{dv}{dt} = cu + dv.$$

The matrix associated with this linearization,

$$A(x_0, y_0) = \begin{pmatrix} a & b \\ c & d \end{pmatrix},$$

is called the **Jacobian matrix** of the system at the equilibrium (x_0, y_0). It seems reasonable that the stability of the equilibrium (x_0, y_0) of the nonlinear system (1.67) is indicated by the stability of $(0, 0)$ of the associated linearized system (1.73). Under suitable conditions, discussed in subsequent sections, this is true. In the next section we analyze linear systems.

EXERCISES

1. In the xy phase plane draw the orbits of a conservative mechanical system governed by $mx'' = F(x)$ if the potential energy function is $V(x) = x^3(1 - x)^2$. For which values of the total energy E does the system oscillate?

2. Redo Exercise 1 if the force is $F(x) = 3x^2 - 1$.

3. Find equations of the orbits in the xy plane for the system $x' = y^2$, $y' = -\frac{2}{3}x$, and sketch the phase diagram. Is the origin stable or unstable?

4. Find the linearization of the system

$$x' = y - x,$$
$$y' = -y + \frac{5x^2}{4 + x^2},$$

about the equilibrium point $(1, 1)$.

1.4.2 Linear Systems

Consider a two-dimensional linear system

$$x' = ax + by, \tag{1.74}$$
$$y' = cx + dy,$$

where a, b, c, and d are constants. We assume that the coefficient matrix is nonsingular, or

$$ad - bc \neq 0. \tag{1.75}$$

This condition guarantees that the algebraic system

$$ax + by = 0,$$
$$cx + dy = 0$$

has an isolated critical point at the origin. Otherwise, there is an entire line of non-isolated critical points. It is easier to examine (1.74) using matrix notation. This is advantageous because the notation generalizes immediately to higher dimensions. Letting

$$\mathbf{x} = \begin{pmatrix} x \\ y \end{pmatrix}, \quad A = \begin{pmatrix} a & b \\ c & d \end{pmatrix},$$

(1.74) can be written

$$\mathbf{x}' = A\mathbf{x}. \tag{1.76}$$

Condition (1.75) is $\det A \neq 0$. Following our experience with linear equations with constant coefficients, we try a solution of (1.76) of the form

$$\mathbf{x} = \mathbf{v}e^{\lambda t}, \tag{1.77}$$

where $\mathbf{v}$ and λ are to be determined. Substituting (1.77) into (1.76) gives, after simplification,

$$A\mathbf{v} = \lambda\mathbf{v}, \tag{1.78}$$

which is the **algebraic eigenvalue problem**. Consequently λ is an eigenvalue of A and $\mathbf{v}$ is an associated eigenvector. The eigenvalues of A are found as roots of the equation

$$\det(A - \lambda I) = 0, \tag{1.79}$$

and the corresponding eigenvector(s) are then determined by solving the homogeneous system

$$(A - \lambda I)\mathbf{v} = 0. \tag{1.80}$$

Because (1.79) is a quadratic equation in λ, it will have two roots. Hence, there are several cases to consider, depending on whether the roots λ_1 and λ_2 are real, equal or unequal, of the same or different signs, are complex, or purely imaginary. We work out one of the cases in detail and indicate the solution for the others, leaving those details to the reader.

– $0 < \lambda_1 < \lambda_2$ (λ_1 and λ_2 are real, distinct, and positive). Let $\mathbf{v}_1$ and $\mathbf{v}_2$ be independent eigenvectors corresponding to λ_1 and λ_2, respectively. Then $\mathbf{v}_1 e^{\lambda_1 t}$ and $\mathbf{v}_2 e^{\lambda_2 t}$ are linearly independent solutions of (1.76), and the general solution is given by

$$\mathbf{x}(t) = c_1 \mathbf{v}_1 e^{\lambda_1 t} + c_2 \mathbf{v}_2 e^{\lambda_2 t}, \tag{1.81}$$

where c_1 and c_2 are arbitrary constants. If we set $c_2 = 0$ then $\mathbf{x} = c_1 \mathbf{v}_1 e^{\lambda_1 t}$ gives two **linear orbits** consisting of two rays emanating from the origin in the directions defined by $\mathbf{v}_1$ and $-\mathbf{v}_1$. Similarly, if $c_1 = 0$, then $\mathbf{x} = c_2 \mathbf{v}_2 e^{\lambda_2 t}$ represents two linear orbits in the directions $\mathbf{v}_2$ and $-\mathbf{v}_2$. These four rays are called **separatrices**. If $c_1 \neq 0$ and $c_2 \neq 0$, then (1.81) represents curved orbits that enter $(0,0)$ as t goes backward in time, i.e., $t \to -\infty$. To see how orbits enter the origin we note that for large negative t the term $e^{\lambda_1 t}$ dominates $e^{\lambda_2 t}$ since $\lambda_2 > \lambda_1$. Thus

$$\mathbf{x}(t) \approx c_1 \mathbf{v}_1 e^{\lambda_1 t}, \quad \text{large } t,$$

and so the orbits enter the origin tangent to $\mathbf{v}_1$. For large positive t all the orbits (1.81) go to infinity with slope asymptotic to $\mathbf{v}_2$, because in that case $e^{\lambda_2 t}$ dominates $e^{\lambda_1 t}$. The origin $(0,0)$ is an **unstable node**.

– $\lambda_2 < \lambda_1 < 0$ (real, distinct, negative). This case is similar to the preceding case, but $(0,0)$ is an **asymptotically stable node**. The direction of the flow is reversed.

– $\lambda_1 < 0 < \lambda_2$ (real, distinct, opposite signs). The general solution is

$$\mathbf{x}(t) = c_1 \mathbf{v}_1 e^{\lambda_1 t} + c_2 \mathbf{v}_2 e^{\lambda_2 t}.$$

As in the first two cases, the eigenvectors $\mathbf{v}_1$ and $\mathbf{v}_2$ and their negatives define linear orbits. One set, corresponding to the eigen-pair $\lambda_1, \mathbf{v}_1$, enter the origin as $t \to \infty$. The other set, corresponding to the eigen-pair $\lambda_2, \mathbf{v}_2$, leave the origin as $t \to \infty$. No orbit other than $c_1 \mathbf{v}_1 e^{\lambda_1 t}$ can enter the origin at $t \to \infty$, and all orbits are asymptotic to the linear orbits as $t \to \pm\infty$. The origin is a **saddle point** in this case, which is unstable.

– $\lambda_1 = \lambda_2$ (real and equal). When there is a single eigenvalue $\lambda \equiv \lambda_1 = \lambda_2$ of multiplicity two, and the solution depends on whether there are one or two linearly independent eigenvectors. (a) If $\mathbf{v}_1$ and $\mathbf{v}_2$ are linearly independent eigenvectors corresponding to λ, then the general solution of (1.76) is

$$\mathbf{x}(t) = c_1 \mathbf{v}_1 e^{\lambda t} + c_2 \mathbf{v}_2 e^{\lambda t}$$
$$= (c_1 \mathbf{v}_1 + c_2 \mathbf{v}_2) e^{\lambda t}.$$

Because $\mathbf{v}_1$ and $\mathbf{v}_2$ are independent, every direction is covered and every orbit is therefore a linear ray entering the origin ($\lambda < 0$) or leaving the origin ($\lambda \dot{<} 0$). Depending upon the sign of λ, the origin is an unstable node or an asymptotically stable node. (b) If $\mathbf{v}$ is the only eigenvector corresponding to λ, then $\mathbf{v}e^{\lambda t}$ is a solution of (1.76). A second linearly independent solution has the form $(\mathbf{w} + \mathbf{v}t)e^{\lambda t}$ for a vector $\mathbf{w}$ satisfying $(A - \lambda I)\mathbf{w} = \mathbf{v}$, and so the general solution to (1.76) is

$$\mathbf{x}(t) = (c_1\mathbf{v} + c_2\mathbf{w} + c_2\mathbf{v}t)e^{\lambda t}.$$

For large t we have $\mathbf{x}(t) \approx c_2 t \mathbf{v}e^{\lambda t}$, and so the orbits enter the origin with direction $\mathbf{v}$ when $\lambda < 0$, or they leave the origin when $\lambda > 0$; the origin is an asymptotically stable **node** or unstable node, respectively.

$-$ $\lambda_1 = \alpha + i\beta, \lambda_2 = \alpha - i\beta$ (complex conjugate roots). Let $\mathbf{w} + i\mathbf{v}$ be an eigenvector corresponding to λ_1. Then a complex solution of (1.76) is

$$(\mathbf{w} + i\mathbf{v})e^{(\alpha+i\beta)t},$$

or, after expanding with Euler's formula,

$$e^{\alpha t}(\mathbf{w}\cos\beta t - \mathbf{v}\sin\beta t) + ie^{\alpha t}(\mathbf{w}\sin\beta t + \mathbf{v}\cos\beta t).$$

The real and imaginary parts of a complex solution are independent real solutions, and therefore the general solution is

$$\mathbf{x}(t) = c_1 e^{\alpha t}(\mathbf{w}\cos\beta t - \mathbf{v}\sin\beta t) + c_2 e^{\alpha t}(\mathbf{w}\sin\beta t + \mathbf{v}\cos\beta t).$$

The other eigenvalue gives the same independent solutions. Notice that the two functions $\mathbf{w}\cos\beta t - \mathbf{v}\sin\beta t$ and $\mathbf{w}\sin\beta t + \mathbf{v}\cos\beta t$ are periodic functions with period $2\pi/\beta$, and the factor $e^{\alpha t}$ is like an amplitude factor. If $\alpha = 0$, then the two components $u(t)$ and $v(t)$ of $\mathbf{x}$ are strictly oscillatory functions with period $2\pi/\beta$. Therefore the orbits are closed curves and $(0,0)$ is a **center**. If $\alpha < 0$, the amplitude of $\mathbf{x}$ decreases and the orbits form spirals that wind into the origin; these are decaying oscillations and the origin is called an asymptotically stable **spiral point**. If $\alpha > 0$, the amplitude of $\mathbf{x}$ increases and the orbits are unstable spirals, representing growing oscillations, and the origin is an unstable spiral point.

The preceding discussion completely characterizes the solution structure and the stability properties of an isolated critical point $(0,0)$ of the linear system (1.76). We summarize key points in the following theorem.

Theorem 1.15

The critical point $(0, 0)$ of the linear system

$$\frac{d\mathbf{x}}{dt} = A\mathbf{x}, \quad \det A \neq 0,$$

is stable if, and only if, the eigenvalues of A have non-positive real parts; it is asymptotically stable if, and only if, the eigenvalues have negative real parts.

Example 1.16

Consider the linear system

$$x' = 3x - 2y,$$
$$y' = 2x - 2y.$$

The coefficient matrix

$$A = \begin{pmatrix} 3 & -2 \\ 2 & -2 \end{pmatrix}$$

has characteristic equation

$$\det \begin{pmatrix} 3 - \lambda & -2 \\ 2 & -2 - \lambda \end{pmatrix} = \lambda^2 - \lambda - 2 = 0,$$

and therefore the eigenvalues are $\lambda = -1, 2$. Therefore the origin is an unstable saddle point. The eigenvectors are found by solving the linear homogeneous system

$$\begin{pmatrix} 3 - \lambda & -2 \\ 2 & -2 - \lambda \end{pmatrix} \begin{pmatrix} v_1 \\ v_2 \end{pmatrix} = \begin{pmatrix} 0 \\ 0 \end{pmatrix}.$$

When $\lambda = -1$,

$$4v_1 - 2v_2 = 0, \quad 2v_1 - v_2 = 0,$$

and an eigenvector corresponding to $\lambda = -1$ is $[v_1, v_2]^{\mathrm{T}} = [1, 2]^{\mathrm{T}}$. When $\lambda = 2$,

$$v_1 - 2v_2 = 0, \quad 2v_1 - 4v_2 = 0,$$

which gives $[v_1, v_2]^{\mathrm{T}} = [2, 1]^{\mathrm{T}}$. The general solution to the system is

$$\mathbf{x}(t) = c_1 \begin{pmatrix} 1 \\ 2 \end{pmatrix} e^{-t} + c_2 \begin{pmatrix} 2 \\ 1 \end{pmatrix} e^{2t}$$

The eigenvectors define the directions of the separatrices. To get a more accurate plot it is helpful to plot the nullclines, or loci of points where $x' = 0$ and where $y' = 0$. A phase diagram is shown in Fig. 1.19.

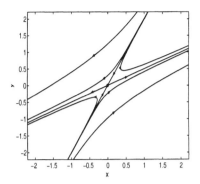

Figure 1.19 Saddle point.

EXERCISES

1. Find the general solution and sketch phase diagrams for the following systems; characterize the equilibria as to type (node, etc.) and stability.

a) $x' = x - 3y$, $y' = -3x + y$.

b) $x' = -x + y$, $y' = y$.

c) $x' = 4y$, $y' = -9x$.

d) $x' = x + y$, $y' = 4x - 2y$.

e) $x' = 3x - 4y$, $y' = x - y$.

f) $x' = -2x - 3y$, $y' = 3x - 2y$.

2. Determine the behavior of solutions near the origin for the system

$$\mathbf{x}' = \begin{pmatrix} 3 & b \\ 1 & 1 \end{pmatrix} \mathbf{x}$$

for different values of b.

3. Prove that the critical point $(0,0)$ of the linear system (1.74) is asymptotically stable if the only if $\operatorname{tr} A < 0$ and $\det A > 0$, where $\operatorname{tr} A = a + d$ is the trace is of A.

4. For the linear system (1.76), let $p = \operatorname{tr} A$ and $q = \det A$. Shade the region in the pq plane representing values where the origin is a saddle point. For which values is the origin a stable spiral? An unstable node? A center?

5. Consider the system

$$\mathbf{x}' = \begin{pmatrix} -a & -1 \\ 1 & -a \end{pmatrix} \mathbf{x}.$$

What are the possible behaviors of solutions to this system, depending upon the parameter a.

6. The equation for a damped harmonic oscillator is

$$\ddot{m}x + a\dot{x} + kx = 0, \quad m, \, a, \, k > 0.$$

Write the equation as a system by introducing $y = x'$ and show that $(0,0)$ is a critical point. Describe the nature and stability of the critical point in the following cases: $a = 0$; $a^2 - 4mk = 0$; $a^2 - 4km < 0$; $a^2 - 4km > 0$. Interpret the results physically.

1.4.3 Nonlinear Systems

We now revisit the question of whether the linearization

$$\mathbf{x}' = A\mathbf{x}, \quad A = \begin{pmatrix} P_x(x_0, y_0) & P_y(x_0, y_0) \\ Q_x(x_0, y_0) & Q_y(x_0, y_0) \end{pmatrix} \tag{1.82}$$

of a nonlinear system

$$x' = P(x, y), \quad y' = Q(x, y) \tag{1.83}$$

at a critical point (x_0, y_0) predicts its type (node, saddle, center, spiral point) and stability. The following result of Poincaré gives a partial answer to this question. The proof can be found in the references.

Theorem 1.17

Let (x_0, y_0) be an isolated critical point for the nonlinear system (1.83) and let $A = A(x_0, y_0)$ be the Jacobian matrix for linearization (1.82), with $\det A \neq 0$. Then (x_0, y_0) is a critical point of the same type as the origin $(0,0)$ for the linearization in the following cases.

(i) The eigenvalues of A are real, either equal or distinct, and have the same sign (node).

(ii) The eigenvalues of A are real and have opposite signs (saddle).

(iii) The eigenvalues of A are complex, but not purely imaginary (spiral).

Therefore, the exceptional case is when the linearization has a center. The orbital structure for the nonlinear system near the critical points mirrors that of the linearization in the non-exceptional cases, with a slight distortion due to the nonlinearity.

Example 1.18

The nonlinear system

$$x' = -y - x^3, \quad y' = x,$$

has a stable spiral at the origin. The linearization has Jacobian

$$A = \begin{pmatrix} 0 & -1 \\ 1 & 0 \end{pmatrix},$$

has purely imaginary eigenvalues $\pm i$, and hence $(0,0)$ is a center for the linearization. This is the exceptional case in the theorem.

The question of stability is answered by the following theorem, which is an immediate corollary of the Poincaré theorem.

Theorem 1.19

If $(0,0)$ is asymptotically stable for (1.82), then it is asymptotically stable for (1.83).

Example 1.20

Consider the nonlinear system

$$\dot{x} = 2x + 3y + xy,$$
$$\dot{y} = -x + y - 2xy^3,$$

which has an isolated critical point at $(0,0)$. The Jacobian matrix A at the origin is

$$A = \begin{pmatrix} -2 & 3 \\ -1 & 1 \end{pmatrix},$$

and it has eigenvalues $-\frac{1}{2} \pm (\sqrt{3}/2)i$. Thus the linearization has an asymptotically stable spiral at $(0,0)$, and thus the nonlinear system has an asymptotically stable spiral at $(0,0)$.

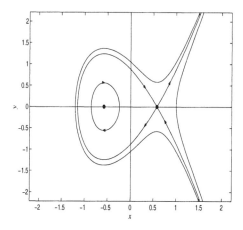

Figure 1.20   Orbits for the conservative system.

Example 1.21

(Mechanics) A particle of mass $m = 1$ moves on the x axis under the influence of a conservative force $f(x) = 3x^2 - 1$. The equations of motion in the phase plane are

$$
\begin{aligned}
x' &= y, \\
y' &= 3x^2 - 1,
\end{aligned}
$$

where the position x and the velocity y are functions of time t. The potential energy is $V(x) = -x^3 + x$, where the zero potential energy level is located at $x = 0$. The orbits are given by the conservation of energy law

$$
\frac{1}{2}y^2 + (-x^3 + x) = E,
$$

or

$$
y = \sqrt{2}\sqrt{x^3 - x + E}, \quad y = -\sqrt{2}\sqrt{x^3 - x + E}.
$$

They are plotted in Fig. 1.20. There are two equilibria, $x = \sqrt{\frac{1}{3}}$, $y = 0$ and $x = -\sqrt{\frac{1}{3}}$, $y = 0$, where the velocity is zero and the force is zero. The equilibrium solution $x = -\sqrt{\frac{1}{3}}$, $y = 0$ has the structure of a center, and for small initial values the system will oscillate. The other equilibrium $x = \sqrt{\frac{1}{3}}$, $y = 0$, has the structure of an unstable saddle point. Because $x' = y$, for $y > 0$ we have $x' > 0$, and the orbits are directed to the right in the upper half-plane.

For $y < 0$ we have $x' < 0$, and the orbits are directed to the left in the lower half plane. The x-nullcline is the y axis, where the vector field is vertical; the y-nullclines are the lines $x = -\sqrt{\frac{1}{3}}$ and $x = \sqrt{\frac{1}{3}}$, where the vector field is horizontal. For large initial energies the system does not oscillate but rather goes to $x = +\infty$, $y = +\infty$; that is, the mass moves farther and farther to the right with faster speed. The Jacobian matrix is

$$A(x, y) = \left(\begin{array}{cc} 0 & 1 \\ 6x & 0 \end{array} \right).$$

Then

$$A\left(-\sqrt{\frac{1}{3}}, 0\right) = \left(\begin{array}{cc} 0 & 1 \\ -2\sqrt{3} & 0 \end{array} \right), \quad A\left(\sqrt{\frac{1}{3}}, 0\right) = \left(\begin{array}{cc} 0 & 1 \\ 2\sqrt{3} & 0 \end{array} \right).$$

The eigenvalues of the first matrix are purely imaginary, so the linearization has a center at $\left(-\sqrt{\frac{1}{3}}, 0\right)$. This is the exceptional case and the theorem gives no information about the nonlinear system. However, for this problem we were able to find equations for the orbits and we found that the nonlinear system does have a center at $\left(-\sqrt{\frac{1}{3}}, 0\right)$. The second matrix has real eigenvalues of opposite sign, and therefore $\left(\sqrt{\frac{1}{3}}, 0\right)$ is a saddle point, confirming the analysis.

Example 1.22

(Ecology) Nonlinear equations play an central role in modeling population dynamics in ecology. Let $x = x(t)$ be the prey population and $y = y(t)$ be the predator population. We can think of rabbits and foxes, food fish and sharks, or any consumer-resource interaction, including herbivores and plants. If there is no predator we assume the prey dynamics is $x' = rx$, or exponential growth, where r is the *per capita* growth rate. In the absence of prey, we assume that the predator dies via $y' = -my$, where m is the *per capita* mortality rate. When there are interactions, we must include terms that decrease the prey population and increase the predator population. To determine the form of the predation term, let us assume that the rate of predation, or the number of prey consumed per unit time, per predator, is proportional to the number of prey. That is the rate of predation is ax. Thus, if there are y predators then the rate that prey is decreased is axy. Note that the interaction term is proportional to xy, the product of the number of predators and the number of prey. For example, if there were 20 prey and 10 predators, there would be 200 possible interactions. Only a fraction of them, a, is assumed to result in a kill. The parameter a depends upon the fraction of encounters and the success of the encounters.

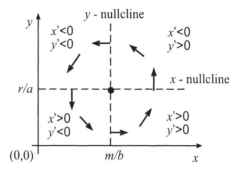

Figure 1.21 Nullclines and vector field for the Lotka–Volterra system.

The prey consumed cause a rate of increase in predators of εaxy, where ε is the conversion efficiency of the predator population (observe that there is not a one-to-one trade of a prey to create one predator). Therefore, we obtain the simplest model of predator–prey interaction, called the **Lotka–Volterra** model[6]

$$x' = rx - axy,$$
$$y' = -my + bxy,$$

where $b = \varepsilon a$. To analyze the Lotka–Volterra model we factor the right sides of the equations to obtain

$$x' = x(r - ay), \quad y' = y(-m + bx). \tag{1.84}$$

Setting the right sides equal to zero gives two equilibria, $(0,0)$ and $(m/b, r/a)$. The origin represents extinction of both species, and the nonzero equilibrium represents a possible coexistent state. The x-nullclines for (1.84), where $x' = 0$, are $x = 0$ and $y = r/a$. Thus, orbits cross these two lines vertically. The y-nullclines, where $y' = 0$, are $y = 0$ and $x = m/b$. The orbits cross these lines horizontally. The equilibria are the intersections of the x and y nullclines. See Fig. 1.21. Along each nullcline we can find the direction of the vector field. For example, on the ray to the right of the equilibrium we have $x > m/b$, $y = r/a$. We know the vector field is vertical so we need only check the sign of y'. We have $y' = y(-m+bx) = (r/a)(-m+bx) > 0$, so the vector field points upward. Similarly we can determine the directions along the other three rays. These are shown in the accompanying figure. Further, we can determine the direction of the vector field in regions between the nullclines either by selecting an arbitrary

---

[6] A. Lotka and V. Volterra developed this model in the 1920s to study fish populations in the Mediterranean Sea after the First World War.

point in that region and calculating x' and y', or by just noting the sign of x' and y' in that region from information obtained from the system. For example, in the quadrant, above and to the right of the nonzero equilibrium, it is easy to see that $x' < 0$ and $y' > 0$; so the vector field points upward and to the left. This task can be performed for each region and obtain the directions shown in the figure. Having the direction of the vector field along the nullclines and in the regions bounded by the nullclines tells us the directions of the orbits. Near $(0, 0)$ the orbits appear to veer away and the equilibrium has a saddle point structure. The Jacobian matrix is

$$A(x, y) = \begin{pmatrix} r - ay & -ax \\ by & -m + bx \end{pmatrix}.$$

Clearly,

$$A(0, 0) = \begin{pmatrix} r & 0 \\ 0 & -m \end{pmatrix}$$

has eigenvalues r and $-m$, and so $(0, 0)$ is indeed a saddle point. It appears that orbits circle around the nonzero equilibrium in a counterclockwise fashion. The Jacobi matrix (C. Jacobi (1804–1851)) is

$$A\left(\frac{m}{b}, \frac{r}{a}\right) = \begin{pmatrix} 0 & -\frac{am}{b} \\ \frac{br}{a} & 0 \end{pmatrix},$$

which has purely imaginary eigenvalues, so the linearization is inconclusive; additional work is required. To obtain the equation of the orbits we divide the two equations in (1.84) to get

$$\frac{y'}{x'} = \frac{y(-m + bx)}{x(r - ay)}.$$

Rearranging and integrating gives

$$r \ln y - ay = bx - m \ln x + C,$$

which is the algebraic equation for the orbits. It is obscure what these curves are because it is not possible to solve for either of the variables. But we can proceed as follows. If we exponentiate we get

$$y^r e^{-ay} = e^C e^{bx} x^{-m}.$$

Now consider the y nullcline where x is fixed at a value m/b, and fix a C value (i.e., an orbit). The right side of the last equation is a positive number γ, and so $y^r = \gamma e^{ay}$. If we plot both sides of this equation (a power function and a growing exponential) we observe that there can be at most two crossings; therefore, this equation can have at most two solutions for y. Hence, along the y nullcline there can be at most two crossings; an orbit cannot spiral into

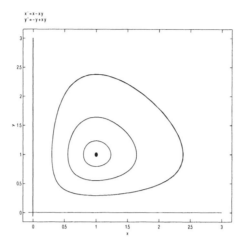

Figure 1.22 Counterclockwise periodic orbits of the Lotka–Volterra system.

or out from the equilibrium point because that would mean many values of y would be possible. We conclude that the nonzero equilibrium is a center with closed, periodic orbits encircling it. A phase diagram is shown in Fig. 1.22. When the prey population is high the predators have a high food source and their numbers start to increase, thereby eventually driving down the prey population. Then the prey population gets low, ultimately reducing the number of predators because of lack of food. Then the process repeats, giving cycles. The Lotka–Volterra model is the simplest model in ecology showing how populations can cycle, and it was one of the first strategic models to explain qualitative observations in natural systems. Note that the nonzero equilibrium is stable. A small perturbation from equilibrium puts the populations on a periodic orbit that stays near the equilibrium. But the system does not return to equilibrium.

A central problem in the theory of nonlinear systems is to determine whether the system admits any cycles, or closed orbits. These orbits represent periodic solutions of the system, or oscillations. In general, for nonlinear systems the existence of closed orbits may be difficult to decide. The following negative result, however, due to Bendixson and Dulac is easy to prove.

Theorem 1.23

For the system (1.83), if $P_x + Q_y$ is of one sign in a region of the phase plane, then system (1.83) cannot have a closed orbit in that region.

By way of contradiction assume the region contains a closed orbit Γ given by

$$x = x(t), \ y = y(t), \quad 0 \le t \le T,$$

and denote the interior of Γ by R. By Green's theorem in the plane,

$$\int_\Gamma P \, dy - Q \, dx = \int \int_R (P_x + Q_y) \, dx \, dy \neq 0.$$

On the other hand,

$$\int_\Gamma P \, dy - Q \, dx = \int_0^T (Py' - Qx') \, dt = \int_0^T (PQ - QP) \, dt = 0,$$

which is a contradiction.

Another negative criterion is due to Poincaré.

Theorem 1.24

A closed orbit of the system (1.83) surrounds at least one critical point of the system.

Positive criteria that ensure the existence of closed orbits for general systems are not abundant, particularly criteria that are practical and easy to apply. For certain special types of equations such criteria do exist. The following fundamental theorem, called the **Poincaré–Bendixson theorem**, is a general theoretical result.

Theorem 1.25

Let R be a closed bounded region in the plane containing no critical points of (1.83). If Γ is an orbit of (1.83) that lies in R for some t_0 and remains in R for all $t > t_0$, then Γ is either a closed orbit or it spirals toward a closed orbit as $t \to \infty$.

The Poincaré–Bendixson theorem guarantees in the plane that either an orbit leaves every bounded set as $t \to \pm\infty$, it is a closed curve, it approaches a critical point or a closed curve as $t \to \pm\infty$, or it is a critical point. Thus, in the plane, the only *attractors* are closed orbits or critical points. Interestingly enough, the Poincaré–Bendixson theorem does not generalize directly to higher dimensions. In three dimensions, for example, there exist *strange attractors* that have neither the character of a point, a curve, or a surface, which act as attractors to all paths.

1.4.4 Bifurcation

Next we examine systems of equations of the form

$$x' = P(x, y, \mu), \quad y' = Q(x, y, \mu), \tag{1.85}$$

depending upon a real parameter μ. As μ varies, the nature of the solution often changes at special values of μ, giving fundamentally different critical point structures. For example, what was once a stable equilibrium can bifurcate into an unstable one. A simple example will illustrate the idea.

Example 1.26

Consider the linear system

$$x' = x + \mu y, \quad y' = x - y,$$

where μ is a parameter. The eigenvalues λ are determined from

$$\det \begin{pmatrix} 1 - \lambda & \mu \\ 1 & -1 - \lambda \end{pmatrix} = 0,$$

or

$$\lambda^2 - (1 + \mu) = 0.$$

Hence

$$\lambda = \pm\sqrt{1 + \mu}.$$

If $\mu > -1$ then the eigenvalues are real and have opposite signs, and therefore the origin is a saddle point. If $\mu = -1$, then the system becomes

$$x' = x - y, \quad y' = x - y$$

and there is a line of equilibrium solutions $y = x$. If $\mu < -1$, then the eigenvalues are purely imaginary and the origin is a center. Consequently, as μ decreases, the nature of the solution changes at $\mu = -1$; the equilibrium state $(0, 0)$ evolves from an unstable saddle to a stable center as μ passes through the critical value. We say a bifurcation occurs at $\mu = -1$.

These types of phenomena occur frequently in physical systems. For example, in a predator–prey interaction in ecology we may ask how coexistent states, or equilibrium populations, depend upon the carrying capacity K of the prey species. The parameter K then acts as a bifurcation parameter. For a given system, we may find a coexistent, asymptotically stable state for small carrying capacities. But, as the carrying capacity increases, the coexistent state can suddenly bifurcate to an unstable equilibrium and the dynamics could change completely. This type of behavior, if present, is an important characteristic of differential equations in physics, engineering, and the natural sciences.

Example 1.27

Consider the system

$$x' = -y - x(x^2 + y^2 - \mu), \qquad (1.86)$$
$$y' = x - y(x^2 + y^2 - \mu).$$

The origin is an isolated equilibrium, and the linearization is

$$x' = \mu x - y, \quad y' = x + \mu y.$$

It is easy to see that the eigenvalues λ of the Jacobian matrix are given by

$$\lambda = \mu \pm i.$$

If $\mu < 0$, then $\mathrm{Re}\,\lambda < 0$, and $(0,0)$ is a stable spiral; if $\mu = 0$, then $\lambda = \pm i$, and $(0,0)$ is a center; if $\mu > 0$, then $\mathrm{Re}\,\lambda > 0$, and $(0,0)$ is an unstable spiral. From results in the preceding sections regarding the relationship between a nonlinear system and its linearization, we know $(0,0)$ is a stable spiral for the nonlinear system when $\mu < 0$ and an unstable spiral when $\mu > 0$. Therefore, there is a bifurcation at $\mu = 0$ where the equilibrium exchanges stability.

The system in the last example deserves a more careful look. It is possible to solve the nonlinear problem (1.86) directly if we transform to polar coordinates $x = r\cos\theta, \quad y = r\sin\theta$. First, it is straightforward to show

$$xx' + yy' = rr', \qquad xy' - yx' = r^2\theta'.$$

If we multiply the first equation by x and the second by y and then add, we obtain

$$\dot{r} = r(\mu - r^2). \qquad (1.87)$$

If we multiply the first equation by $-y$ and the second by x and then add, we obtain

$$\dot{\theta} = 1. \qquad (1.88)$$

The equivalent system (1.87)–(1.88) in polar coordinates may be integrated directly. Clearly

$$\theta = t + t_0, \qquad (1.89)$$

where t_0 is a constant. Therefore, the polar angle θ winds around counterclockwise as time increases.

Next we examine the r-equation when $\mu > 0$. By separating variables and using the partial fraction decomposition

$$\frac{1}{r(r^2 - \mu)} = \frac{-1/\mu}{r} + \frac{1/2\mu}{r - \sqrt{\mu}} + \frac{1/2\mu}{r + \sqrt{\mu}},$$

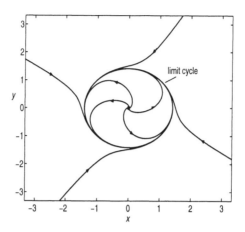

Figure 1.23 limit cycles.

we obtain

$$r = \frac{\sqrt{\mu}}{\sqrt{1 + c\ \exp(-2\mu t)}}, \tag{1.90}$$

where c is a constant of integration. When $c = 0$ we obtain the solution

$$r = \sqrt{\mu},$$
$$\theta = t + t_0,$$

which represents a periodic solution whose orbit is the circle $r = \sqrt{\mu}$ in the phase plane. If $c < 0$, then (1.90) represents an orbit that spirals counterclockwise toward the circle $r = \sqrt{\mu}$ from the outside, and if $c > 0$ the solution represents an orbit that spirals counterclockwise toward the circle from the inside. This means that the origin is an unstable spiral as indicated by the linear analysis, and the periodic solution $r = \sqrt{\mu}$ is approached by all other solutions. When a periodic orbit, or cycle, is approached by orbits from both the inside and outside, we say that the orbit is a stable **limit cycle**. See Fig. 1.23.

In the case $\mu = 0$ the r-equation becomes

$$\frac{dr}{r^3} = -dt,$$

and direct integration gives the solution

$$r = \frac{1}{\sqrt{2t + c}}, \quad \theta = t + t_0.$$

This implies that the origin is a stable spiral.

Finally, in the case $\mu < 0$, let $k^2 = -\mu > 0$. Then the radial r-equation is

$$\frac{dr}{r(r^2 + k^2)} = -dt. \tag{1.91}$$

The partial fraction decomposition is

$$\frac{1}{r(r^2 + k^2)} = \frac{1/k^2}{r} - \frac{(1/k^2)r}{r^2 + k^2},$$

and (1.91) integrates to

$$r^2 = \frac{ck^2 e^{-2k^2 t}}{1 - ce^{-2k^2 t}}.$$

It is easy to verify that this solution represents spirals that approach the origin at $t \to +\infty$. See Fig. 1.23.

Now let us review the results. As μ passes through the critical value $\mu = 0$ from negative to positive, the origin changes from a stable spiral to an unstable spiral and there appears a new time periodic solution. This type of bifurcation is common and it is an example of a **Hopf bifurcation**. A graph in the complex plane of the eigenvalues $\lambda = \mu + i$ and $\lambda = \mu - i$ of the linearized system, as functions of μ, shows that the complex conjugate pair crosses the imaginary axis at $\mu = 0$. Generally this signals bifurcation to a time periodic solution.

Another important consequence of the Poincaré–Bendixson theorem that shows the existence of limit cycles is the following. Suppose there is a closed region R in the plane where at each point (x, y) on the boundary of R the vector field $(P(x, y), Q(x, y))$ for the system points inward into R. We call R a **basin of attraction**. If R contains no critical points, then there must be a periodic orbit in R.

EXERCISES

1. Determine the nature and stability properties of the critical points of the systems, and sketch the phase diagram:

   a) $x' = x + y - 2x^2$, $y' = -2x + y + 3y^2$.

   b) $x' = -x - y - 3x^2 y$, $y' = -2x - 4y + y \sin x$.

   c) $x' = y^2$, $y' = -\frac{2}{3}x$.

   d) $x' = x^2 - y^2$, $y' = x - y$.

   e) $x' = x^2 + y^2 - 4$, $y' = y - 2x$.

2. Consider the system

$$x' = 4x + 4y - x(x^2 + y^2),$$
$$x' = -4x + 4y - y(x^2 + y^2).$$

a) Show there is a closed orbit in the region $1 \leq r \leq 3$, where $r^2 = x^2 + y^2$.

b) Find the general solution.

3. Determine if the following systems admit periodic solutions.

a) $x' = y$, $\quad y' = (x^2 + 1)y - x^5$.

b) $x' = x + x^3 - 2y$, $\quad y' = -3x + y^5$.

c) $x' = y$, $\quad y' = y^2 + x^2 + 1$.

d) $x' = y$, $\quad y' = 3x^2 - y - y^5$.

e) $x' = 1 + x^2 + y^2$, $\quad y' = (x - 1)^2 + 4$.

4. Find the values of μ where solutions bifurcate and examine the stability of the origin in each case.

a) $x' = x + \mu y$, $\quad y' = \mu x + y$.

b) $x' = y$, $\quad y' = -2x + \mu y$.

c) $x' = x + y$, $\quad y' = \mu x + y$.

d) $x' = 2y$, $\quad y' = 2x - \mu y$.

e) $x' = y$, $\quad y' = \mu x + x^2$.

f) $x' = y$, $\quad y' = x^2 - x + \mu y$.

g) $x' = \mu y + xy$, $\quad y' = -\mu x + \mu y + x^2 + y^2$.

h) $x' = y$, $\quad y' = -y - \mu + x^2$.

i) $x' = \mu x - x^2 - 2xy$, $\quad y' = (\mu - \frac{5}{3})y + xy + y^2$.

5. For a nonlinear system (1.83) assume there is a region D in the plane where the vector field points inward, into the region, at all points on the boundary of D. If there is an unstable critical point in the interior of D, explain why there must be a limit cycle inside D.

6. Discuss the dependence on the sign of μ of the critical point at the origin of the system
$$x' = -y + \mu x^3, \quad y' = x + \mu y^3.$$

7. Show that the orbits of $x'' + f(x) = 0$ in the xy phase plane (where $y = x'$) are periodic if $f(x)$ is an increasing function and $f(0) = 0$ (Hint: Show $\frac{1}{2}y^2 + V(x) = $ constant, where $V(x) = \int_0^x f(u)\,du$.)

8. Prove a generalization of the Bendixson–Dulac theorem : For the system (1.83), if there is a smooth function $h(x, y)$ for which $(hP)_x + (hQ)_y$ is of one sign in a region R of the phase plane, then system (1.83) cannot have a closed orbit in R. Use this result to show that the system

$$x' = x(A - ax + by), \quad y' = y(B - cy + dx), \quad a, c > 0,$$

has no periodic orbits. (Hint: Take $h(x, y) = 1/xy$.)

9. If $p(0, 0)$ and $h(0)$ are positive, prove that the origin is an asymptotically stable critical point for the dynamical equation

$$x'' + p(x, x')x' + xh(x) = 0.$$

10. Examine the stability of the equilibrium solutions of the system

$$\frac{dx}{dt} = \gamma - y^2 - vx, \quad \frac{dy}{dt} = -vy + xy,$$

where γ and v are positive constants.

11. Let P denote the carbon biomass of plants in an ecosystem, H be the carbon biomass of herbivores, and ϕ the rate of primary carbon production in plants due to photosynthesis. A model of plant–herbivore dynamics is given by

$$P' = \phi - aP - bHP,$$
$$H' = \varepsilon bHP - cH,$$

where a, b, c, and ε are positive parameters.

a) Explain the various terms and parameters in the model and determine the dimensions of each parameter.

b) Non-dimensionalize the model and find the equilibrium solutions.

c) Analyze the dynamics in the cases of high primary production ($\phi > ac/\varepsilon b$) and low primary production ($\phi < ac/\varepsilon b$). Explain what happens in the system if primary production is slowly increased from a low value to a high value.

12. An RCL circuit with a nonlinear resistance, where the voltage drop across the resistor is a nonlinear function of current, can be modeled by the Van der Pol equation
$$x'' + \rho(x^2 - 1)x' + x = 0,$$
where ρ is a positive constant, and $x(t)$ is the current.

a) In the phase plane, show that the origin is an unstable equilibrium.

b) Sketch the nullclines and the vector field. What are the possible dynamics? Is there a limit cycle?

c) Use software to sketch the phase diagram when $\rho = 1$, and plot $x = x(t)$ if $x(0) = 0.05$ and $x'(0) = 0$. Describe the dynamics in this case.

13. Show that period orbits for the mechanical system

$$\begin{aligned} x' &= y, \\ y' &= -ky - V'(x), \end{aligned}$$

are possible only if $k = 0$.

14. Analyze the following nonlinear system, which models cell differentiation.

$$x' = y - x, \quad y' = -y + \frac{5x^2}{4 + x^2}.$$

15. Consider the scaled predator–prey model

$$\begin{aligned} x' &= \frac{2}{3}x\left(1 - \frac{x}{4}\right) - \frac{xy}{1+x}, \\ y' &= ry\left(1 - \frac{y}{x}\right), \end{aligned}$$

where the carrying capacity of the predator depends upon the prey population. Examine the dynamics of the model as the predator growth rate r decreases from $\frac{1}{2}$ to a small value.

16. (Project) Consider two competing species where one of the species immigrates or emigrates at a constant rate h. The populations are governed by

$$\begin{aligned} x' &= x(1 - ax) - xy, \\ y' &= y(b - y) - xy + h, \end{aligned}$$

where a and b are positive. Give an analysis of the dynamics in the cases $h = 0$, $h < 0$, and $h > 0$. In particular, in each case indicate the behavior of the system in (a,b)–parameter space.

REFERENCES AND NOTES

There are many outstanding books on mathematical modeling. The classic, which everyone should own and read, is Lin & Segel (1974). Several books devoted to advanced models and problems can serve as a resource for problems, projects, and advanced mathematical methods, and we mention Fowler (1997) and Howison (2005). Excellent, brief introductions are contained in Mooney & Swift (1999) and Illner *et al* (2005).

A concise account of elementary differential equations can be found in the author's text, Logan (2006); a more detailed treatment is in Ledder (2004). At the next level we refer to Hirsch, Smale, & Devaney (2004) and Strogatz (1994), both of which discuss applications to physics, biology, engineering, and chemistry.

Birkhoff, G. 1950. *Hydrodynamics: A Study in Logic, Fact, and Similitude*, Princeton University Press, Princeton, NJ.

Fowler, A. C. 1997. *Mathematical Models in the Applied Sciences*, Cambridge University Press, Cambridge.

Hirsh, M. W., Smale, S. & Devaney. R. L. 2004. *Differential Equations, Dynamical Systems, & An Introduction to Chaos*, 2nd ed., Academic Press, San Diego.

Howison, S. 2005. *Practical Applied Mathematics*, Cambridge University Press, Cambridge.

Illner, R., Bohun, C. S., McCollum, S. & van Roode, T. 2005. *Mathematical Modelling: A Case Studies Approach*, American Mathematical Society, Providence.

Ledder, G. 2004. *Differential Equations: A Modeling Approach*, McGraw-Hill, New York.

Lin, C. C. & Segel, L. A. 1974. *Mathematics Applied to Deterministic Problems in the Natural Sciences*, Macmillan, New York [reprinted by SIAM (Soc. Industr. & Appl. Math.), Philadelphia, 1988].

Logan, J. D. 2006. *A First Course in Differential Equations*, Springer-Verlag, New York.

Mooney, D. & Swift, 1999. *A Course in Mathematical Modeling*, Mathematical Association of America, Washington DC.

Strogatz, S. H. 1994. *Nonlinear Dynamics and Chaos*, Addison-Wesley, Reading, MA.

Perturbation Methods

Equations arising from mathematical models usually cannot be solved in exact form. Therefore, we must resort to approximation and numerical methods. Foremost among approximation techniques are perturbation methods. Perturbation methods lead to an approximate solution to a problem when the model equations have terms that are small. These terms arise because the underlying physical process has small effects. For example, in a fluid flow problem the viscosity may be small compared to advection, or in the motion of a projectile the force caused by air resistance may be small compared to gravity. These low-order effects are represented by terms in the model equations that, when compared to the other terms, are negligible. When scaled properly, the order of magnitude of these terms is represented by a small coefficient parameter, say ε. By a perturbation solution we mean an approximate solution that is the first few terms of a Taylor-like expansion in the parameter ε.

To fix the idea, consider a differential equation

$$F(t, y, y', y'', \varepsilon) = 0, \quad t \in I, \tag{2.1}$$

where t is the independent variable, I is the time interval, and y is the dependent variable. The appearance of a small parameter ε is shown explicitly. In general, initial or boundary conditions may accompany the equation, but for the present we ignore auxiliary conditions. To denote that ε is a small parameter we write

$$\varepsilon \ll 1.$$

This means that ε is *small compared to unity*, a glib phrase in which no specific cutoff is explicitly defined; certainly $0.001 \ll 1$, but 0.75 is not small compared

85

to unity. The previous remarks also apply to equations with a large parameter λ, since in that case we may introduce $\varepsilon = 1/\lambda$, which is small.

By a **perturbation series** we understand a power series in ε of the form

$$y_0(t) + \varepsilon y_1(t) + \varepsilon^2 y_2(t) + \cdots . \tag{2.2}$$

The basis of the **regular perturbation** method is to assume a solution of (2.1) of the form (2.2) where the functions $y_0, y_1, y_2, \ldots$ are found by substitution of (2.2) into (2.1). The first few terms of such a series form an approximate solution, called a **perturbation solution** or **approximation**; usually no more than two or three terms are taken. Generally, the method will be successful if the approximation is uniform; that is, the difference between the approximate solution and the exact solution converges to zero at some well-defined rate as ε approaches zero, uniformly on I. These ideas are made precise later. We emphasize in this context that there is no advantage in taking *a priori* a specific value for ε. Rather, we consider ε to be an arbitrary small number, so that the analysis will be valid for any choice of ε. Of particular interest in many problems is the behavior of the solutions as $\varepsilon \to 0$. Generally, the perturbation series is a valid representation for $\varepsilon < \varepsilon_0$ for some undetermined ε_0.

The term y_0 in perturbation series is called the *leading order* term. The terms $\varepsilon y_1, \varepsilon^2 y_2, \ldots$ are regarded as higher-order correction terms that are expected to be small. If the method is successful, y_0 will be the solution of the **unperturbed problem**

$$F(t, y, y', y'', 0) = 0, \quad t \in I,$$

in which ε is set to zero. In this context (2.1) is called the **perturbed problem**. Therefore, when a model equation is obtained in which a small parameter appears, it is often regarded as a perturbed equation where the term(s) containing the small parameter represent small perturbations or changes from some basic unperturbed problem. It is implicit that the unperturbed equation should always be solvable so that the leading-order behavior of the solution can be found. One frequently encounters problems of this sort whose solutions are facilitated by the fact that the governing equation, the boundary condition, or the shape of the region in higher-dimensional problems, is not too different from those of a simpler problem.

For a variety of reasons, sometimes the naive perturbation approach described above fails. This failure often occurs when the leading order problem is not well-posed, or its solution is invalid on all parts of the domain, for example, when there are multiple time or spatial scales. Then we must modify the approach. These modifications, which are myriad, belong to the class of **singular perturbation** methods.

Perturbation methods apply to differential equations (ordinary and partial), algebraic equations, integral equations, and all of the other types of equations we encounter in applied mathematics. Their importance cannot be overstated.

2.1 Regular Perturbation

The basic idea of the perturbation method is easily illustrated by investigating a simple algebraic equation.

Example 2.1

Consider the quadratic equation

$$x^2 + 2\varepsilon x - 3 = 0,$$

where ε is a small positive parameter. Assume a solution in the form of a perturbation series $x = x_0 + x_1\varepsilon + x_2\varepsilon^2 + \cdots$, where the dots represent higher-order terms. Substituting into the equation gives

$$(x_0 + x_1\varepsilon + x_2\varepsilon^2 + \cdots)^2 + 2\varepsilon(x_0 + x_1\varepsilon + x_2\varepsilon^2 + \cdots) - 3 = 0.$$

Expanding out and collecting terms in powers of ε gives

$$x_0^2 - 3 + 2x_0(x_1 + 1)\varepsilon + (x_1^2 + 2x_0x_2 + 2x_1)\varepsilon^2 + \cdots = 0.$$

Equating the coefficients to zero yields

$$x_0^2 = 3, \; x_1 = -1, \; x_1^2 + 2x_0x_2 + 2x_1 = 0, \ldots.$$

Solving,

$$x_0 = \pm\sqrt{3}, \; x_1 = -1, \; x_2 = \pm\frac{1}{2\sqrt{3}}, \ldots$$

Therefore we obtain have two approximate solutions

$$x = \sqrt{3} - \varepsilon + \frac{1}{2\sqrt{3}}\varepsilon^2 + \cdots,$$

$$x = -\sqrt{3} - \varepsilon - \frac{1}{2\sqrt{3}}\varepsilon^2 + \cdots.$$

We can calculate as many terms as we wish, and often computer algebra systems can be quite useful in performing all the tedious algebra. The approximate solution can be compared to the exact solution obtained by the quadratic formula. We have

$$x = \frac{1}{2}\left(-2\varepsilon \pm \sqrt{4\varepsilon^2 + 12}\right) = -\varepsilon \pm \sqrt{3 + \varepsilon}.$$

The radical can be expanded by the binomial formula to get

$$\sqrt{3 + \varepsilon^2} = \sqrt{3}\left(1 + \frac{\varepsilon^2}{\sqrt{3}}\right)^{1/2} = \sqrt{3}\left(1 + \frac{\varepsilon^2}{2\sqrt{3}} + \cdots\right).$$

Therefore the quadratic formula leads to

$$x = -\varepsilon \pm \sqrt{3}\left(1 + \frac{\varepsilon^2}{2\sqrt{3}} + \cdots\right),$$

which is the same as the perturbation approximation.

2.1.1 Motion in a Resistive Medium

Next consider a differential equation. Suppose a body of mass m, initially with velocity V_0, moves in a straight line in a medium that offers a resistive force of magnitude $av - bv^2$, where $v = v(\tau)$ is the velocity of the object as a function of time τ, and a and b are positive constants with $b \ll a$ (b is much smaller than a). Therefore the nonlinear part of the force is assumed to be small compared to the linear part. The constants a and b have units of force per velocity and force per velocity-squared, respectively. By Newton's second law, the equation of motion is

$$m\frac{dv}{d\tau} = -av + bv^2, \quad v(0) = V_0.$$

We first non-dimensionlize the problem. A velocity scale is the maximum velocity, which is clearly V_0, because the object must slow down in the resistive medium. If the small nonlinear term bv^2 were not present, then the velocity would decay like $e^{-a\tau/m}$; therefore, the characteristic time is m/a. Introducing dimensionless variables

$$y = \frac{v}{V_0}, \quad t = \frac{\tau}{m/a},$$

the problem becomes

$$\frac{dy}{dt} = -y + \varepsilon y^2, \quad t > 0, \tag{2.3}$$

$$y(0) = 1, \tag{2.4}$$

where

$$\varepsilon \equiv \frac{bV_0}{a} \ll 1.$$

Equation (2.3) is a Bernoulli equation and can be solved exactly by making the substitution $w = y^{-1}$, and then integrating the resulting linear equation to obtain

$$y_{\text{ex}} = \frac{e^{-t}}{1 + \varepsilon(e^{-t} - 1)}.$$

Equation (2.3) is a slightly altered or perturbed form of the linear equation

$$\frac{dy}{dt} = -y, \quad y(0) = 1, \tag{2.5}$$

which is easily solved by $y = e^{-t}$. Because the nonlinear term εy^2 is small, the function e^{-t} appears to be a good approximate solution to the problem, provided the scaling was performed correctly. The exact solution can be expanded in a Taylor series in powers of ε as

$$y_{\text{ex}} = e^{-t} + \varepsilon(e^{-t} - e^{-2t}) + \varepsilon^2(e^{-t} - 2e^{-2t} + e^{-3t}) + \cdots. \tag{2.6}$$

If ε is small, the leading order term $y_0 = e^{-t}$ is a reasonable approximation, but it does not include any effects of the nonlinear term in the original equation.

To obtain an approximate solution by a perturbation method we employ a perturbation series. That is, we assume that the solution of (2.3) is representable as

$$y = y_0(t) + \varepsilon y_1(t) + \varepsilon^2 y_2(t) + \cdots, \tag{2.7}$$

which is a series in powers of ε. The functions $y_0, y_1, y_2, \ldots$ are determined by substituting (2.7) into both the differential equation (2.3) and initial condition, and then equating coefficients of like powers of ε. Thus,

$$y_0' + \varepsilon y_1' + \varepsilon^2 y_2' + \cdots = -(y_0 + \varepsilon y_1 + \varepsilon^2 y_2 + \cdots) + \varepsilon(y_0 + \varepsilon y_1 + \varepsilon^2 y_2 + \cdots)^2,$$

which, when coefficients are collected, gives a sequence of linear differential equations

$$\begin{aligned} y_0' &= -y_0, \\ y_1' &= -y_1 + y_0^2, \\ y_2' &= -y_2 + 2y_0 y_1, \ldots. \end{aligned}$$

The initial condition leads to

$$y_0(0) + \varepsilon y_1(0) + \varepsilon^2 y_2(0) + \cdots = 1,$$

or the sequence of initial conditions

$$y_0(0) = 1, \quad y_1(0) = y_2(0) = \cdots = 0.$$

Therefore we have obtained a set of linear initial value problems for y_0, y_1, $y_2, \ldots$. These are easily solved in sequence to obtain

$$y_0 = e^{-t},$$
$$y_1 = e^{-t} - e^{-2t},$$
$$y_2 = e^{-t} - 2e^{-t} + e^{-3t}, \ldots.$$

Notice that y_1 and y_2 are the first- and second-order correction terms to the leading-order approximation $y_0 = e^{-t}$, which is in agreement with (2.6). Therefore we have obtained a three-term perturbation solution

$$y_a = e^{-t} + \varepsilon(e^{-t} - e^{-2t}) + \varepsilon^2(e^{-t} - 2e^{-2t} + e^{-3t}), \qquad (2.8)$$

which is an approximation of y_{ex} and includes nonlinear effects due to the term εy^2 in the original differential equation. The approximate solution is the first three terms of the Taylor expansion of the exact solution. In this problem the exact solution is known (this is rare), and we can compare it to the approximate solution. The error in the approximation (2.8) is given by the difference

$$y_{ex} - y_a = m_1(t)\varepsilon^3 + m_2(t)\varepsilon^4 + \cdots, \quad t > 0,$$

for bounded functions $m_1, m_2, \ldots$. For a fixed positive t the error approaches zero as $\varepsilon \to 0$ as the same rate as ε^3 goes to zero. In fact one can show that the convergence is uniform as $\varepsilon \to 0$ in the interval $0 \leq t < \infty$. Convergence notions are explored further after an additional example.

2.1.2 Nonlinear Oscillations

In the last example the regular perturbation method led to a satisfactory result, consistent with intuition and that compared favorably with the exact solution. In the next example the procedure is the same, but the result does not turn out favorable; it is the first signal of the need to modify the regular perturbation method and develop a singular perturbation method. In this example, the perturbation approximation will be valid only if certain restrictions are placed on the interval of time that the solution evolves.

Consider a spring–mass oscillator where a mass m is connected to a spring whose restoring force has magnitude $ky + ay^3$, where y is the replacement of the mass, measured positively from equilibrium, and k and a are positive constants characterizing the stiffness properties of the spring. Assume that the nonlinear portion of the restoring force is small in magnitude compared to the linear part, or $a \ll k$. If, initially, the mass is released from a positive displacement A, then the motion is given by a function $y = y(\tau)$ of time τ, which satisfies Newton's second law

$$m\frac{d^2y}{d\tau^2} = -ky - ay^3, \quad \tau > 0, \tag{2.9}$$

and the initial conditions

$$y(0) = A, \quad \frac{dy}{d\tau}(0) = 0. \tag{2.10}$$

Because of the presence of the nonlinear term ay^3, this problem cannot be solved exactly. Because $a \ll k$, however, a perturbation method is suggested.

From the projectile problem in Chapter 1 the reader should have gained enough skepticism not to just neglect the term ay^3 and proceed without rescaling. Therefore, to properly analyze the problem we seek appropriate time and length scales that reduce the problem to dimensionless form. The constants in the problem are k, a, m, and A have dimensions

$$[k] = \frac{\text{mass}}{\text{time}^2}, \quad [a] = \frac{\text{mass}}{\text{length}^2 \cdot \text{time}^2}, \quad [m] = \text{mass}, \quad [A] = \text{length}.$$

An obvious choice is to scale y by the amplitude A of the initial displacement. To scale time τ we argue as follows. If the small term ay^3 is neglected, then the differential equation is $my'' = -ky$, which has periodic solutions of the form $\cos\sqrt{k/mt}$ with a period proportional to $\sqrt{m/k}$. Hence, we choose $\sqrt{m/k}$ to be the characteristic time and introduce dimensionless variables t and u via

$$t = \frac{\tau}{\sqrt{m/k}}, \quad u = \frac{y}{A}.$$

Under this change of variables the differential equation (2.9) and initial conditions become (*overdots* denote derivatives with respect to τ)

$$\ddot{u} + u + \varepsilon u^3 = 0, \quad t > 0,$$
$$u(0) = 1, \quad \dot{u}(0) = 0, \tag{2.11}$$

where

$$\varepsilon \equiv \frac{aA^2}{k} \ll 1$$

is a dimensionless small parameter. It now becomes clear exactly what the assumption of a small nonlinear restoring force means; the precise assumption is $aA^2 \ll k$. Equation (2.11) is the *Duffing equation*.

We attempt a solution of the form

$$u(t) = u_0(t) + \varepsilon u_1(t) + \varepsilon^2 u_2(t) + \cdots,$$

where u_0, u_1, ... are to be determined. Substituting into the differential equation and initial conditions, and equating coefficients of like powers of ε, gives a sequence of linear initial value problems

$$\ddot{u}_0 + u_0 \;=\; 0, \quad u_0(0) = 1, \quad \dot{u}_0(0) = 0, \tag{2.12a}$$

$$\ddot{u}_1 + u_1 \;=\; -u_0^3, \quad u_1(0) = 0, \quad \dot{u}_0(0) = 0, \ldots \tag{2.12b}$$

The leading-order solution to (2.12a) is easily found to be

$$u_0(t) = \cos t.$$

Then the problem (2.12b) for the first correction term becomes

$$\ddot{u}_1 + u_1 = -\cos^3 t, \quad u_1(0) = \dot{u}_1(0) = 0.$$

We employ the trigonometric identity $\cos 3t = 4\cos^3 t - 3\cos t$ to write the differential equation as

$$\ddot{u}_1 + u_1 = -\frac{1}{4}(3\cos t + \cos 3t), \tag{2.13}$$

which can be solved by standard methods. The general solution of the homogeneous equation $\ddot{u}_1 + u_1 = 0$ is $c_1 \cos t + c_2 \sin t$. A particular solution can be found by the method of undetermined coefficients and is of the form

$$u_p = C\cos 3t + Dt\cos t + Et\sin t.$$

Substituting u_p into the differential equation and equating like terms gives $C = \frac{1}{32}, D = 0$, and $E = -\frac{3}{8}$. Thus the general solution of (2.13) is

$$u_1 = c_1\cos t + c_2\sin t + \frac{1}{32}\cos 3t - \frac{3}{8}t\sin t.$$

Applying the initial conditions on u_1 gives

$$u_1 = \frac{1}{32}(\cos 3t - \cos t) - \frac{3}{8}t\sin t.$$

Therefore, in scaled variables, a 2-term approximate solution takes the form

$$u_{\mathrm{a}} = \cos t + \varepsilon\left[\frac{1}{32}(\cos 3t - \cos t) - \frac{3}{8}t\sin t\right]. \tag{2.14}$$

The leading-order behavior of the approximate solution is $\cos t$, which is an oscillation. The second term, or the correction term, however, is not necessarily small. For a fixed value of time t the term goes to zero as $\varepsilon \to 0$, but if t itself is of the order of ε^{-1} or larger as $\varepsilon \to 0$, then the term $-\frac{3}{8}t\sin t$ will be large. Such a term is called a **secular term**. Therefore, we expect that the amplitude of the approximate solution will grow with time, which is not consistent with the physical situation or with the exact solution. Indeed, it is easy to show that the latter is bounded for all $t > 0$. In this approximation, the correction term cannot be made arbitrarily small for t in $(0, \infty)$ by choosing ε small enough. And, it is not possible to improve on (2.14) by calculating additional higher-order terms; they too will contain secular terms that will not cancel the effects of the lower-order terms. One may legitimately ask, therefore, if there is any value at all in the approximate solution (2.14). In a limited sense the answer is yes. If we restrict the independent variable t to a *finite* interval $[0, T]$, then the correction term $\varepsilon \left[\frac{1}{2}(\cos 3t - \cos t) - \frac{3}{8}t \sin t \right]$ can be made arbitrarily small by choosing ε sufficiently small, for any $t \in [0, T]$. So, as long as the coefficient $3\varepsilon t/8$ is kept small by limiting t and taking ε small, the leading-order term $\cos t$ is a reasonable approximate solution.

2.1.3 The Poincaré–Lindstedt Method

A straightforward application of the regular perturbation method on the initial value problem (2.11) for the Duffing equation led to a secular term in the first correction that ruined the approximation unless t was restricted. In this section we present a method to remedy this type of singular behavior, which occurs in all cases of periodic motion. The key to the analysis is to recognize that not only does the correction term grow in amplitude, but it also does not correct for the difference between the exact period of oscillation and the approximate period 2π of the leading order term $\cos t$. Over several oscillations the error in the period increases until the approximation and the actual solution are completely out of phase. To confirm this, the reader should use a software system to solve (2.11) numerically, and then plot the numerical solution vs. $\cos t$.

The idea of the Poincaré–Lindstedt method is to introduce a distorted time scale in the perturbation series. In particular, we let

$$u(\tau) = u_0(\tau) + \varepsilon u_1(\tau) + \varepsilon^2 u_2(\tau) + \cdots , \qquad (2.15)$$

where

$$\tau = \omega t, \qquad (2.16)$$

with

$$\omega = 1 + \varepsilon \omega_1 + \varepsilon^2 \omega_2 + \cdots . \qquad (2.17)$$

Here ω_0 has been chosen to be unity, the frequency of the solution of the unperturbed problem. Under the scale transformation (2.16) the initial value problem (2.11) becomes

$$\omega^2 u'' + u + \varepsilon u^3 = 0, \quad \tau > 0, \qquad (2.18)$$

$$u(0) = 1, \quad u'(0) = 0, \qquad (2.19)$$

where $u = u(\tau)$ and prime denotes differentiation with respect to τ. Substituting (2.15) and (2.17) into (2.18) and (2.19) gives

$$(1 + 2\varepsilon\omega_1\omega_0 + \cdots)(u_0'' + \varepsilon u_1'' + \cdots) + (u_0 + \varepsilon u_1 + \cdots) + \varepsilon(u_0^3 + 3\varepsilon u_0^2 u_1 + \cdots) = 0$$

and

$$u_0(0) + \varepsilon u_1(0) + \cdots = 1, \quad u_0'(0) + \varepsilon u_1'(0) + \cdots = 0.$$

Collecting coefficients of powers of ε gives

$$u_0'' + u_0 = 0, \quad u_0(0) = 1, \quad u_0'(0) = 0, ..., \qquad (2.20)$$
$$u_1'' + u_1 = -2\omega_1 u_0'' - u_0^3, \quad u_1(0) = u_1'(0) = 0, ... \qquad (2.21)$$

The solution of (2.20) is

$$u_0(\tau) = \cos\tau.$$

Then the differential equation in (2.21) becomes

$$u_1'' + u_1 = 2\omega_1\cos\tau - \cos^3\tau = \left(2\omega_1 - \frac{3}{4}\right)\cos\tau - \frac{1}{4}\cos 3\tau. \qquad (2.22)$$

Because $\cos\tau$ is a solution to the homogeneous equation, the $\cos\tau$ term on the right side of the equation leads to a particular solution with a term $\tau\cos\tau$, which is a secular term. The key observation is that we can avoid a secular term if we choose

$$\omega_1 = \frac{3}{8}.$$

Then (2.22) becomes

$$u_1'' + u_1 = -\frac{1}{4}\cos 3\tau,$$

which has the general solution

$$u_1(\tau) = c_1\cos\tau + c_2\sin\tau + \frac{1}{32}\cos 3\tau.$$

The initial conditions on u_1 lead to

$$u_1(\tau) = \frac{1}{32}(\cos 3\tau - \cos\tau).$$

Therefore, a first-order, uniformly valid, perturbation solution of (2.11) is

$$u(\tau) = \cos\tau + \frac{1}{32}\varepsilon(\cos 3\tau - \cos\tau) + \cdots,$$

where

$$\tau = t + \frac{3}{8}\varepsilon t + \cdots.$$

The Poincaré–Lindstedt method is successful on a number of similar problems, and it is one of a general class of multiple scale methods. Generally, it works successfully on some (not all) equations of the form

$$u'' + \omega_0^2 u = \varepsilon F(t, u, u'), \ 0 < \varepsilon \ll 1.$$

These are problems whose leading order is oscillatory with frequency ω_0. The basic technique is to change variables to one with a different frequency, $\tau = (\omega_0 + \omega_1 \varepsilon + \ldots)t$, and then assume $u = u(\tau)$ is a perturbation series in ε. The constants $\omega_0, \omega_1, \ldots$ are chosen at each step to avoid the presence of secular terms in the expansion.

2.1.4 Asymptotic Analysis

With insight from the previous examples we define some notions regarding convergence and uniformity. We observed that substitution of a perturbation series into a differential equation does not always lead to a valid approximate solution. Ideally we would like to say that a few terms in a truncated perturbation series provides, for a given ε, an approximate solution for the entire range of the independent variable t. Unfortunately, as we have seen, this is not always the case. Failure of this regular perturbation method is the rule rather than the exception.

To aid the analysis of approximate solutions we introduce some basic notation and terminology that permits the comparison of two functions as their common argument approaches some fixed value. These comparisons are called order relations.

Let $f(\varepsilon)$ and $g(\varepsilon)$ be defined in some neighborhood (or punctured neighborhood) of $\varepsilon = 0$. We write

$$f(\varepsilon) = o(g(\varepsilon)) \quad \text{as} \quad \varepsilon \to 0, \tag{2.23}$$

if

$$\lim_{\varepsilon \to 0} \left| \frac{f(\varepsilon)}{g(\varepsilon)} \right| = 0,$$

and we write

$$f(\varepsilon) = O(g(\varepsilon)) \quad \text{as} \quad \varepsilon \to 0, \tag{2.24}$$

if there exists a positive constant M such that

$$|f(\varepsilon)| \leq M|g(\varepsilon)|$$

for all ε in some neighborhood (punctured neighborhood) of zero. The comparison function g is called a **gauge function**.

In this definition, $\varepsilon \to 0$ may be replaced by a one-sided limit or by $\varepsilon \to \varepsilon_0$, where ε_0 is any finite or infinite number, with the domain of f and g defined appropriately. If (2.23) holds, we say f is *little oh* of g as $\varepsilon \to 0$, and if (2.24) holds we say f is *big oh* of g as $\varepsilon \to 0$. Common gauge functions are $g(\varepsilon) = \varepsilon^n$ for some exponent n, and $g(\varepsilon) = \varepsilon^n (\ln \varepsilon)^m$ for exponents m and n. The statement $f(\varepsilon) = O(1)$ means f is bounded in a neighborhood of $\varepsilon = 0$, and $f(\varepsilon) = o(1)$ means $f(\varepsilon) \to 0$ as $\varepsilon \to 0$. If $f = o(g)$, then f goes to zero faster than g goes to zero as $\varepsilon \to 0$, and we write $f(\varepsilon) \ll f(\varepsilon)$. The following examples illustrate a few methods to prove order relations.

Example 2.2

Verify $\varepsilon^2 \ln \varepsilon = o(\varepsilon)$ as $\varepsilon \to 0^+$. By L'Hôpital's rule,

$$\lim_{\varepsilon \to 0^+} \frac{\varepsilon^2 \ln \varepsilon}{\varepsilon} = \lim_{\varepsilon \to 0^+} \frac{\ln \varepsilon}{(1/\varepsilon)} = \lim_{\varepsilon \to 0^+} \frac{(1/\varepsilon)}{(-1/\varepsilon^2)} = 0.$$

Example 2.3

Verify $\sin \varepsilon = O(\varepsilon)$ as $\varepsilon \to 0^+$. By the mean value theorem there is a number c between 0 and ε such that

$$\frac{\sin \varepsilon - \sin 0}{\varepsilon - 0} = \cos c.$$

Hence $|\sin \varepsilon| = |\varepsilon \cos c| \leq |\varepsilon|$, because $|\cos c| \leq 1$. An alternate argument is to note that $(\sin \varepsilon)/\varepsilon \to 1$ as $\varepsilon \to 0^+$. Because the limit exists, the function $(\sin \varepsilon)/\varepsilon$ must be bounded for $0 < \varepsilon < \varepsilon_0$, for some ε_0. Therefore $|(\sin \varepsilon)/\varepsilon| \leq M$, for some constant M and $\sin \varepsilon = O(\varepsilon)$.

The order definitions may be extended to functions of ε and another variable t lying in an interval I. First we review the notion of uniform convergence. Let $h(t, \varepsilon)$ be a function defined for ε in a neighborhood of $\varepsilon = 0$, possibly not including the value $\varepsilon = 0$ itself, and for t in some interval I, either finite or infinite. We say

$$\lim_{\varepsilon \to 0} h(t, \varepsilon) = 0 \text{ uniformly on } I,$$

if the convergence to zero is at the same rate for each $t \in I$; that is, if for any positive number η there can be chosen a positive number ε_0, independent of t, such that $|h(t, \varepsilon)| < \eta$ for all $t \in I$, whenever $|\varepsilon| < \varepsilon_0$. In other words, if $h(t, \varepsilon)$ can be made arbitrarily small over the entire interval I by choosing ε small enough, then the convergence is uniform. If merely $\lim_{\varepsilon \to 0} h(t_0, \varepsilon) = 0$ for each fixed $t_0 \in I$, then we say that the convergence is *pointwise* on I.

To prove $\lim_{\varepsilon \to 0} h(t, \varepsilon) = 0$ uniformly on I it is sufficient to find a function $H(\varepsilon)$ such that $|h(t, \varepsilon)| \leq H(\varepsilon)$ holds for all $t \in I$, with $H(\varepsilon) \to 0$ as $\varepsilon \to 0$. To prove that convergence is not uniform on I, it is sufficient to produce a $\bar{t} \in I$ such that $|h(\bar{t}, \varepsilon)| \geq \eta$ for some positive η, regardless of how small ε is chosen.

Let $f(t, \varepsilon)$ and $g(t, \varepsilon)$ be defined for all $t \in I$ and all ε in a (punctured) neighborhood of $\varepsilon = 0$. We write

$$f(t, \varepsilon) = o(g(t, \varepsilon)) \quad \text{as} \quad \varepsilon \to 0,$$

if

$$\lim_{\varepsilon \to 0} \left| \frac{f(t, \varepsilon)}{g(t, \varepsilon)} \right| = 0 \tag{2.25}$$

pointwise on I. If the limit is uniform on I, we write $f(t, \varepsilon) = o(g(t, \varepsilon))$ as $\varepsilon \to 0$, uniformly on I. If there exists a positive function $M(t)$ on I such that

$$|f(t, \varepsilon)| \leq M(t)|g(t, \varepsilon)|$$

for all $t \in I$ and ε in some neighborhood of zero, then we write

$$f(t, \varepsilon) = O(g(t, \varepsilon)) \quad \text{as} \quad \varepsilon \to 0, \quad t \in I.$$

If $M(t)$ is a bounded function on I, we write

$$f(t, \varepsilon) = O(g(t, \varepsilon)) \quad \text{as} \quad \varepsilon \to 0, \quad \text{uniformly on } I.$$

The big oh and little oh notations permit us to make quantitative statements about the error in a given approximation. We can make the following definition. A function $y_a(t, \varepsilon)$ is a *uniformly valid asymptotic approximation* to a function $y(t, \varepsilon)$ on an interval I as $\varepsilon \to 0$ if the error $E(t, \varepsilon) \equiv y(t, \varepsilon) - y_a(t, \varepsilon)$ converges to zero as $\varepsilon \to 0$, uniformly for $t \in I$. We often express the fact that $E(t, \varepsilon)$ is little oh or big oh of ε^n (for some n) as $\varepsilon \to 0$ to make an explicit statement regarding the rate that the error goes to zero, and whether or not the convergence is uniform.

Example 2.4

Let

$$y(t, \varepsilon) = e^{-t\varepsilon}, \quad t > 0, \quad \varepsilon \ll 1.$$

The first three terms of the Taylor expansion in powers of ε provide an approximation

$$y_a(t, \varepsilon) = 1 - t\varepsilon + \tfrac{1}{2}t^2\varepsilon^2.$$

The error is

$$E(t,\varepsilon) = e^{-t\varepsilon} - 1 + t\varepsilon - \tfrac{1}{2}t^2\varepsilon^2 = -\frac{1}{3!}t^3\varepsilon^3 + \cdots.$$

For a fixed t the error can be made as small as desired by choosing ε sufficiently small. Thus $E(t,\varepsilon) = O(\varepsilon^2)$ as $\varepsilon \to 0$. If ε is fixed, however, regardless of how small, t may be chosen large enough so that the approximation is totally invalid. Thus, the approximation is not uniform on $I = [0,\infty)$. Clearly, by choosing $t = 1/\varepsilon$ we have $E(1/\varepsilon,\varepsilon) = e^{-1} - \tfrac{1}{2}$, which is not small. We may *not* write $E(t,\varepsilon) = O(\varepsilon^2)$ as $\varepsilon \to 0$, uniformly on $[0,\infty)$.

The difficulty of these definitions with regard to differential equations is that the exact solution to the equation is seldom known, and therefore a direct error estimate cannot be made. Consequently, we require some notion of how well an approximate solution satisfies the differential equation and the auxiliary conditions. For definiteness consider the differential equation in (2.1). We say that an approximate solution $y_a(t,\varepsilon)$ *satisfies the differential equation* (2.1) *uniformly for* $t \in I$ *as* $\varepsilon \to 0$ if

$$r(t,\varepsilon) \equiv F(t, y_a(t,\varepsilon), \dot{y}_a(t,\varepsilon), \ddot{y}_a(t,\varepsilon), \varepsilon) \to 0$$

uniformly on I as $\varepsilon \to 0$. The quantity $r(t,\varepsilon)$ is the residual error, which measures how well the approximate solution $y_a(t,\varepsilon)$ satisfies the equation.

Example 2.5

Consider the initial value problem

$$\ddot{y} + \dot{y}^2 + \varepsilon y = 0, \quad t > 0, \quad 0 < \varepsilon \ll 1,$$
$$y(0) = 0, \quad \dot{y}(0) = 1.$$

Substituting the perturbation series $y = y_0 + \varepsilon y_1 + \cdots$ gives the initial value problem

$$\ddot{y}_0 + \dot{y}_0^2 = 0, \quad t > 0,$$
$$y_0(0) = 0, \quad \dot{y}_0(0) = 1,$$

for the leading-order term y_0. It is easily found that $y_0(t) = \ln(t+1)$, and hence

$$r(t,\varepsilon) \equiv \ddot{y}_0 + \dot{y}_0^2 + \varepsilon y_0 = \varepsilon \ln(t+1).$$

Thus $r(t,\varepsilon) = O(\varepsilon)$ as $\varepsilon \to 0$, but not uniformly on $[0,\infty)$. On any finite interval $[0,T]$, however, we have $|\varepsilon \ln(t+1)| \le \varepsilon \ln(T+1)$, and so $r(t,\varepsilon) = O(\varepsilon)$ as $\varepsilon \to 0$, uniformly on $[0,T]$.

Specifically, the regular perturbation method produces an asymptotic expansion

$$y_0(t) + y_1(t)\varepsilon + y_2(t)\varepsilon^2 + \cdots,$$

for which an approximate solution can be obtained by taking the first few terms. Such an expansion in the integral powers of ε, that is, $1, \varepsilon, \varepsilon^2, \ldots$, is called an asymptotic power series. In some problems the expansion is of the form

$$y_0(t) + y_1(t)\sqrt{\varepsilon} + y_2(t)\varepsilon + y_3(t)\varepsilon^{3/2} + \cdots,$$

in terms of the sequence $1, \varepsilon^{1/2}, \varepsilon, \varepsilon^{3/2}, \ldots$. In yet other problems the required expansion must have the form

$$y_0(t)\ln\varepsilon + y_1(t)\varepsilon\ln\varepsilon + y_2(t)\varepsilon + y_3(t)\varepsilon^2\ln^2\varepsilon + y_4(t)\varepsilon^2\ln\varepsilon + y_5(t)\varepsilon^2 + \cdots.$$

The type of expansion depends on the problem. In general, we say a sequence of gauge functions $\{g_n(t, \varepsilon)\}$ is an **asymptotic sequence** as $\varepsilon \to 0$, $t \in I$, if

$$g_{n+1}(t, \varepsilon) = o(g_n(t, \varepsilon)) \quad \text{as} \quad \varepsilon \to 0,$$

for $n = 0, 1, 2, \ldots$. That is, each term in the sequence tends to zero faster than its predecessor, as $\varepsilon \to 0$. Given a function $y(t, \varepsilon)$ and an asymptotic sequence $\{g_n(t, \varepsilon)\}$ as $\varepsilon \to 0$, the formal series

$$\sum_{n=0}^{\infty} a_n g_n(t, \varepsilon) \tag{2.26}$$

is said to be an **asymptotic expansion** of $y(t, \varepsilon)$ as $\varepsilon \to 0$, if

$$y(t, \varepsilon) - \sum_{n=0}^{N} a_n g_n(t, \varepsilon) = o(g_N(t, \varepsilon)), \quad \text{as} \quad \varepsilon \to 0,$$

for every N. In other words, for any partial sum the remainder is little oh of the last term. If the limits just cited are uniform for $t \in I$, then we speak of a uniform asymptotic sequence and uniform asymptotic expansion. In most cases the sequence $\{g_n(t, \varepsilon)\}$ is of the form of a product, $g_n(t, \varepsilon) = y_n(t)\phi_n(\varepsilon)$, as in the previous examples. The notation

$$y \sim \sum_{n=0}^{\infty} a_n g_n(t, \varepsilon)$$

often denotes an asymptotic expansion.

The formal series (2.26) need not converge to be valuable. The value of such expansions, although perhaps divergent, is that often only a few terms are required to obtain an accurate approximation, whereas a convergent Taylor series may yield an accurate approximation only if many terms are calculated.

A rather obvious question arises. If a given approximate solution $y_a(t, \varepsilon)$
satisfies the differential equation uniformly on $t \in I$, is it in fact a uniformly
valid approximation to the exact solution $y(t, \varepsilon)$? A complete discussion of this
question is beyond our scope, but a few remarks are appropriate to caution
the reader regarding the nature of this problem. Probably more familiar is
the situation in linear algebra where we consider a linear system of equations
$A\mathbf{x} = \mathbf{b}$. Let $\mathbf{x}_a$ be an approximate solution. A measure of how well it satisfies
the system is the magnitude $|\mathbf{r}|$ of the *residual vector* $\mathbf{r}$ defined by $\mathbf{r} = A\mathbf{x}_a -$
$\mathbf{b}$. If $\mathbf{r} = \mathbf{0}$, then $\mathbf{x}_a$ must be the exact solution $\bar{\mathbf{x}}$. But if $|\mathbf{r}|$ is small, it
does not necessarily follow that the magnitude $|\mathbf{e}|$ is small, where $\mathbf{e} = \bar{\mathbf{x}} -$
$\mathbf{x}_a$ is the error. In *ill-conditioned systems,* where $\det A$ is close to zero, a
small residual may not imply a small error. A similar state of affairs exists for
differential equations. Therefore one must proceed cautiously in interpreting the
validity of a perturbation solution. Numerical calculations or the computation
of additional correction terms may aid in the interpretation. Often a favorable
comparison with experiment leads one to conclude that an approximation is
valid.

EXERCISES

1. In a spring–mass problem assume that the restoring force is $-ky$ and that
   there is a resistive force numerically equal to $a\dot{y}^2$, where k and a are constants with appropriate units. With initial conditions $y(0) = A$, $\dot{y}(0) = 0$
   determine the correct time and displacement scales for small damping and
   show that the problem can be written in dimensionless form as

$$\bar{y}'' + \varepsilon(\bar{y}')^2 + \bar{y} = 0,$$
$$\bar{y}(0) = 1, \quad \bar{y}'(0) = 0,$$

   where $\varepsilon \equiv aA/m$ is a dimensionless parameter and prime denotes the
   derivative with respect to the scaled time $\bar{t}$.

2. Consider the initial value problem

$$u'' - u = \varepsilon tu, \quad t > 0, \quad u(0) = 1, \quad u'(0) = -1.$$

   Find a 2-term perturbation approximation for $0 < \varepsilon \ll 1$ and compare it
   graphically to a 6-term Taylor series approximation (centered at $t = 0$)
   where $\varepsilon = 0.04$. Use a numerical differential equation solver to find an
   accurate numerical solution and compare.

3. Write down an order relation (big O or little o) that expresses the fact that
   e^{-t} decays to zero faster that $1/t^2$ as $t \to \infty$, and prove your assertion.

4. Let
$$f(y, \varepsilon) = \frac{1}{(1 + \varepsilon y)^{3/2}}, \quad y = y_0 + y_1 \varepsilon + \cdots.$$

Expand f in powers of ε up to $O(\varepsilon^2)$.

5. Verify the following order relations:

a) $t^2 \tanh t = O(t^2)$ as $t \to \infty$.

b) $\exp(-t) = o(1)$ as $t \to \infty$.

c) $\sqrt{\varepsilon(1 - \varepsilon)} = O(\sqrt{\varepsilon})$ as $\varepsilon \to 0^+$.

d) $\frac{\sqrt{\varepsilon}}{1 - \cos \varepsilon} = O(\varepsilon^{-3/2})$ as $\varepsilon \to 0^+$.

e) $t = O(t^2)$ as $t \to \infty$.

f) $\exp(\varepsilon) - 1 = O(\varepsilon)$ as $\varepsilon \to 0$.

g) $\int_0^\varepsilon \exp(-x^2) dx = O(\varepsilon)$ as $\varepsilon \to 0^+$.

h) $\exp(\tan \varepsilon) = O(1)$ as $\varepsilon \to 0$.

i) $e^{-\varepsilon} = O(\varepsilon^{-p})$ as $\varepsilon \to \infty$ for all $p > 0$.

j) $\ln \varepsilon = o(\varepsilon^{-p})$ as $\varepsilon \to 0^+$ for all $p > 0$.

6. Consider the algebraic equation $\phi(x, \varepsilon) = 0, 0 < \varepsilon \ll 1$, where ϕ is a function having derivatives of all order. Assuming $\phi(x, 0) = 0$ is solvable to obtain x_0, show how to find a 3-term perturbation approximation of the form $x = x_0 + x_1 \varepsilon + x_2 \varepsilon^2$. What condition on ϕ is required to determine x_1 and x_2? Find a 3-term approximation to the roots of $\exp(\varepsilon x) = x^2 - 1$.

7. Find a 3-term approximation to the real solution of $(x + 1)^3 = \varepsilon x$.

8. Use the Poincaré—Lindstedt method to obtain a 2-term perturbation approximation to the following problems:

a) $y'' + y = \varepsilon y y'^2$, $y(0) = 1$, $y'(0) = 0$.

b) $y'' + 9y = 3\varepsilon y^3$, $y(0) = 0$, $y'(0) = 1$.

c) $y'' + y = \varepsilon y(1 - y'^2)$, $y(0) = 1$, $y'(0) = 0$.

9. To find approximations to the roots of the cubic equation

$$x^3 - 4.001x + 0.002 = 0,$$

why is it easier to examine the equation

$$x^3 - (4 + \varepsilon)x + 2\varepsilon = 0?$$

Find a 2-term approximation to this equation.

10. Consider the algebraic system

$$0.01x + y = 0.1, \quad x + 101y = 11.$$

In the first equation ignore the apparently small first term $0.01x$ and obtain an approximate solution for the system. Is the approximation a good one? Analyze what went wrong by examining the solution of the system

$$\varepsilon x + y = 0.1, \quad x + 101y = 11.$$

11. In Section 1.3 we obtained the initial value problem

$$\frac{d^2 h}{dt^2} = -\frac{1}{(1 + \varepsilon h)^2}, \quad h(0) = 0, \quad h'(0) = 1, \quad 0 < \varepsilon \ll 1,$$

governing the motion of a projectile. Use regular perturbation theory to obtain a 3-term perturbation approximation. Up to the accuracy of ε^2 terms, determine the value t_m when h is maximum. Find $h_{\max} \equiv h(t_m)$ up through order ε^2 terms.

12. A mass m is attached to an aging spring that exerts a force $F_s = -kye^{-rt}$, where k and r are positive constants, and $y = y(t)$ is the displacement from equilibrium. The system is also immersed in a medium that exerts a drag force $F_d = -ay'$, where a is a positive constant. Initially, the mass is displaced to position y_0 and released.

a) Formulate an initial value problem that governs the motion of the mass on the spring.

b) Make a table of the parameters in the problem and state their dimensions.

c) Non-dimensionlize the problem in the simplest way in the case that the constant k is very small. Use y_0 as the characteristic length. Denote the small dimensionless parameter by ε.

d) Use a perturbation method to obtain a 2-term approximation (up to $O(\varepsilon)$) to the scaled problem.

e) In mks units, take $m = 0.01$, $a = 0.01$, $r = 0.07$, $k = 0.0001$. Plot an accurate numerical approximation and the 2-term approximation on the same set of axes, and comment on the accuracy of the approximation.

13. Consider the boundary value problem

$$y'' = 3y \cos\left(\frac{\pi x}{2}\right), \quad y(0) = 0, \ y(1) = 1.$$

Obtain an approximation by considering the equation

$$y'' = \varepsilon y \cos \left(\frac{\pi x}{2} \right)$$

for small ε and then setting $\varepsilon = 3$. How accurate is the approximation?

14. Consider the problem

$$\frac{dy}{dt} = e^{-\varepsilon/y}, \quad y(0) = 1.$$

Find the leading order approximation and derive initial value problems for the $O(\varepsilon)$ and $O(\varepsilon^2)$ terms. (Do not solve.) Write down the exact solution of the problem.

15. Consider the scaled pendulum problem

$$\frac{d^2\theta}{d\tau^2} + \frac{\sin \varepsilon\theta}{\varepsilon} = 0, \quad \tau > 0, \quad 0 < \varepsilon \ll 1,$$

$$\theta(0) = 1, \quad \frac{d\theta}{d\tau}(0) = 0.$$

a) Apply the regular perturbation method to find a 2-term expansion. Show that the correction term is secular and comment on the validity of the approximation.

b) Show that the exact period of oscillation is

$$P = \frac{4\varepsilon}{\sqrt{2}} \int_0^1 \frac{dx}{\sqrt{\cos(\varepsilon x) - \cos \varepsilon}}.$$

c) Show that $P = 2\pi + \frac{\pi^2}{8}\varepsilon^2 + \cdots$. Find an expansion for the frequency ω.

16. Consider the initial value problem $y'' = \varepsilon t y$, $0 < \varepsilon \ll 1$, $y(0) = 0$, $y'(0) = 1$. Using regular perturbation theory obtain a 3-term approximate solution on $t \geq 0$. Does the approximation satisfy the differential equation uniformly on $t \geq 0$ as $\varepsilon \to 0^+$?

17. Consider the boundary value problem

$$Ly \equiv t^2 y'' + \varepsilon t^2 y' + \frac{1}{4}y = 0, \quad 1 \leq t \leq e, \quad 0 < \varepsilon \ll 1,$$

$$y(1) = 1, \quad y(e) = 0.$$

a) Use regular perturbation to find the leading-order behavior $y_0(t)$.

b) Compute an upper bound for $|Ly_0|$ on $1 \leq t \leq e$ when $\varepsilon = 0.01$. Can you conclude that y_0 is a good approximation to the exact solution?

18. Find a 2-term perturbation solution of

$$u' + u = \frac{1}{1 + \varepsilon u}, \quad u(0) = 0, \quad 0 < \varepsilon \ll 1.$$

19. A wildebeest starts at $t = 0$ from a point (a, b) in the xy-plane and moves along the line $x = a$. A lion, starting from the origin at $t = 0$ begins the chase, always running in a direction pointing at the wildebeest. The differential equation for the lion's path $y(x)$ is

$$(a - x)y'' = \varepsilon \left(1 + y'^2\right),$$

where ε is a small parameter giving the ratio of the speed of the wildebeest to the speed of the lion.

a) Find a leading order perturbation approximation. Based upon your answer, do you expect the lion to catch the wildebeest?

b) Find the equation for a correction term in the perturbation approximation, but do not solve.

2.2 Singular Perturbation

2.2.1 Algebraic Equations

As we observed in Section 2.1 for the Duffing equation, a straightforward application of the regular perturbation method failed because secular terms appeared in the expansion, making the approximation nonuniform for large times. We remedied this by adjusting the frequency of oscillations using a scale transformation. Now we indicate a different type of singular behavior, first for a simple algebra problem.

Example 2.6

Solve the quadratic equation

$$\varepsilon x^2 + 2x + 1 = 0, \quad 0 < \varepsilon \ll 1.$$

This equation can be solved exactly, of course, but our goal is to illustrate the failure of the regular perturbation method. Observe that the equation is a slight alteration of the unperturbed equation

$$2x + 1 = 0,$$

which has solution $x = -\frac{1}{2}$, and the unperturbed problem (linear) is fundamentally different from the original problem (quadratic). If we attempt regular perturbation by substituting the perturbation series

$$x = x_0 + x_1\varepsilon + x_2\varepsilon^2 + \cdots,$$

then we obtain, after setting the coefficients of like powers of ε equal to zero, the sequence of equations

$$2x_0 + 1 = 0,$$
$$x_0^2 + 2x_1 = 0,$$
$$2x_1x_0 + 2x_2 = 0, \ldots$$

Hence, $x_0 = -\frac{1}{2}, x_1 = -\frac{1}{8}, x_2 = -\frac{1}{16}, \ldots$, and we obtain a single perturbation solution

$$x = -\frac{1}{2} - \frac{1}{8}\varepsilon - \frac{1}{16}\varepsilon^2 - \cdots$$

What happened to the other solution? Regular perturbation assumed a leading term of order unity, and it is not surprising that it recovered only one root, of order unity. The other root could be different order, either large or small. To find the second root we examine the three terms, εx^2, $2x$, and 1, of the equation more closely. Discarding the εx^2 gave the root close to $x = -\frac{1}{2}$, and in that case, the term εx^2 is small compared to $2x$ and 1. So it is reasonable to ignore εx^2; the apparent small term is indeed small. For the second root, εx^2 may not be small (because x is large). Two cases are possible if one term is to be neglected from the equation to make a simplification. (i) εx^2 and 1 are the same order and $2x \ll 1$, or (ii) εx^2 and $2x$ are the same order and both are large compared to 1.

In case (i) we have $x = O(1/\sqrt{\varepsilon})$; but then $2x \ll 1$ could not hold because $2x$ would be large. This in inconsistent. In case (ii) we have $x = O(1/\varepsilon)$; hence εx^2 and $2x$ both are of order $1/\varepsilon$ and are large compared to 1. Therefore case (ii) is consistent and the second root is of order $1/\varepsilon$, which is large. This provides a clue to a rescaling. Let us choose a new variable y of order 1 defined by

$$y = \frac{x}{1/\varepsilon} = \varepsilon x.$$

With this change of variables the original equation becomes

$$y^2 + 2y + \varepsilon = 0.$$

Now the principle of proper scaling holds—each term has a magnitude defined by its coefficient. Assuming a regular perturbation series

$$y = y_0 + y_1\varepsilon + y_2\varepsilon^2 + \cdots$$

and substituting into the last equation, we obtain

$$y_0^2 + 2y_0 = 0,$$
$$2y_0 y_1 + 2y_1 + 1 = 0, \ldots$$

Hence $y_0 = -2$, $y_1 = \frac{1}{2}, \ldots$, giving

$$y = -2 + \frac{1}{2}\varepsilon + \cdots,$$

or

$$x = -\frac{2}{\varepsilon} + \frac{1}{2} + \cdots,$$

as the second root. In summary, the roots are different order, and one expansion does not reveal both.

The reasoning we used in this example is called **dominant balancing**. We examine each term carefully and determine which ones combine to give a dominant balance. In the next section we show type of balancing argument applies equally well to differential equations.

Example 2.7

Find a leading order approximation of the four roots to the equation

$$\varepsilon x^4 - x - 1 = 0,$$

where $0 < \varepsilon \ll 1$. When $\varepsilon = 0$ we get only a single root $x = -1$, which is order 1. To determine the leading order of the other roots we use dominant balancing. If the first and third terms balance then $x = O(\varepsilon^{-1/4})$, which is large; this is inconsistent. If the first and second terms balance, then $x = O(\varepsilon^{-1/3})$, which is large compared to 1; this case is consistent. Therefore we re-scale according to $y = \varepsilon^{1/3} x$, which gives

$$y^4 - y - \varepsilon^{1/3} = 0.$$

To leading order $y^4 - y = 0$, giving $y = 1$, $e^{2\pi i/3}$, $e^{-2\pi i/3}$. We discard $y = 0$ because that corresponds to $x = 0$, which is clearly not a valid approximation to the equation. Consequently, the leading order behavior of the four roots is

$$x = -1, \ \varepsilon^{-1/3}, \ \varepsilon^{-1/3}e^{2\pi i/3}, \ \varepsilon^{-1/3}e^{-2\pi i/3}.$$

Three of them are large. We can obtain higher-order correction terms using the perturbation series

$$y = y_0 + \varepsilon^{1/3} y_1 + \varepsilon^{2/3} + \cdots.$$

2.2.2 Differential Equations

Now we encounter a different type of singular behavior from that caused by the appearance of secular terms in the perturbation series. Often, with a perturbation series, some problems do not even permit the calculation of the leading-order term because the perturbed problem is of totally different character from the unperturbed problem.

Consider the boundary problem

$$\varepsilon y'' + (1+\varepsilon)y' + y = 0, \quad 0 < x < 1, \quad 0 < \varepsilon \ll 1,$$
$$y(0) = 0, \quad y(1) = 1, \tag{2.27}$$

and assume a naive perturbation series of the form

$$y = y_0(x) + \varepsilon y_1(x) + \varepsilon^2 y_2(x) + \cdots .$$

Substitution into the differential equation gives

$$\varepsilon(y_0'' + \varepsilon y_1'' + \varepsilon^2 y_2'' + \cdots) + (y_0' + \varepsilon y_1' + \varepsilon^2 y_2' + \cdots)$$
$$+ (\varepsilon y_0' + \varepsilon^2 y_1' + \varepsilon^3 y_2' + \cdots) + (y_0 + \varepsilon y_1 + \varepsilon^2 y_2 + \cdots) = 0.$$

Equating the coefficients of like powers of ε to zero gives a sequence of problems

$$y_0' + y_0 = 0,$$
$$y_1' + y_1 = -y_0'' - y_0', \dots .$$

The boundary conditions force

$$y_0(0) = 0, \quad y_0(1) = 1,$$
$$y_1(0) = 0, \quad y_1(1) = 0, \dots$$

Therefore, we have obtained a sequence of boundary value problems for y_0, $y_1, \dots$. The leading-order problem is

$$y_0' + y_0 = 0, \quad y_0(0) = 0, \quad y_0(1) = 1. \tag{2.28}$$

Already we see a difficulty. The differential equation is first-order, yet there are two conditions to be satisfied. The general solution of the equation is

$$y_0(t) = ce^{-x}. \tag{2.29}$$

Application of the boundary condition $y_0(0) = 0$ gives $c = 0$, and so $y_0(x) = 0$. This function cannot satisfy the boundary condition at $x = 1$. Conversely, application of the boundary condition $y_0(1) = 1$ in (2.29) gives $c = e$, and so

$$y_0(t) = e^{1-x}. \tag{2.30}$$

This function cannot satisfy the condition at $x = 0$. Therefore we are at an impasse; regular perturbation fails at the first step.

2.2.3 Boundary Layers

Careful examination of the preceding problem shows what went wrong and points the way toward a correct, systematic method to obtain an approximate solution. First of all, we should have been suspicious from the beginning. The unperturbed problem, found by setting $\varepsilon = 0$, is

$$y' + y = 0, \quad y(0) = 0, \quad y(1) = 1. \tag{2.31}$$

This problem is of a different character than the perturbed problem (2.27) in that it is first order rather than second. Because the small parameter multiplied the highest derivative in (2.27), the second derivative term disappeared when ε was set to zero. This type of phenomenon signals in almost all cases the failure of the regular perturbation method.

Equation (2.27), a linear equation with constant coefficients, can be solved exactly to get

$$y(x) = \frac{1}{e^{-1} - e^{-1/\varepsilon}}(e^{-x} - e^{-x/\varepsilon}). \tag{2.32}$$

A look at the graph of the solution of (2.27) reveals the reason for the difficulty. See Fig. 2.1. We observe that $y(x)$ is changing very rapidly in a narrow interval, called the **boundary layer**, near the origin, and more slowly in the larger interval, called the **outer layer** away from the origin. One spatial scale does not describe the variations in both layers. Rather, two spatial scales are indicated, one for each zone. It is instructive and revealing to compute the derivatives of y and estimate the size of the terms in the equation. An easy calculation shows

$$y' = \frac{1}{e^{-1} - e^{-1/\varepsilon}}\left(-e^{-x} + \frac{1}{\varepsilon}e^{-x/\varepsilon}\right),$$

$$y'' = \frac{1}{e^{-1} - e^{-1/\varepsilon}}\left(e^{-x} - \frac{1}{\varepsilon^2}e^{-x/\varepsilon}\right).$$

First we examine the second derivative. Suppose ε is small and x is in the narrow boundary layer near $x = 0$. For definiteness assume $x = \varepsilon$. Then

$$y''(\varepsilon) = \frac{1}{e^{-1} - e^{-1/\varepsilon}}\left(e^{-\varepsilon} - \frac{1}{\varepsilon^2}e^{-1}\right) = O(\varepsilon^{-2})$$

Thus y'' is very large inside this narrow band, and therefore the term $\varepsilon y''$ is not small, as would be anticipated in a regular perturbation calculation; in fact $\varepsilon y'' = O(\varepsilon^{-1})$. We see again that terms that appear small in a differential equation are not always necessarily small. The problem is that the differential equation (2.27), as it stands, is not scaled properly for small x, and a rescaling must occur in the boundary layer if correct conclusions are to be drawn. For

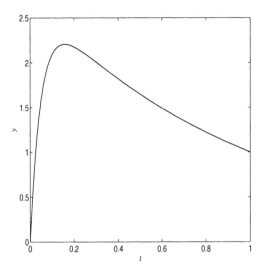

Figure 2.1 The exact solution (2.32) showing a narrow layer where rapid changes are occurring.

values of x away from the boundary layer, the term $\varepsilon y''$ is indeed small. For example, if $x = \frac{1}{2}$, then

$$y''\left(\tfrac{1}{2}\right) = \frac{1}{e^{-1} - e^{-1/\varepsilon}}\left(e^{-1/2} - \frac{1}{\varepsilon^2}e^{-1/2\varepsilon}\right) = O(1)$$

Consequently, in the outer layer $\varepsilon y''$ is small and may be safely neglected. The reader may deduce similar conclusions regarding the first derivative term.

This calculation suggests that in the outer region, the region away from the boundary layer, the leading-order problem (2.31) obtained by setting $\varepsilon = 0$ in the original problem is a valid approximation provided we take only the right $(x = 1)$ boundary condition. Thus,

$$y_0(x) = e^{1-x} \tag{2.33}$$

This is consistent with the exact solution (2.32). For, if ε is small, then $e^{-1} - e^{-1/\varepsilon} \approx e^{-1}$ and y may be approximated by

$$y(x) \approx e^{1-x} - e^{1-x/\varepsilon}. \tag{2.34}$$

For x of order 1 we have $e^{1-x/\varepsilon} \approx 0$, and so

$$y(t) \approx e^{1-t}, \quad x = O(1).$$

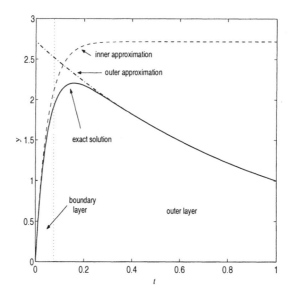

Figure 2.2 The exact solution (2.32) and inner and outer approximations.

The approximate solution (2.33), which is valid in the outer region, is called the **outer approximation** and is plotted in Fig. 2.2.

What about an approximate solution in the narrow boundary layer near $x = 0$? From (2.34), if x is small, then

$$y(x) \approx e - e^{1-x/\varepsilon}, \quad x \text{ small} \tag{2.35}$$

This approximate solution, which is valid only in the boundary layer where rapid changes are taking place, is called the **inner approximation** and is denoted by $y_i(x)$; it is also plotted in Fig. 2.2.

In this problem we have the advantage of knowing the exact solution, and as yet we have little clue how to determine the inner solution if the exact solution is unknown. The key to the analysis in the boundary layer is re-scaling, as in the polynomial equations examined earlier. In the layer, the term $\varepsilon y''$ in the differential equation is not small, as it appears. Consequently, we must re-scale the independent variable x in the boundary layer by selecting a spatial scale that will reflect rapid and abrupt changes and will force each term in the equation into its proper form in the re-scaled variables, namely as the product of a coefficient representing the magnitude of the term and a term of order unity. In the next section we develop a general procedure for dealing with boundary layers.

Before embarking on an analysis of the boundary layer phenomena we summarize our observations and make some general remarks. We have noted that a naive regular perturbation expansion does not always produce an approximate solution. In fact, there are several indicators that often suggest its failure.

1. When the small parameter multiplies the highest derivative in the problem.

2. When setting the small parameter equal to zero changes the character of the problem, as in the case of a partial differential equation changing type (elliptic to parabolic, for example), or an algebraic equation changing degree. In other words, the solution for $\varepsilon = 0$ is fundamentally different in character from the solutions for ε close to zero.

3. When problems occur on infinite domains, giving, for example, secular terms.

4. When singular points are present in the interval of interest.

5. When the equations that model physical processes have multiple time or spatial scales.

Such perturbation problems fall in the general category of singular perturbation problems. For ordinary differential equations, problems involving boundary layers are common. The procedure is to determine whether there is a boundary layer and where it is located. If there is a boundary layer, then the leading-order perturbation term found by setting $\varepsilon = 0$ in the equation often provides a valid approximation in a large outer layer. The inner approximation in the boundary layer is found by re-scaling, which we discuss below. We show that the inner and outer approximations can be *matched* to obtain a uniformly valid approximation over the entire interval of interest. The singular perturbation method applied in this context is also called the method of *matched asymptotic expansions* or *boundary layer theory*.

EXERCISES

1. Use a dominant balancing method to determine the leading-order behavior of the roots of the following algebraic equations. In all cases $0 < \varepsilon \ll 1$.

   a) $\varepsilon x^4 + \varepsilon x^3 - x^2 + 2x - 1 = 0$.

   b) $\varepsilon x^3 + x - 2 = 0$.

   c) $\varepsilon^2 x^6 - \varepsilon x^4 - x^3 + 8 = 0$.

   d) $\varepsilon x^5 - x^3 - 1 = 0$.

   e) $\varepsilon^2 x^5 - x^3 + 1 = 0$.

2. In (1b) find a first-order correction for the leading behavior (Hint: for the scaled equation assume $\bar{x} = x_0 + x_1\sqrt{\varepsilon} + x_2\varepsilon + \cdots$.)

3. Find a 2-term approximation to the three roots $x^3 + \varepsilon x + 2\varepsilon^2 = 0$, $0 < \varepsilon \ll 1$.

4. Show that regular perturbation fails on the boundary value problem

$$\varepsilon y'' + y' + y = 0, \quad 0 < x < 1, \quad 0 < \varepsilon \ll 1,$$
$$y(0) = 0, \quad y(1) = 1.$$

Find the exact solution and sketch it for $\varepsilon = 0.05$ and $\varepsilon = 0.005$. If $x = O(\varepsilon)$, show that $\varepsilon y''(x)$ is large; if $x = O(1)$, show that $\varepsilon y''(x) = O(1)$. Find the inner and outer approximations from the exact solution.

2.3 Boundary Layer Analysis

2.3.1 Inner and Outer Approximations

We return to the boundary value problem

$$\varepsilon y'' + (1 + \varepsilon)y' + y = 0, \quad 0 < x < 1, \tag{2.36}$$
$$y(0) = 0, \quad y(1) = 1,$$

where $0 < \varepsilon \ll 1$. By examining the exact solution near $x = 0$ we found that rapid changes were occurring in y, y', and y'', suggesting a small characteristic length; the term $\varepsilon y''$ was not small as it appears to be in the equation. Away from the boundary layer, in the region where $x = O(1)$, it was noted that $\varepsilon y''$ and $\varepsilon y'$ are small, and so the solution could be approximated accurately by setting $\varepsilon = 0$ in the equation to obtain

$$y' + y = 0,$$

and selecting the boundary condition $y(1) = 1$. This gives the outer approximation

$$y_0(t) = e^{1-x}. \tag{2.37}$$

To analyze the behavior in the boundary layer, we notice that significant changes in y take place on a very short spatial interval, which suggests a length scale on the order of a function of ε, say $\delta(\varepsilon)$. If we change variables via

$$\xi = \frac{x}{\delta(\varepsilon)}, \quad Y(\xi) = y(\delta(\varepsilon)\xi), \tag{2.38}$$

the differential equation (2.36) becomes

$$\frac{\varepsilon}{\delta(\varepsilon)^2} Y''(\xi) + \frac{(1+\varepsilon)}{\delta(\varepsilon)} Y'(\xi) + Y(\xi) = 0, \tag{2.39}$$

where prime denotes derivatives with respect to ξ. Another way of looking at the rescaling is to regard (2.38) as a scale transformation that permits examination of the boundary layer close up, as under a microscope.

The coefficients of the four terms in the differential equation are

$$\frac{\varepsilon}{\delta(\varepsilon)^2}, \quad \frac{1}{\delta(\varepsilon)}, \quad \frac{\varepsilon}{\delta(\varepsilon)}, \quad 1. \tag{2.40}$$

If the scaling is correct, each will reflect the order of magnitude of the term in which it appears. To determine the scale factor $\delta(\varepsilon)$ we estimate these magnitudes by considering all possible dominant balances between pairs of terms in (2.40). In the pairs we include the first term because it was deleted in the outer layer, and it is known that it plays a significant role in the boundary layer. Because the goal is to make a simplification in the problem, we do not consider dominant balancing of three terms. If all four terms are equally important, no simplification can be made at all. Therefore there are three cases to consider. For notation we will sometimes use the symbol $\sim$ to denote "of the same order."

(i) The terms $\varepsilon/\delta(\varepsilon)^2$ and $1/\delta(\varepsilon)$ are of the same order and $\varepsilon/\delta(\varepsilon)$ and 1 are small in comparison.

(ii) The terms $\varepsilon/\delta(\varepsilon)^2$ and 1 are of the same order and $1/\delta(\varepsilon)$ and $\varepsilon/\delta(\varepsilon)$ are small in comparison.

(iii) The terms $\varepsilon/\delta(\varepsilon)^2$ and $\varepsilon/\delta(\varepsilon)$ are of the same order and $1/\delta(\varepsilon)$ and 1 are small in comparison.

Only case (i) is possible. For, in case (ii), $\varepsilon/\delta(\varepsilon)^2 \sim 1$ implies $\delta(\varepsilon) = O(\sqrt{\varepsilon})$; but then $1/\delta(\varepsilon)$ is not small compared to 1. This case is inconsistent. In case (iii), $\varepsilon/\delta(\varepsilon)^2 \sim \varepsilon/\delta(\varepsilon)$ implies $\delta(\varepsilon) = O(1)$, which leads to the outer approximation. In case (i), $\varepsilon/\delta(\varepsilon)^2 \sim 1/\delta(\varepsilon)$ forces $\delta(\varepsilon) = O(\varepsilon)$; then $\varepsilon/\delta(\varepsilon)^2$ and $1/\delta(\varepsilon)$ are both order $1/\varepsilon$, which is large compared to $\varepsilon/\delta(\varepsilon)$ and 1. Therefore, a consistent scaling is possible if we select $\delta(\varepsilon) = O(\varepsilon)$; hence, we take

$$\delta(\varepsilon) = \varepsilon. \tag{2.41}$$

Therefore, the scaled differential equation (2.30) becomes

$$Y'' + Y' + \varepsilon Y' + \varepsilon Y = 0. \tag{2.42}$$

Now, (2.42) is amenable to regular perturbation. Because we are interested only in the leading-order approximation, which we denote by Y_i, we set $\varepsilon = 0$ in (2.42) to obtain

$$Y_i'' + Y_i' = 0. \tag{2.43}$$

The general solution is

$$Y_i(\tau) = C_1 + C_2 e^{-\xi}.$$

Because the boundary layer is located near $x = 0$, we apply the boundary condition $y(0) = 0$, or $Y_i(0) = 0$. This yields $C_2 = -C_1$, and so

$$Y_i(\xi) = C_1(1 - e^{-\xi}). \tag{2.44}$$

In terms of y and x,

$$y_i(t) = C_1(1 - e^{-x/\varepsilon}). \tag{2.45}$$

This is the inner approximation for $x = O(\varepsilon)$.

In summary, we have an approximation solution

$$y_0(x) = e^{1-x}, \quad x = O(1)$$
$$y_i(x) = C_1(1 - e^{-x/\varepsilon}), \quad x = O(\varepsilon), \tag{2.46}$$

each valid for an appropriate range of x. There remains to determine the constant C_1, which is accomplished by the process of matching.

2.3.2 Matching

By matching we do *not* mean selecting a specific value of ε, say ε_0, and requiring that $y_0(\varepsilon_0) = y_i(\varepsilon_0)$ to obtain the constant C_1. This process is nothing more than patching the two approximations together so as to be continuous at some fixed $x = \varepsilon_0$. Rather, the goal is to construct a single composite expansion in ε that is uniformly valid on the entire interval $[0, 1]$ as $\varepsilon \to 0$. Thus, there is little gain in pinpointing the edge of the boundary layer, since it becomes narrower as $\varepsilon \to 0$. We figuratively state that the boundary layer for this problem has width ε. In general, it is the scaling factor $\delta(\varepsilon)$ that defines the **width** of the boundary layer. One can proceed as follows, however. It seems reasonable that the inner and outer expansions should agree to some order in an overlap domain, intermediate between the boundary layer and outer region (see Fig. 2.3). If $x = O(\varepsilon)$, then x is in the boundary layer, and if $x = O(1)$, then x is in the outer region; therefore, this overlap domain could be characterized as values of x for which $x = O(\sqrt{\varepsilon})$, for example, because $\sqrt{\varepsilon}$ is between ε and

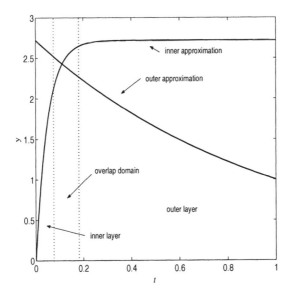

Figure 2.3 The overlap domain where inner and outer approximations are matched.

1 ($\sqrt{\varepsilon}$ goes to zero slower than ε does, or $\varepsilon \ll \sqrt{\varepsilon}$). This intermediate scale suggests a new scaled independent variable η in the overlap domain defined by

$$\eta = \frac{x}{\sqrt{\varepsilon}} \tag{2.47}$$

The condition for matching is that the inner approximation, written in terms of the *intermediate variable* η, should agree with the outer approximation, written in terms of the intermediate variable η, in the limit as $\varepsilon \to 0^+$. In symbols, for matching we require that for fixed η

$$\lim_{\varepsilon \to 0^+} y_0(\sqrt{\varepsilon}\eta) = \lim_{\varepsilon \to 0^+} y_i(\sqrt{\varepsilon}\eta). \tag{2.48}$$

For the present problem,

$$\lim_{\varepsilon \to 0^+} y_0(\sqrt{\varepsilon}\eta) = \lim_{\varepsilon \to 0^+} e^{1-\sqrt{\varepsilon}\eta} = e$$

and

$$\lim_{\varepsilon \to 0^+} y_i(\sqrt{\varepsilon}\eta) = \lim_{\varepsilon \to 0^+} C_1(1 - e^{-\eta/\sqrt{\varepsilon}}) = C_1.$$

Therefore, matching requires $C_1 = e$ and the inner approximation becomes

$$y_i(t) = e(1 - e^{-x/\varepsilon}). \tag{2.49}$$

Because we are seeking only a leading order approximation, we can avoid introducing an intermediate variable and just write the **matching condition** (2.48) simply as

$$\lim_{x \to 0^+} y_0(x) = \lim_{\xi \to \infty} Y_i(\xi). \tag{2.50}$$

This states that the outer approximation, as the outer variable moves into the inner region, must equal the inner approximation, as the inner variable moves into the outer region. Finally, we point out that we have only introduced approximations of leading order. Higher order approximations can be obtained by more elaborate matching schemes, and we refer to the references for discussions of these methods. For example, see Bender & Orszag (1978). In the sequel we usually use (2.50) instead of introducing an intermediate variable.

2.3.3 Uniform Approximations

To obtain a composite expansion that is uniformly valid throughout $[0, 1]$, we note the sum of the outer and inner approximations is

$$y_0(x) + y_i(x) = e^{1-x} + e - e^{1-x/\varepsilon}$$

$$\approx \begin{cases} e^{1-x} + e, & x = O(1) \\ 2e - e^{1-x/\varepsilon}, & x = O(\varepsilon) \end{cases}$$

By subtracting the common limit (2.48), which is e, we have

$$y_u(x) \equiv y_0(x) + y_i(x) - e = e^{1-x} - e^{1-x/\varepsilon}. \tag{2.51}$$

When x is in the outer region the second term is small and $y_u(x)$ is approximately e^{1-x}, which is the outer approximation. When x is in the boundary layer the first term is nearly e and $y_u(x)$ is approximately $e(1 - e^{-x/\varepsilon})$, which is the inner approximation. In the intermediate or overlap region both the inner and outer approximations are approximately equal to e. Therefore, in the overlap domain the sum of $y_0(x)$ and $y_i(x)$ gives $2e$, or twice the contribution. This is why we must subtract the common limit from the sum. In summary, $y_u(x)$ provides a uniform approximate solution throughout the interval $[0, 1]$.

Substituting $y_u(x)$ into the differential equation gives

$$\varepsilon y_u'' + (1 + \varepsilon)y_u' + y_u = 0,$$

so $y_u(x)$ satisfies the differential equation exactly on $(0, 1)$. Checking the boundary conditions,

$$y_u(0) = 0, \quad y_u(1) = 1 - e^{1-1/\varepsilon}.$$

The left boundary condition is satisfied exactly and the right boundary condition holds up to $O(\varepsilon^n)$, for any $n > 0$, because

$$\lim_{\varepsilon \to 0+} \frac{e^{1-1/\varepsilon}}{\varepsilon^n} = 0,$$

for any $n > 0$. Consequently, y_u is a uniformly valid approximation on $[0, 1]$.

Example 2.8

We work through another simple example without the detailed exposition of the preceding paragraphs. We determine an approximate solution of the boundary value problem

$$\varepsilon y'' + y' = 2x, \quad 0 < x < 1, \quad 0 < \varepsilon \ll 1,$$
$$y(0) = 1, \quad y(1) = 1,$$

using singular perturbation methods. Clearly, regular perturbation will fail since the unperturbed problem is

$$y' = 2x,$$

which has the general solution

$$y(x) = x^2 + C.$$

Such a function cannot satisfy both boundary conditions. Consequently we assume a boundary layer at $x = 0$ and impose the boundary condition $y(1) = 1$ (since $x = 1$ is in the outer layer) to get the outer approximation

$$y_0(x) = x^2, \quad x = O(1).$$

To determine the width $\delta(\varepsilon)$ of the boundary layer we re-scale near $x = 0$ via

$$\xi = \frac{x}{\delta(\varepsilon)}, \quad Y(\xi) = y(x).$$

In scaled variables the differential equation becomes

$$\frac{\varepsilon}{\delta(\varepsilon)^2} Y'' + \frac{1}{\delta(\varepsilon)} Y' = 2\delta(\varepsilon)\xi.$$

If $\varepsilon/\delta(\varepsilon)^2 \sim 2\delta(\varepsilon)$ is the dominant balance, then $\delta(\varepsilon) = O(\varepsilon^{1/3})$ and the second term $1/\delta(\varepsilon)$ would be $O(\varepsilon^{-1/3})$, which is not small compared to the assumed dominant terms. Therefore assume $\varepsilon/\delta(\varepsilon)^2 \sim 1/\delta(\varepsilon)$ is the dominant balance. In that case $\delta(\varepsilon) = O(\varepsilon)$ and the term $2\delta(\varepsilon)$ has order $O(\varepsilon)$, which is small compared to $\varepsilon/\delta(\varepsilon)^2$ and $1/\delta(\varepsilon)$, both of which are $O(\varepsilon^{-1})$. Therefore $\delta(\varepsilon) = \varepsilon$ is consistent and the scaled differential equation is

$$Y'' + Y' = 2\varepsilon^2 \xi.$$

The inner approximation to first-order satisfies

$$Y_i'' + Y_i' = 0,$$

whose general solution is

$$Y(\xi) = C_1 + C_2 e^{-\xi}.$$

In terms of y and x

$$y(x) = C_1 + C_2 e^{-x/\varepsilon}.$$

Applying the boundary condition $y(0) = 1$ in the boundary layer gives $C_1 = 1 - C_2$, and therefore the inner approximation is

$$y_i(x) = (1 - C_2) + C_2 e^{-x/\varepsilon}.$$

To find C_2 we introduce an overlap domain of order $\sqrt{\varepsilon}$ and an appropriate intermediate scaled variable

$$\eta = \frac{x}{\sqrt{\varepsilon}}.$$

Then $x = \sqrt{\varepsilon}\eta$ and the matching condition becomes (with η fixed)

$$\lim_{\varepsilon \to 0^+} y_0(\sqrt{\varepsilon}\eta) = \lim_{\varepsilon \to 0^+} y_i(\sqrt{\varepsilon}\eta),$$

or

$$\lim_{\varepsilon \to 0^+} \varepsilon\eta^2 = \lim_{\varepsilon \to 0^+} [(1 - C_2) + C_2 e^{-\eta/\sqrt{\varepsilon}}].$$

This gives $C_2 = 1$, and thus the inner approximation is

$$y_i(x) = e^{-x/\varepsilon}.$$

A uniform composite approximation $y_u(x)$ is found by adding the inner and outer approximations and subtracting the common limit in the overlap domain, which is zero in this case. Consequently,

$$y_u(x) = x^2 + e^{-x/\varepsilon}.$$

As an exercise, the reader should plot this approximation for different values of ε. Again, as a shortcut, the matching condition may be written

$$\lim_{x \to 0} y_0(x) = \lim_{\xi \to \infty} Y_i(\xi),$$

which gives the same result.

2.3.4 General Procedures

In the preceding examples the boundary layer occurred at $x = 0$, or at the left endpoint. In the general case boundary layers can occur at any point in the interval, at the right endpoint or an interior point; in fact, multiple boundary layers can occur in the same problem. Boundary layers also appear in initial value problems, which we discuss in the next section. The Exercises point out different boundary layer phenomena. When solving a problem one should assume a boundary layer at $x = 0$ (or the left endpoint) and then proceed. If this assumption is in error, then the procedure will break down when trying to match the inner and outer approximations. At that time make an assumption of a boundary layer at the right endpoint. The analysis is exactly the same, but the scale transformation to define the inner variable in the boundary layer becomes

$$\xi = \frac{x_0 - x}{\delta(\varepsilon)},$$

where $x = x_0$ is the right endpoint, and $Y(\xi) = y(x_0 - \delta(\varepsilon)\xi)$. Even though preceding examples exhibit a boundary layer width of $\delta(\varepsilon) = \varepsilon$, this is not the rule. Also, we have only matched the leading-order behavior of the inner and outer approximations. Refined the matching procedures can include higher order terms. Finally we point out that this method is not a universal technique. For certain classes of problems it works well, but for other problems significant modifications must be made. Singular perturbation theory is an active area of research in applied mathematics and a well-developed rigorous theory is only available for restricted classes of differential equations.

We conclude with a theorem for linear equations with variable coefficients. For this class of problems, the boundary layer can be completely characterized, provided suitable restrictions are placed on the coefficients.

Theorem 2.9

Consider the boundary value problem

$$\varepsilon y'' + p(x)y' + q(x)y = 0, \quad 0 < x < 1, \quad 0 < \varepsilon \ll 1, \tag{2.52}$$

$$y(0) = a, \quad y(1) = b,$$

where p and q are continuous functions on $0 \leq x \leq 1$, and $p(x) > 0$ for $0 \leq x \leq 1$. Then there exists a boundary layer at $x = 0$ with inner and outer approximations given by

$$y_i(x) = C_1 + (a - C_1)e^{-p(0)x/\varepsilon}, \tag{2.53}$$

$$y_0(x) = b \exp \left(\int_x^1 \frac{q(s)}{p(s)} \, ds \right), \tag{2.54}$$

where

$$C_1 = b \exp\left(\int_0^1 \frac{q(s)}{p(s)}\, ds\right). \tag{2.55}$$

To demonstrate the theorem, we show that the assumption of a boundary layer at $x = 0$ is consistent and leads to the approximations given above. If the boundary layer is at $x = 0$, then the outer solution $y_0(x)$ will satisfy

$$p(x)y_0' + q(x)y_0 = 0$$

and the condition $y_0(1) = b$. Separating variables and applying the condition at $x = 1$ gives (2.54). In the boundary layer we introduce a scaled variable ξ defined by $\xi = x/\delta(\varepsilon)$, where $\delta(\varepsilon)$ is to be determined. If $Y(\xi) = y(\delta(\varepsilon)\xi)$, then the differential equation becomes

$$\frac{\varepsilon}{\delta(\varepsilon)^2}Y'' + \frac{p(\delta(\varepsilon)\xi)}{\delta(\varepsilon)}Y' + q(\delta(\varepsilon)\xi)Y = 0. \tag{2.56}$$

As $\varepsilon \to 0^+$ the coefficients behave like

$$\frac{\varepsilon}{\delta(\varepsilon)^2}, \quad \frac{p(0)}{\delta(\varepsilon)}, \quad q(0).$$

It is easy to see that the dominant balance is $\varepsilon/\delta(\varepsilon)^2 \sim p(0)/\delta(\varepsilon)$, and therefore the boundary layer has thickness $\delta(\varepsilon) = O(\varepsilon)$. For definiteness take $\delta(\varepsilon) = \varepsilon$. Equation (2.56) then becomes

$$Y'' + p(\varepsilon\xi)Y' + \varepsilon q(\varepsilon\xi)Y = 0,$$

which to leading order is

$$Y_i'' + p(0)Y_i' = 0.$$

The general solution is

$$Y(\xi) = C_1 + C_2 e^{-p(0)\xi}.$$

Applying the boundary condition $Y_i(0) = a$ yields $C_2 = a - C_1$, and thus the inner approximation is

$$y_i(x) = C_1 + (a - C_1)e^{-p(0)x/\varepsilon}. \tag{2.57}$$

To match we introduce the intermediate variable $\eta = x/\sqrt{\varepsilon}$ and require that for fixed η,

$$\lim_{\varepsilon \to 0^+} y_i(\sqrt{\varepsilon}\eta) = \lim_{\varepsilon \to 0^+} y_0(\sqrt{\varepsilon}\eta).$$

In this case the matching condition becomes

$$\lim_{\varepsilon \to 0^+} \{C_1 + (a - C_1)e^{-p(0)\eta/\sqrt{\varepsilon}}\} = \lim_{\varepsilon \to 0^+} \left\{b \exp\left(\int_{\sqrt{\varepsilon}\eta}^1 \frac{q(s)}{p(s)}\, ds\right)\right\},$$

which forces

$$C_1 = b \exp \left(\int_0^1 \frac{q(s)}{p(s)} \, ds \right).$$

Consequently, the inner approximation is given as stated in (2.53) and (2.55), and the proof is complete. A uniform composite approximation $y_u(x)$ is given by

$$y_u(x) = y_0(x) + y_i(x) - C_1.$$

It can be shown that $y_u(x) - y(x) = O(\varepsilon)$ as $\varepsilon \to 0^+$, uniformly on $[0,1]$, where $y(x)$ is the exact solution to (2.52).

If $p(x) < 0$ on $0 \leq x \leq 1$, then no match is possible because $y_i(x)$ would grow exponentially, unless $C_1 = a$. On the other hand, a match is possible if $p(x) < 0$ if the boundary layer is at $x = 1$. In summary, one can show that the boundary layer is at $x = 0$ if $p(x) > 0$ and at $x = 1$ if $p(x) < 0$. As a final observation, there can be no boundary layer at an interior point $0 < x < 1$ in either case. If p changes sign in the interval, then interior layers are possible; these problems are called *turning point* problems.

EXERCISES

1. Use singular perturbation methods to obtain a uniform approximate solution to the following problems. In each case assume $0 < \varepsilon \ll 1$ and $0 < x < 1$.

   a) $\varepsilon y'' + 2y' + y = 0$, $\quad y(0) = 0$, $\quad y(1) = 1$.

   b) $\varepsilon y'' + y' + y^2 = 0$, $\quad y(0) = \frac{1}{4}$, $\quad y(1) = \frac{1}{2}$.

   c) $\varepsilon y'' + (1+x)y' = 1$, $\quad y(0) = 0, y(1) = 1 + \ln 2$.

   d) $\varepsilon y'' + (x+1)y' + y = 0$, $\quad y(0) = 0$, $\quad y(1) = 1$

   e) $\varepsilon y'' + x^{1/3} y' + y = 0$, $\quad y(0) = 0$, $\quad y(1) = \exp\left(-\frac{3}{2}\right)$.

   f) $\varepsilon y'' + xy' - xy = 0$, $\quad y(0) = 0$, $\quad y(1) = e$.

   g) $\varepsilon y'' + 2y' + e^y = 0$, $\quad y(0) = y(1) = 0$.

   h) $\varepsilon y'' - (2 - x^2)y = -1$, $\quad y'(0) = 0$, $\quad y(1) = 1$.

   i) $\varepsilon y'' - b(x)y' = 0$, $\quad y(0) = \alpha$, $y(1) = \beta$, where $\alpha \neq \beta$, $b(x) > 0$.

   j) $\varepsilon y'' - 4(\pi - x^2)y = \cos x$, $\quad y(0) = 0$, $\quad y(\pi/2) = 1$.

2. Examine the exact solution to show why singular perturbation methods fail on the boundary value problem

$$\varepsilon u'' + u = 0, \quad 0 < x < 1,$$
$$u(0) = 1, \quad u(1) = 2.$$

3. Determine values of a for which the problem

$$\varepsilon y'' + y' + ae^y = 0, \quad y(0) = 0, \ y(1) = 0$$

has a solution with a boundary layer structure.

4. Obtain a uniform approximation to the problem

$$\varepsilon u'' - (2x + 1)u' + 2u = 0, \quad 0 < x < 1,$$
$$u(0) = 1, \quad u(1) = 0.$$

5. Is the problem

$$\varepsilon y'' + \frac{1}{x} y' + y = 0, \quad x > 0,$$
$$y(0) = 1, \quad y'(0) = 0$$

a singular perturbation problem? Discuss.

6. Attempt singular perturbation methods on the boundary value problem

$$\varepsilon y'' + \left(x - \tfrac{1}{2}\right) y = 0, \quad 0 < x < 1,$$
$$y(0) = 1, \quad y(1) = 2.$$

Give a complete discussion.

7. Find a leading order approximation to

$$\varepsilon y'' + e^x y' = 1, \quad y(0) = 1, \ y(\infty) = 0.$$

8. Let f be a smooth function, monotonic on $[0, 1]$, and let $f'(0) = b \neq 0$. Either find a uniformly valid approximation to the problem

$$\varepsilon y'' - \frac{1}{f'(x)} y' = -1, \quad y'(0) = 1, \ y(1) = 0,$$

or show that such an approximation does not exist. Does the answer depend upon the value b?

9. Find a uniformly valid approximation to

$$-\varepsilon u'' + \cos(b - x)u = 1, \quad 0 < x < 1,$$
$$u(0) = 0, \quad u'(b) = 0,$$

where $0 < \varepsilon \ll 1$ and $b > 0$. For what values of b is the approximation valid?

10. Find a uniformly valid approximation to

$$\varepsilon u'' - a(x)u = f(t), \quad 0 < x < 1,$$
$$u(0) = 0, \quad u(b) = -f(1)/a(1),$$

where $0 < \varepsilon \ll 1$ and $a > 0$, and a and f have infinitely many derivatives on $\mathbb{R}$.

11. Find a uniform approximation to the boundary value problem

$$\varepsilon y'' + \frac{1}{1+x}y' + \varepsilon y = 0, \quad 0 < x < 1, \quad 0 < \varepsilon \ll 1,$$
$$y(0) = 0, \quad y(1) = 1.$$

Draw a graph of the approximation.

12. Find uniform asymptotic approximations to the solutions of the following problems on $0 \le x \le 1$ with $0 < \varepsilon \ll 1$.

a) $\varepsilon y'' + (x^2 + 1)y' - x^3 y = 0, \quad y(0) = y(1) = 1.$

b) $\varepsilon y'' + (\cosh x)y' - y = 0, \quad y(0) = y(1) = 1.$

c) $\varepsilon y'' + 2\varepsilon x^{-1}y' - y = 0, \quad y(0) = 0, \quad y'(1) = 1.$

d) $\varepsilon(y'' + y') - y = 0, \quad y(0) = y(1) = 0.$

e) $\varepsilon y'' + (1+x)^{-1}y' + \varepsilon y = 0, \quad y(0) = 0, \quad y(1) = 1.$

2.4 Initial Layers

2.4.1 Damped Spring–Mass System

Initial value problems containing a small parameter can have a layer near $t = 0$. For evolution problems these layers are called initial layers. They are a zones in which there are rapid temporal changes. We will illustrate the idea on a simple, solvable initial value problem. Consider a damped spring-mass system with y denoting the positive displacement of the mass m from equilibrium, and with the forces given by

$$F_{\text{spring}} = -ky, \quad F_{\text{damping}} = -a\dot{y}, \quad (k, a > 0),$$

where k and a are the spring constant and damping constant, respectively. By Newton's second law the governing differential equation is

$$m\ddot{y} + a\dot{y} + ky = 0, \quad t > 0. \tag{2.58}$$

Initially, we assume the displacement is zero and that the mass is put into motion by imparting to it (say, with a hammer blow) a positive impulse I. Therefore, the initial conditions are given by

$$y(0) = 0, \quad m\dot{y}(0) = I. \tag{2.59}$$

We want determine the leading-order behavior of the system in the case that the mass has very small magnitude.

The first step is recast the problem in dimensionless form. The independent and dependent variables are t and y having dimensions of time T and length L, respectively. The constants m, a, k, and I have dimensions

$$[m] = M, \quad [a] = MT^{-1}, \quad [k] = MT^{-2}, \quad [I] = MLT^{-1},$$

where M is a mass dimension. We can identify three possible time scales,

$$\frac{m}{a}, \quad \sqrt{\frac{m}{k}}, \quad \frac{a}{k}. \tag{2.60}$$

These correspond to balancing inertia and damping terms, the inertia and the spring terms, and the damping and the spring terms, respectively. Possible length scales are

$$\frac{I}{a}, \quad \frac{I}{\sqrt{km}}, \quad \frac{aI}{km}. \tag{2.61}$$

By assumption $m \ll 1$, and so $m/a \ll 1$, $\sqrt{m/k} \ll 1$, $I/\sqrt{km} \gg 1$, and $aI/km \gg 1$. Such relations are often important in determining appropriate time and length scales, because scales should be chosen so that the dimensionless dependent and independent variables are of order one.

Before proceeding further, we us use our intuition regarding the motion of the mass. The positive impulse given to the mass will cause a rapid displacement to some maximum value, at which time the force due to the spring will attempt to restore it to its equilibrium position. Because the mass is small there will be very little inertia, and therefore it will probably not oscillate about equilibrium; the system is strongly over-damped. Consequently we expect the graph of $y = y(t)$ to exhibit the features shown in Fig. 2.4—a quick rise and then a gradual decay. Such a function is indicative of a multiple time-scale description. In the region near $t = 0$, where there is an abrupt change, a short time scale seems appropriate; we call this the **initial layer**. Away from the initial layer an order 1 time scale seems appropriate. In all, the problem has the characteristics of a singular perturbation problem, namely multiple time scales and an apparent small quantity m multiplying the highest derivative term. Of the length scales given in (2.61), it appears that only I/a is suitable. The remaining two are large, which violates our intuition regarding the maximum displacement. Of

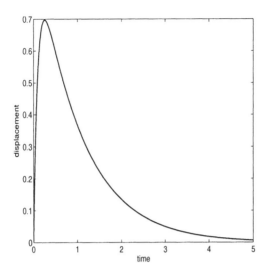

Figure 2.4 Displacement vs. time.

the time scales, only a/k is first-order. The remaining two are small, and one of them may be suitable in the assumed initial layer near $t = 0$. As a guess, since m/a depends on the mass and the damping, and $\sqrt{m/k}$ depends on the mass and the spring, we predict the former. The reason lies in intuition concerning the dominant processes during the early stages of motion. The high initial velocity should influence the damping force more than the force due to the spring. Therefore, the terms $m\ddot{y}$ and $a\dot{y}$ should dominate during the initial phase, whereas $m\ddot{y}$ should be low-order in the later phase.

With these remarks as motivation, we introduce scaled variables

$$\bar{t} = \frac{t}{a/k}, \quad \bar{y} = \frac{y}{I/a}, \tag{2.62}$$

which seem appropriate for the outer layer, away from the initial layer at $t = 0$. Note that both $\bar{t}$ and $\bar{y}$ are of order 1. In terms of these variables the initial value problem becomes

$$\varepsilon\bar{y}'' + \bar{y}' + \bar{y} = 0,$$
$$\bar{y}(0) = 0, \quad \varepsilon\bar{y}'(0) = 1, \tag{2.63}$$

where prime denotes the derivative with respect to $\bar{t}$, and the dimensionless constant ε is given by

$$\varepsilon = \frac{mk}{a^2} \ll 1. \tag{2.64}$$

Notice, by proper scaling, the small term occurs where we expect it, namely in the inertia term. Using singular perturbation techniques we can obtain an approximate solution to (2.63). The leading order approximation in the outer layer is

$$\overline{y}' + \overline{y} = 0,$$

where we have set $\varepsilon = 0$. Then,

$$\overline{y}_0(\overline{t}) = Ce^{-\overline{t}} \tag{2.65}$$

is the outer approximation. Neither initial condition is appropriate to apply because both are at $\overline{t} = 0$, which is in the assumed initial layer. To obtain an inner approximation we re-scale according to

$$\tau = \frac{\overline{t}}{\delta(\varepsilon)}, \quad Y = \overline{y}. \tag{2.66}$$

Then (2.63) becomes

$$\frac{\varepsilon}{\delta(\varepsilon)^2}Y'' + \frac{1}{\delta(\varepsilon)}Y' + Y = 0. \tag{2.67}$$

The dominant balance is $\varepsilon/\delta(\varepsilon)^2 \sim 1/\delta(\varepsilon)$, which gives the width of the initial layer, $\delta(\varepsilon) = \varepsilon$. (The balance $\varepsilon/\delta(\varepsilon)^2 \sim 1$ implies $\delta(\varepsilon) = O(\sqrt{\varepsilon})$, but then $1/\delta(\varepsilon)$ is not small, giving inconsistency.) Consequently, the initial layer has width ε, or $O(m)$. The scale transformation (2.66) is then

$$\tau = \frac{\overline{t}}{\varepsilon}, \quad Y = \overline{y},$$

and the differential equation (2.67) becomes

$$Y'' + Y' + \varepsilon Y = 0.$$

Setting $\varepsilon = 0$ and solving the resulting equation gives the inner approximation

$$Y_i(\tau) = A + Be^{-\tau},$$

or

$$\overline{y}_i(\overline{t}) = A + Be^{-\overline{t}/\varepsilon}.$$

The initial condition $\overline{y}_i(0) = 0$ forces $B = -A$, and the condition $\varepsilon\overline{y}_i'(0) = 1$ forces $A = 1$. Therefore, the inner approximation is completely determined by

$$\overline{y}_i(\overline{t}) = 1 - e^{-\overline{t}/\varepsilon}. \tag{2.68}$$

Next we use matching to determine the constant C. We can introduce an overlap domain with time scale $\sqrt{\varepsilon}$ and scaled dimensionless variable $\eta = \frac{\bar{t}}{\sqrt{\varepsilon}}$, but we prefer to apply the simple condition (2.50), or

$$\lim_{\bar{t} \to 0} \bar{y}_0(\bar{t}) = \lim_{\tau \to \infty} Y_i(\tau).$$

Both limits give $C = 1$. A uniformly valid approximation is therefore

$$\bar{y}_u(\bar{t}) = \bar{y}_0(\bar{t}) + \bar{y}_i(\bar{t}) - \lim_{\varepsilon \to 0^+} y_0(\bar{t}),$$

or

$$\bar{y}_u(\bar{t}) = e^{-\bar{t}} - e^{-\bar{t}/\varepsilon}.$$

Here the limits are taken with η fixed. In terms of the original dimensioned variables t and y,

$$y_u(t) = \frac{I}{a}(e^{-kt/a} - e^{-at/m}). \tag{2.69}$$

Intuition proved to be correct; for small t we have

$$y_u(t) \cong \frac{I}{a}(1 - e^{-at/m}),$$

which describes a rapidly rising function with time scale m/a. For larger t the first term in (2.69) dominates and

$$y_u(t) \cong \frac{I}{a}e^{-kt/a},$$

which represents a function that is decaying exponentially on a time scale a/k.

This example shows how physical reasoning can complement mathematical analysis in solving problems arising out of empirics. The constant interplay between the physics and the mathematics leads to a good understanding of a problem and gives insight into its structure, as well as strategies for obtaining solutions.

2.4.2 Chemical Reaction Kinetics

In chemical processes it is important to know the concentrations of the chemical species when reactions take place, and reaction kinetics provide a rich source of singular perturbation phenomena because reactions in a chain often occur on different time scales. For example, some intermediary chemical species may be short-lived, and reactions creating those species may be able to be ignored, thereby leading to simplification of the system.

The differential equations that govern chemical reactions are based on the **law of mass action**. It states that the *rate* r of the chemical reaction

$$m\mathbf{A} + n\mathbf{B} + \cdots \xrightarrow{k} \text{products},$$

where m molecules of $\mathbf{A}$ combine with n molecules of $\mathbf{B}$, and so on, is proportional to the concentrations $a, b, \ldots$ of the reactants entering the reaction, each raised to the integer powers $m, n, \ldots$, respectively, or

$$r = ka^m b^n \cdots.$$

The constant k is the rate constant, and it often appears in the chemical equation as shown. Actually, k depends upon temperature, but under isothermal assumptions it is taken to be constant. Several examples illustrate the law of mass action.

Example 2.10

The reaction rates of the reactions $\mathbf{A} \xrightarrow{k}$ products, $\mathbf{A} + \mathbf{B} \xrightarrow{k}$ products, $2\mathbf{A} + \mathbf{B} \xrightarrow{k}$ products, and $\mathbf{A} + \mathbf{B} + \mathbf{C} \xrightarrow{k}$ products are, respectively, $r = ka$, $r = kab$, $r = ka^2 b$, and $r = kabc$.

The differential equation(s) governing a chemical reaction state that the rate that a chemical reactant is consumed or created is equal to the reaction rate, with the appropriate number of molecules either consumed or created by the reaction. For several coupled reactions, or a chain, the rate of change of a reactant concentration is the sum of the rates for each reaction where the chemical species is involved.

Example 2.11

Consider the reaction $\mathbf{A} + \mathbf{B} \xrightarrow{k}$ products. We have

$$\frac{da}{dt} = \frac{db}{dt} = -kab.$$

For the reaction $3\mathbf{A} + \mathbf{B} \xrightarrow{k} \mathbf{A} + \mathbf{C}$, we have

$$\frac{da}{dt} = -2a^3 b.$$

For the reversible reaction $\mathbf{A} + \mathbf{B} \xrightarrow{k_1} \mathbf{C}$, $\mathbf{C} \xrightarrow{k_2} \mathbf{A} + \mathbf{B}$, we have

$$\frac{da}{dt} = -k_1 a + k_2 c.$$

Example 2.12

(Enzyme kinetics) To model a certain classes of enzyme kinetics we consider an idealized reaction

$$A + B \rightleftarrows C \rightarrow P + B, \tag{2.70}$$

where reactant molecules A (a substrate) and B (an enzyme) combine to form a complex molecule C. The molecule C breaks up to form a product molecule P and the original enzyme B,or the reverse reaction, C disassociating into the original reactants A and B, can occur. The goal is to formulate a set of equations that govern the concentrations a, b, c, and p of A, B, C, and P, as functions of time. We measure concentration in, for example, number of molecules per unit volume. There are three reactions in (2.70), namely $A + B \overset{k_1}{\rightarrow} C$, $C \overset{k_2}{\rightarrow} A + B$, and $C \overset{k_3}{\rightarrow} P + B$. By the law of mass action, the equations for the kinetics are

$$\frac{da}{dt} = -k_1 ab + k_2 c, \tag{2.71a}$$

$$\frac{db}{dt} = -k_1 ab + k_2 c + k_3 c, \tag{2.71b}$$

$$\frac{dc}{dt} = k_1 ab - k_2 c - k_3 c, \tag{2.71c}$$

$$\frac{dp}{dt} = k_3 c. \tag{2.71d}$$

At $t = 0$ suppose the initial concentrations are

$$a(0) = \hat{a}, \quad b(0) = \hat{b}, \quad c(0) = 0, \quad p(0) = 0.$$

A substantial reduction can be made by adding (2.71b) and (2.71c) to obtain $b + c = \text{constant} = \hat{b}$. Furthermore, the first three equations do not involve p. Therefore the system of four equations can be reduced to two, namely

$$\frac{da}{dt} = -k_1 a(\hat{b} - c) + k_2 c, \tag{2.72}$$

$$\frac{dc}{dt} = k_1 a(\hat{b} - c) - (k_2 + k_3)c. \tag{2.73}$$

Once these are solved for a and c, then p can be found from (2.71d) by simple quadrature, and b is given by $\hat{b} - c$. We give a thorough analysis of these equations in the sequel.

Next we investigate the dynamical equations (2.72)–(2.73) subject to the initial conditions, where the initial concentration $\hat{b}$ of the enzyme B is assumed to be small compared to the initial substrate concentration $\hat{a}$ of A. Before tackling the problem directly, let us use intuition to discover some relevant features of the reaction. In the early stages we expect **A** and **B** to combine rapidly to produce the complex **C**. This is because it is relatively easy for the few molecules of **B** to bond with the abundant molecule **A**. In the long run we expect **C** to decrease to zero because eventually it goes to form the product **P** through an irreversible reaction. Under these assumptions the concentration of **A** would decrease quickly at first, then ultimately level off at some positive value. Therefore, it appears that the problem has singular perturbation character with an initial layer near $t = 0$ where rapid dynamics occurs.

We select dimensionless variables consistent with the scaling principle; that is, the magnitude of each term is correctly estimated by the coefficient that precedes it. It is clear that a should be scaled by its value at $t = 0$, namely $\hat{a}$. To scale c we choose the concentration $\hat{b}$, since eventually all of **B** will go into forming the complex **C**. There are time scales for the process, but it is not clear at this point what the scales are. For the moment we choose an unknown time scale $\hat{t}$. Consequently, under the scale transformation

$$\bar{t} = \frac{t}{\hat{t}}, \quad \bar{a} = \frac{a}{\hat{a}}, \quad \bar{c} = \frac{c}{\hat{b}}, \tag{2.74}$$

equations (2.72)–(2.73) become

$$\frac{\hat{a}}{\hat{t}}\frac{d\bar{a}}{d\bar{t}} = -k_1\hat{a}\hat{b}\bar{a}(1 - \bar{c}) + k_2\hat{b}\bar{c}, \tag{2.75a}$$

$$\frac{\hat{b}}{\hat{t}}\frac{d\bar{c}}{d\bar{t}} = k_1\hat{a}\hat{b}\bar{a}(1 - \bar{c}) - (k_2 + k_3)\hat{b}\bar{c}. \tag{2.75b}$$

In (2.75a) the term $-k_1\hat{a}\hat{b}\bar{a}(1 - \bar{c})$ defines the rate at which **A** is depleted, while the term $k_2\hat{b}\bar{c}$ defines the rate at which **A** is formed from the breakup of C. If the back reaction $\mathbf{C} \to \mathbf{A} + \mathbf{B}$ dominated the forward reaction $\mathbf{A} + \mathbf{B} \to \mathbf{C}$, then little would be accomplished in the total reaction. Hence we assume the dominant balance in (2.75a) is between the two terms

$$\frac{\hat{a}}{\hat{t}}\frac{d\bar{a}}{d\bar{t}} \quad \text{and} \quad -k_1\hat{a}\hat{b}\bar{a}(1 - \bar{c}),$$

which suggests that the time scale is

$$\hat{t} = 1/k_1\hat{b}. \tag{2.76}$$

With this choice the dimensionless equations (2.75a)–(2.75b) become

$$\frac{d\bar{a}}{d\bar{t}} = -\bar{a} + (\bar{a} + \lambda)\bar{c}, \tag{2.77a}$$

$$\varepsilon\frac{d\bar{c}}{d\bar{t}} = \bar{a} - (\bar{a} + \mu)\bar{c}, \tag{2.77b}$$

where the dimensionless parameters λ and μ are given by

$$\lambda = \frac{k_2}{\hat{a}k_1}, \quad \mu = \frac{k_2 + k_3}{\hat{a}k_1},$$

and $\varepsilon = \hat{b}/\hat{a} \ll 1$. The initial conditions are

$$\bar{a}(0) = 1, \quad \bar{c}(0) = 0.$$

At this point we can make a naive calculation that is used routinely in enzyme kinetics. If the $\bar{c}$-equation equilibrates fast, then $\frac{d\bar{c}}{dt} = 0$ and $\bar{a} = (\bar{a} + \mu)\bar{c}$. The $\bar{c}$-equation is said to be in *quasi-equilibrium*. Then the $\bar{a}$ equation (2.77a) becomes, approximately,

$$\frac{d\bar{a}}{dt} = -\frac{(\mu - \lambda)\bar{a}}{\mu + \bar{a}}.$$

This equation is called the **Michaelis–Menten** reaction rate, and it is the usual expression for enzyme kinetics. Below we will see how this classical approximation arises as the outer approximation using singular perturbation methods.

Assume that the outer approximation can be found by setting $\varepsilon = 0$ in (2.77b). If $\bar{a}_0(\bar{t})$ and $\bar{c}_0(\bar{t})$ denote these outer approximations, then

$$\frac{d\bar{a}_0}{dt} = -\bar{a}_0 + (\bar{a}_0 + \lambda)\bar{c}_0,$$

$$0 = \bar{a}_0 - (\bar{a}_0 + \mu)\bar{c}_0.$$

The second equation is an algebraic equation that yields

$$\bar{c}_0 = \frac{\bar{a}_0}{\bar{a}_0 + \mu}. \tag{2.78}$$

Substitution into the first equation then yields a single first-order equation for $\bar{a}_0$ that, when variables are separated, gives

$$\left(1 + \frac{\mu}{\bar{a}_0}\right) d\bar{a}_0 = -(\mu - \lambda)\,d\bar{t}.$$

Integrating,

$$\bar{a}_0 + \mu \ln \bar{a}_0 = -(\mu - \lambda)\bar{t} + K, \tag{2.79}$$

where K is a constant of integration and $\mu - \lambda > 0$. Equation (2.79) defines $\bar{a}_0(\bar{t})$ implicitly. Because it is assumed to hold in the outer region, neither initial condition applies; the constant K will be determined by matching. In summary, (2.78) and (2.79) determine the outer approximation.

To find the inner approximation in the layer near $t = 0$ we change scale via

$$\tau = \frac{\bar{t}}{\delta(\varepsilon)}, \quad A = \bar{a}, \quad C = \bar{c}.$$

Equations (2.77a)–(2.77b) become

$$\frac{1}{\delta(\varepsilon)} \frac{dA}{d\tau} = -A + (A + \lambda)C, \tag{2.80a}$$

$$\frac{\varepsilon}{\delta(\varepsilon)} \frac{dC}{d\tau} = A - (A + \mu)C, \tag{2.80b}$$

where $A = A(\tau)$ and $C = C(\tau)$. In the boundary layer we expect the term involving $dC/d\tau$ to be important. Thus, from (2.80b), we choose $\delta(\varepsilon) = \varepsilon$, which gives

$$\frac{dA}{d\tau} = [-A + (A + \lambda)C]\varepsilon, \tag{2.81a}$$

$$\frac{dC}{d\tau} = A - (A + \mu)C. \tag{2.81b}$$

Setting $\varepsilon = 0$, we find the leading order inner approximation $A_i(\tau)$ is constant. From the initial condition $A_i(0) = 1$, we have

$$A_i(\tau) = 1. \tag{2.82}$$

Then, (2.81b) becomes

$$\frac{dC_i}{d\tau} = 1 - (1 + \mu)C_i,$$

which is a linear first-order equation for C. Solving, we obtain

$$C_i(\tau) = \frac{1}{\mu + 1} + M e^{-(\mu+1)\tau},$$

where M is a constant of integration. Using $C_i(0) = 0$ gives $M = -1/(\mu + 1)$, and so the inner approximation is

$$C_i(\tau) = \frac{1}{\mu + 1}(1 - e^{-(\mu+1)\tau}). \tag{2.83}$$

Finally, we match the outer approximation with the inner approximation. The matching conditions are

$$\lim_{\bar{t} \to 0} \bar{a}_0(\bar{t}) = \lim_{\tau \to \infty} A_i(\tau) \tag{2.84}$$

and

$$\lim_{\bar{t} \to 0} \bar{c}_0(\bar{t}) = \lim_{\tau \to \infty} C_i(\tau). \tag{2.85}$$

The left-hand side of (2.84) cannot be evaluated directly, since $\bar{a}_0(\bar{t})$ is defined only implicitly by (2.79). But we may still evaluate the limit to get

$$\lim_{\bar{t}\to 0} \bar{a}_0(\bar{t}) + \mu \lim_{\bar{t}\to 0} \ln \bar{a}_0(\bar{t}) = K,$$

or

$$1 + \mu \ln(1) = K.$$

Whence

$$K = 1.$$

Equation (2.84) is satisfied identically, and from (2.83) it is easy to see that

$$\lim_{\tau\to\infty} C_i(\tau) = \frac{1}{\mu + 1}.$$

Equation (2.85) gives no new information, but it is consistent.

In summary, uniform leading order approximations for $\bar{a}$ and $\bar{c}$ are

$$\bar{a}_u(\bar{t}) = \bar{a}_0(\bar{t}) + \bar{a}_i(\bar{t}) - 1,$$

$$\bar{c}_u(\bar{t}) = \bar{c}_0(\bar{t}) + \bar{c}_i(\bar{t}) - \frac{1}{1+\mu},$$

or

$$\bar{a}_u(\bar{t}) = \bar{a}_0(\bar{t}), \tag{2.86}$$

$$\bar{c}_u(\bar{t}) = \bar{c}_0(\bar{t}) + \bar{c}_i(\bar{t}) - \frac{1}{1+\mu}, \tag{2.87}$$

where $\bar{a}_0(\bar{t})$ is defined implicitly by the algebraic equation

$$\bar{a}_0(\bar{t}) + \mu \ln \bar{a}_0(\bar{t}) = -(\mu - \lambda)\bar{t} + 1.$$

Numerical calculations are now required to determine solution curves.

EXERCISES

1. Use a singular perturbation method to find a leading-order approximation to the initial value problem

$$\varepsilon y' + y = e^{-t}, \quad y(0) = 2,$$

on $t \geq 0$. Calculate the residual and show that it converges to zero uniformly on $t \geq 0$ as $\varepsilon \to 0$.

2. Find a uniformly valid approximation to the problem

$$\varepsilon y'' + b(t)y' + y = 0, \quad 0 < t < T,$$
$$y(0) = 1, \quad y'(0) = \beta/\varepsilon + \gamma,$$

as $\varepsilon \to 0$. Here, b is smooth and $b(t) > 0$, and T, β, $\gamma > 0$.

3. Find a uniformly valid approximation to the problem

$$\varepsilon y'' + (t+1)^2 y' = 1, \quad t > 0, \ 0 < \varepsilon \ll 1,$$
$$y(0) = 1, \quad \varepsilon y'(0) = 1.$$

4. Consider a nonlinear damped oscillator whose dynamics is governed by the initial value problem

$$m\ddot{y} + a\dot{y} + ky^3 = 0, \quad t > 0,$$
$$y(0) = 0, \quad m\dot{y}(0) = I,$$

where the mass m is small.

a) Non-dimensionalize the problem with a correct scaling.

b) Find a leading-order perturbation approximation.

5. Find approximations to the following systems where $0 < \varepsilon \ll 1$:

a) $\frac{dx}{dt} = y - \varepsilon \sin x$, $\quad \varepsilon \frac{dy}{dt} = x^2 y + \varepsilon y^3$, $\quad x(0) = k$, $\quad y(0) = 0$.

b) $\frac{du}{dt} = v$, $\quad \varepsilon \frac{dv}{dt} = -u^2 - v$, $\quad u(0) = 1$, $\quad v(0) = 0$.

6. Consider the reversible reaction $\mathbf{A} \rightleftarrows \mathbf{B}$, where k_f and k_b are the rate constants for the forward and backward reactions, respectively. If $a(0) = a_0$ and $b(0) = 0$, find the concentration $a = a(t)$.

7. For the bimolecular reaction $\mathbf{X} + \mathbf{Y} \xrightarrow{k} \mathbf{Z}$, show that the concentration $x = x(t)$ of $\mathbf{X}$ satisfies the logistics equation.

8. Consider an exothermic reaction $\mathbf{X} \to$ products that releases heat. Initially, the concentration and temperature are $x(0) = x_0$ and $T(0) = T_0$, respectively. The rate $k = k(T)$ is a function of temperature and is given by the Arrhenius form

$$k(T) = re^{E/RT_0} e^{-E/RT},$$

where E is the activation energy, R is the gas constant, and r is a multiplicative constant. As the reaction proceeds, T is related to x via the constitutive equation $T = T_0 + h(x_0 - x)$, where h is the increase in temperature when one gram of x is consumed by reaction.

a) If $y = (x_0 - x)/x_0$ is a progress variable that measures the extent of the reaction, find a differential equation for y.

b) Define a scaled temperature and time by $\theta = T/T_0$ and $\tau = rt$, respectively, and show that the model can be written

$$\frac{dy}{d\tau} = e^A e^{-A/\theta}(1 - y), \quad \theta = 1 + \beta y,$$

for appropriately chosen constants A and β.

c) Find an initial value problem for the scaled temperature θ.

d) Assume the parameter A is small and obtain approximations for θ and y.

e) Does a perturbation method work for A large? Perform some numerical experiments to support your answer to this question.

9. Choosing $\mu = 1$, $\lambda = 0.625$, and $\varepsilon = 0.1$, sketch graphs of the uniform approximations $\bar{a}_u(\bar{t})$ and $\bar{c}_u(\bar{t})$ given by (2.86)–(2.87).

2.5 The WKB Approximation

The WKB method (Wentzel–Kramers–Brillouin) is a perturbation method that applies to a variety of problems—in particular, to linear differential equations of the form

$$\varepsilon^2 y'' + q(x)y = 0, \quad 0 < \varepsilon \ll 1, \tag{2.88}$$

$$y'' + (\lambda^2 p(x) - q(x))y = 0, \quad \lambda \gg 1, \tag{2.89}$$

$$y'' + q(\varepsilon x)^2 y = 0, \quad 0 < \varepsilon \ll 1. \tag{2.90}$$

We note that (2.88) and (2.90) have a small parameter, while (2.89) has a large parameter; the latter may be changed to the former by letting $\varepsilon = \lambda^{-1}$.

It is not possible, in general, to obtain a closed-form solution, in terms of elementary functions, to differential equations with variable coefficients like equations (2.88)–(2.90). So approximation methods are often necessary. The WKB method was developed by several individuals. The first work was by Liouville[1] in the mid 1800s; in the early 1900s Liouville's method was extended by others, for example, Rayleigh (in the context of sound wave reflection) and Jeffrey (Mathieu's equation); the method now bears the initials of Wentzel, Kramers, and Brillouin because of their important application of the method in quantum mechanics. In the next few paragraphs we briefly develop some of the basic concepts of quantum mechanics, which will give a physical context in which to examine the WKB approximation.

For simplicity, consider a particle of mass m moving on the x axis under the influence of a conservative force given by $F(x)$. The potential $V(x)$ is defined by the equation $F(x) = -V'(x)$. According to the canon of classical particle mechanics, the motion $x = x(t)$ is governed by the dynamical equation

---

[1] J. Liouville (1809–1882) was one of the great mathematicians of the 19th century. His name is attached to key results in differential equations, complex analysis, and many other areas of mathematics.

$$m\frac{d^2x}{dt^2} = F(x), \tag{2.91}$$

which is Newton's second law of motion. If the initial position $x(0)$ and velocity $dx/dt(0)$ are specified, then one can, in theory, determine the state of the particle (position and velocity) for all times $t > 0$. In this sense, classical mechanics is a *deterministic* model.

In the early 1900s it was found that this classic model of motion failed on the atomic scale. Out of this revolution quantum mechanics was developed. Quantum theory dictates that the particle has no definite position or velocity; rather, one postulates a statistical or probabilistic interpretation of the state of the particle in terms of a **wave function** $\Psi(x,t)$. The square of the wave function is the probability density for the position, which is a random variable X. Therefore,

$$\Pr(a < X \le b) = \int_a^b |\Psi(x,t)|^2 \, dx$$

is the probability of the particle being in the interval $a < x \le b$ at time t. Also, $\int_{-\infty}^{\infty} |\Psi(x,t)|^2 \, dx = 1$, because the particle is located somewhere on the x axis. From elementary probability theory it is known that the probability density contains all of the statistical information for a given problem (for example, the mean and variance, of position). The question is how to find the wave function. The equation that governs the evolution of a quantum mechanical system (the analog of Newton's law for the classic system) is the **Schrödinger[2] equation**, a second-order partial differential equation having the form

$$i\hbar\Psi_t = -\frac{\hbar^2}{2m}\Psi_{xx} + V(x)\Psi, \tag{2.92}$$

where V is the potential energy, m is the mass, and where $\hbar = h/2\pi$, where $h = 6.625 \cdot 10^{-34}$ kg m^2/ sec is Planck's constant.

Assuming solutions have the form of a product $\Psi(x,t) = y(x)\phi(t)$, we obtain, upon substitution into (2.92) and after rearrangement,

$$\frac{i\hbar\frac{d\phi}{dt}}{\phi} = \frac{\frac{-\hbar^2}{2m}y'' + V(x)y}{y}.$$

The left side of this equation depends only on t, and the right side depends only on x. The only way equality can occur for all t and x is if both sides are equal to the same constant, which we shall call E. Therefore we obtain two equations, one for $\phi(t)$,

$$\frac{d\phi}{dt} = (-iE/\hbar)\phi,$$

---

[2] E. Schrödinger was one of the major scientists in the early 20th century who developed the theory underlying quantum mechanics.

and one for $y(x)$, namely

$$-\frac{\hbar^2}{2m}y'' + (V(x) - E)y = 0. \tag{2.93}$$

The solution of the time equation is $\phi = C\exp(-iEt/\hbar)$, where C is a constant. Equation (2.93), whose solution $y(x)$ gives the spatial part of the wave function, is called the *time-independent* Schrödinger equation, and it is one of the fundamental equations of mathematical physics. Because $\hbar$ is extremely small, we may set $\varepsilon = \hbar/\sqrt{2m} \ll 1$ and obtain a problem of the form (2.88), where $q(x) = E - V(x)$.

2.5.1 The Non-oscillatory Case

We first consider equation (2.88) in the case that $q(x)$ is strictly negative on the interval of interest. For definiteness, let us write $q(x) = -k(x)^2$, where $k(x) > 0$, and consider the equation

$$\varepsilon^2 y'' - k(x)^2 y = 0. \tag{2.94}$$

If k were a constant, say $k = k_0$, then (2.94) would have real exponential solutions of the form $\exp(\pm k_0 x/\varepsilon)$. This suggests making in (2.94) the substitution

$$y = e^{u(x)/\varepsilon}.$$

Then (2.94) becomes

$$\varepsilon v' + v^2 - k(x)^2 = 0, \quad v = u'. \tag{2.95}$$

Now we try a regular perturbation expansion

$$v(x) = v_0(x) + \varepsilon v_1(x) + O(\varepsilon^2).$$

Substituting this expansion into (2.95) gives

$$v_0^2 = \pm k(x)^2, \quad 2v_0 v_1 = -v_0',$$

at $O(1)$ and $O(\varepsilon)$, respectively. Therefore

$$v_0 = \pm k(x), \quad v_1 = -\frac{k'(x)}{2k(x)},$$

and the expansion for v becomes

$$v(x) = \pm k(x) - \varepsilon \frac{k'(x)}{2k(x)} + O(\varepsilon^2).$$

Consequently, u is given by

$$u(x) = \pm \int_a^x k(\xi)\,d\xi - \frac{\varepsilon}{2}\ln k(x) + O(\varepsilon^2),$$

where a is an arbitrary constant; the other constants are incorporated into the indefinite integral. Finally, the expansion for $y(x)$ is

$$y(x) = e^{u(x)/\varepsilon} = \frac{1}{\sqrt{k(x)}}\exp\left(\pm\frac{1}{\varepsilon}\int_a^x k(\xi)\,d\xi\right)(1 + O(\varepsilon)), \quad \varepsilon \ll 1$$

This equation defines two linearly independent approximations (one with the plus sign, and one with the minus sign) to the linear equation (2.94); taking the linear combination of these gives the WKB approximation to (2.94) in the non-oscillatory case:

$$y_{\mathrm{WKB}}(x) = \frac{c_1}{\sqrt{k(x)}}\exp\left(\frac{1}{\varepsilon}\int_a^x k(\xi)\,d\xi\right) + \frac{c_2}{\sqrt{k(x)}}\exp\left(-\frac{1}{\varepsilon}\int_a^x k(\xi)\,d\xi\right).$$

$$(2.96)$$

The arbitrary constants c_1 and c_2 can be determined by boundary data; the arbitrary lower limit of integration a should be taken at the point where the data is given.

Example 2.13

Find the WKB approximation to the equation

$$\varepsilon^2 y'' - (1 + x)^2 y = 0, \quad x > 0$$

Here, $k(x) = 1 + x$ and $\int_0^x k(\xi)\,d\xi = x + x^2/2$. Therefore the WKB approximation is

$$y_{\mathrm{WKB}}(x) = \frac{c_1}{\sqrt{1 + x}}e^{(x + x^2/2)/\varepsilon} + \frac{c_2}{\sqrt{1 + x}}e^{-(x + x^2/2)/\varepsilon}.$$

If we append the boundary conditions $y(0) = 0$ and $\lim_{x \to \infty} y(x) = 0$, then $c_1 = 0$ to ensure the latter condition at infinity. Then the condition at $x = 0$ forces $c_2 = 1$. So

$$y_{\mathrm{WKB}}(x) = \frac{1}{\sqrt{1 + x}}e^{-(x + x^2/2)/\varepsilon}.$$

2.5.2 The Oscillatory Case

In the case that $q(x)$ is strictly positive, we expect oscillatory solutions of (2.88). So, we consider the equation

$$\varepsilon^2 y'' + k(x)^2 y = 0, \tag{2.97}$$

where $k(x) > 0$. Nearly the same calculation can be made as for the non-oscillatory case. Now we make the substitution $y(x) = \exp(iu(x)/\varepsilon)$ into equation (2.97) and proceed exactly as before to obtain the WKB approximation in the oscillatory case as

$$y_{\text{WKB}}(x) = \frac{c_1}{\sqrt{k(x)}} \exp\left(\frac{i}{\varepsilon} \int_a^x k(\xi)\, d\xi\right) + \frac{c_2}{\sqrt{k(x)}} \exp\left(-\frac{i}{\varepsilon} \int_a^x k(\xi)\, d\xi\right).$$

This is the same result except for the presence of the complex number i in the exponential. Using Euler's formula, $e^{i\theta} = \cos\theta + i\sin\theta$, we can rewrite the approximation to (2.97) in terms of sines and cosines as

$$y_{\text{WKB}}(x) = \frac{c_1}{\sqrt{k(x)}} \sin\left(\frac{1}{\varepsilon} \int_a^x k(\xi)\, d\xi\right) + \frac{c_2}{\sqrt{k(x)}} \cos\left(-\frac{1}{\varepsilon} \int_a^x k(\xi)\, d\xi\right). \tag{2.98}$$

Example 2.14

(Schrödinger Equation) In the case that $E > V(x)$, which is the oscillatory case, the WKB approximation (2.98) to the Schrö dinger equation (2.93) is often written in terms of an amplitude A and phase ϕ (which play the role of arbitrary constants) as

$$y_{\text{WKB}} = \frac{A}{(E - V(x))^{1/4}} \cos\left(\frac{\sqrt{2m}}{\hbar} \int_a^x \sqrt{E - V(\xi)}\, d\xi + \phi\right).$$

Example 2.15

The WKB approximation can be used to determine the large eigenvalues for simple differential operators. (A detailed discussion of eigenvalue problems is presented in Chapter 4.) For the present, consider the boundary value problem

$$y'' + \lambda q(x) y = 0, \quad 0 < x < \pi; \ y(0) = y(\pi) = 0. \tag{2.99}$$

We assume $q(x) > 0$. A number λ is called an **eigenvalue** of the boundary value problem (2.99) if there exists a nontrivial solution of (2.99) for that particular value of λ; the corresponding nontrivial solutions are called **eigenfunctions**. It is an important problem to determine all the eigenvalues and eigenfunctions for a given boundary value problem. In the present case, if we define $\varepsilon = 1/\sqrt{\lambda}$ and $k(x) = \sqrt{q(x)}$, then the differential equation in (2.99) has the form

$$\varepsilon^2 y'' + k(x)^2 y = 0,$$

for which the WKB method applies, provided ε is small or, equivalently, if λ is large in (2.99). The WKB approximation is given by

$$y_{\mathrm{WKB}}(x) = \frac{1}{q(x)^{1/4}} \left(c_1 \sin\left(\sqrt{\lambda}\int_0^x \sqrt{q(\xi)}\,d\xi\right) + c_2 \cos\left(\sqrt{\lambda}\int_0^x \sqrt{q(\xi)}\,d\xi\right) \right).$$

But $y(0) = 0$ forces $c_2 = 0$, and $y(\pi) = 0$ forces

$$\frac{c_1}{q(\pi)^{1/4}}\sin\left(\sqrt{\lambda}\int_0^\pi \sqrt{q(\xi)}\,d\xi\right) = 0.$$

Choosing $c_1 = 0$ gives the trivial solution, which we are trying to avoid; so we force

$$\sin\left(\sqrt{\lambda}\int_0^\pi \sqrt{q(\xi)}\,d\xi\right) = 0,$$

which can be satisfied by choosing λ such that

$$\sqrt{\lambda}\int_0^\pi \sqrt{q(\xi)}\,d\xi = n\pi,$$

where n is a large integer (recall, the approximation is valid only for large values of λ). Thus, the large eigenvalues of problem (2.99) are given by

$$\lambda = \lambda_n = \left(n\pi \bigg/ \int_0^\pi \sqrt{q(\xi)}\,d\xi \right)^2,$$

for large n. The corresponding eigenfunctions are

$$y_{\mathrm{WKB}}(x) = \frac{c_1}{q(x)^{1/4}}\sin\left(n\pi \frac{\int_0^x \sqrt{q(\xi)}\,d\xi}{\int_0^\pi \sqrt{q(\xi)}\,d\xi} \right),$$

for large n.

If the function $q(x)$ has zeros in the interval of interest, then the problem of determining asymptotic approximations of (2.88) is significantly more difficult. On one side of a simple zero of q the solution is oscillatory, and on the other side it is exponential; therefore these two solutions must be connected, or matched, in a layer located at that root. The points x where $q(x) = 0$ are called *turning points*; in terms of quantum mechanics and the Schrödinger equation (2.93), the turning points are the values of x where $E = V(x)$, or where the total energy equals the potential energy. We refer the reader to Holmes (1995) for an elementary treatment of turning point problems. Murray (1984) shows how to apply the WKB method to the more general Liouville equation

$$y'' + (\lambda^2 q_0(x) + \lambda q_1(x) + q_2(x) + \cdots)y = 0, \quad \lambda \gg 1.$$

Interestingly enough, even though the WKB approximation is valid for $\lambda \gg 1$ or $\varepsilon \ll 1$, it often provides a good approximation even when λ and ε are $O(1)$. Several numerical examples are discussed in Bender & Orszag (1978).

EXERCISES

1. Use the WKB method to find an approximate solution to the initial value problem

$$y'' - \lambda(1 + x^2)^2 y = 0, \quad y(0) = 0, \quad y'(0) = 1.$$

   for large λ.

2. Show that the large eigenvalues of the problem

$$y'' + \lambda(x + \pi)^4 y = 0, \quad y(0) = y(\pi) = 0,$$

   are given approximately by

$$\lambda = \frac{9n^2}{49\pi^4}$$

   for large integers n, and find the corresponding eigenfunctions.

3. Consider the eigenvalue problem

$$-\frac{1}{x}y'' = \lambda y, \quad 1 < x < 4,$$
$$y(1) = y(4) = 0.$$

   Find an approximation for the large eigenvalues and eigenfunctions.

4. Show that the differential equation $y'' + p(x)y' + q(x)y = 0$ can be transformed into the equation $u'' + r(x)u = 0$, $r = q - p'/2 - p^2/4$, without a first derivative, by the Liouville transformation $u = y \exp(\frac{1}{2}\int_a^x p(\xi)\,d\xi)$.

5. Find a formula in terms of the sinh function for the WKB approximation to the initial value problem

$$\varepsilon^2 y'' - q(x)y = 0, \quad y(0) = 0, \quad y'(0) = 1; \quad q(x) > 0.$$

6. The *slowly varying oscillator* equation is

$$y'' + q(\varepsilon x)^2 y = 0,$$

   where $\varepsilon \ll 1$ and q is strictly positive. Find an approximate solution. Why is the equation called "slowly varying"?

7. Find an approximation for the large eigenvalues of

$$y'' + \lambda e^{4x} y = 0, \quad 0 < x < 1, \quad y(0) = y(1) = 1.$$

8. Find an approximation for the general solution of

$$y'' + (\lambda^2 x^2 + x)y = 0, \quad x > 0, \quad \lambda \gg 1.$$

Hint: let $y = \exp(iu/\varepsilon)$, $\varepsilon = 1/\lambda$.

2.6 Asymptotic Expansion of Integrals

The solution of differential equations often leads to formulas involving complicated integrals that cannot be evaluated in closed form. For example, the initial value problem

$$y'' + 2\lambda t y' = 0, \quad y(0) = 0, \quad y'(0) = 1,$$

has solution

$$y = y(t; \lambda) = \int_0^t e^{-\lambda s^2}\, ds,$$

but we cannot calculate the integral because there is no antiderivative in terms of simple functions. For some problems, however, we may want to know the behavior of y for large t (with λ fixed), or we may want to determine the value of y for a fixed value of t in the case that λ is a large parameter. Problems like this are common in applied mathematics, and this section we introduce standard techniques to approximate certain kinds of integrals.

2.6.1 Laplace Integrals

One type of integral we wish to study has the form

$$I(\lambda) = \int_a^b f(t)e^{-\lambda g(t)}\, dt, \quad \lambda \gg 1, \tag{2.100}$$

where g is a strictly increasing function on $[a, b]$ and the derivative g' is continuous. Here, $a < b \leq +\infty$, and $\lambda \gg 1$ is another notation for the asymptotic statement "as $\lambda \to \infty$." An example is the Laplace transform

$$U(\lambda) = \int_0^\infty f(t)e^{-\lambda t}\, dt.$$

Actually, it is sufficient to examine integrals of the form

$$I(\lambda) = \int_0^b f(t)e^{-\lambda t}\, dt, \quad \lambda \gg 1. \tag{2.101}$$

To observe this, we can make a change of variables $s = g(t) - g(a)$ in (2.100) to obtain

$$I(\lambda) = e^{-\lambda g(a)} \int_0^{g(b)-g(a)} \frac{f(t(s))}{g'(t(s))} e^{-\lambda s} \, ds,$$

where $t = t(s)$ is the solution of the equation $s = g(t) - g(a)$ for t. The integral on the right has the form of (2.101).

The fundamental idea in obtaining an approximation for (2.101) is to determine what subinterval gives the dominant contribution. The function $\exp(-\lambda t)$ is a rapidly decaying exponential; thus, if f does not grow too fast at infinity and if it is reasonably well behaved at $t = 0$, then it appears that the main contribution comes from a neighborhood of $t = 0$. A detailed example will illustrate our reasoning; the method is called **Laplace's method**.

Example 2.16

Consider the integral

$$I(\lambda) = \int_0^\infty \frac{\sin t}{t} e^{\lambda t} \, dt, \quad \lambda \gg 1. \tag{2.102}$$

Figure 2.8, which shows a graph of the integrand for different values of λ, confirms that the main contribution to the integral occurs near $t = 0$. To proceed we partition the interval of integration into two parts, $[0, T]$ and $[T, \infty]$, to obtain

$$I(\lambda) = \int_0^T \frac{\sin t}{t} e^{-\lambda t} \, dt + \int_T^\infty \frac{\sin t}{t} e^{-\lambda t} \, dt, \tag{2.103}$$

where T is any positive real number. For any T, the second integral is an exponentially small term (EST); that is, it has order $\lambda^{-1} \exp(-\lambda T)$ as $\lambda \to \infty$. A simple calculation confirms this fact:

$$|I(\lambda)| \leq \int_T^\infty \left| \frac{\sin t}{t} \right| e^{-\lambda t} \, dt \leq \int_T^\infty e^{-\lambda t} \, dt = \frac{1}{\lambda} e^{-\lambda T}.$$

We note that an exponentially small term is $o(\lambda^{-m})$, for any positive integer m, which means that it decays much faster than any power of λ. Thus,

$$I(\lambda) = \int_0^T \frac{\sin t}{t} e^{-\lambda t} \, dt + \text{EST}.$$

In the finite interval $[0, T]$ we can replace $\sin t / t$ by its Taylor series about $t = 0$,

$$\frac{\sin t}{t} = 1 - \frac{t^2}{3!} + O(t^4).$$

Thus

$$I(\lambda) = \int_0^T \left(1 - \frac{t^2}{3!} + \mathrm{O}(t^4)\right) e^{-\lambda t}\,dt + \mathrm{EST}.$$

Now change variables via $u = \lambda t$ to obtain

$$I(\lambda) = \frac{1}{\lambda} \int_0^{\lambda T} \left(1 - \frac{u^2}{3!\lambda^2} + \mathrm{O}\left(\frac{1}{\lambda^4}\right)\right) e^{-u}\,du + \mathrm{EST}.$$

But now the upper limit can be replaced by ∞ since the error incurred by doing this is exponentially small as $\lambda \to \infty$. (Since we are already discarding algebraically small terms, there is no worry about adding exponentially small terms to the integral.) Consequently, we obtain

$$I(\lambda) = \frac{1}{\lambda} \int_0^{\infty} \left(1 - \frac{u^2}{3!\lambda^2} + \mathrm{O}\left(\frac{1}{\lambda^4}\right)\right) e^{-u}\,du.$$

We can now use the integration formula

$$\int_0^{\infty} u^m e^{-u}\,du = m!, \quad m = 0, 1, 2, \ldots, \tag{2.104}$$

to obtain the asymptotic approximation of (2.102) given by

$$I(\lambda) \sim \frac{1}{\lambda} - \frac{2!}{3!\lambda^3} + \mathrm{O}\left(\frac{1}{\lambda^5}\right), \quad \lambda \gg 1.$$

Before discussing general problems, we pause to define a special function, the *gamma function*, which plays an important role in subsequent calculations. It is defined by

$$\Gamma(x) = \int_0^{\infty} u^{x-1} e^{-u}\,du, \quad x > 0.$$

Because the gamma function has the property

$$\Gamma(x+1) = x\Gamma(x),$$

it generalizes the idea of a factorial function to non-integer values of the argument; note that $\Gamma(n) = (n-1)!$ for positive integers n, and $\Gamma(\frac{1}{2}) = \sqrt{\pi}$. Other values of the gamma function are tabulated in handbooks on special functions.

We now formulate a general result known as **Watson's lemma**, which gives the asymptotic behavior of a large number of integrals. We shall only outline the proof, and those interested in a more detailed discussion should consult, for example, Murray (1984).

Theorem 2.17

Consider the integral

$$I(\lambda) = \int_0^b t^\alpha h(t) e^{-\lambda t}\, dt, \tag{2.105}$$

where $\alpha > -1$, where $h(t)$ has a Taylor series expansion about $t = 0$, with $h(0) \neq 0$, and where $|h(t)| < k e^{ct}$, $0 < t < b$, for some positive constants k and c. Then

$$I(\lambda) \sim \sum_{n=0}^{\infty} \frac{h^{(n)}(0)\Gamma(\alpha + n + 1)}{n!\lambda^{\alpha+n+1}}, \qquad \lambda \gg 1. \tag{2.106}$$

The condition on α will guarantee the convergence of the improper integral at $t = 0$ and the exponential boundedness of h is required to ensure that the integral converges as t goes to infinity. The argument is similar to the one in the previous example. We split up the integral (2.105) for any $T > 0$, where T lies in the radius of convergence of the power series for h. Then

$$I(\lambda) = \int_0^T t^\alpha h(t) e^{-\lambda t}\, dt + \int_T^b t^\alpha h(t) e^{-\lambda t}\, dt.$$

The second integral is exponentially small. Now we substitute the Taylor series expansion for h about $t = 0$ in the first integral to obtain

$$I(\lambda) = \int_0^T t^\alpha \left(h(0) + h'(0)t + \frac{1}{2!}h''(0)t^2 + \cdots \right) e^{-\lambda t}\, dt + \text{EST}.$$

Making the substitution $u = \lambda t$ and then replacing the upper limit of integration by ∞ gives

$$I(\lambda) \sim \frac{1}{\lambda} \int_0^\infty \left(h(0) \left(\frac{u}{\lambda}\right)^\alpha + h'(0) \left(\frac{u}{\lambda}\right)^{\alpha+1} \right.$$
$$\left. + \frac{1}{2!}h''(0) \left(\frac{u}{\lambda}\right)^{\alpha+2} + \cdots \right) e^{-\lambda t}\, dt.$$

The result (2.106) now follows from the definition of the gamma function.

Example 2.18

Some integrals that do not have the form (2.105) can be transformed into an integral that does. For example, consider the complementary error function erfc defined by

$$\text{erfc}(\lambda) = \frac{2}{\sqrt{\pi}} \int_\lambda^\infty e^{-s^2}\, ds. \tag{2.107}$$

Letting $t = s - \lambda$ gives

$$\text{erfc}(\lambda) = \frac{2}{\sqrt{\pi}} e^{-\lambda^2} \int_0^\infty e^{-t^2} e^{-2\lambda t} \, dt = \frac{1}{\sqrt{\pi}} e^{-\lambda^2} \int_0^\infty e^{-\tau^2/4} e^{-\lambda \tau} \, d\tau.$$

Watson's lemma applies, and we leave it as an exercise to show that

$$\text{erfc}(\lambda) \sim \frac{2}{\sqrt{\pi}} e^{-\lambda^2} \left(\frac{1}{2\lambda} - \frac{\Gamma(3)}{(2\lambda)^3} + \frac{\Gamma(5)}{2!(2\lambda)^5} - \frac{\Gamma(7)}{3!(2\lambda)^7} + \cdots \right). \tag{2.108}$$

2.6.2 Integration by Parts

Sometimes an asymptotic expansion can be found by a simple, successive, integration by parts procedure. However, this method is not often as applicable as the Laplace method, and it usually involves more work. Let us consider the erfc function. Using integration by parts, we have

$$\text{erfc}(\lambda) = \frac{2}{\sqrt{\pi}} \int_\lambda^\infty e^{-t^2} \, dt$$

$$= \frac{2}{\sqrt{\pi}} \int_\lambda^\infty \frac{-2te^{-t^2}}{-2t} \, dt$$

$$= \frac{2}{\sqrt{\pi}} \left(\frac{e^{-t^2}}{-2t} \bigg|_\lambda^\infty - \int_\lambda^\infty \frac{e^{-t^2}}{2t^2} \, dt \right)$$

$$= \frac{2}{\sqrt{\pi}} \left(\frac{e^{-\lambda^2}}{2\lambda} - \int_\lambda^\infty \frac{e^{-t^2}}{2t^2} \, dt \right).$$

The integral on the right may be integrated by parts to get

$$\int_\lambda^\infty \frac{e^{-t^2}}{2t^2} \, dt = \frac{1}{2} \int_\lambda^\infty \frac{-2te^{-t^2}}{-2t^3} \, dt$$

$$= \frac{1}{2} \left(\frac{e^{-\lambda^2}}{2\lambda^3} - \int_\lambda^\infty \frac{3e^{-t^2}}{2t^4} \, dt \right).$$

Thus

$$\text{erfc}(\lambda) = \frac{2}{\sqrt{\pi}} e^{-\lambda^2} \left(\frac{1}{2\lambda} - \frac{1}{4\lambda^3} \right) + \frac{1}{\sqrt{\pi}} \int_\lambda^\infty \frac{3e^{-t^2}}{2t^4} \, dt.$$

By repeated integration by parts we can eventually reproduce (2.108). It is important in this method to ensure that each succeeding term generated in the expansion is asymptotically smaller than the preceding term; this is required by the definition of asymptotic expansion in Section 2.1.

Using the ratio test, we can easily observe that for fixed λ the series in (2.108) for erfc is not convergent. One may legitimately ask, therefore, how a divergent series can be useful in approximating values of erfc(λ); for fixed λ we cannot obtain a good approximation by taking more and more terms. For calculations, the idea is to *fix* the number of terms n and apply the resulting approximation for large values of λ. The error in this approximation is about the magnitude of the first discarded term. Practically, then, to approximate erfc(λ), we compute the magnitude of the first few terms of the series and then terminate the series when the magnitude begins to increase.

To illustrate this strategy, we approximate erfc(2). We have

$$\text{erfc}(2) = \frac{e^{-4}}{\sqrt{\pi}} \left(\frac{1}{2} - \frac{1}{2^4} + \frac{1 \cdot 3}{2^7} - \frac{1 \cdot 3 \cdot 5}{2^{10}} + \frac{1 \cdot 3 \cdot 5 \cdot 7}{2^{13}} \right.$$
$$\left. - \frac{1 \cdot 3 \cdot 5 \cdot 7 \cdot 9}{2^{16}} + \cdots \right).$$

After the fifth term, the magnitude of the terms begins to increase. So we terminate after five terms and obtain erfc(2) $\approx$ 0.004744; the tabulated value is 0.004678 and thus our approximation has an error of 0.000066.

The reader might inquire why one shouldn't just write

$$\text{erfc}(\lambda) = \frac{2}{\sqrt{\pi}} \int_0^\infty e^{-t^2} \, dt - \frac{2}{\sqrt{\pi}} \int_0^\lambda e^{-t^2} \, dt$$
$$= 1 - \frac{2}{\sqrt{\pi}} \int_0^\lambda e^{-t^2} \, dt,$$

and then replace $\exp(-t^2)$ by its Taylor series expansion and integrate term-by term. This procedure would yield a convergent series. So, adopting this strategy we get

$$\text{erfc}(\lambda) = 1 - \frac{2}{\sqrt{\pi}} \int_0^\lambda \left(1 - \frac{t^2}{2} + \frac{1}{2!} \left(\frac{t^2}{2} \right)^2 - \cdots \right) dt$$
$$= 1 - \frac{2}{\sqrt{\pi}} \left(\lambda - \frac{\lambda^3}{3} + \frac{\lambda^5}{5 \cdot 2!} - \frac{\lambda^7}{7 \cdot 3!} + \cdots \right)$$

The reader should check that it takes about 20 terms to get the same accuracy that the divergent, asymptotic series gave in only five terms. Therefore, asymptotic expansions can be significantly more efficient than traditional Taylor series expansions. For small λ, however, the Taylor series in preferable.

2.6.3 Other Integrals

We can extend the previous ideas to approximate integrals of the form

$$I(\lambda) = \int_a^b f(t)e^{\lambda g(t)}\, dt, \quad \lambda \gg 1, \tag{2.109}$$

where f is continuous and g is sufficiently smooth and has a unique maximum at the point $t = c$ in (a, b). Specifically, $g'(c) = 0$, $g''(c) < 0$. Again we expect the main contribution of (2.109) to come from where g assumes its maximum. To obtain a leading order approximation we replace $f(t)$ by $f(c)$, and we replace g by the first three terms of its Taylor series (the g' term is zero) to obtain

$$I(\lambda) \sim \int_a^b f(c)e^{\lambda(g(c)+g''(c)(t-c)^2/2)}\, dt \tag{2.110}$$

$$\sim f(c)e^{\lambda g(c)} \int_a^b e^{\lambda/2 g''(c)(t-c)^2}\, dt. \tag{2.111}$$

Letting $s = (t - c)\sqrt{-\lambda g''(c)/2}$, we get

$$I(\lambda) \sim f(c)e^{\lambda g(c)} \sqrt{\frac{-2}{\lambda g''(c)}} \int_{(a-c)\sqrt{-\lambda g''(c)/2}}^{(b-c)\sqrt{-\lambda g''(c)/2}} e^{-s^2}\, ds. \tag{2.112}$$

We may now replace the limits of integration by $-\infty$ and $+\infty$ and use $\int_{-\infty}^{+\infty} \exp(-s^2)ds = \sqrt{\pi}$ to obtain

$$I(\lambda) \sim f(c)e^{\lambda g(c)} \sqrt{\frac{-2\pi}{\lambda g''(c)}}, \quad \lambda \gg 1. \tag{2.113}$$

Remark 2.19

If the maximum of g occurs at an endpoint, then $c = a$ or $c = b$ and one of the limits of integration in (2.110) is zero; thus we will obtain one-half of the value indicated in (2.111).

Example 2.20

Approximate the integral

$$I(\lambda) = \int_0^\pi e^{\lambda \sin t}\, dt$$

for large λ. Here $f(t) = 1$, and $g(t) = \sin t$ has its maximum at $t = \pi/2$ with $g''(\pi/2) = -1$. Thus, by (2.111), $I(\lambda) \sim e^\lambda \sqrt{2\pi/\lambda}$.

EXERCISES

1. Verify (2.108).

2. Find a 2-term approximation for large λ for

$$I(\lambda) = \int_0^{\pi/2} e^{-\lambda \tan^2 \theta}\, d\theta.$$

3. Consider the integral in (2.109), where the maximum of g occurs at an endpoint of the interval with a nonzero derivative. To fix the idea assume that the maximum occurs at $t = b$ with $g'(t) > 0$ for $a \le t \le b$. Derive the formula

$$I(\lambda) \sim \frac{f(b)e^{\lambda g(b)}}{\lambda g'(b)}, \quad \lambda \gg 1.$$

4. Verify the following approximations for large λ.

   a) $\int_0^\infty e^{-\lambda t} \ln(1+t^2)dt \sim \frac{2!}{\lambda^3} - \frac{1}{2}\frac{4!}{\lambda^5} + \cdots.$

   b) $\int_0^1 \sqrt{1+t}\,e^{\lambda(2t-t^2)}dt \sim \sqrt{\pi/2\lambda}\,e^\lambda.$

   c) $\int_1^2 \sqrt{3+t}\,e^{\lambda/(t+1)}dt \sim \frac{8}{\lambda}e^{\lambda/2}.$

5. Use integration by parts to show

$$\Gamma(x+1) = x\Gamma(x).$$

6. Find a 3-term asymptotic expansion of

$$\int_0^\infty e^{-\lambda t^2} \sin t\, dt$$

   for large λ.

7. Find a leading order asymptotic approximation for the following integrals, $\lambda \gg 1$.

   a) $\int_0^\pi e^{-\lambda t}\frac{\cos t}{\sqrt{t}}\, dt.$

   b) $\int_0^1 e^{-\lambda t}\frac{1}{\sqrt{t}}\frac{1}{\sqrt{1-t/2}}\, dt.$

   c) $\int_0^2 e^{\lambda(2-t)t} \ln(t+1)\, dt.$

   d) $\int_0^1 e^t e^{\lambda(3t^2+2t^3)}dt.$

8. The modified *Bessel* function $I_n(x)$ has integral representation

$$I_n(x) = \frac{1}{\pi}\int_0^\pi e^{x\cos\theta}\cos n\theta\, d\theta, \quad n = 1, 2, 3, \ldots$$

   Find the leading-order asymptotic approximation of $I_n(x)$ as $x \to +\infty$.

9. Find the first two terms in the asymptotic expansion of the integral

$$I(\lambda) = \int_0^1 e^{-\lambda(1+s)} \frac{1}{\sqrt{2 + 3s + 3s^2 + s^3}} \, ds, \quad \text{as} \quad \lambda \to \infty.$$

10. Show that

$$\int_0^{\pi/2} e^{-x \sin^2 t} \, dt \sim \frac{\sqrt{\pi}}{2\sqrt{x}} \left(1 + \frac{1}{4x} + \cdots \right), \quad \lambda \gg 1.$$

11. Use integration by parts to find a 2-term asymptotic approximation of the function

$$\text{Ci}(\lambda) = \int_\lambda^\infty \frac{\cos x}{x} \, dx$$

for large λ.

12. Consider the real exponential integral

$$\text{Ei}(\lambda) = \int_\lambda^\infty \frac{e^{-t}}{t} \, dt$$

for large values of λ.

a) Use integration by parts to show that

$$\text{Ei}(\lambda) = e^{-\lambda} \left(\frac{1}{\lambda} - \frac{1}{\lambda^2} + \frac{2!}{\lambda^3} - \frac{3!}{\lambda^4} + \cdots + \frac{(-1)^{n-1}(n-1)!}{\lambda^n} \right) + r_n(\lambda),$$

where

$$r_n(\lambda) = (-1)^n n! \int_\lambda^\infty \frac{e^{-t}}{t^{n+1}} \, dt.$$

b) Show that for fixed n, $|r_n(\lambda)| \to 0$ as $\lambda \to \infty$.

c) Show that for fixed n, $r_n(\lambda) = o\left(\frac{(n-1)! e^{-\lambda}}{\lambda^n} \right)$ as $\lambda \to \infty$, and conclude that $\text{Ei}(\lambda)$ has asymptotic expansion

$$\text{Ei}(\lambda) \sim e^{-\lambda} \left(\frac{1}{\lambda} - \frac{1}{\lambda^2} + \frac{2!}{\lambda^3} - \frac{3!}{\lambda^4} + \cdots \right).$$

d) Show that the asymptotic series in part (c) diverges for fixed λ.

e) Approximate $\text{Ei}(10)$.

13. Use integration by parts to obtain an asymptotic expansion for large λ of

$$I(\lambda) = \int_0^\infty \frac{e^{-t}}{(t + \lambda)^2} \, dt.$$

REFERENCES AND NOTES

The four books listed below cover a large amount of material on perturbation methods and asymptotic analysis.

Bender, C. M. & and Orszag, S. A. 1978. *Advanced Mathematical Methods for Scientists and Engineers*, McGraw-Hill, New York (reprinted by Springer–Verlag, New York 1998).

Holmes, M. 1995. *Introduction to Perturbation Methods*, Springer-Verlag, New York (1995).

Lin, C. C. & Segel, L. A. 1974. *Mathematics Applied to Deterministic Problems in the Natural Sciences*, Macmillan, New York (reprinted by SIAM, Philadelphia 1988).

Murray, J. D. 1984. *Asymptotic Analysis*, Springer–Verlag, New York (1984).

3

Calculus of Variations

Generally, the calculus of variations is concerned with the optimization of variable quantities called functionals over some admissible class of competing objects. Many of its methods were developed over two hundred years ago by L. Euler[1], L. Lagrange[2], and others. It continues to the present day to bring important techniques to many branches of engineering and physics, and it introduces many important methods to applied analysis.

This chapter is independent of the remainder of the text.

3.1 Variational Problems

3.1.1 Functionals

To motivate the basic concepts of the calculus of variations we first review some simple notions in elementary differential calculus. One of the central problems in the calculus is to maximize or minimize a given real valued function of a single variable. If f is a given function defined in an open interval I, then f

---

[1] Leonard Euler (1701–1783), a Swiss mathematician who spent much of his professional life in St. Petersburg, was perhaps one of the most prolific contributors to mathematics and science of all time. Some may rank him at the top of all mathematicians. His name is attached to major results in nearly every area of study in mathematics.

[2] Joseph Louis Lagrange (1736–1813) was one of the great mathematicians in the 18th century whose work had a deep influence on subsequent research.

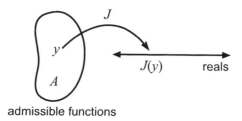

admissible functions

Figure 3.1 Schematic of a functional mapping a function space into the real numbers.

has a **local** (or *relative*) **minimum** at a point x_0 in I if $f(x_0) \leq f(x)$ for all x, satisfying $|x - x_0| < \delta$ for some δ. If f has a local minimum at x_0 in I and f is differentiable in I, then it is well known that

$$f'(x_0) = 0, \tag{3.1}$$

where the prime denotes the ordinary derivative of f. Similar statements can be made if f has a local maximum at x_0.

Condition (3.1) is called a **necessary condition** for a local minimum; that is, if f has a local minimum at x_0, then (3.1) necessarily follows. Equation (3.1) is not sufficient for a local minimum, however; that is, if (3.1) holds, it does not guarantee that x_0 provides an actual minimum. The following conditions are **sufficient conditions** for f to have a local minimum at x_0,

$$f'(x_0) = 0 \quad \text{and} \quad f''(x_0) > 0,$$

provided f'' exists. Again, similar conditions can be formulated for local maxima. If (3.1) holds, we say f is **stationary** at x_0 and that x_0 is an critical point of f. The calculus of variations deals with generalizations of this problem from the calculus. Rather than find conditions under which functions have extreme values, the calculus of variations deals with extremizing general quantities called functionals. A **functional** is a rule that assigns a real number to each function $y(t)$ in a well-defined class. Like a function, a functional is a rule or association, but its domain is some set of functions rather than a subset of real numbers. To be more precise let A be a set of functions $y, z, \ldots$; then a functional J on A is a rule that associates to each $y \in A$ a real number denoted by $J(y)$. Figure 3.1 illustrates this notion and it is symbolically represented as $J: A \to \mathbb{R}$. We say J *maps* A into $\mathbb{R}$. A fundamental problem of the calculus of variations can be stated as follows: Given a functional J and a well-defined set of functions A, determine which functions in A afford a minimum (or maximum) value to J. The word minimum can be interpreted as a local minimum (if A is equipped with some device to measure closeness of its elements) or an absolute minimum, that is, a minimum relative to all elements in A. The set A is called the set of **admissible functions**. It is those functions that are the competing functions for extremizing J. For example, the set of admissible functions might be the set of all continuous functions on an interval $[a, b]$, the set of all continuously differentiable functions on $[a, b]$ satisfying the condition $f(a) = 0$, or whatever, as long as the set is well defined. Later we shall impose some general conditions on A. For the most part the classical calculus of variations restricts itself to functionals that are defined by certain integrals and to the determination of both necessary and sufficient conditions for extrema. The problem of extremizing a functional J over the set A is called a variational problem. Several examples are presented in the next section.

To a certain degree the calculus of variations could be characterized as the calculus of functionals. In the present discussion we restrict ourselves to an analysis of necessary conditions for extrema. Applications of the calculus of variations are found in geometry, physics, engineering, numeral analysis, and optimal control theory.

3.1.2 Examples

We now formulate a few classical variational problems to illustrate the notions of a functional and the corresponding admissible functions.

Example 3.1

(Arclength) Let A be the set of all continuously differentiable functions on an interval $a \leq x \leq b$ that satisfy the boundary conditions $y(a) = y_0$, $y(b) = y_1$. Let J be the arclength functional on A defined by

$$J(y) = \int_a^b \sqrt{1 + y'(x)^2}\, dx.$$

To each y in A the functional J associates a real number that is the arc-length of the curve $y = y(x)$ between the two fixed points $P\colon (a, y_0)$ and $Q\colon (b, y_1)$. The associated calculus of variations problem is to minimize J, that is, the arc-length, over the set A. Clearly the minimizing arc is the straight line between P and Q given by

$$y(x) = \frac{y_1 - y_0}{b - a}(x - a) + y_0.$$

The general methods of the calculus of variations allows us to prove that this is the case, and we will therefore have a demonstration of the well known geometric fact that the shortest distance between two points is a straight line.

Example 3.2

(Brachistochrone problem) A bead of mass m with initial velocity zero slides with no friction under the force of gravity g from a point $(0, b)$ to a point $(a, 0)$ along a wire defined by a curve $y = y(x)$ in the xy plane. Which curve leads to the fastest time of descent? Historically this problem was important in the development of the calculus of variations. J. Bernoulli posed it in 1696 and several well-known mathematicians of that day proposed solutions; Euler's solution led to general methods that were useful in solving a wide variety of such problems. The solution curve itself is an arc of a cycloid, the curve traced out by a point on the rim of a rolling wheel; so the *brachistochrone*, or the curve of shortest time, is a cycloid. To formulate this problem analytically we compute the time of descent T for a fixed curve $y = y(x)$ connecting the points $(0, b)$ and $(a, 0)$ (see Fig. 3.2). Letting s denote the arc-length along the curve measured from the initial point $(0, b)$, we have

$$T = \int_0^T dt = \int_0^S \frac{dt}{ds}\, ds = \int_0^S \frac{1}{v}\, ds,$$

where S is the total arc-length of the curve and v denotes the velocity. Because $ds = (1 + y'(x)^2)^{1/2}\, dx$ we have

$$T = \int_0^a \frac{\sqrt{1 + y'(x)^2}}{v}\, dx.$$

To obtain an expression for v in terms of x we use the fact that energy is conserved; that is,

(Kinetic energy at $t > 0$) + (potential energy at $t > 0$)

$$= (\text{kinetic energy at } t = 0) + (\text{potential energy at } t = 0).$$

Expressed in terms of our notation,

$$\frac{1}{2}mv^2 + mgy = 0 + mgb.$$

Solving for v gives

$$v = \sqrt{2g(b - y(x))}.$$

Therefore the time required for the bead to descend is

$$T(y) = \int_0^a \frac{\sqrt{1 + y'(x)^2}}{\sqrt{2g(b - y(x))}}\, dx, \tag{3.2}$$

where we have explicitly noted that T depends on the curve y. Here, the set of admissible functions A is the set of all continuously differentiable functions y

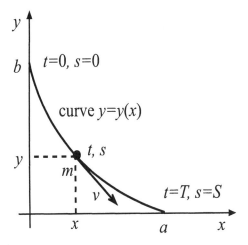

Figure 3.2 The brachistochrone problem.

defined on $0 \leq x \leq a$ that satisfy the boundary conditions $y(0) = b$, $y(a) = 0$, and the conditions

$$y \leq b, \quad \int_0^a (b - y)^{-1/2} \, dx < +\infty.$$

Equation (3.2) defines a functional on A. A real number is associated to each curve, namely the time required for the bead to descend the curve. The associated variational problem is to minimize $T(y)$, where $y \in A$, that is, to find which curve y in A minimizes the time. Because of the fixed boundary conditions this problem, like Example 3.1, is called a **fixed-endpoint problem**.

Example 3.3

Let A be the set of all nonnegative continuous functions on an interval $x_1 \leq x \leq x$, and define the functional J on A by

$$J(y) = \int_{x_1}^{x_2} y(x) \, dx,$$

which gives the area under the curve $y = y(x)$. Clearly $y(x) = 0$ provides an absolute minimum for J and there is no maximum.

The types of functionals that are of interest in the classic calculus of variations are ones defined by expressions of the form

$$J(y) = \int_a^b L(x, y(x), y'(x))\, dx, \quad y \in A, \tag{3.3}$$

where $L = L(x, y, y')$ is some given function and A is a well-defined class of functions. In the last three examples the functionals are each of the form (3.3) with

$$L = \sqrt{1 + (y')^2}, \tag{3.4}$$
$$L = \sqrt{1 + (y')^2}/\sqrt{2g(y_1 - y)}, \tag{3.5}$$
$$L = y, \tag{3.6}$$

respectively. The function L is called the **Lagrangian** (after Lagrange). We assume L to be twice continuously differentiable in each of its three arguments. It is customary to drop the independent variable x in parts of the integrand in (3.3) and just write

$$J(y) = \int_a^b L(x, y, y')\, dx, \quad y \in A.$$

In problems where time is an independent variable we write

$$J(y) = \int_a^b L(t, y, \dot{y})\, dt, \quad y \in A,$$

where the overdot on y denotes differentiation with respect to t.

EXERCISES

1. Show that the functional

$$J(y) = \int_0^1 \{\dot{y} \sin \pi y - (t + y)^2\}\, dt,$$

   where y is continuously differentiable on $0 \le t \le 1$, actually assumes its maximum value $2/\pi$ for the function $y(t) = -t$.

2. Consider the functional $J(y) = \int_0^1 (1 + x)(y')^2\, dx$, where y is twice continuously differentiable and $y(0) = 0$, $y(1) = 1$. Of all functions of the form $y(x) = x + c_1 x(1 - x) + c_2 x^2(1 - x)$, where c_1 and c_2 are constants, find the one that minimizes J.

3.2 Necessary Conditions for Extrema

3.2.1 Normed Linear Spaces

As stated in Section 3.1, a necessary condition for a function f to have a local minimum at some point x_0 is $f'(x_0) = 0$. A similar condition holds for functionals, provided the concept of derivative can be defined. In elementary calculus the derivative of a function f at x_0 is defined by a limit process, that is

$$f'(x_0) = \lim_{\Delta x \to 0} \frac{f(x_0 + \Delta x) - f(x_0)}{\Delta x},$$

if the limit exists. For a functional $J : A \to \mathbb{R}$, this formula makes no sense if f is replaced by J, and x_0 and Δx are replaced by functions in A. The essence of the derivative, namely the limit process, however, can be carried over to functionals in a restricted manner that we now describe.

To define the derivative of a functional J at some element y_0 in its domain A, we require a notion of closeness in the set A so that the value $J(y_0)$ can be compared to $J(y)$ for y *close to* y_0. To carry out this idea in a fairly general manner we assume that the set of admissible functions A is a subset of a **normed linear space**. This is a set of functions where addition and multiplication by scalars (real numbers) are defined, and there is some measuring device to give definiteness to the idea of the size of an object. Actually, the objects in the normed linear space may be vectors (in a geometric sense), functions, numbers, tensors, matrices, and so on. Normed linear spaces whose elements are functions, are often called function spaces. We make precise definition in two parts—first we define a linear space, and then we define a norm.

A collection V of objects $u, v, w, \ldots$ is called a *real* **linear space** if the following conditions are satisfied.

1. There is a binary operation $+$ on V such that $u + v \in V$ whenever $u, v \in V$. That is, V is *closed* under the operation $+$, which is called *addition*.

2. Addition is *commutative* and *associative*; that is, $u + v = v + u$ and $u + (v + w) = (u + v) + w$.

3. There is *zero element* 0 in V that has the property that $u + 0 = u$ for all $u \in V$. The zero element is also called the identity under addition and it is unique.

4. For each $u \in V$ there is a $v \in V$ such that $u + v = 0$. The element v is called the *inverse* and u and is denoted by $-u$.

5. For each $u \in V$ and $\alpha \in \mathbb{R}$, *scalar multiplication* αu is defined and $\alpha u \in V$.

6. Scalar multiplication is associative and distributive; that is,

$$\alpha(\beta u) = (\alpha\beta)u,$$
$$(\alpha + \beta)u = \alpha u + \beta u,$$
$$\alpha(u + v) = \alpha u + \alpha v,$$

for all $\alpha, \beta \in \mathbb{R}$, and $u, v \in V$.

7. $1u = u$ for all $u \in V$.

These conditions, or axioms, put algebraic structure on the set V. The first four conditions state that there is an addition defined on V that satisfies the usual rules of arithmetic; the last three conditions state that the objects of V can be multiplied by real numbers and that reasonable rules governing that multiplication hold true. The reader is already familiar with some examples of linear spaces.

Example 3.4

(The set $\mathbb{R}^n$) Here an object is an n-tuple of real numbers $(x_1, x_2, \ldots, x_n)$, where $x_i \in \mathbb{R}$. Addition and scalar multiplication are defined by

$$(x_1, x_2, \ldots, x_n) + (y_1, y_2, \ldots, y_n) = (x_1 + y_1, x_2 + y_2, \ldots, x_n + y_n)$$

and

$$\alpha(x_1, x_2, \ldots, x_n) = (\alpha x_1, \alpha x_2, \ldots, \alpha x_n).$$

The zero element is $(0, 0, \ldots, 0)$ and the inverse of $(x_1, \ldots, x_n)$ is $(-x_1, \ldots, -x_n)$. It is easily checked that the remaining properties hold true. The objects of $\mathbb{R}^n$ are called vectors.

Example 3.5

(The function space $C[a, b]$) The objects in this set are the continuous functions $f, g, h, \ldots$ defined on and interval $a \leq x \leq b$. Addition and scalar multiplication are defined pointwise by

$$(f + g)(x) = f(x) + g(x),$$
$$(\alpha f)(x) = \alpha f(x).$$

From elementary calculus we know that the sum of two continuous functions is a continuous function, and a continuous function multiplied by a constant is also a continuous function. The zero element is the 0 function, that is, the function that is constantly zero on $[a, b]$. The inverse of f is $-f$, which is the reflection of $f(x)$ through the x axis. Again it is easily checked that the other conditions are satisfied. Geometrically the sum of the two functions f and g is the graphical sum, and multiplication of f by α stretches the graph of f vertically by a factor α.

Example 3.6

The function space $C^2[a, b]$ is the set of all continuous functions on an interval $[a, b]$ whose second derivative is continuous. The elements are functions. Addition and scalar multiplication are defined as in the last example. It is clear that $C^2[a, b]$ satisfies the other conditions for a linear space. If $y \in C^2[a, b]$, we say y is a function *of class* C^2 on $[a, b]$.

The linear spaces, function spaces, in the last two examples are typical of those occurring in the calculus of variations. The most common function spaces occurring in applied analysis are the spaces $C^n[a, b]$, that is, the set of all continuous functions on the interval $a \leq x \leq b$ whose n th derivative exists and is continuous. The operations in $C^n[a, b]$ are the same as those defined in $C[a, b]$.

A linear space is said to have a norm if there is a rule that uniquely determines the size of a given element in the space. In particular if V is a linear space, then a **norm** on V is a mapping that associates to each $y \in V$ a nonnegative real number denoted by $||y||$, called the norm of y, and that satisfies the following conditions:

1. $||y|| = 0$ if, and only if, $y = 0$.

2. $||\alpha y|| = |\alpha| ||y||$ for all $\alpha \in \mathbb{R}$, $y \in V$.

3. $||y_1 + y_2|| \leq ||y_1|| + ||y_2||$ for all $y_1, y_2 \in V$.

A **normed linear space** is a linear space V on which there is defined a norm $|| \cdot ||$. The number $||y||$ is interpreted as the magnitude or size of y. Thus, a norm puts geometric structure on V.

Example 3.7

The linear space $\mathbb{R}^n$ is a normed linear space if we define a norm by

$$||\mathbf{y}|| = \sqrt{y_1^2 + \cdots + y_n^2}, \quad \mathbf{y} = (y_1, y_2, \ldots, y_n).$$

It is not difficult to show that this (Euclidean) norm satisfies the conditions 1 through 3. A norm on a linear space is not unique. In $\mathbb{R}^n$ the quantity

$$||\mathbf{y}|| = \max\{|y_1|, |y_2|, \ldots, |y_n|\}$$

also defines a norm.

Example 3.8

$C[a, b]$ becomes a normed linear space if we define a norm of a function by

$$||y||_M = \max_{a \leq x \leq b} |y(x)|. \tag{3.7}$$

This norm is called the **strong norm**. The quantity

$$||y||_1 = \int_a^b |y(x)| \, dx \tag{3.8}$$

also defines a norm on $C[a, b]$, called the L^1-norm. Both these norms provide a way to define the magnitude of a continuous function on $[a, b]$, and both satisfy the axioms 1 through 3.

Example 3.9

In the function space $C^1[a, b]$ the norm defined by

$$||y||_w = \max_{a \leq x \leq b} |y(x)| + \max_{a \leq x \leq b} |y'(x)| \tag{3.9}$$

is called the **weak norm** and it plays an important role in the calculus of variations. The same norm may also be defined in $C^n[a, b]$, $n \geq 2$.

If V is a normed linear space with norm $|| \cdot ||$, then the **distance** between the elements y_1 and y_2 in V is defined by

$$||y_1 - y_2||.$$

The motivation for this definition comes from the set of real numbers $\mathbb{R}$, where the absolute value $|x|$ is a norm. The distance between two real numbers a and b is $|a - b|$. Another example is the distance between two functions y_1 and y_2 in $C^1[a, b]$ measured in the weak norm (3.9),

$$||y_1 - y_2||_w = \max_{a \leq x \leq b} |y_1(x) - y_2(x)| + \max_{a \leq x \leq b} |y_1'(x) - y_2'(x)|. \tag{3.10}$$

Having the notion of distance allows us to talk about nearness. Two elements of a normed linear space are close if the distance between them is small. Two functions y_1 and y_2 in $C^1[a, b]$ are close in the weak norm (3.9) if the quantity $||y_1 - y_2||_w$ is small; this means that y_1 and y_2 have values that do not differ greatly, and their derivatives do not differ greatly. The two functions y_1 and y_2 shown in Fig. 3.3 are not close in the weak norm, but they are close in the strong norm because $\max |y_1(x) - y_2(x)|$ is small. Hence, closeness in one norm does not imply closeness in another.

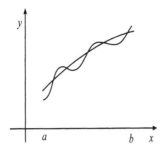

Figure 3.3 Two functions close in the strong norm but not close in the weak norm.

3.2.2 Derivatives of Functionals

For a variational problem we assume the set of admissible functions A is a subset of a normed linear space V. Then the set A can inherit the norm and algebraic properties of the space V. Therefore the notions of local maxima and minima make sense. If $J: A \to \mathbb{R}$ is a functional on A, where $A \subseteq V$ and V is a normed linear space with norm $||\cdot||$, then J has a **local minimum** at $y_0 \in A$, provided $J(y_0) \leq J(y)$ for all $y \in A$ with $||y - y_0|| < d$, for some positive number d. We call y_0 an **extremal** of J and say J is **stationary** at y_0. In the special case that A is a set of functions and $A \subseteq C^1[a,b]$ with the weak norm, we say J has a weak relative minimum at y_0. If the norm on $C^1[a,b]$ is the strong norm, then we say J has a strong relative minimum at y_0. In the calculus of variations the norm comes into play only in the theory of sufficient conditions for extrema. In this text we focus only on necessary conditions.

To motivate the definition of a derivative of a functional, let us rewrite the limit definition of derivative of a function f at x_0 as

$$f(x_0 + \Delta x) - f(x_0) = f'(x_0)\Delta x + o(\Delta x).$$

The **differential** of f at x_0, defined by $df(x_0, \Delta x) = f'(x_0)\Delta x$, is the linear part in the increment Δx of the total change $\Delta f \equiv f(x_0 + \Delta x) - f(x_0)$. That is,

$$\Delta f = df(x_0, \Delta x) + o(\Delta x).$$

Now consider a functional $J: A \to \mathbb{R}$, where A is a subset of a normed linear space V. Specifically, V may be a function space. Let $y_0 \in A$. To increment y_0 we fix an element $h \in V$ such that $y_0 + \varepsilon h$ is in A for all real numbers ε sufficiently small. The increment εh is called the **variation** of the function y_0 and is often denoted by δy_0; that is, $\delta y_0 \equiv \varepsilon h$. Then we define the total change in the functional J due to the change εh in y_0 by

$$\Delta J = J(y_0 + \varepsilon h) - J(y_0).$$

Our goal is to calculate the linear part of this change. To this end, we define the ordinary function

$$\mathcal{J}(\varepsilon) \equiv J(y_0 + \varepsilon h),$$

which is a function of a real variable ε defined by evaluating the functional J on the one parameter family of functions $y_0 + \varepsilon h$. Assuming $\mathcal{J}$ is sufficiently differentiable in ε, we have

$$\Delta J = \mathcal{J}(\varepsilon) - \mathcal{J}(0) = \mathcal{J}'(0)\varepsilon + \mathrm{o}(\varepsilon).$$

Therefore, $\mathcal{J}'(0)\varepsilon$ is like a differential for the functional J, being the linear part, or lowest order, of the increment ΔJ Thus, we are led to the following definition.

Definition 3.10

Let $J \colon A \to \mathbb{R}$ be a functional on A, where $A \subseteq V$, V a normed linear space. Let $y_0 \in A$ and $h \in V$ such that $y_0 + \varepsilon h \in A$ for all ε sufficiently small. Then the **first variation** (also the **Gâteaux derivative**) of J at y_0 in the direction of h is defined by

$$\delta J(y_0, h) \equiv \mathcal{J}'(0) \equiv \frac{d}{d\varepsilon} J(y_0 + \varepsilon h)|_{\varepsilon = 0}, \tag{3.11}$$

provided the derivative exists. Such a direction h for which (3.11) exists is called an **admissible variation** at y_0.

There is a analogy of δJ with a directional derivative. Let $f(\mathbf{X})$ be a real valued function defined for $\mathbf{X} = (x, y)$ in $\mathbb{R}^2$, and let $\mathbf{n} = (n_1, n_2)$ be a unit vector. We recall from calculus that the directional derivative of f at $\mathbf{X}_0 = (x_0, y_0)$ in the direction $\mathbf{n}$ is defined by

$$D_{\mathbf{n}} f(\mathbf{X}_0) \equiv \lim_{\varepsilon \to 0} \frac{f(\mathbf{X}_0 + \varepsilon \mathbf{n}) - f(\mathbf{X}_0)}{\varepsilon}.$$

When (3.11) is written out in limit form, it becomes

$$\delta J(y_0, h) = \lim_{\varepsilon \to 0} \frac{J(y_0 + \varepsilon h) - J(y_0)}{\varepsilon},$$

and the analogy stands out clearly.

The recipe for calculating the first variation $\delta J(y_0, h)$ is to evaluate J at $y_0 + \varepsilon h$, where h is chosen such that $y_0 + \varepsilon h \in A$; then take the derivative with respect to ε, afterwards setting $\varepsilon = 0$. Examples are in the next section.

3.2.3 Necessary Conditions

Let $J\colon A \to \mathbb{R}$ be a functional on A, where $A \subseteq V$, and let $y_0 \in A$ be a local minimum for J relative to a norm $||\cdot||$ in V. Guided by results from the calculus, we expect that the first variation δJ of J should vanish at y_0. It is not difficult to see that this is indeed the case. For, let $h \in V$ be such that $y_0 + \varepsilon h$ is in A for sufficiently small values of ε. Then the function $\mathcal{J}(\varepsilon)$ defined by

$$\mathcal{J}(\varepsilon) = J(y_0 + \varepsilon h)$$

has a local minimum at $\varepsilon = 0$. Hence from ordinary calculus we must have $\mathcal{J}'(0) = 0$ or $\delta J(y_0, h) = 0$. This follows regardless which $h \in V$ is chosen. Hence we have proved the following theorem that gives a necessary condition for a local minimum.

Theorem 3.11

Let $J\colon A \to \mathbb{R}$ be a functional, $A \subseteq V$. If $y_0 \in A$ provides a local minimum for J relative to the norm $||\cdot||$, then

$$\delta J(y_0, h) = 0 \tag{3.12}$$

for all admissible variations h.

The fact that (3.12) holds for all *admissible variations* h often allows us to eliminate h from the condition and obtain an equation just in terms of y_0, which can then be solved for y_0. In the calculus of variations, the equation for y_0 is a differential equation. Since (3.12) is a necessary condition we are not guaranteed that solutions y_0 actually will provide a minimum. Therefore, all we know is that y_0 satisfying (3.12) are extremals, which are the candidates for maxima and minima. In other words, $\delta J(y_0, h) = 0$ is *stationary* at y_0 in the direction h.

Example 3.12

Consider the functional $J(y) = \int_0^1 (1 + y'(x)^2)dx$, where $y \in C^1[0, 1]$ and $y(0) = 0$, $y(1) = 1$. Let $y_0(x) = x$ and $h(x) = x(1-x)$. A few members of the family of curves $y_0 + \varepsilon h = x + \varepsilon x(1-x)$ are sketched in Fig. 3.4. These are the variations in $y_0(x) = x$. We evaluate J on the family $y_0 + \varepsilon h$ to get

$$\mathcal{J}(\varepsilon) = J(y_0 + \varepsilon h) = \int_0^1 [1 + (y_0'(x) + \varepsilon h'(x))^2]\,dx$$

$$= \int_0^1 [1 + (1 + \varepsilon(1 - 2x))^2]\,dx = 2 + \frac{\varepsilon^2}{3}.$$

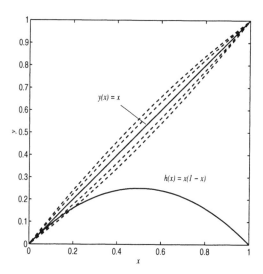

Figure 3.4 The 1–parameter family of curves $x + \varepsilon x(1 - x)$ (dashed) for different values of ε.

Then $\mathcal{J}'(\varepsilon) = \frac{2}{3}\varepsilon$ and $\mathcal{J}'(0) = 0$. Hence we conclude that $\delta J(y_0, h) = 0$ and $y_0 = x$ is an extremal; J is stationary at $y_0 = x$ in the direction $h = x(1 - x)$.

EXERCISES

1. Determine whether the given set constitutes a real linear space under the usual operations associated with elements of the set.

   a) The set of all polynomials of degree ≤ 2.

   b) The set of all continuous functions on $[0, 1]$ satisfying $f(0) = 0$.

   c) The set of all continuous functions on $[0, 1]$ satisfying $f(1) = 1$.

2. Prove that (3.7) and (3.8) both satisfy properties for a norm on $C[a, b]$.

3. In $C^1[0, 1]$ compute the distances between the two functions $y_1(x) = 0$ and $y_2(x) = \frac{1}{100}\sin(1000x)$ in both the weak and strong norms.

4. Show that the first variation $\delta J(y_0, h)$ satisfies the homogeneity condition

$$\delta J(y_0, \alpha h) = \alpha \delta J(y_0, h), \quad \alpha \in \mathbb{R}.$$

5. A functional $J\colon V \to \mathbb{R}$, where V is a normed linear space, is *linear* if

$$J(y_1 + y_2) = J(y_1) + J(y_2), \quad y_1, y_2 \in V$$

and

$$J(\alpha y_1) = \alpha J(y_1), \quad \alpha \in \mathbb{R}, \quad y_1 \in V$$

Which of the following functionals on $C^1[a, b]$ are linear?

a) $J(y) = \int_a^b yy' \, dx.$

b) $J(y) = \int_a^b ((y')^2 + 2y) \, dx.$

c) $J(y) = e^{y(a)}.$

d) $J(y) = \int_a^b \int_a^b K(x, t) y(x) y(t) \, dx \, dt.$

e) $J(y) = \int_a^b y \sin x \, dx.$

f) $J(y) = \int_a^b (y')^2 dx + G(y(b))$, where G is a given differentiable function

6. A functional $J \colon V \to \mathbb{R}$, where V is a normed linear space, is continuous at $y_0 \in V$ if for any $\varepsilon > 0$ there is a $\delta > 0$ such that $|J(y) - J(y_0)| < \varepsilon$ whenever $\|y - y_0\| < \delta.$

   a) Let $V = C^1[a, b]$. Prove that if J is continuous in the strong norm, then it is continuous in the weak norm.

   b) Give a specific example to show that continuity in the weak norm does not imply continuity in the strong norm.

7. Compute the first variation of the functionals on $C^1[a, b]$ given in Exercise 5(a)–(f).

8. Let $J \colon V \to \mathbb{R}$ be a linear functional and V a normed linear space. Prove that if J is continuous at $y_0 = 0$, then it is continuous at each $y \in V$.

9. The second variation of a functional $J \colon A \to \mathbb{R}$ at $y_0 \in A$ in the direction h is defined by

$$\delta^2 J(y_0, h) \equiv \frac{d^2}{d\varepsilon^2} J(y_0 + \varepsilon h)|_{\varepsilon=0}.$$

Find the second variation of the functional

$$J(y) = \int_0^1 (xy'^2 + y \sin y') \, dx,$$

where $y \in C^2[0, 1].$

10. If $J(y) = \int_0^1 (x^2 - y^2 + (y')^2) dx$, $y \in C^2[0, 1]$, calculate ΔJ and $\delta J(y, h)$ when $y(x) = x$ and $h(x) = x^2.$

11. Consider the functional $J(y) = \int_0^1 (1+x)(y')^2 dx$ where $y \in C^2$ and $y(0) = 0$ and $y(1) = 1$. Show that $\delta J(y_0, h) = 0$ for all $h \in C^2$ and $h(0) = h(1) = 0$, where $y_0(x) = \ln(1 + x)/\ln 2.$

12. Consider the functional $J(y) = \int_0^{2\pi} (y')^2 \, dx$. Sketch the function $y_0(x) = x$ and the family $y_0(x) + \varepsilon h(x)$, where $h(x) = \sin x$. Compute $\mathcal{J}(\varepsilon) \equiv J(y_0 + \varepsilon h)$ and show that $\mathcal{J}'(0) = 0$. Deduce that J is stationary at x in the direction $\sin x$.

13. Let $J(y) = \int_0^1 (3y^2 + x)dx + \frac{1}{2}y(0)^2$, where y is a continuous function on $0 \leq x \leq 1$. Let $y = x$ and $h = x + 1$, $0 \leq x \leq 1$. Compute $\delta J(x, h)$.

3.3 The Simplest Problem

3.3.1 The Euler Equation

The *simplest problem* in the calculus of variations is to find a local minimum for the functional

$$J(y) = \int_a^b L(x, y, y') \, dx, \tag{3.13}$$

where $y \in C^2[a, b]$ and $y(a) = y_0$, $y(b) = y_1$. Here L is a given, twice continuously differentiable function on $[a, b] \times \mathbb{R}^2$, and $|| \cdot ||$ is some norm in $C^2[a, b]$. Actually, our assumption that y is of class C^2 could be weakened to $y \in C^1[a, b]$. The increased smoothness assumption, however, will make the work a little easier. In this discussion we assume that the competing functions are class C^2. The references at the end of the chapter can be consulted to see how the analysis changes if smoothness conditions are relaxed.

We seek a necessary condition. Guided by the last two sections we compute the first variation of J. Let y be a local minimum and h a twice continuously differentiable function satisfying $h(a) = h(b) = 0$. Then $y + \varepsilon h$ is an admissible function and

$$J(y + \varepsilon h) = \int_a^b L(x, y + \varepsilon h, y' + \varepsilon h') \, dx.$$

Therefore

$$\frac{d}{d\varepsilon} J(y + \varepsilon h) = \int_a^b \frac{\partial}{\partial \varepsilon} L(x, y + \varepsilon h, y' + \varepsilon h') \, dx$$

$$= \int_a^b [L_y(x, y + \varepsilon h, y' + \varepsilon h')h + L_{y'}(x, y + \varepsilon h, y' + \varepsilon h')h'] \, dx,$$

where L_y denotes $\partial L/\partial y$ and $L_{y'}$ denotes $\partial L/\partial y'$. Hence

$$\delta J(y, h) = \frac{d}{d\varepsilon} J(y + \varepsilon h)|_{\varepsilon = 0} = \int_a^b \{L_y(x, y, y')h + L_{y'}(x, y, y')h'\} \, dx.$$

Therefore, a necessary condition for $y(x)$ to be a local minimum is

$$\int_a^b \left\{ L_y(x, y, y')h + L_{y'}(x, y, y')h' \right\} dx = 0 \tag{3.14}$$

for all $h \in C^2[a, b]$ with $h(a) = h(b) = 0$.

Condition (3.14) is not useful as it stands for determining $y(x)$. Using the fact that it must hold for all h, however, we can eliminate h and thereby obtain a condition for y alone. First we integrate the second term in (3.14) by parts to obtain

$$\int_a^b \left(L_y(x, y, y') - \frac{d}{dx} L_{y'}(x, y, y') \right) h \, dx + L_{y'}(x, y, y')h|_{x=a}^{x=b} = 0. \tag{3.15}$$

Because h vanishes at a and b, the boundary terms vanish and the necessary condition becomes

$$\int_a^b \left(L_y(x, y, y') - \frac{d}{dx} L_{y'}(x, y, y') \right) h \, dx = 0 \tag{3.16}$$

for all $h \in C^2[a, b]$ with $h(a) = h(b) = 0$. The following lemma, due to Lagrange, provides the final step in our calculation by effectively allowing us to eliminate h from condition (3.16). This lemma is one version of what is called the *fundamental lemma* of the calculus of variations.

Lemma 3.13

If $f(x)$ is continuous on $[a, b]$ and if

$$\int_a^b f(x)h(x) \, dx = 0 \tag{3.17}$$

for every twice continuously differentiable function h with $h(a) = h(b) = 0$, then $f(x) = 0$ for $x \in [a, b]$.

Proof

The proof is by contradiction. Assume for some x_0 in (a, b) that $f(x_0) > 0$. Because f is continuous, $f(x) > 0$ for all x in some interval (x_1, x_2) containing x_0. For $h(x)$ choose

$$h(x) = \begin{cases} (x - x_1)^3 (x_2 - x)^3, & x_1 \leq x \leq x_2 \\ 0, & \text{otherwise} \end{cases}$$

a graph of which is shown in Fig. 3.5. The cubic factors $(x - x_1)^3$ and $(x_2 - x)^3$

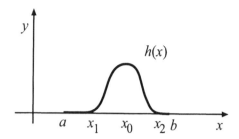

Figure 3.5 Plot of the C^2 function $h(x)$.

appear so that h is smoothed out at x_1 and x_2 and is therefore of class C^2. Then

$$\int_a^b f(x)h(x)\, dx = \int_{x_1}^{x_2} f(x)(x - x_1)^3 (x_2 - x)^3\, dx > 0,$$

because f is positive on (x_1, x_2). This contradicts (3.17) and the lemma is proved. □

Applying the lemma to (3.16) with $f(x) = L_y(x, y, y') - (d/dx)L_{y'}(x, y, y')$, which is continuous because L is twice continuously differentiable, we obtain the following important theorem.

Theorem 3.14

If a function y provides a local minimum for the functional

$$J(y) = \int_a^b L(x, y, y')\, dx,$$

where $y \in C^2[a, b]$ and

$$y(a) = y_0, \quad y(b) = y_1,$$

then y must satisfy the differential equation

$$L_y(x, y, y') - \frac{d}{dx}L_{y'}(x, y, y') = 0, \quad x \in [a, b]. \tag{3.18}$$

Equation (3.18) is called the **Euler equation** or **Euler–Lagrange equation**. It is a second-order ordinary differential equation that is, in general, nonlinear. It represents a *necessary condition* for a local minimum and it is analogous to the derivative condition $f'(x) = 0$ in differential calculus. Therefore its solutions are not necessarily local minima. The solutions of (3.18) are (local) **extremals**. If y is an extremal, then $\delta J(y, h) = 0$ for all h, and J is stationary at y.

By writing out the total derivative in (3.18) using the chain rule, the Euler equation becomes

$$L_y - L_{y'x}(x, y, y') - L_{y'y}(x, y, y')y' - L_{y'y'}(x, y, y')y'' = 0.$$

Thus it is evident that the Euler equation is second-order, provided $L_{y'y'} \neq 0$.

3.3.2 Solved Examples

We now determine the extremals for some specific variational problems.

Example 3.15

Find the extremals of the functional

$$J(y) = \int_0^1 ((y')^2 + 3y + 2x)\, dx, \quad y(0) = 0, \quad y(1) = 1.$$

The Lagrangian is $L = (y')^2 + 3y + 2x$. Then $L_y = 3$, $L_{y'} = 2y'$ and the Euler equation is

$$3 - \frac{d}{dx}(2y') = 0.$$

Integrating twice gives the extremals

$$y = \frac{3}{4}x^2 + C_1 x + C_2.$$

The boundary conditions readily give $C_2 = 0$ and $C_1 = \frac{1}{4}$. Therefore the extremal of J satisfying the boundary conditions is

$$y = \frac{3}{4}x^2 + \frac{1}{4}x.$$

Further calculations are required to determine whether this maximizes or minimizes J.

Example 3.16

The arclength functional is

$$J(y) = \int_a^b \sqrt{1 + (y')^2}\, dx,$$

where $y \in C^2[a, b]$, $y(a) = y_0$ and $y(b) = y_1$. A necessary condition for J to have a local minimum at y is that y satisfy the Euler equation

$$L_y - \frac{d}{dx} L_{y'} = 0,$$

or

$$\frac{d}{dx}\left(\frac{y'}{\sqrt{1 + (y')^2}}\right) = 0.$$

Hence

$$\frac{y'}{\sqrt{1 + (y')^2}} = C,$$

where C is a constant. Solving for y' gives

$$y' = K,$$

where K is a constant. Therefore

$$y = Kx + M,$$

and the extremals are straight lines. Applying the boundary conditions determines the constants K and M,

$$K = \frac{y_1 - y_0}{b - a}, \quad M = \frac{by_0 - ay_1}{b - a}.$$

Therefore the unique extremal is $y = \frac{y_1 - y_0}{b-a} x + \frac{by_0 - ay_1}{b-a}$, which is the equation of the straight line connecting (a, y_0) and (b, y_1).

3.3.3 First Integrals

The Euler equation is a second-order ordinary differential equation and the solution will generally contain two arbitrary constants. In special cases, when the Lagrangian does not depend explicitly on one of its variables x, y, or y', it is possible to make an immediate simplification of the Euler equation. We consider three cases:

1. If $L = L(x, y)$, then the Euler equation is $L_y(x, y) = 0$, which is an algebraic equation.

2. If $L = L(x, y')$, then the Euler equation is

$$L_{y'}(x, y') = C, \tag{3.19}$$

where C is any constant.

3. If $L = L(y, y')$, then the Euler equation is

$$L(y, y') - y'L_{y'}(y, y') = C, \qquad (3.20)$$

where C is any constant.

Propositions (1) and (2) follow obviously from the Euler equation. To prove (3) we note that

$$\frac{d}{dx}[L - y'L_{y'}] = \frac{dL}{dx} - y'\frac{d}{dx}L_{y'} - L_{y'}y''$$

$$= L_y y' + L_{y'}y'' - y'\frac{d}{dx}L_{y'} - L_{y'}y''$$

$$= y'\left(L_y - \frac{d}{dx}L_{y'}\right) = 0.$$

Equations (3.19) and (3.20) are first integrals of the Euler equation. A **first integral** of a second-order differential equation $F(x, y, y', y'') = 0$ is an expression of the form $g(x, y, y')$, involving one lower derivative, which is constant whenever y is a solution of the original equation $F(x, y, y', y'') = 0$. Hence

$$g(x, y, y') = C$$

represents an integration of the second-order equation. In the calculus of variations, if L is independent of x, then $L - y'L_{y'}$ is a first integral of the Euler equation and we say $L - y'L_{y'}$ is constant on the extremals. In mechanics, first integrals are called **conservation laws**.

Example 3.17

(Brachistochrone) We determine the extremals of the brachistochrone problem. The functional is

$$J(y) = \int_{x_1}^{x_2} \frac{\sqrt{1 + (y')^2}}{\sqrt{2g(y_1 - y)}} \, dx,$$

subject to the boundary conditions $y(0) = b$, $y(a) = 0$. The Lagrangian is independent of x, and therefore a first integral of the Euler equation is given by (3.20). Then

$$\frac{\sqrt{1 + (y')^2}}{\sqrt{b - y}} - (y')^2 \frac{(1 + (y')^2)^{-1/2}}{\sqrt{b - y}} = C,$$

which simplifies to

$$\left(\frac{dy}{dx}\right)^2 = \frac{1 - C^2(b - y)}{C^2(b - y)}.$$

Taking the square root of both sides and separating variables gives

$$dx = -\frac{\sqrt{b-y}}{\sqrt{C_1 - (b-y)}}\, dy, \quad C_1 = C^{-2},$$

where the minus sign is taken because $dy/dx < 0$. The last equation can be integrated by making the trigonometric substitution

$$b - y = C_1 \sin^2 \frac{\phi}{2}. \tag{3.21}$$

One obtains

$$dx = C_1 \sin^2 \frac{\phi}{2}\, d\phi = \frac{C_1}{2}(1 - \cos\phi)\, d\phi,$$

which yields

$$x = \frac{C_1}{2}(\phi - \sin\phi) + C_2. \tag{3.22}$$

Equations (3.21) and (3.22) are parametric equations for a cycloid. Here, in contrast to the problem of finding the curve of shortest length between two points, it is not clear that the cycloids just obtained actually minimize the given functional. Further calculations would be required for confirmation.

Example 3.18

(Fermat's principle) In the limit of geometric optics the principle of Fermat (1601–1665) states that the time elapsed in the passage of light between two fixed points in a medium is an extremum with respect to all possible paths connecting the points. For simplicity we consider only light rays that lie in the xy plane. Let $c = c(x,y)$ be a positive continuously differentiable function representing the velocity of light in the medium. Its reciprocal $n = c^{-1}$ is called the index of refraction of the medium. If $P\colon (x_1, y_1)$ and $Q\colon (x_2, y_2)$ are two fixed points in the plane, then the time required for light to travel along a given path $y = y(x)$ connecting the two points is

$$T(y) = \int_P^Q \frac{ds}{c} = \int_{x_1}^{x_2} n(x,y)\sqrt{1 + (y')^2}\, dx.$$

Therefore, the actual light path connecting P and Q is the one that extremizes the integral $T(y)$. The differential equation for the extremal path is the Euler equation

$$n_y(x,y)\sqrt{1 + (y')^2} - \frac{d}{dx}\left(\frac{n(x,y)y'}{\sqrt{1 + (y')^2}}\right) = 0.$$

EXERCISES

1. Find the extremal(s).

   a) $J(y) = \int_0^1 y' \, dx$, $\quad y \in C^2[0,1]$, $\quad y(0) = 0$, $\quad y(1) = 1$.

   b) $J(y) = \int_0^1 yy' \, dx$, $\quad y \in C^2[0,1]$, $\quad y(0) = 0$, $\quad y(1) = 1$.

   c) $J(y) = \int_0^1 xyy' \, dx$, $\quad y \in C^2[0,1]$, $\quad y(0) = 0$, $\quad y(1) = 1$.

2. Find extremals for the following functionals:

   a) $\int_a^b \frac{(y')^2}{x^3} \, dx$.

   b) $\int_a^b (y^2 + (y')^2 + 2ye^x) \, dx$.

3. Show that the Euler equation for the functional

$$J(y) = \int_a^b f(x,y)\sqrt{1 + (y')^2} \, dx$$

   has the form

$$f_y - f_x y' - \frac{f y''}{1 + (y')^2} = 0.$$

4. Find an extremal for

$$J(y) = \int_1^2 \frac{\sqrt{1 + (y')^2}}{x} \, dx, \quad y(1) = 0, \quad y(2) = 1.$$

5. Prove that the functional $J(y) = \int_a^b (p(x)(y')^2 + q(x)y^2) dx$, $y \in C^2[a,b]$, $y(a) = y_a$, $y(b) = y_b$, where p and q are positive, assumes its absolute minimum value for the function $y = Y(x)$, where Y is the solution to the Euler equation.

6. Obtain a necessary condition for a function $y \in C[a,b]$ to be a local minimum of the functional

$$J(y) = \int\int_R K(s,t)y(s)y(t) \, ds \, dt + \int_a^b y(t)^2 \, dt - 2\int_a^b y(t)f(t) \, dt,$$

   where $K(s,t)$ is a given continuous function of s and t on the square $R: \{(s,t) \,|\, a \le s \le b, \, a \le t \le b\}$, $K(s,t) = K(t,s)$, and $f \in C[a,b]$. (The Euler equation is an integral equation.)

7. Find the extremal for $J(y) = \int_0^1 (1 + x)(y')^2 dx$, $y \in C^2$ and $y(0) = 0$, $y(1) = 1$. What is the extremal if the boundary condition at $x = 1$ is changed to $y'(1) = 0$?

8. Describe the paths of light rays in the plane where the medium has index of refraction given by

   a) $n = kx$.

   b) $n = ky^{-1}$.

   c) $n = ky^{1/2}$.

   d) $n = ky^{-1/2}$, where $k > 0$.

9. Show that the Euler equation for the problem

$$J(y) = \int_a^b L(t, y, \dot{y})\, dt$$

can be written in the form

$$L_t - \frac{d}{dt}(L - \dot{y}L_{\dot{y}}) = 0.$$

10. Show that the minimal area of a surface of revolution is a catenoid, that is, the surface found by revolving a catenary

$$y = c_1 \cosh\left(\frac{x + c_2}{c_1}\right)$$

about the x axis.

11. Find the extremals for

$$J(y) = \int_a^b (x^2 y'^2 + y^2)\, dx, \quad y \in C^2[a, b].$$

12. A Lagrangian has the form

$$L(x, y, y') = \frac{a^2}{12} y'^4 + ay'^2 G(y) - G(y)^2,$$

where G is a given differentiable function. Find Euler's equation and a first integral.

13. Find extremals of the functional

$$J(y) = \int_0^1 (y^2 + yy' + (y' - 2)^2)\, dx$$

over the domain $A = \{y \in C^2[0, 1] : y(0) = y(1) = 0\}$. Show that J does not assume a maximum value at these extremals. Explain why it does, or does not, follow that the extremals yield minimum values of J.

14. (Economics) Let $y = y(t)$ be a individual's total capital at time t and let $r = r(t)$ be the rate that capital is spent. If $U = U(r)$ is the rate of enjoyment, then his total enjoyment over a lifetime $0 \le t \le T$ is

$$E = \int_0^T e^{-\beta t} U(r(t)) \, dt,$$

where the exponential factor reflects the fact that future enjoyment is discounted over time. Initially, his capital is Y, and he desires $y(T) = 0$. Because his capital gains interest at rate α,

$$y' = \alpha y - r(t).$$

Assume $\alpha < 2\beta < 2\alpha$. Determine $r(t)$ and $y(t)$ for which the individual's total enjoyment is maximized if his enjoyment function is $U(r) = 2\sqrt{r}$.

3.4 Generalizations

3.4.1 Higher Derivatives

In the simplest variational problem the Lagrangian depends on x, y, and y'. An obvious generalization is to include higher derivatives in the Lagrangian. Thus we consider the *second-order problem*

$$J(y) = \int_a^b L(x, y, y', y'') \, dx, \quad y \in A, \tag{3.23}$$

where A is the set of all $y \in C^4[a, b]$ that satisfy the boundary conditions

$$y(a) = A_1, \quad y'(a) = A_2, \quad y(b) = B_1, \quad y'(b) = B_2. \tag{3.24}$$

The function L is assumed to be twice continuously differentiable in each of its four arguments.

Again we seek a necessary condition. Let $y \in A$ provide a local minimum for J with respect to some norm in $C^4[a, b]$, and choose $h \in C^4[a, b]$ satisfying

$$h(a) = h'(a) = h(b) = h'(b) = 0. \tag{3.25}$$

To compute the first variation, we form the function $J(y + \varepsilon h)$. Then

$$
\begin{aligned}
\delta J(y, h) &= \frac{d}{d\varepsilon} J(y + \varepsilon h)|_{\varepsilon = 0} \\
&= \frac{d}{d\varepsilon} \int_a^b L(x, y + \varepsilon h, y' + \varepsilon h', y'' + \varepsilon h'') \, dx|_{\varepsilon = 0} \\
&= \int_a^b [L_y h + L_{y'} h' + L_{y''} h''] \, dx,
\end{aligned}
$$

where L_y, and $L_{y'}$, and $L_{y''}$ are evaluated at (x, y, y', y''). In this calculation two integrations by parts are required on the term $L_{y''}h''$. Proceeding,

$$\delta J(y, h) = \int_a^b \left[L_y - \frac{d}{dx} L_{y'} + \frac{d^2}{dx^2} L_{y''} \right] h \, dx$$
$$+ \left[L_{y''}h' - \frac{d}{dx} L_{y''}h + h L_{y'} \right]_{x=a}^{x=b} . \quad (3.26)$$

By (3.25) the terms on the boundary at $x = a$ and $x = b$ vanish. Therefore,

$$\int_a^b \left[L_y - \frac{d}{dx} L_{y'} + \frac{d^2}{dx^2} L_{y''} \right] h \, dx = 0 \quad (3.27)$$

for all $h \in C^4[a, b]$ satisfying (3.25). To complete the argument, another version of the fundamental lemma is needed.

Lemma 3.19

If f is a continuous function on $[a, b]$, and if

$$\int_a^b f(x)h(x) \, dx = 0$$

for all $h \in C^4[a, b]$ satisfying (3.25), then $f(x) \equiv 0$ for $x \in [a, b]$.

The proof is nearly the same as before and is left as an exercise. Applying this lemma to (3.27) gives a necessary condition for y to be a local minimum, that is,

$$L_y - \frac{d}{dx} L_{y'} + \frac{d^2}{dx^2} L_{y''} = 0, \quad a \leq x \leq b. \quad (3.28)$$

This is the Euler equation for the second-order problem defined by (3.23). Equation (3.28) is a fourth-order ordinary differential equation for y, and its general solution involves four arbitrary constants that can be determined from the boundary conditions.

The *nth-order variational problem* is defined by the functional

$$J(y) = \int_a^b L(x, y, y', y'', \ldots, y^{(n)}) \, dx,$$

$y \in C^{2n}[a, b]$, with

$$y(a) = A_1, \quad y'(a) = A_2, \ldots, y^{(n-1)}(a) = A_n,$$
$$y(b) = B_1, \quad y'(b) = B_2, \ldots, y^{(n-1)}(b) = B_n.$$

The Euler equation in this case is

$$L_y - \frac{d}{dx}L_{y'} + \frac{d^2}{dx^2}L_{y''} - \cdots + (-1)^n \frac{d^n}{dn^n}L_{y^{(n)}} = 0,$$

which is an ordinary differential equation of order $2n$.

3.4.2 Several Functions

The functional J may depend on several functions $y_1, \ldots, y_n$. To fix the notion, let $n = 2$ and consider the functional

$$J(y_1, y_2) = \int_a^b L(x, y_1, y_2, y_1', y_2') \, dx, \tag{3.29}$$

where $y_1, y_2 \in C^2[a, b]$ with boundary conditions given by

$$\begin{aligned} y_1(a) = A_1, \quad y_1(b) = B_1, \\ y_2(a) = A_2, \quad y_2(b) = B_2. \end{aligned} \tag{3.30}$$

In this case suppose the pair y_1 and y_2 provide a local minimum for J. We vary each independently by choosing h_1 and h_2 in $C^2[a, b]$, satisfying

$$h_1(a) = h_2(a) = h_1(b) = h_2(b) = 0, \tag{3.31}$$

and forming a one-parameter admissible pair of functions $y_1 + \varepsilon h_1$ and $y_2 + \varepsilon h_2$. Then

$$\mathcal{J}(\varepsilon) = \int_a^b L(x, y_1 + \varepsilon h_1, y_2 + \varepsilon h_2, y_1' + \varepsilon h_1', y_2' + \varepsilon h_2') \, dx,$$

and $\mathcal{J}$ has a local minimum at $\varepsilon = 0$. Now

$$\mathcal{J}'(0) = \int_a^b (L_{y_1} h_1 + L_{y_2} h_2 + L_{y_1'} h' + L_{y_2'} h_2') \, dx.$$

Integrating the last two terms in the integrand by parts and applying conditions (3.31) gives

$$\int_a^b \left\{ \left(L_{y_1} - \frac{d}{dx} L_{y_1'} \right) h_1 + \left(L_{y_2} - \frac{d}{dx} L_{y_2'} \right) h_2 \right\} dx = 0 \tag{3.32}$$

for all $h_1, h_2 \in C^2[a, b]$. Picking $h_2 = 0$ on $[a, b]$ gives

$$\int_a^b \left(L_{y_1} - \frac{d}{dx} L_{y_1'} \right) h_1 \, dx = 0$$

for all $h_1 \in C^2[a, b]$. By the fundamental lemma,

$$L_{y_1} - \frac{d}{dx} L_{y_1'} = 0. \tag{3.33}$$

Now in (3.32) pick $h_1 = 0$ on $[a, b]$. Then

$$\int_a^b \left(L_{y_2} - \frac{d}{dx} L_{y_2'} \right) h_2 \, dx = 0$$

for all $h_2 \in C^2[a, b]$. Applying the fundamental lemma again yields

$$L_{y_2} - \frac{d}{dx} L_{y_2'} = 0. \tag{3.34}$$

Consequently, if the pair y_1, y_2 provides a local minimum for J, then y_1 and y_2 must satisfy the system of two ordinary differential equations (3.33) and (3.34). These are the Euler equations for the problem. The four arbitrary constants appearing in the general solution can be determined from the boundary data.

In general, if J depends on n functions,

$$J(y_1, \ldots, y_n) = \int_a^b L(x, y_1, \ldots, y_n, y_1', \ldots, y_n') \, dx, \tag{3.35}$$

where $y_i \in C^2[a, b]$, and

$$y_i(a) = A_i, \quad y_i(b) = B_i, \quad i = 1, \ldots, n,$$

then a necessary condition for $y_1, \ldots, y_n$ to provide a local minimum for J is that $y_1, \ldots, y_n$ satisfy the Euler system of n ordinary differential equations

$$L_{y_i} - \frac{d}{dx} L_{y_i'} = 0, \quad i = 1, \ldots, n. \tag{3.36}$$

If the Lagrangian L is independent of explicit dependence on x, that is, $L_x = 0$, then it is left as an exercise to show that

$$L - \sum_{i=1}^n y_i' L_{y_i'} = C$$

is a first integral of (3.36).

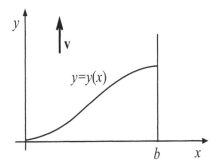

Figure 3.6 Path $y = y(x)$ across a river with velocity field $\mathbf{v}(x, y) = v(x)\mathbf{j}$.

3.4.3 Natural Boundary Conditions

Let us consider the following problem: A river with parallel straight banks b units apart has stream velocity given by $\mathbf{v}(x, y) = v(x)\mathbf{j}$, where $\mathbf{j}$ is the unit vector in the y direction (see Fig. 3.6). Assuming that one of the banks is the y axis and that the point $(0, 0)$ is the point of departure, what route should a boat take to reach the opposite bank in the shortest possible time? Assume that the speed of the boat in still water is c, where $c > v$.

This problem differs from those in earlier sections in that the right-hand endpoint, the point of arrival on the line $x = b$, is not specified; it must be determined as part of the solution to the problem. It can be shown (see the Exercises) that the time required for the boat to cross the river along a given path $y = y(x)$ is

$$J(y) = \int_0^b \frac{\sqrt{c^2(1 + (y')^2) - v^2} - vy'}{c^2 - v^2} \, dx. \tag{3.37}$$

The variational problem is to minimize $J(y)$ subject to the conditions

$$y(0) = 0, \quad y(b) \text{ free}.$$

Such a problem is called a **free endpoint problem,** and if $y(x)$ is an extremal then a certain condition must hold at $x = b$. Conditions of these types, called **natural boundary conditions**, are the subject of this section. Just as common are problems where both endpoints are unspecified.

To fix the notion we consider the problem

$$J(y) = \int_a^b L(x, y, y') \, dx, \tag{3.38}$$

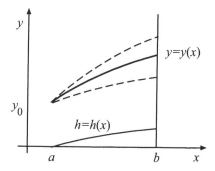

Figure 3.7 Variations $y = y(x) + \varepsilon h(x)$ of $y(x)$ satisfying $h(a) = 0$.

where $y \in C^2[a, b]$ and

$$y(a) = y_0, \quad y(b) \text{ free.} \tag{3.39}$$

Let y be a local minimum. Then the variations h must be C^2 functions that satisfy the single condition

$$h(a) = 0. \tag{3.40}$$

Figure 3.7 shows several variations $y + \varepsilon h$ of the extremal y. No condition on h at $x = b$ is required, because the admissible functions are unspecified at the right endpoint.

The first variation is

$$\delta J(y, h) = \frac{d}{d\varepsilon} \int_a^b L(x, y + \varepsilon h, y' + \varepsilon h')\, dx \big|_{\varepsilon=0}$$

$$= \int_a^b \left(L_y - \frac{d}{dx} L_{y'} \right) h\, dx + L_{y'} h \big|_{x=a}^{x=b}.$$

Because $h(a) = 0$, a necessary condition for y to be a local minimum is

$$\int_a^b \left(L_y - \frac{d}{dx} L_{y'} \right) h\, dx + L_{y'}(b, y(b), y'(b)) h(b) = 0 \tag{3.41}$$

for every $h \in C^2(a, b)$ satisfying $h(a) = 0$. Because (3.41) holds for these h, it must hold for h also satisfying the condition $h(b) = 0$. Hence

$$\int_a^b \left(L_y - \frac{d}{dx} L_{y'} \right) h\, dx = 0, \tag{3.42}$$

and by the fundamental lemma, y must satisfy the Euler equation

$$L_y - \frac{d}{dx} L_{y'} = 0. \tag{3.43}$$

Then, substituting (3.43) into (3.41) gives

$$L_{y'}(b, y(b), y'(b))h(b) = 0. \tag{3.44}$$

Now, because (3.44) holds for all choices of $h(b)$, we must have

$$L_{y'}(b, y(b), y'(b)) = 0, \tag{3.45}$$

which is a condition on the extremal y at $x = b$. Equation (3.45) is called a **natural boundary condition**. The Euler equation (3.43), the fixed boundary condition $y(a) = A$, and the natural boundary condition (3.45) are enough to determine the extremal for the variational problem (3.38) and (3.39). By similar arguments if the left endpoint $y(a)$ is unspecified, then the natural boundary condition on an extremal y at $x = a$ is $L_{y'}(a, y(a), y'(a)) = 0$.

Example 3.20

We find the natural boundary condition at $x = b$ for the river crossing problem (3.37). In this case

$$L_{y'} = \frac{1}{c^2 - v^2} \left(\frac{c^2 y'}{(c^2(1 + y'^2) - v^2)^{1/2}} - v \right).$$

The boundary condition (3.45) becomes

$$\frac{c^2 y'(b)}{(c^2(1 + y'(b)^2) - v(b)^2)^{1/2}} - v(b) = 0,$$

which can be simplified to

$$y'(b) = \frac{v(b)}{c}.$$

Thus, the slope that the boat enters the bank at $x = b$ is the ratio of the water speed at the bank to the boat velocity in still water.

Example 3.21

Find the differential equation and boundary conditions for the extremal of the variational problem

$$J(y) = \int_0^1 (p(x)y'^2 - q(x)y^2) \, dx,$$

$$y(0) = 0, \quad y(1) \text{ free,}$$

where p and q are positive smooth functions on $[0, 1]$. In this case $L_y = -2q(x)y$, $L_{y'} = 2p(x)y'$, and the Euler equation is

$$\frac{d}{dx}(p(x)y') + q(x)y = 0, \quad 0 < x < 1,$$

which is a Sturm–Liouville type equation . The natural boundary condition at $x = 1$ is given by (3.45), or

$$2p(1)y'(1) = 0.$$

Because $p > 0$, the boundary conditions are

$$y(0) = 0, \quad y'(1) = 0.$$

EXERCISES

1. Find extremals for the following functionals.

   a) $J(y_1, y_2) = \int_0^{\pi/4}[4y_1^2 + y_2^2 + y_1'y_2']\,dx$, $\quad y_1(0) = 1$, $y_1\left(\frac{\pi}{4}\right) = 0$, $y_2(0) = 0$, $y_2\left(\frac{\pi}{4}\right) = 1$.

   b) $J(y) = \int_0^1 (1 + (y'')^2)\,dx$, $\quad y(0) = 0$, $y'(0) = 1$, $y(1) = 1$, $y'(1) = 1$.

   c) $J(y) = \int_a^b (y^2 + 2\dot{y}^2 + (y'')^2)\,dt$.

   d) $J(y) = \int_0^1 (yy' + (y'')^2)\,dx$, $y(0) = 0$, $y'(0) = 1$, $y(1) = 2$, $y'(1) = 4$.

   e) $J(y) = \int_0^\infty [y^2 + y'^2 + (y'' + y')^2]\,dx$, $\quad y(0) = 1$, $y'(0) = 2$, $y(\infty) = 0$, $y'(\infty) = 0$.

2. Prove that the Euler equation for $J(y) = \int_a^b L(x, y, y', y'')\,dx$ has first integral $L_{y'} - (d/dx)L_{y''} = C$ when L is independent of y. If L is independent of x, show that

$$L - y'\left(L_{y'} - \frac{d}{dx}L_{y''}\right) - y''L_{y''} = C$$

   is a first integral.

3. One version of the Ramsey growth model in economics involves minimizing the "total product"

$$J(M) = \int_0^T (aM - M' - b)^2 dt, \quad a, b > 0,$$

   over a fixed planning period $[0, T]$, where $M = M(t)$ is the capital at time t and $M(0) = M_0$ is the initial capital. If $M(t)$ minimizes J, find the capital $M(T)$ at the end of the planning period.

4. Find the extremals for a functional of the form

$$J(y, z) = \int_a^b L(\dot{y}, \dot{z}) \, dt,$$

given that L satisfies the condition $L_{\dot{y}\dot{y}} L_{\dot{z}\dot{z}} - L_{\dot{y}\dot{z}}^2 \neq 0.$

5. Find the extremals for

a) $J(y) = \int_0^1 (y'^2 + y^2) \, dx$, $y(0) = 1$, $y(1)$ free.

b) $J(y) = \int_0^3 e^{2x}(y'^2 - y^2) \, dx$, $y(0) = 1$, $y(3)$ free.

c) $J(y) = \int_0^1 (\frac{1}{2}y'^2 + y'y + y' + y) \, dx$, $y(0) = \frac{1}{2}$, $y(1)$ free.

d) $J(y) = \int_1^e (\frac{1}{2}x^2 y'^2 - \frac{1}{8}y^2) \, dx$, $y(1) = 1$, $y(e)$ free.

e) $J(y) = \int_0^1 y'^2 \, dx + y(1)^2$, $y(0) = 1$, $y(1)$ free.

6. Determine the natural boundary condition at $x = b$ for the variational problem defined by

$$J(y) = \int_a^b L(x, y, y') \, dx + G(y(b)), \quad y \in C^2[a, b], \ y(a) = y_0,$$

where G is a given differentiable function on $\mathbb{R}$.

7. Find the extremal(s) of the functional

$$J(y) = \int_0^b (\sqrt{1 - k^2 + y'^2} - ky') \, dx,$$

in the class of all $C^2[0, b]$ functions with $y(0) = 0$ and $y(b)$ free. Take $0 < k < 1$.

8. Find the natural boundary condition at $x = 2$ associated with the functional

$$J(y) = (1 - 3y(2))^2 + \int_0^2 [(x + 1)^2 y'^2 + xy' - y^2] \, dx.$$

3.5 The Canonical Formalism

3.5.1 Hamilton's Principle

According to the doctrine of classical dynamics, one associates with the system being described a set of quantities or dynamical variables, each of which has a well-defined value at each instant of time and which defines the state of the dynamical system at that instant. Further, it is assumed that the time evolution of the system is completely determined if its state is known at some given instant. Analytically this doctrine is expressed by the fact that the dynamical variables satisfy a set of differential equations (the equations of motion of the system) as functions of time, along with initial conditions. The program of classical dynamics consists of listing the dynamical variables and formulating the equations of motion that predict the system's evolution in time. Newton's second law of motion describes the dynamics of a mechanical system.

Another method of obtaining the equations of motion is from a variational principle. This method is based on the idea that a system should evolve along a path of least resistance. Principles of this sort have a long history in physical theories dating back to antiquity when Hero of Alexandria stated a minimum principle concerning the path of reflected light rays. In the 17th century, Fermat's principle, that light rays travel along the path of shortest time, was put forth. For mechanical systems, Maupertuis's principle of least action stated that a system should evolve from one state to another in such a way that the *action* (a vaguely defined term with the energy × time) is smallest. Lagrange and Gauss were also advocates of similar principles. In the early part of the 19th century, however, W. R. Hamilton (1805–1865) stated what has become an encompassing, aesthetic principle that can be generalized to embrace many areas of physics.

Hamilton's principle states that the time evolution of a mechanical system occurs in such a manner that the integral of the difference between kinetic and potential energy is stationary. To be more precise, let $y_1, \ldots, y_n$ denote a set of *generalized coordinates* of a given dynamical system. That is, regarded as functions of time, we assume that $y_1, \ldots, y_n$ completely specify the state of the system at any instant. Further we assume that there are no relations among the y_i, so that they may be regarded as independent. In general, the y_i may be lengths, angles, or whatever. The time derivatives $\dot{y}_1, \ldots, \dot{y}_n$ (we will use the common physics convention of an overdot to denote time derivatives) are called the *generalized velocities*. The kinetic energy T is, in the most general case, a quadratic form in the $\dot{y}_i$, that is,

$$T = \sum_{i=1}^{n} \sum_{j=1}^{n} a_{ij}(y_1, \ldots, y_n) \dot{y}_i \dot{y}_j, \tag{3.46}$$

where the a_{ij} are known functions of the coordinates $y_1, \ldots, y_n$. The potential energy V may be a function of t, y_i, and $\dot{y}_i$, or

$$V = V(t, y_1, \ldots, y_n, \dot{y}_1, \ldots, \dot{y}_n). \tag{3.47}$$

For systems where such a scalar potential energy exists we define the *Lagrangian* by

$$L = L(t, y_1, \ldots, y_n, \dot{y}_1, \ldots, \dot{y}_n) \equiv T - V. \tag{3.48}$$

Hamilton's principle for these systems may then be stated as follows: Consider a mechanical system described by generalized coordinates $y_1, \ldots, y_n$ with Lagrangian given by (3.48). Then the motion of the system from time t_0 to t_1 is such that the functional

$$J(y_1, \ldots, y_n) = \int_{t_0}^{t_1} L(t, y_1, \ldots, y_n, \dot{y}_1, \ldots, \dot{y}_n)\, dt \tag{3.49}$$

is stationary for the functions $y_1(t), \ldots, y_n(t)$, which describe the actual time evolution of the system. If we regard the set of coordinates $y_1, \ldots, y_n$ as coordinates in n dimensional space, then the equations

$$y_i = y_i(t), \quad i = 1, \ldots, n, \quad (t_0 \leq t \leq t_1)$$

are parametric equations of a curve C that joins two states $S_0: (y_1(t_0), \ldots, y_n(t_0))$ and $S_1: (y_1(t_1), \ldots, y_n(t_1))$. Hamilton's principle then states that among all paths in configuration space connecting the initial state S_0 to the final state S_1, the actual motion takes place along the path that affords an extreme value to the integral (3.49). The actual path is an extremal. In physics and engineering, Hamilton's principle is often stated concisely as

$$\delta \int_{t_0}^{t_1} L(t, y_1, \ldots, y_n, \dot{y}_1, \ldots, \dot{y}_n)\, dt = 0.$$

The functional $\int L\, dt$ is called the **action integral** for the system.

Because the curve $y_i = y_i(t)$, $i = 1, \ldots, n$, along which the motion occurs, makes the functional J stationary, it follows from the calculus of variations that the $y_i(t)$ must satisfy the Euler equations

$$L_{y_i} - \frac{d}{dt} L_{\dot{y}_i} = 0, \quad i = 1, \ldots, n. \tag{3.50}$$

In mechanics, the Euler equations (3.50) are called **Lagrange's equations**. They form the equations of motion, or governing equations, for the system. We say that the governing equations follow from a *variational principle* if we can find an L such that $\delta \int L\, dt = 0$ gives those governing equations as necessary conditions for an extremum.

If the Lagrangian L is independent of time t, that is, $L_t = 0$, then a first integral is

$$L - \sum_{i=1}^{n} \dot{y}_i L_{\dot{y}_i} = \text{const.}$$

This equation is a **conservation law**. The quantity $-L + \sum_{i=1}^{n} \dot{y}_i L_{\dot{y}_i}$ is called the **Hamiltonian** of the system, and it frequently represents the total energy. Thus, if L is independent of time, then energy is conserved.

Example 3.22

(Harmonic oscillator) Consider the motion of a mass m attached to a spring obeying Hooke's law, where the restoring force is $F = -ky$. The variable y is a generalized coordinate representing the positive displacement of the mass from equilibrium, and $k > 0$ is the spring constant. The kinetic and potential energies are

$$T = \tfrac{1}{2} m \dot{y}^2, \quad V = \tfrac{1}{2} k y^2.$$

The Lagrangian is therefore

$$L = T - V = \tfrac{1}{2} m \dot{y}^2 - \tfrac{1}{2} k y^2,$$

and Hamilton's principle states that the motion takes place in such a way that

$$J(y) = \int_{t_0}^{t_1} \left(\tfrac{1}{2} m \dot{y}^2 - \tfrac{1}{2} k y^2 \right) dt$$

is stationary. Lagrange's equation is

$$L_y - \frac{d}{dt} L_{\dot{y}} = -ky - \frac{d}{dt}(m\dot{y}) = 0,$$

or

$$\ddot{y} + \frac{k}{m} y = 0,$$

which expresses Newton's second law that force equals mass times acceleration. Its solution gives the extremals

$$y(t) = C_1 \cos \sqrt{\frac{k}{m}} t + C_2 \sin \sqrt{\frac{k}{m}} t,$$

where C_1 and C_2 are constants that can be determined from boundary conditions.

Example 3.23

(Pendulum) Consider a simple pendulum of length l and bob mass m suspended from a frictionless support. To describe the state at any time t we choose the generalized coordinate θ measuring the angle displaced from the vertical equilibrium position. If s denotes the actual displacement of the bob along a circular arc measured from equilibrium, then the kinetic energy is

$$T = \tfrac{1}{2} m \dot{s}^2 = \tfrac{1}{2} m l^2 \dot{\theta}^2,$$

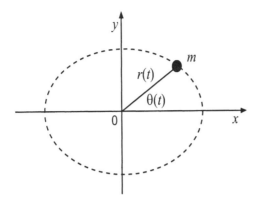

Figure 3.8 The motion of a mass in a inverse-square, central force field.

where we have used $s = l\theta$. The potential energy of the bob is mg times the height above its equilibrium position, or

$$V = mg(l - l\cos\theta).$$

By Hamilton's principle the motion takes place so that

$$J(\theta) = \int_{t_0}^{t_1} \left(\tfrac{1}{2}ml^2\dot\theta^2 - mg(l - l\cos\theta)\right)\,dt$$

is stationary. Lagrange's equation is

$$L_\theta - \frac{d}{dt}L_{\dot\theta} = -mgl\sin\theta - \frac{d}{dt}(ml^2\dot\theta) = 0,$$

or

$$\ddot\theta + \frac{g}{l}\sin\theta = 0,$$

which describes the evolution the system. For small displacements $\sin\theta \approx \theta$ and the equation becomes $\ddot\theta + (g/l)\theta = 0$, whose solution exhibits simple harmonic motion.

Example 3.24

(Motion in a central force field) Consider the planar motion of a mass m that is attracted to the origin with a force inversely proportional to the square of the distance from the origin (see Fig. 3.8). For generalized coordinates we take the polar coordinates r and θ of the position of the mass. The kinetic energy is

$$T = \tfrac{1}{2}mv^2 = \tfrac{1}{2}m(\dot{x}^2 + \dot{y}^2)$$

$$= \tfrac{1}{2}m\left[\left(\frac{d}{dt}(r\cos\theta)\right)^2 + \left(\frac{d}{dt}(r\sin\theta)\right)^2\right]$$

$$= \tfrac{1}{2}m[(\dot{r}\cos\theta - r\sin\theta\dot{\theta})^2 + (\dot{r}\sin\theta + r\cos\theta\dot{\theta})^2]$$

$$= \tfrac{1}{2}m(\dot{r}^2 + r^2\dot{\theta}^2).$$

The force is $-k/r^2$ for some constant k, and thus the potential energy is

$$V = -\int -\frac{k}{r^2}\,dr = -\frac{k}{r}.$$

Hamilton's principle requires that

$$J(r,\theta) = \int_{t_0}^{t_1}\left(\frac{m}{2}(\dot{r}^2 + r^2\dot{\theta}^2) + \frac{k}{r}\right)dt$$

is stationary. Lagrange's equations are

$$L_\theta - \frac{d}{dt}L_{\dot{\theta}} = 0, \quad L_r - \frac{d}{dt}L_{\dot{r}} = 0,$$

or

$$-\frac{d}{dt}(mr^2\dot{\theta}) = 0, \quad mr\dot{\theta}^2 - \frac{k}{r^2} - \frac{d}{dt}(m\dot{r}) = 0.$$

These can be rewritten as

$$mr^2\dot{\theta} = \text{constant}, \quad m\ddot{r} - mr\dot{\theta}^2 + \frac{k}{r^2} = 0,$$

which is a coupled system of ordinary differential equations. The Exercises request a solution.

In summary, Hamilton's principle gives us a procedure for finding the equations of motion of a system if we can write down the kinetic and potential energies. This offers an alternative approach to writing down Newton's second law for a system, which requires that we know the forces. Because Hamilton's principle only results in writing the equations of motion, why not just directly determine the governing equations and forgo the variational principle altogether? Actually, this may be a legitimate objection, particularly in view of the fact that the variational principle is usually derived *a posteriori*, that is, from the known equations of motion and not conversely, as would be relevant from the point of view of the calculus of variations. Moreover, if a variational principle is given as the basic principle for the system, then there are complicated sufficient conditions for extrema that must be considered, and they seem to have little or no role in physical problems. Finally, although variational principles do to some extent represent a unifying concept for physical theories, the extent is by no means universal; it is impossible to state such principles for some systems with constraints or dissipative forces.

On the other hand, aside from the aesthetic view, the *ab initio* formulation of the governing law by a variational principle has arguments on its side. The action integral plays a fundamental role in the development of numerical methods for solving differential equations (Rayleigh–Ritz method, Galerkin methods, etc.); it also plays a decisive role in the definition of Hamilton's characteristic function, the basis for the Hamilton–Jacobi theory. Furthermore, many variational problems occur in geometry and other areas apart from physics; in these problems the action or fundamental integral *is* an *a priori* notion. In summary, the calculus of variations provides a general context in which to study wide classes of problems of interest in many areas of science, engineering, and mathematics.

3.5.2 Hamilton's Equations

The Euler equations for the variational problem

$$J(y_1, \ldots, y_n) = \int_{t_0}^{t_1} L(t, y_1, \ldots, y_n, \dot{y}_1, \ldots, \dot{y}_n)\, dt \tag{3.51}$$

form a system of n second-order ordinary differential equations. We now introduce a canonical method for reducing these equations to a system of $2n$ first-order equations. For simplicity we examine the case $n = 1$,

$$J(y) = \int_{t_0}^{t_1} L(t, y, \dot{y})\, dt. \tag{3.52}$$

The Euler equation is

$$L_y - \frac{d}{dt}L_{\dot{y}} = 0. \tag{3.53}$$

First we define a new variable p, called the **canonical momentum** by

$$p \equiv L_{\dot{y}}(t, y, \dot{y}). \tag{3.54}$$

If $L_{\dot{y}\dot{y}} \neq 0$, then the implicit function theorem guarantees that (3.54) can be solved for $\dot{y}$ in terms of t, y, and p to get

$$\dot{y} = \phi(t, y, p). \tag{3.55}$$

Now we define the **Hamiltonian** H by

$$H(t, y, p) \equiv -L(t, y, \phi(t, y, p)) + \phi(t, y, p)p. \tag{3.56}$$

Notice that H is just the expression $-L + \dot{y}L_{\dot{y}}$, where $\dot{y}$ is given by (3.55). In many systems H is the total energy.

Example 3.25

For a particle of mass m moving in one dimension with potential energy $V(y)$, the Lagrangian is given by $L = \frac{1}{2}m\dot{y}^2 - V(y)$. Hence $p = m\dot{y}$, which is the momentum of the particle, and

$$H = -\frac{1}{2}m\left(\frac{p}{m}\right)^2 + V(y) + \frac{p^2}{m} = \frac{1}{2}\frac{p^2}{m} + V(y)$$

which is the total energy of the particle written in terms of position and momentum.

Now, from (3.56),

$$\frac{\partial H}{\partial p} = -\frac{\partial L}{\partial \dot{y}}\frac{\partial \phi}{\partial p} + \phi + p\frac{\partial \phi}{\partial p} = \phi = \dot{y},$$

and

$$\frac{\partial H}{\partial y} = -\frac{\partial L}{\partial y} - \frac{\partial L}{\partial \dot{y}}\frac{\partial \phi}{\partial y} + p\frac{\partial \phi}{\partial y} = -L_y = -\frac{d}{dt}L_{\dot{y}} = -\dot{p}.$$

Therefore, we have shown that the Euler equation (3.53) can be written as an equivalent system of equations

$$\dot{y} = \frac{\partial H}{\partial p}(t, y, p), \quad \dot{p} = -\frac{\partial H}{\partial y}(t, y, p). \tag{3.57}$$

Equations (3.57) are called **Hamilton's equations**. They form a system of first-order differential equations for y and p.

Example 3.26

Consider the harmonic oscillator whose Lagrangian is

$$L(t, y, \dot{y}) = \frac{1}{2}m\dot{y}^2 - \frac{1}{2}ky^2. \tag{3.58}$$

The canonical momentum is

$$p = L_{\dot{y}} = m\dot{y}.$$

Solving for $\dot{y}$ gives

$$\dot{y} = \frac{p}{m}, \quad (= \phi(t, y, p)),$$

and therefore the Hamiltonian is

$$H = -L + \dot{y}L_{\dot{y}} = -\left(\frac{1}{2}m\left(\frac{p}{m}\right)^2 - \frac{1}{2}ky^2\right) + \left(\frac{p}{m}\right)p$$

$$= \frac{1}{2}\frac{p^2}{m} + \frac{1}{2}ky^2,$$

which is the sum of the kinetic and potential energy, or the total energy of the system. Now

$$\frac{\partial H}{\partial y} = ky, \quad \frac{\partial H}{\partial p} = \frac{p}{m},$$

and so Hamilton's equations are

$$\dot{y} = \frac{p}{m}, \quad \dot{p} = -ky. \tag{3.59}$$

One way to solve these equations in the yp (phase) plane is to divide them to obtain

$$\frac{dp}{dy} = \frac{-ky}{p/m}.$$

Separating variables and integrating yields

$$p^2 + kmy^2 = C,$$

which is a family of ellipses in the yp plane.

For the general action integral

$$J(y_1, \ldots, y_n) = \int_{t_0}^{t_1} L(t, y_1, \ldots, y_n, \dot{y}_1, \ldots, \dot{y}_n) \, dt$$

depending on n functions $y_1, \ldots, y_n$, the canonical momenta are

$$p_i = L_{\dot{y}_i}(t, y_1, \ldots, y_n, \dot{y}_1, \ldots, \dot{y}_n), \quad i = 1, \ldots, n. \tag{3.60}$$

Assuming

$$\det(L_{\dot{y}_i \dot{y}_j}) \neq 0,$$

we can solve the system of n equations (3.60) for $\dot{y}_1, \ldots, \dot{y}_n$ to obtain

$$\dot{y}_i = \phi_i(t, y_1, \ldots, y_n, p_1, \ldots, p_n), \quad i = 1, \ldots, n.$$

The Hamiltonian H is defined by

$$H(t, y_1, \ldots, y_n, p_1, \ldots, p_n) = -L + \sum_{i=1}^{n} \dot{y}_i L_{\dot{y}_i}$$

$$= -L(t, y_1, \ldots, y_n, \phi_1, \ldots, \phi_n)$$

$$+ \sum_{i=1}^{n} \phi_i(t, y_1, \ldots, y_n, p_1, \ldots, p_n) p_i.$$

Using an argument exactly like the one used for the case $n = 1$, we obtain the canonical form of the Euler equations,

$$\dot{y}_i = \frac{\partial H}{\partial p_i}(t, y_1, \ldots, y_n, p_1, \ldots, p_n), \tag{3.61a}$$

$$\dot{p}_i = -\frac{\partial H}{\partial y_i}(t, y_1, \ldots, y_n, p_1, \ldots, p_n), \tag{3.61b}$$

$i = 1, \ldots, n$, which are Hamilton's equations. They represent $2n$ first-order ordinary differential equations for the $2n$ functions $y_1, \ldots, y_n, p_1, \ldots, p_n$. A complete discussion of the role of the canonical formalism in the calculus of variations, geometry, and physics can be found in Rund (1966).

3.5.3 The Inverse Problem

In general, a variational principle exists for a given physical system if there is a Lagrangian L such that the Euler equations

$$L_{y_i} - \frac{d}{dt} L_{\dot{y}_i} = 0, \quad i = 1, \ldots, n,$$

corresponding to the action integral $\int L \, dt$ coincide with the governing equations of the system. For conservative mechanical systems Hamilton's principle tells us that $L = T - V$. How can we determine the Lagrangian L for other systems if we know the equations of motion? This problem is known as the **inverse problem** of the calculus of variations. Let us formulate the inverse problem for $n = 1$. Given a second-order ordinary differential equation

$$\ddot{y} = F(t, y, \dot{y}), \tag{3.62}$$

find a Lagrangian $L(t, y, \dot{y})$ such that (3.62) is the Euler equation

$$L_y - \frac{d}{dt} L_{\dot{y}} = 0. \tag{3.63}$$

Generally, the problem has infinitely many solutions. To determine L we write the Euler equation as

$$L_y - L_{\dot{y}t} - L_{\dot{y}y}\dot{y} - L_{\dot{y}\dot{y}}\ddot{y} = 0,$$

or

$$L_y - L_{\dot{y}t} - L_{\dot{y}y}\dot{y} - F L_{\dot{y}\dot{y}} = 0.$$

There is a general procedure for characterizing all such Lagrangians, but it is often simpler to proceed directly by matching terms in the Euler equation to terms in the given differential equation.

Example 3.27

Consider a damped spring–mass oscillator where a mass m is suspended from a spring with spring constant k and there is a dashpot offering a resistive force numerically equal to $a\dot{y}$, where a is the damping constant, and y is the displacement of the mass from equilibrium. By Newton's law

$$m\ddot{y} = -ky - a\dot{y},$$

or

$$m\ddot{y} + a\dot{y} + ky = 0. \tag{3.64}$$

Equation (3.64) is the damped harmonic oscillator equation. The system is not conservative because of the damping, and there is not a scalar potential. We cannot apply Hamilton's principle directly. We seek a Lagrangian $L(t, y, \dot{y})$ such that (3.64) is the Euler equation. Multiplying (3.64) by a nonnegative function $f(t)$ we obtain

$$mf(t)\ddot{y} + af(t)\dot{y} + kf(t)y = 0. \tag{3.65}$$

Equation (3.65) must coincide with the Euler equation

$$-L_y + L_{\dot{y}t} + L_{\dot{y}y}\dot{y} + L_{\dot{y}\dot{y}}\ddot{y} = 0.$$

An obvious choice is to take $L_{\dot{y}\dot{y}} = mf(t)$, so that

$$L_{\dot{y}} = mf(t)\dot{y} + M(t, y),$$

and

$$L = mf(t)\frac{\dot{y}^2}{2} + M(t, y)\dot{y} + N(t, y),$$

where M and N are arbitrary functions. It follows that

$$-L_y + L_{\dot{y}t} + L_{\dot{y}y}\dot{y} = af(t)\dot{y} + kf(t)y,$$

or

$$-(M_y\dot{y} + N_y) + mf'(t)\dot{y} + M_t + M_y\dot{y} = af(t)\dot{y} + kf(t)y.$$

Thus $mf'(t) = af(t)$, or $f(t) = e^{at/m}$. Hence

$$-N_y + M_t = kye^{at/m}.$$

Selecting $M = 0$ gives $N = -(k/2)y^2 e^{at/m}$, and therefore a Lagrangian for the damped harmonic oscillator is

$$L(t, y, \dot{y}) = \left(\frac{m}{2}\dot{y}^2 - \frac{k}{2}y^2\right)e^{at/m}.$$

This expression is the Lagrangian for the undamped oscillator times a time-dependent factor.

EXERCISES

1. Consider the functional $J(y) = \int_a^b [r(t)\dot{y}^2 + q(t)y^2]\, dt$. Find the Hamiltonian $H(t, y, p)$ and write down Hamilton's equations for the problem.

2. Write down Hamilton's equations for the functional

$$J(y) = \int_a^b \sqrt{(t^2 + y^2)(1 + \dot{y}^2)}\, dt.$$

Solve the equations and sketch the solution curves in the yp plane.

3. Derive Hamilton's equations for the functional

$$J(\theta) = \int_{t_1}^{t_2} \left(\frac{m}{2} l^2 \dot{\theta}^2 + mgl\cos\theta - mgl \right) dt,$$

representing the motion of a pendulum in a plane.

4. A particle of unit mass moves along a y axis under the influence of a potential given by

$$f(y) = -\omega^2 y + ay^2,$$

where ω and a are positive constants.

a) What is the potential energy $V(y)$? Determine the Lagrangian and write down the equation of motion.

b) Find the Hamiltonian $H(y, p)$ and show it coincides with the total energy. Write down Hamilton's equations. Is energy conserved? Is momentum conserved?

c) If the total energy E is $\omega^2/10$, and at time $t = 0$ the particle is at $y = 0$, what is the initial velocity?

d) Sketch the possible phase trajectories in phase space when the total energy in the system is given by $E = \omega^6/12a^2$. [Hint: note that $p = \pm\sqrt{2}\sqrt{E - V(y)}$.) What is the value of E above which oscillatory motion is not possible?

5. Give an alternate derivation of Hamilton's equations by rewriting the action integral

$$\int_{t_0}^{t_1} L(t, y, \dot{y})\, dt$$

as

$$J(p, y) = \int_{t_0}^{t_1} F(t, y, \dot{y}, p, \dot{p})\, dt,$$

where $F = \dot{y}p - H(t, y, p)$, treating y and p as independent and finding the Euler equations.

6. A particle of mass m moves in one dimension under the influence of a force $F(y, t) = ky^{-2}e^t$, where y is position, t is time, and k is a constant. Formulate Hamilton's principle for this system and derive the equation of motion. Determine the Hamiltonian and compare it with the total energy. Is energy conserved?

7. A particle of mass m is falling under the action of gravity with resistive forces neglected. If y denotes the distance measured downward, find the Lagrangian for the system and determine the equation of motion from a variational principle.

8. Consider a system of n particles where m_i is the mass of the ith particle and (x_i, y_i, z_i) is its position in space. The kinetic energy of the system is

$$T = \frac{1}{2} \sum_{i=1}^{n} m_i(\dot{x}_i^2 + \dot{y}_i^2 + \dot{z}_i^2),$$

and assume that the system has a potential energy

$$V = V(t, x_1, \ldots, x_n, y_1, \ldots, y_n, z_1, \ldots, z_n)$$

such that the force acting on the ith particle has components

$$F_i = -\frac{\partial V}{\partial x_i}, \quad G_i = -\frac{\partial V}{\partial y_i}, \quad H_i = -\frac{\partial V}{\partial z_i}.$$

Show that Hamilton's principle applied to this system yields

$$m_i\ddot{x}_i = F_i, \quad m_i\ddot{y}_i = G_i, \quad m_i\ddot{z}_i = H_i,$$

which are Newton's equations for a system of n particles.

9. Consider a simple plane pendulum with a bob of mass m attached to a string of length l. After the pendulum is set in motion the string is shortened by a constant rate $dl/dt = -\alpha = $ const. Formulate Hamilton's principle and determine the equation of motion. Compare the Hamiltonian to the total energy. Is energy conserved?

10. Consider the differential equations

$$r^2\dot{\theta} = C, \quad \ddot{r} - r\dot{\theta}^2 + \frac{k}{m}r^{-2} = 0,$$

governing the motion of a mass in an inverse square central force field.

a) Using the chain rule show that

$$\dot{r} = Cr^{-2}\frac{dr}{d\theta}, \qquad \ddot{r} = C^2 r^{-4}\frac{d^2r}{d\theta^2} - 2C^2 r^{-5}\left(\frac{dr}{d\theta}\right)^2,$$

and therefore the equations of motion may be written

$$\frac{d^2r}{d\theta^2} - 2r^{-1}\left(\frac{dr}{d\theta}\right)^2 - r + \frac{k}{C^2 m}r^2 = 0.$$

b) Let $r = u^{-1}$ and show that

$$\frac{d^2u}{d\theta^2} + u = \frac{k}{C^2 m}.$$

c) Solve the differential equation in part (b) to obtain

$$u = r^{-1} = \frac{k}{C^2 m}(1 + \varepsilon\cos(\theta - \theta_0)),$$

where ε and θ_0 are constants of integration.

d) Show that elliptical orbits are obtained when $\varepsilon < 1$.

11. Find a Lagrangian $L(t, y, \dot{y})$ such that the Euler equation coincides with the Emden–Fowler equation

$$\ddot{y} + \frac{2}{t}\,\dot{y} + y^5 = 0.$$

12. A particle moving in one dimension in a constant external force field with frictional force proportional to its velocity has equation of motion

$$\ddot{y} + a\dot{y} + b = 0, \qquad a, b > 0.$$

Find a Lagrangian $L(t, y, \dot{y})$.

13. Using the electrical–mechanical analogy and the result for the damped harmonic oscillator, determine a Lagrangian for an RCL electrical circuit whose differential equation is

$$L\ddot{I} + R\dot{I} + \frac{1}{C}I = 0,$$

where $I(t)$ is the current, L the inductance, C the capacitance, and R the resistance.

14. Consider the nonlinear equation

$$y'' + p(t)y' + f(y) = 0,$$

where p and f are continuous functions with antiderivatives $P(t)$ and $F(y)$, respectively. Find a Lagrangian $L = L(t, y, y')$ such that the Euler equation associated with the functional $J(y) = \int_a^b L\,dt$ is equivalent to the given equation. Hint: Multiply the equation by $e^{P(t)}$.

3.6 Isoperimetric Problems

In differential calculus the problem of minimizing a given function $f(x,y)$ subject to a subsidiary condition $g(x,y) = $ constant is a common one. Now we consider classes of variational problems in which the competing functions are required to conform to certain restrictions, in addition to the normal endpoint conditions. Such restrictions are called *constraints* and they may take the form of integral relations, algebraic equations, or differential equations. In this section we limit the discussion to integral constraints.

One method for solving the constraint problem in calculus is the **Lagrange multiplier rule**. We state this necessary condition precisely because it is used in the subsequent discussion.

Theorem 3.28

Let f and g be differentiable functions with $g_x(x_0, y_0)$ and $g_y(x_0, y_0)$ not both zero. If (x_0, y_0) provides an extreme value to f subject to the constraint $g = C = $ constant, then there exists a constant λ such that

$$f_x^*(x_0, y_0) = 0, \tag{3.66}$$
$$f_y^*(x_0, y_0) = 0, \tag{3.67}$$

and

$$g(x_0, y_0) = C, \tag{3.68}$$

where

$$f^* \equiv f + \lambda g.$$

The proof can be found in most calculus books. For practical calculations the three conditions in the theorem can be used to determine x_0, y_0 , and the Lagrange multiplier λ.

A variational problem having an integral constraint is known as an **isoperimetric problem**. In particular, consider the problem of minimizing the functional

$$J(y) = \int_a^b L(x, y, y')\, dx \tag{3.69}$$

subject to

$$W(y) \equiv \int_a^b G(x, y, y')\, dx = k, \tag{3.70}$$

where $y \in C^2[a, b]$ and

$$y(a) = y_0, \quad y(b) = y_1, \tag{3.71}$$

and k is a fixed constant. The given functions L and G are assumed to be twice continuously differentiable. The subsidiary condition (3.70) is called an **isoperimetric constraint**.

In essence, we follow the procedure in earlier sections for problems without constraints. We embed an assumed local minimum $y(x)$ in a family of admissible functions. A *one-parameter* family $y(x) + \varepsilon h(x)$ is not, however, a suitable choice, because those curves may not maintain the constancy of W. Therefore we introduce a *two-parameter* family

$$z(x) = y(x) + \varepsilon_1 h_1(x) + \varepsilon_2 h_2(x),$$

where h_1, $h_2 \in C^2[a, b]$,

$$h_1(a) = h_1(b) = h_2(a) = h_2(b) = 0, \tag{3.72}$$

and ε_1 and ε_2 are real parameters ranging over intervals containing the origin. We assume that W does not have an extremum at y. Then for any choice of h_1 and h_2 there will be values of ε_1 and ε_2 in the neighborhood of $(0,0)$ for which $W(z) = k$. Evaluating J and W at z gives

$$\mathcal{J}(\varepsilon_1, \varepsilon_2) = \int_a^b L(x, z, z') \, dx,$$

$$\mathcal{W}(\varepsilon_1, \varepsilon_2) = \int_a^b G(x, z, z') \, dx = k.$$

Because y is the local minimum for (3.69) subject to the constraint (3.70), the point $(\varepsilon_1, \varepsilon_2) = (0,0)$ must be a local minimum for $\mathcal{J}(\varepsilon_1, \varepsilon_2)$ subject to the constraint $\mathcal{W}(\varepsilon_1, \varepsilon_2) = k$. This is just a differential calculus problem and so the Lagrange multiplier rule may be applied. There must exist a constant λ such that

$$\frac{\partial \mathcal{J}^*}{\partial \varepsilon_1} = \frac{\partial \mathcal{J}^*}{\partial \varepsilon_2} = 0 \quad \text{at} \quad (\varepsilon_1, \varepsilon_2) = (0,0), \tag{3.73}$$

where $\mathcal{J}^*$ is defined by

$$\mathcal{J}^* = \mathcal{J} + \lambda \mathcal{W} = \int_a^b L^*(x, z, z') \, dx,$$

with

$$L^* = L + \lambda G. \tag{3.74}$$

We now calculate the derivatives in (3.73), afterward setting $\varepsilon_1 = \varepsilon_2 = 0$. Accordingly,

$$\frac{\partial \mathcal{J}^*}{\partial \varepsilon_i}(0,0) = \int_a^b (L_y^*(x, y, y')h_i + L_{y'}^*(x, y, y')h_i') \, dx, \quad i = 1, 2.$$

Integrating the second term by parts and applying conditions (3.72) give

$$\frac{\partial J^*}{\partial \varepsilon_i}(0,0) = \int_a^b \left(L_y^*(x,y,y') - \frac{d}{dx} L_{y'}^*(x,y,y') \right) h_i \, dx, \quad i = 1, 2.$$

From (3.73), and because of the arbitrary character of h_1 or h_2, the fundamental lemma implies

$$L_y^*(x,y,y') - \frac{d}{dx} L_{y'}^*(x,y,y') = 0, \tag{3.75}$$

which is a necessary condition for an extremum.

In practice, the solution of the second-order differential equation (3.75) will yield a function $y(x)$ that involves two arbitrary constants and the unknown multiplier λ. These may be evaluated by the two boundary conditions (3.71) and by substituting $y(x)$ into the isoperimetric constraint (3.70).

Example 3.29

(Shape of a Hanging Rope) A rope of length l with constant density ρ hangs from two fixed points (a, y_0) and (b, y_1) in the plane. Let $y(x)$ be an arbitrary configuration of the rope with the y axis adjusted so that $y(x) > 0$. A small element of length ds at (x, y) has mass ρds and potential energy $\rho g y \, ds$ relative to $y = 0$. Therefore the total potential energy of the rope hanging in the arbitrary configuration $y = y(x)$ is given by the functional

$$J(y) = \int_0^l \rho g y \, ds = \int_a^b \rho g y \sqrt{1 + y'^2} \, dx. \tag{3.76}$$

It is known that the actual configuration minimizes the potential energy. Therefore, we are faced with minimizing (3.76) subject to the isoperimetric condition

$$W(y) = \int_a^b \sqrt{1 + y'^2} \, dx = l. \tag{3.77}$$

Hence, we form the auxiliary function

$$L^* = L + \lambda G = \rho g y \sqrt{1 + y'^2} + \lambda \sqrt{1 + y'^2},$$

and write the associated Euler equation (3.75). In this case L^* does not depend explicitly on x, and thus a first integral is

$$L^* - y' L_{y'}^* = C,$$

or

$$(\rho \, g y + \lambda) \left(\sqrt{1 + y'^2} - \frac{y'^2}{\sqrt{1 + y'^2}} \right) = C.$$

Solving for y' and separating variables yields

$$\frac{dy}{\sqrt{(\rho\, gy + \lambda)^2 - C^2}} = \frac{1}{C}\, dx.$$

Letting $u = \rho\, gy + \lambda$ and using the antiderivative formula

$$\int \frac{du}{\sqrt{u^2 - C^2}} = \cosh^{-1}\frac{u}{C}$$

gives

$$\frac{1}{\rho\, g}\cosh^{-1}\frac{u}{C} = \frac{1}{C}x + C_1.$$

Then

$$y = -\frac{\lambda}{\rho g} + \frac{C}{\rho g}\cosh\left(\frac{\rho g x}{C} + C_2\right).$$

Therefore the shape of a hanging rope is a **catenary**. The constants C, C_2, and λ may be determined from (3.77) and the endpoint conditions $y(a) = y_0$ and $y(b) = y_1$. In practice this calculation is difficult, and there may not be a smooth solution for large values of l (Why?)

Example 3.30

Consider the problem of minimizing

$$J(\psi) = \int_{-\infty}^{\infty}\left(\frac{\hbar^2}{2m}(\psi')^2 + V(x)\psi^2\right)dx,$$

subject to the constraint

$$\int_{-\infty}^{\infty}\psi^2(x)\,dx = 1. \tag{3.78}$$

Here $\hbar$ and m are constants, and $V = V(x)$ is a given function. The auxiliary Lagrangian

$$L^* = \frac{\hbar^2}{2m}(\psi')^2 + V(x)\psi^2 - E\psi^2,$$

where $-E$ is a Lagrange multiplier. The Euler equation is

$$L_\psi^* - \frac{d}{dx}L_{\psi'}^* = 0,$$

or

$$-\frac{\hbar^2}{2m}\psi'' + V(x)\psi = E\psi. \tag{3.79}$$

Equation (3.79) is the **Schrödinger equation** *for the wave function ψ in quantum mechanics for a particle of mass m under the influence of a potential V.* Equation (3.78) is a normalization condition for the wave function ψ, whose square is a probability density. In general, solutions ψ of (3.79) will exist only for discrete values of the multiplier E, which are identified with the possible energy levels of the particle and are the eigenvalues. These problems are discussed in Chapter 4.

EXERCISES

1. Find extremals of the isoperimetric problem

$$J(y) = \int_0^\pi y'^2\, dx, \quad y(0) = y(\pi) = 0,$$

subject to

$$\int_0^\pi y^2\, dx = 1.$$

2. Find extremals for the isoperimetric problem

$$J(y) = \int_0^1 (y'^2 + x^2)\, dx, \quad y(0) = 0, \quad y(1) = 0$$

$$\int_0^1 y^2\, dx = 2$$

3. Derive a necessary condition for an extremum for the isoperimetric problem

   Minimize

$$J(y_1, y_2) = \int_a^b L(x, y_1, y_2, y_1', y_2')\, dx$$

   subject to

$$\int_a^b G(x, y_1, y_2, y_1', y_2')\, dx = C$$

   and

$$y_1(a) = A_1, \quad y_2(a) = A_2, \quad y_1(b) = B_1, \quad y_2(b) = B_2$$

   where C, A_1, A_2, B_1, and B_2 are constants.

4. Use the result of the preceding exercise to maximize

$$J(x, y) = \int_{t_0}^{t_1} (xy' - yx')\, dt,$$

   subject to

$$\int_{t_0}^{t_1} \sqrt{x'^2 + y'^2}\, dt = l.$$

   Show that J represents the area enclosed by a curve with parametric equations $x = x(t)$, $y = y(t)$, and the constraint fixes the length of the curve.

Thus, this is the problem of determining which curve of a specified perimeter encloses the maximum area. The term *isoperimetric*, meaning *same perimeter*, originated in this context.

Answer: Circles $(x - c_1)^2 + (y - c_2)^2 = \lambda^2$.

5. Determine the equation of the shortest arc in the first quadrant that passes through $(0,0)$ and $(1,0)$ and encloses a prescribed area A with the x axis, where $0 < A \leq \pi/8$.

6. Write down equations that determine the solution of the isoperimetric problem

$$\int_a^b (p(x)y'^2 + q(x)y^2)\, dx \rightarrow \min,$$

subject to

$$\int_a^b r(x)y^2\, dx = 1,$$

where p, q, and r are given functions and $y(a) = y(b) = 0$. The result is a *Sturm–Liouville eigenvalue problem* (Chapter 4).

7. Consider the problem of minimizing the functional

$$J(y, z) = \int_a^b L(t, y, z, \dot{y}, \dot{z})\, dt, \tag{3.80}$$

subject to the algebraic constraint

$$G(t, y, z) = 0, \quad G_z \neq 0 \tag{3.81}$$

and boundary conditions

$$y(a) = y(b) = z(a) = z(b) = 0,$$

with all of the functions possessing the usual differentiability conditions. If $y(t)$ and $z(t)$ provide a local minimum, show that there exists a function $\lambda(t)$ such that

$$L_y - \frac{d}{dt}L_{\dot{y}} = \lambda(t)G_y, \quad L_z - \frac{d}{dt}L_{\dot{z}} = \lambda(t)G_z.$$

(*Hint:* Solve (3.81) to get $z = g(t, y)$, and substitute into (3.80) to obtain a functional depending only upon y. Find the Euler equation and use the condition $(\partial/\partial y)G(t, y, g(t, y)) = 0$.)

REFERENCES AND NOTES

The calculus of variations is an old subject dating back to the late 17th century. As such, the literature is vast, and there are many excellent treatments of the theory for both necessary and sufficient conditions. The references below represent a few of the recent books that complement this text. A classical reference is Gelfand & Fomin (1963). More recent are Pinch (1993) and van Brunt (2004). A definitive treatise on the canonical formalism is in Rund (1966), and a treatment of invariance of functionals under local Lie groups of transformations, leading to conservation laws and Noether's theorem, is contained in Logan (1977). This latter material is also covered in the first edition of this text (J. D. Logan, 1987). An important generalization of the calculus of variations is optimal control theory, which is not covered in present text; see Pinch (1993) for an accessible introduction.

Gelfand, I. M. & Fomin, S. V. 1963. *Calculus of Variations*, Prentice-Hall, Englewood Cliffs, NJ (reprinted by Dover Publications, Inc., 2000).

Logan, J. D. 1987. *Applied Mathematics: A Contemporary Approach*, Wiley-Interscience, New York.

Logan, J. D. 1977. *Invariant Variational Principles*, Academic Press, New York

Pinch, E. R. 1993. *Optimal Control and the Calculus of Variations*, Oxford University Press, Oxford.

Rund, H. 1966. *The Hamilton-Jacobi Theory in the Calculus of Variations*, Van Nostrand, London.

van Brunt, B. 2004. *The Calculus of Variations*, Springer-Verlag, New York.

4

Eigenvalue Problems, Integral Equations, and Green's Functions

The topics in this chapter form the traditional core of applied mathematics—orthogonal expansions, boundary value problems, integral equations, and Green's functions. All of these subjects are interrelated in one of the most aesthetic theories in all of mathematics.

4.1 Orthogonal Expansions

4.1.1 Orthogonality

From calculus the reader is familiar with the idea of expanding a given function f in an infinite series of simpler functions. For example, if f has infinitely many derivatives at $x = x_0$, then its **Taylor series** about $x = x_0$ is

$$f(x) = \sum_{n=0}^{\infty} a_n (x - x_0)^n,$$

where the coefficients a_n, the Taylor coefficients, are given by $a_n = f^{(n)}(x_0)/n!$. For example, the Taylor series about $x_0 = 0$ for the exponential function is

$$e^x = 1 + x + \frac{1}{2!}x^2 + \frac{1}{3!}x^3 + \cdots,$$

and the series converges for all real numbers x. Another type of series is a Fourier series. If f is defined and integrable on the interval $[-\pi, \pi]$, then its **Fourier series**[1] is an infinite series of the form

$$f(x) \sim \frac{a_0}{2} + \sum_{n=1}^{\infty}(a_n \cos nx + b_n \sin nx),$$

where the coefficients are calculated by the formulae

$$a_n = \frac{1}{\pi}\int_{-\pi}^{\pi} f(x)\cos nx\, dx, \quad n = 0, 1, 2 \ldots,$$

$$b_n = \frac{1}{\pi}\int_{-\pi}^{\pi} f(x)\sin nx\, dx, \quad n = 1, 2 \ldots.$$

(The tilde in the formula denotes that the function f has the right side as its Fourier series, and it implies nothing at this point about convergence.) In the case of Taylor series the set of power functions $1, x, x^2, \ldots$ form the basis of the expansion, and in the Fourier series the trigonometric functions $1, \cos x, \cos 2x$, $\ldots$, $\sin x, \sin 2x, \ldots$ form the basis of the expansion. Each has its advantages. The first few terms of the Taylor series can provide a good approximation to f locally near $x = x_0$. The first few terms of a Fourier series can, on the other hand, provide a good global approximation to periodic signals, which are common in science and engineering.

In this section we expose the underlying idea of a Fourier expansion by allowing the basis functions to be any set of functions with an *orthogonality* property. The result is a general theory that incorporates Fourier series as a special case. Orthogonality (perpendicularity) is a notion that has its origins in vector analysis, but it has meaning in more general settings, such as with respect to sets of functions. We define the set of functions $L^2 \equiv L^2[a, b]$ to be the set all the real-valued functions f defined on the bounded interval $[a, b]$ that satisfy

$$\int_a^b |f(x)|^2 dx < \infty.$$

When this property holds, we say that f is **square integrable**[2] on the interval $[a, b]$, and we write $f \in L^2 = L^2[a, b]$. It turns out that L^2 is an infinite-dimensional linear space. In this function space we define a inner product,

---

[1] Joseph Fourier (1768–1930), a French mathematician, was one of the stars of the early 19th century. His fundamental work on heat conduction motivated the study of partial differential equations, and questions about the convergence of Fourier series spawned much of the analysis developed in that century.

[2] The integral here is, appropriately, the Lebesgue (1875–1941) integral studied in advanced analysis courses rather than the Riemann (1826–1866) integral, which is studied in elementary calculus; however, there will be no harm in interpreting the integrals in this text as a Riemann integrals since all of the functions that we examine will be Riemann integrable (and therefore Lebesgue integrable).

which is a generalization of the scalar product for vectors. If f, $g \in L^2$, then the **inner product** of f and g is a real number, denoted by (f, g), and is defined by

$$(f, g) = \int_a^b f(x)g(x)\, dx.$$

It is easy to check, by the properties of integrals, that the inner product satisfies the same properties of dot products of vectors:

(i) $(f, g) = (g, f)$.

(ii) $(f, g + h) = (f, g) + (f, h)$.

(iii) $(f, \alpha g) = \alpha(f, g), \quad \alpha \in \mathbb{R}$.

The scalar product induces a natural norm in the function space L^2. If $f \in L^2$, then we define the **norm** of f to be the nonnegative real number

$$\|f\| \equiv \sqrt{(f, f)} = \sqrt{\int_a^b |f(x)|^2\, dx}.$$

Observe that this definition is exactly analogous to the definition of the length of a vector in Euclidean space. We say that a set of functions f_1, f_2, $f_3, \ldots$ in $L^2[a, b]$ is **orthogonal** if $(f_i, f_j) = 0$ for $i \neq j$. For example, the set 1, $\cos x$, $\cos 2x, \ldots, \sin x, \sin 2x, \ldots$ is orthogonal on $[-\pi, \pi]$ because if we integrate the product of any two of these over the interval $[-\pi, \pi]$, we get zero. The reader should verify this statement using formulas from a calculus book. As another example, one can verify that

$$\int_0^\pi \sin mx \sin nx\, dx = 0, \quad n \neq m.$$

Thus the set of functions $\sin x, \sin 2x, \ldots$ is orthogonal on the interval $[0, \pi]$. The reader can check that the power functions 1, x, $x^2, \ldots$ that appear in the Taylor series do *not* form an orthogonal set on any interval $[a, b]$. Functions can be *normalized* in the same way that a vector can be turned into a unit vector, namely by dividing by its norm. For example, if $f \in L^2$, then the function h defined by $h(x) = f(x)/\|f\|$ has unit norm. If each function in an orthogonal set of functions has unit norm, then the set is called **orthonormal**. Thus, orthogonal sets of functions can always be turned into orthonormal sets.

In the case of complex-valued functions we make the following definitions. If u and v are functions, then we define their inner product by

$$(f, v) = \int_a^b f(x)\overline{v(x)}\, dx,$$

where the overbar denotes complex conjugation. The inner product is linear in each variable, and it has the properties $(f, g) = \overline{(g, f)}$, and $(\beta f, g) = \beta(f, g)$, where β is a scalar. The real number $\|f\| = \sqrt{(f, \overline{f})}$ is called the norm of f.

Example 4.1

The set of functions

$$\sin \frac{n\pi x}{l}, \quad n = 1, 2, \ldots,$$

is orthogonal on the interval $[0, l]$, and $\|\sin \frac{n\pi x}{l}\|^2 = l/2$. Therefore, the functions

$$\sqrt{\frac{2}{l}} \sin \frac{n\pi x}{l}$$

form an orthonormal set of functions on $[0, l]$. These functions clearly are square integrable.

The question that we ask is as follows. Can we expand an arbitrary function $f \in L^2$ in a series of given orthogonal functions $f_1, f_2, \ldots$? That is, can we write

$$f(x) = \sum_{n=1}^{\infty} c_n f_n(x) \tag{4.1}$$

for some set of constants c_n, and in what sense can we interpret the expansion? Intuitively, we think of f as a signal and $f_1, f_2, \ldots$ as fundamental modes. So, can we break a signal down into fundamental modes? In some instances, when f represents a signal, we will think of x as time, or we use the variable t.

When we write a series like (4.1), a natural question is, "What is meant by convergence of the infinite series?" There are several possible answers to this question, depending on how we measure the error of an N-term approximation $s_N(x) \equiv \sum_{n=1}^{N} c_n f_n(x)$ to $f(x)$. On one hand, by the **pointwise error** $E_N(x)$ we mean the expression

$$E_N(x) \equiv f(x) - s_N(x) \equiv f(x) - \sum_{n=1}^{N} c_n f_n(x).$$

Therefore, for a fixed number N (giving the number of terms), $E_N(x)$ is a *function* whose value gives the error at each point x. On the other hand, we can define an *integrated* error; the number e_N defined by

$$e_N \equiv \int_a^b |f(x) - s_N(x)|^2 \, dx \equiv \int_a^b |f(x) - \sum_{n=1}^{N} c_n f_n(x)|^2 \, dx$$

is called the **mean-square error** in the approximation. Note that e_N is just the square of the pointwise error, integrated over the interval $[a, b]$. Thus, the mean-square error is a type of average over all the pointwise errors. Each error expressions leads to a definition of convergence.

If for each *fixed* x in $[a, b]$ we have $\lim_{N \to \infty} E_N(x) = 0$, then we say that the infinite series (4.1) converges **pointwise** to f on the interval $[a, b]$; in this case, for each fixed x the series is a numerical series that converges to the numerical value $f(x)$, at that x; pointwise convergence means that the error goes to zero at each point. This is the type of convergence first encountered in calculus. We say that the infinite series (4.1) converges to f in the **mean-square** sense if

$$\lim_{N \to \infty} e_N = 0.$$

Mean-square convergence is also called *convergence in* $L^2[a, b]$. It requires that the integrated pointwise error-squared go to zero as more and more terms are taken. In advanced courses it is shown that pointwise convergence does not imply mean-square convergence, and conversely.

There is a stronger definition of convergence, called uniform convergence , that implies both mean-square and pointwise convergence. We say that the series (4.1) converges *uniformly* to $f(x)$ on $[a, b]$ if for any given tolerance $\varepsilon > 0$, we can find a number N (representing the number of terms) for which $|E_N(x)| < \varepsilon$ *for all* x in $[a, b]$. That is, the error can be made uniformly small over the entire interval by choosing the number of terms large enough. For uniform convergence, N (the number of terms) depends only on the tolerance ε and not on where in the interval the error is taken.

A novice may not appreciate the subtle differences in convergence and wonder why we have defined three types (actually, there are others). Suffice it to say that the error may not go to zero in a pointwise or uniform manner, but that does not mean the approximation is not useful; a weaker form of convergence, like mean-square convergence, may be all that is required.

The mean-square convergence criterion can be written simply in terms of the norm of a function. The series (4.1) converges to f in the mean-square sense if and only if $\lim_{N \to \infty} \|E_N\| = 0$. By the way, all of the definitions above are valid whether or not the functions $f_n(x)$ are mutually orthogonal.

Next suppose (4.1) holds in the sense of mean-square convergence, where f is a given function in L^2 and f_n is a given orthogonal sequence in L^2. How can we compute the coefficients c_n? Orthogonality is crucial. We will assume in this argument that the series in (4.1) can be integrated term by term (it can be proved). If we multiply both sides of (4.1) by f_m and integrate over the interval $[a, b]$, we get, using the inner product notation,

$$(f, f_m) = \int_a^b \sum_{n=1}^{\infty} c_n f_n(x) f_m(x)\, dx = \sum_{n=1}^{\infty} c_n (f_n, f_m) = c_m (f_m, f_m) = c_m \|f_m\|^2.$$

Observe that the infinite sum collapsed to a single term because of the orthogonality of the f_n. So the c_n (change the subscript m to n) are given by

$$c_n = \frac{1}{\|f_n\|^2}(f, f_n), \quad n = 1, 2, \dots. \tag{4.2}$$

Therefore, if (4.1) holds in the mean-square sense, then the coefficients c_n are given by formula (4.2). The c_n are called the **Fourier coefficients**, and when the coefficients are computed in this way, the orthogonal series (4.1) is called the **generalized Fourier series** for f.

If the series expansion (4.1) is valid in the mean-square sense and the f_n are *orthonormal*, then the Fourier coefficients (4.2) are given simply by

$$c_n = (f, f_n). \tag{4.3}$$

There is another useful property of Fourier coefficients. If $f(x)$ is a given function and $f_1, \dots, f_N$ is an orthonormal set of functions, we can ask how to best approximate f by a linear combination of the f_n. If a_n is any other sequence of coefficients, then

$$\left\| f - \sum_{n=1}^{N} c_n f_n \right\|^2 \le \left\| f - \sum_{n=1}^{N} a_n f_n \right\|^2. \tag{4.4}$$

That is, the mean-square error is minimized for the approximation with the Fourier coefficients c_n given by (4.3).

The demonstration of (4.4) is straightforward using the definition of norm and properties of the inner product. We have

$$
\begin{aligned}
\left\| f - \sum_{n=1}^{N} a_n f_n \right\|^2 &= \left(f - \sum_{n=1}^{N} a_n f_n, f - \sum_{n=1}^{N} a_n f_n \right) \\
&= (f, f) - 2\sum_{n=1}^{N} a_n (f, f_n) + \sum_{n=1}^{N} a_n^2 \\
&= (f, f) - 2\sum_{n=1}^{N} a_n c_n + \sum_{n=1}^{N} a_n^2 + \sum_{n=1}^{N} c_n^2 - \sum_{n=1}^{N} c_n^2 \\
&= (f, f) - \sum_{n=1}^{N} c_n^2 + \sum_{n=1}^{N} (a_n - c_n)^2 \\
&= \left\| f - \sum_{n=1}^{N} c_n f_n \right\|^2 + \sum_{n=1}^{N} (a_n - c_n)^2.
\end{aligned}
$$

Because the second term on the right is nonnegative, (4.4) is valid. If the f_n are not orthonormal, and just orthogonal, then the same result is true with the c_n given by (4.3).

In the last line of the preceding proof we used the fact that

$$\left\| f - \sum_{n=1}^{N} c_n f_n \right\|^2 = \|f\|^2 - \sum_{n=1}^{N} c_n^2,$$

which the reader should verify. This equality leads to another interesting inequality. Because the left side is nonnegative, we have

$$\sum_{n=1}^{N} c_n^2 \leq \|f\|^2.$$

This inequality is valid for each N, and so

$$\sum_{n=1}^{\infty} c_n^2 \leq \|f\|^2,$$

which is **Bessel's inequality**. This shows the series of squared Fourier coefficients converges, and thus $c_n \to 0$ as $n \to \infty$. Therefore the Fourier coefficients get smaller and smaller.

If the f_n are an orthonormal set and we form the generalized Fourier series (4.1) with c_n the Fourier coefficients given by (4.3), does the series automatically converge to f in one sense or another? The answer to the question is yes, in the mean-square sense, provided that the orthonormal sequence f_n is complete. An orthonormal system f_n is said to be *complete* if $(f, f_n) = 0$ for all n implies that $f \equiv 0$. Thus, an orthonormal sequence f_n is complete if the only function having all its Fourier coefficients vanish is the zero function. Sometimes it is difficult to show completeness, and we shall usually just state whether a given orthonormal sequence in complete. It is shown in advanced treatments that completeness is equivalent to the property that *any* function $f \in L^2$ can be expanded in a generalized Fourier series (4.1), where the c_n are the Fourier coefficients, and that the series converges to f in the mean-square sense. Furthermore, it can be proved that completeness is equivalent to strict equality holding in the Bessel inequality; that is, f_n is complete if and only if for each $f \in L^2$,

$$\sum_{n=1}^{\infty} c_n^2 = \|f\|^2.$$

This equation is **Parseval's equality**. Pointwise convergence results and stronger uniform convergence results are more difficult to obtain, and they usually require continuity and smoothness assumptions on the function f.

We can carry the modal interpretation of the generalized Fourier series still further. Thinking of f as a signal and the orthonormal set f_n as the fundamental modes, the Fourier coefficient c_n determines the contribution of the nth mode, and the generalized Fourier series is the decomposition of the signal into fundamental modes. The sequence of squared coefficients, c_1^2, c_2^2, $c_3^2, \ldots$, is called the **energy spectrum**, and c_n^2 is called the energy of the nth mode; by Parseval's equality, the *total energy* in the signal is $\|f\|^2$.

EXERCISES

1. Verify that the set of functions $\cos(n\pi x/l)$, $n = 0, 1, 2 \ldots$, form an orthogonal set on the interval $[0, l]$. If

$$f(x) = \sum_{n=0}^{\infty} c_n \cos\left(\frac{n\pi x}{l}\right)$$

   in the mean-square on $[0, l]$, what are the formulae for the c_n? (This series is called the **Fourier cosine series** on $[0, l]$. Find the Fourier cosine series for $f(x) = 1 - x$ on $[0, 1]$.

2. Let f be defined and integrable on $[0, l]$. The orthogonal expansion

$$f(x) = \sum_{n=1}^{\infty} b_n \sin \frac{n\pi x}{l}, \quad b_n = \frac{2}{l} \int_0^l f(x) \sin\left(\frac{n\pi x}{l}\right) dx,$$

   is called the **Fourier sine series** for f on $[0, l]$. Find the Fourier sine series for $f(x) = \cos x$ on $[0, \pi/2]$. What is the Fourier sine series of $f(x) = \sin x$ on $[0, \pi]$?

3. For $f, g \in L^2[a, b]$, prove the **Cauchy–Schwarz inequality**

$$|(f, g)| \leq \|f\| \, \|g\|.$$

   Hint: Define $q(t) = (f + tg, f + tg)$ for any real number t. Using the rules of inner product, expand this expression to obtain a quadratic polynomial in t; finally, look at the discriminant of the polynomial.

4. (Gram–Schmidt orthogonalization) A result from linear algebra is that any set of linearly independent vectors may be turned into a set of orthogonal vectors by the **Gram–Schmidt** orthogonalization process. The same process works for functions in $L^2[a, b]$. Let $f_1, f_2, f_3, \ldots$ be an independent set of functions in L, and define the set g_n by

$$g_1 = f_1, \quad g_2 = f_2 - \frac{(f_2, g_1)}{\|g_1\|^2} g_1, \quad g_3 = f_3 - \frac{(f_3, g_2)}{\|g_2\|^2} g_2 - \frac{(f_3, g_1)}{\|g_1\|^2} g_1, \ldots.$$

   Show that g_n is an orthogonal set.

5. (Lengendre polynomials) The functions 1, x, x^2, x^3 are independent functions on the interval $[-1, 1]$. Use the Gram–Schmidt method to generate a sequence of four orthogonal polynomials $P_0(x), \ldots, P_3(x)$ on $[-1, 1]$. Find an approximation of e^x on $[-1, 1]$ of the form

$$c_0 P_0(x) + c_1 P_1(x) + c_2 P_2(x) + c_3 P_3(x)$$

that is best in the mean-square sense, and plot e^x and the approximation. Compute the pointwise error, the maximum pointwise error over $[-1, 1]$, and the mean-square error.

6. (Haar wavelets) Let ϕ be a function defined by $\phi(x) = 1$ for $x \in [0, 1)$, and $\phi(x) = 0$ otherwise. Let $\psi(x) = \phi(2x) - \phi(2x - 1)$. Then the Haar wavelets are the functions
$$\psi_{mn}(x) = 2^{m/2} \psi(2^m x - n),$$
for $m, n = 0, \pm 1, \pm 2, \ldots$. Sketch a graph of $\psi(x)$, and then sketch a graph of $\psi_{mn}(x)$ for $m, n = 0, \pm 1, \pm 2$. Generally, what is the graph of $\psi_{mn}(x)$? If f is square integrable and
$$f(x) = \sum_{m=-\infty}^{\infty} \sum_{n=-\infty}^{\infty} c_{mn} \psi_{mn}(x),$$
find coefficients c_{mn}.

7. (Hermite polynomials) Consider the differential equation
$$-y'' + x^2 y = Ey, \quad x \in \mathbb{R}, \tag{4.5}$$
and the Hermite polynomials
$$H_n(x) = (-1)^n e^{x^2} \frac{d^n}{dx^n} e^{-x^2}, \; n = 0, 1, 2, \ldots.$$

a) Find $H_0(x), \ldots, H_4(x)$. Why is $H_n(x)$ always a polynomial?

b) Verify that $Y_n(x) = H_n(x) e^{-x^2/2}$ is a solution of (4.5) when $E = 2n+1$. Hint: You will need to show $H_n' = 2nH_{n-1}$.

c) Show that $\int_{-\infty}^{\infty} Y_n Y_m dx = 0$, $m \neq n$, and thus the Y_n are orthogonal on the interval $(-\infty, \infty)$. Hint: Use the fact that both Y_n and Y_m satisfy (3.16) for appropriate E.

d) If a function $f(x)$ is represented by $f(x) = \sum_{n=0}^{\infty} c_n Y_n(x)$, how would you expect to find the c_n? (Assume uniform convergence.) Take $f(x) = 1/\sqrt{1 + x^4}$ and find $c_0, \ldots, c_4$.

e) The differential equation (4.5) is a time-independent Schrödinger equation. Thus $E = 2n + 1$ is interpreted as the energy associated with the probability density $Y_n(x)^2$, appropriately normalized so that $\int Y_n^2 dx = 1$. Sketch the normalized probability densities for $n = 0, 1, 2$.

4.1.2 Classical Fourier Series

We introduced the idea of representing a given function $f(x)$ in terms of an infinite series of orthogonal functions $f_n(x)$. Now we focus those concepts on a special set of orthogonal functions where the f_n are given by sines and cosines. The resulting series is called a (classical) Fourier series. Classical Fourier series are a special case of the orthogonal series discussed in the last section. They were introduced by Joseph Fourier in the early 1800s to solve partial differential equations in heat transfer, and.we discuss this important application in Chapter 5. Questions about convergence of Fourier series spawned much of the development of classical analysis in the 1800s, and the role of Fourier analysis is key in signal processing, remote sensing, and other engineering areas.

We work on an arbitrary symmetric interval $[-l, l]$ about the origin. If f is an integrable function on $[-l, l]$, then its **Fourier series** is the trigonometric polynomial

$$\frac{a_0}{2} + \sum_{n=1}^{\infty} \left(a_n \cos \frac{n\pi x}{l} + b_n \sin \frac{n\pi x}{l} \right), \tag{4.6}$$

where the **Fourier coefficients** a_n and b_n are given by the formulas

$$a_n = \frac{1}{l} \int_{-l}^{l} f(x) \cos \frac{n\pi x}{l} dx, \quad n = 0, 1, 2, \ldots,$$

$$b_n = \frac{1}{l} \int_{-l}^{l} f(x) \sin \frac{n\pi x}{l} dx, \quad n = 1, 2, \ldots.$$

The set of functions

$$1, \cos \frac{n\pi x}{l}, \sin \frac{n\pi x}{l}, \quad n = 1, 2, \ldots, \tag{4.7}$$

are orthogonal on the interval $[-l, l]$. They play the role of the $f_n(x)$ in formula 4.1), and a_n and b_n are the coefficients c_n. It is shown in more advanced texts that the set of functions (4.7) is complete, and therefore the Fourier series (4.6) converges in the mean-square sense to f when $f \in L^2[-l, l]$.

Example 4.2

Let $f(x) = x$ on $-l \le x \le l$. The Fourier coefficients are easily computed to be

$$a_n = \frac{1}{l} \int_{-l}^{l} x \cos \frac{n\pi x}{l} dx = 0, \quad n = 0, 1, 2, \ldots,$$

$$b_n = \frac{1}{l} \int_{-l}^{l} x \sin \frac{n\pi x}{l} dx = \frac{2l}{n\pi} (-1)^{n+1}, \quad n = 1, 2, \ldots.$$

Therefore the Fourier series for f is

$$\frac{2l}{\pi} \left(\sin \frac{\pi x}{l} - \frac{1}{2} \sin \frac{2\pi x}{l} + \frac{1}{3} \sin \frac{3\pi x}{l} - \cdots \right).$$

We make two observations. First, at $x = \pm l$ the series converges to zero and thus does not converge to $f(x)$ at these two points. So the series does not converge pointwise to $f(x)$ on the interval $[-l, l]$. It does, however, converge pointwise to $f(x) = x$ on $(-l, l)$. We do know that the series converges to f in the mean-square sense. Second, the derived series, obtained by differentiating term by term, does not converge at all, much less to the derivative $f'(x) = 1$. So the series cannot be differentiated term by term.

For the Fourier series (4.6) the nth **harmonic** is

$$a_n \cos \frac{n\pi x}{l} + b_n \sin \frac{n\pi x}{l}.$$

Each harmonic in the series has a higher frequency, and thus more oscillations, than the preceding harmonic. The **frequency spectrum** of the series is the sequence of numbers ω_n defined by

$$\omega_0 = \frac{|a_0|}{\sqrt{2}}, \quad \omega_n = \sqrt{a_n^2 + b_n^2} \quad (n \geq 1).$$

The frequency spectrum measures the contribution of the various harmonics in the decomposition of f. The numbers ω_n^2, $n \geq 0$, form the **energy spectrum**. One can show that that Parseval's formula takes the form

$$\frac{a_0^2}{2} + \sum_{n=1}^{\infty} (a_n^2 + b_n^2) = \frac{1}{l} ||f||^2.$$

A Fourier series simplifies considerably if f is an even or an odd function. Observe that $\sin \frac{n\pi x}{l}$ is an odd function and $\cos \frac{n\pi x}{l}$ is an even function; moreover, an even function times an odd function is odd. Therefore, if f is an even function, then the product $f(x) \sin \frac{n\pi x}{l}$ is an odd function; so all of the coefficients b_n are zero because an odd function integrated over a symmetric interval about the origin is zero. Hence, if f is an even function, then its Fourier series reduces to a cosine series of the form

$$\frac{a_0}{2} + \sum_{n=1}^{\infty} a_n \cos \frac{n\pi x}{l}.$$

Similarly, if f is an odd function, then the coefficients a_n vanish, and the Fourier series reduces to a sine series

$$\sum_{n=1}^{\infty} b_n \sin \frac{n\pi x}{l}.$$

Fourier series are a valuable tool in signal processing and data storage. Let us explore this idea in more detail. For this discussion it is helpful to think of the variable x as representing time. We observe that the Fourier series (4.6), although computed only on the interval $[-l, l]$, is a $2l$-periodic function[3] because the sines and cosines are $2l$-periodic. Thus, it repeats its values in every interval of length $2l$. Therefore, if f is defined on all of $\mathbb{R}$ and is $2l$-periodic, we can represent it by its Fourier series on all of $\mathbb{R}$. Hence, if f is some given signal or data set $2l$-periodic in time (say a signal from an electrical circuit, an electrocardiogram, or an extra-terrestrial), then we could digitize the signal and save it by storing the Fourier coefficients, or some finite subset of them; knowledge of those coefficients allows us reproduce the signal via (4.6) and the Fourier coefficients.

Finally, we make some brief comments about convergence of Fourier series. We have already noted that if f is square integrable, then mean-square convergence is automatic. Pointwise convergence requires additional assumptions about the smoothness of f. The case when the graph of f is made up of finitely many continuous, smooth pieces will cover most of the interesting functions encountered in science and engineering. A function f is **piecewise continuous** on $[a, b]$ if it is continuous except possibly at a finite number of of points in $[a, b]$ where it has simple jump discontinuities; f has a simple jump discontinuity at $x = c$ if both one-sided limits of $f(x)$ exist and are finite at c. The function may or may not be defined at a jump discontinuity. We say f is **piecewise smooth** on $[a, b]$ if both f and f' are piecewise continuous on $[a, b]$, and we say f is piecewise smooth on $(-\infty, \infty)$ if it is piecewise smooth on each bounded subinterval $[a, b]$ of the real line. Then the basic pointwise convergence theorem can be stated as follows.

Theorem 4.3

If f is piecewise smooth on $[-l, l]$ and otherwise $2l$-periodic, then its Fourier series (4.6) converges pointwise for all $x \in \mathbb{R}$ to the value $f(x)$ if f is continuous at x, and to the average value of its left and right limits at x, namely $\frac{1}{2}(f(x-) + f(x+))$, if f is discontinuous at x.

To get stronger convergence results, like uniform convergence, additional smoothness conditions on f are required. Observe that continuity of f is not enough to guarantee pointwise convergence. Incredible as it may appear, there are continuous functions whose Fourier series diverge at every point!

---

[3] A function f is P-periodic, or periodic of period P, if it repeats itself in every interval of length P; that is, $f(x + P) = f(x)$ for all x.

EXERCISES

1. Consider a triangular wave given analytically by: $f(x) = x + 1$ if $-1 < x \leq 0$; $f(x) = 1 - x$ if $0 < x \leq 1$; and otherwise 2-periodic. Compute its Fourier series, and plot f and two-term, three-term, and four-term Fourier approximations.

2. Find the Fourier series for $f(x) = x^2$ on $[-\pi, \pi]$. Sketch a graph of the function defined on $\mathbb{R}$ to which the Fourier series converges pointwise. Use the Fourier series to show

$$\frac{\pi^2}{12} = 1 - \frac{1}{4} + \frac{1}{9} - \frac{1}{16} + \cdots .$$

Graph the frequency spectrum.

3. Consider the signal defined by $f(t) = t + 1$ for $-2\pi < t \leq 0$ and $f(t) = t$ for $0 < t \leq 2\pi$, and otherwise 4π-periodic. Plot the signal, and find its Fourier series. Plot the frequency spectrum. To what value does the Fourier series converge at $t = 0$, $t = 2\pi$, and $t = \pi$? Plot the sum of the first four harmonics and compare it to f.

4. Show that the Fourier series for $f(x) = \cos ax$ on $[-\pi, \pi]$, where a is not an integer, is

$$\cos ax = \frac{2a \sin a\pi}{\pi} \left(\frac{1}{2a^2} + \sum_{n=1}^{\infty} (-1)^n \frac{\cos nx}{a^2 - n^2} \right).$$

Show

$$\csc a\pi = \frac{1}{a\pi} + \frac{2a}{\pi} \sum_{n=1}^{\infty} (-1)^n \frac{1}{n^2 - a^2}.$$

5. Let $f(x) = -\frac{1}{2}$ on $-\pi < x \leq 0$ and $f(x) = \frac{1}{2}$ on $0 \leq x \leq \pi$. Show that the Fourier series for f is

$$\sum_{n=1}^{\infty} \frac{2}{(2n-1)\pi} \sin(2n-1)x.$$

If $s_N(x)$ denotes the sum of the first N terms, sketch graphs of $s_1(x)$, $s_3(x)$, $s_7(x)$, and $s_{10}(x)$ and compare with $f(x)$. Observe that the approximations overshoot $f(x)$ in a neighborhood of $x = 0$, and the overshoot is not improved regardless of how many terms are taken in the approximation. This overshoot behavior of Fourier series near a discontinuity is called the **Gibbs' phenomenon**.

4.2 Sturm–Liouville Problems

An ordinary differential equation of the form

$$-(p(x)y')' + q(x)y = \lambda y, \quad a < x < b, \tag{4.8}$$

where λ is constant and p and q are functions is called a **Sturm–Liouville** type differential equation. On a bounded interval, the equation is usually accompanied by boundary conditions on $y(x)$ of the form

$$\alpha_1 y(a) + \alpha_2 y'(a) = 0, \quad \beta_1 y(b) + \beta_2 y'(b) = 0. \tag{4.9}$$

Here the constants α_1 and α_2 are not both zero, and the constants β_1 and β_2 are not both zero (i.e., the boundary condition at an endpoint does not collapse). Two special cases of the boundary conditions (4.9) are

$$y(a) = 0, \quad y(b) = 0, \quad \text{(Dirichlet conditions)}$$

and

$$y'(a) = 0, \quad y'(b) = 0 \quad \text{(Neumann conditions)} \,.$$

If the interval $[a, b]$ is bounded, if p, p', and q are continuous functions on the interval $[a, b]$, and p is never zero in $[a, b]$, then the boundary value problem (4.8)–(4.9) is called a **regular Sturm–Liouville problem** (SLP) . Otherwise it is called **singular**. Singular problems arise when the interval $[a, b]$ is infinite or when $p(x_0) = 0$ for some $x_0 \in [a, b]$. SLPs are named after J.C.F. Sturm and J. Liouville, who studied such problems in the mid-1800s.

A regular SLP will not have a nontrivial solution for every value of the constant λ. (Clearly, the zero function $y(x) \equiv 0$ is always a trivial solution.) A value of λ for which there is a nontrivial solution of (4.8)–(4.9) is called an **eigenvalue**, and the corresponding solution is called an **eigenfunction**. Notice that any constant multiple of an eigenfunction gives another (not independent) eigenfunction. The interesting fact about regular SLPs is that they have an infinite number of eigenvalues, and the corresponding eigenfunctions form a complete, orthogonal set, which makes orthogonal expansions possible. This is a key idea in applied mathematics, physics, and engineering.

Example 4.4

The SLP

$$\begin{aligned} -y''(x) &= \lambda y(x), \quad 0 < x < \pi, \\ y(0) &= 0, \quad y(\pi) = 0, \end{aligned}$$

has eigenvalues $\lambda = \lambda_n = n^2$ and corresponding eigenfunctions $y = y_n(x) = \sin nx$, $n = 1, 2, \ldots$. Thus the problem has solution $y = \sin x$ when $\lambda = 1$, $y = \sin 2x$ when $\lambda = 4$, and so on. One way to find the eigenvalues and eigenfunctions is to go through a case argument by separately considering $\lambda = 0$, $\lambda > 0$, and $\lambda < 0$. We prove later that λ cannot be a complex number. We examine different cases because the solution of the differential equation has a different form depending on the sign of λ. If $\lambda = 0$, then the equation has the form $y'' = 0$, whose general solution is the linear function $y(x) = Ax + B$. But $y(0) = 0$ implies $B = 0$, and $y(\pi) = 0$ implies $A = 0$. Thus we get only the trivial solution in this case and so $\lambda = 0$ is *not* an eigenvalue. If $\lambda < 0$, say for definiteness $\lambda = -k^2$, then the equation has the form $y'' - k^2 y = 0$ with general solution

$$y(x) = Ae^{kx} + Be^{-kx}.$$

The boundary conditions require

$$\begin{aligned} y(0) &= A + B = 0, \\ y(\pi) &= Ae^{k\pi} + Be^{-k\pi} = 0, \end{aligned}$$

which in turn force $A = B = 0$, and therefore we obtain only the trivial solution in this case. Thus, there are no negative eigenvalues. Finally, consider the case $\lambda > 0$, or $\lambda = k^2$. Then the differential equation takes the form

$$y'' + k^2 y = 0,$$

which has general solutions

$$y(x) = A \cos kx + B \sin kx.$$

First, $y(0) = 0$ forces $A = 0$. Thus $y(x) = B \sin kx$. The right boundary condition yields

$$y(\pi) = B \sin k\pi = 0.$$

But now we are not required to take $B = 0$; rather, we can select k to make this equation hold. Because the sine function vanishes at multiples of π, we can choose $k = n$, a nonzero integer (we have already considered the case when $k = 0$), or

$$\lambda = \lambda_n = n^2, \quad n = 1, 2, \ldots.$$

The corresponding solutions, or eigenfunctions, are

$$y = y_n(x) = \sin nx, \quad n = 1, 2, \ldots.$$

Here we have arbitrarily selected the constant $B = 1$ for each eigenfunction; we can always multiply eigenfunctions by a constant to get another (not independent) eigenfunction. Alternately, we could write the eigenfunctions as, for example, $y_n(x) = B_n \sin nx$, where a different constant is chosen for each eigenfunction. Frequently the constants B_n are chosen to normalize the eigenfunction, i.e., force $||y_n|| = 1$. Observe that the set of eigenfunctions are orthogonal on the interval $[0, \pi]$; they are also complete, and any square integrable function can be expanded as a Fourier series in the eigenfunctions, and the convergence is mean-square.

We can always carry out a case argument as in the preceding example to determine the eigenvalues and eigenfunctions of an SLP. This calculation can be tedious, and so it is advantageous to have some general results that allow us to reject certain cases immediately, without detailed calculation. We discuss such matters in the sequel. For the present we state, without proof, the central result regarding SLPs.

Theorem 4.5

The regular Sturm–Liouville problem (4.8)–(4.9) has infinitely many eigenvalues λ_n, $n = 1, 2, \ldots$. The eigenvalues are real, and $\lim_{n \to \infty} |\lambda_n| = +\infty$. The eigenfunctions corresponding to distinct eigenvalues are orthogonal, and the set of all eigenfunctions $y_n(x)$ are complete in the sense that every square-integrable function f on $[a, b]$ can be expanded as

$$f(x) = \sum_{n=1}^{\infty} c_n y_n(x)$$

in the mean-square sense.

For the proof of existence of the eigenvalues we refer to the references (see, for example, Birkhoff and Rota, 1978). But it is straightforward to demonstrate the orthogonality property. To this end, let λ_1 and λ_2 be distinct eigenvalues with corresponding eigenfunctions $y_1(x)$ and $y_2(x)$. This means

$$-(py_1')' + qy_1 = \lambda_1 y_1, \quad -(py_2')' + qy_2 = \lambda_2 y_2.$$

Multiplying the first equation by y_2, multiplying the second by y_1, subtracting, and then integrating the result over the interval $[a, b]$ gives

$$(\lambda_1 - \lambda_2) \int_a^b y_1 y_2 \, dx = \int_a^b [-y_2 (py_1')' + y_1 (py_2')'] \, dx.$$

The integral on the right can be calculated by integrating each term by parts, or by noticing that the integrand is an exact derivative. That is,

$$\frac{d}{dx} [p(y_1 y_2' - y_2 y_1')] = -y_2 (py_1')' + y_1 (py_2')'.$$

Then the fundamental theorem of calculus gives

$$(\lambda_1 - \lambda_2) \int_a^b y_1 y_2 \, dx = [p(y_1 y_2' - y_2 y_1')] \, |_a^b .$$

For simple Dirichlet or Neumann boundary conditions, it is clear that the right side is zero, since both eigenfunctions satisfy the boundary conditions. For the general, mixed boundary conditions (4.9) the right side is zero as well, and we leave this verification to the reader. Therefore it follows that

$$\int_a^b y_1 y_2 \, dx = 0,$$

which means that y_1 and y_2 are orthogonal functions.

The eigenvalues of a regular SLP cannot be complex. We will demonstrate this fact for the special SLP (a Schrödinger-type equation) with Dirichlet boundary conditions

$$-y'' + q(x)y = \lambda y, \quad a < x < b, \tag{4.10}$$

$$y(a) = y(b) = 0. \tag{4.11}$$

Let λ be an eigenvalue and $y(x)$ a corresponding eigenfunction. Let an *overbar* denote complex conjugation. Then, taking the complex conjugate of the differential equation gives

$$-\overline{y}'' + q(x)\overline{y} = \overline{\lambda}\,\overline{y}.$$

We are assuming that q is a real-valued function. Now multiply the given differential equation by $\overline{y}$, multiply the last equation by y, and subtract the results and integrate to obtain

$$-\int_a^b [y''\overline{y} - \overline{y}''y] \, dx = (\lambda - \overline{\lambda}) \int_a^b y\overline{y} \, dx.$$

The integral on the left side is zero because

$$\int_a^b [y''\overline{y} - \overline{y}''y] \, dx = (y'\overline{y} - \overline{y}'y) \, |_a^b = 0.$$

We have used $\overline{y}(a) = \overline{y}(b) = 0$, which comes from taking the complex conjugate of the boundary conditions. Therefore,

$$(\lambda - \overline{\lambda}) \int_a^b y\overline{y} \, dx = 0.$$

Because the integral is positive (the integrand $y\overline{y} = |y|^2$ is positive), we must have $\lambda = \overline{\lambda}$. A number that equals its complex conjugate is real, and so λ is real.

Another issue concerns the sign of eigenvalues. The following **energy argument** is sometimes useful in showing that the eigenvalues are of one sign. We examine the problem (4.10)–(4.11) where q is a positive continuous function. Multiplying the differential equation by y and integrating gives

$$-\int_a^b yy''dx + \int_a^b qy^2\,dx = \lambda\int_a^b y^2dx.$$

The first integral can be integrated by parts to obtain

$$-yy'\big|_a^b + \int_a^b y'^2dx + \int_a^b qy^2dx = \lambda\int_a^b y^2dx.$$

Under Dirichlet boundary conditions (4.11) the boundary terms in the last equation are zero. Hence

$$\int_a^b y'^2dx + \int_a^b qy^2dx = \lambda\int_a^b y^2dx.$$

Because q is positive, the second integral on the left side is positive. Thus, the eigenvalues λ for (4.10)–(4.11) must be positive.

EXERCISES

1. Show that the SLP

$$\begin{aligned}-y''(x) &= \lambda y(x), \quad 0 < x < l,\\ y(0) &= 0, \ \ y(l) = 0,\end{aligned}$$

has eigenvalues $\lambda_n = n^2\pi^2/l^2$ and corresponding eigenfunctions $y_n(x) = \sin(n\pi x/l)$, $n = 1, 2, \ldots$.

2. Show that the SLP

$$\begin{aligned}-y''(x) &= \lambda y(x), \quad 0 < x < l,\\ y'(0) &= 0, \ \ y(l) = 0,\end{aligned}$$

with mixed Dirichlet and Neumann boundary conditions has eigenvalues

$$\lambda_n = \left(\frac{(1+2n)\pi}{2l}\right)^2$$

and corresponding eigenfunctions

$$y_n(x) = \cos\frac{(1+2n)\pi x}{2l}$$

for $n = 0, 1, 2, \ldots$.

3. Find the eigenvalues and eigenfunctions for the problem with *periodic* boundary conditions:

$$-y''(x) = \lambda y(x), \quad 0 < x < l,$$
$$y(0) = y(l), \quad y'(0) = y'(l).$$

4. Consider the SLP

$$-y'' = \lambda y, \quad 0 < x < 1,$$
$$y(0) + y'(0) = 0, \quad y(1) = 0.$$

Is $\lambda = 0$ an eigenvalue? Are there any negative eigenvalues? Show that there are infinitely many positive eigenvalues by finding an equation whose roots are those eigenvalues, and show graphically that there are infinitely many roots.

5. Show that the SLP

$$-y'' = \lambda y, \quad 0 < x < 2,$$
$$y(0) + 2y'(0) = 0, \quad 3y(2) + 2y'(2) = 0,$$

has exactly one negative eigenvalue. Is zero an eigenvalue? How many positive eigenvalues are there?

6. For the SLP

$$-y'' = \lambda y, \quad 0 < x < l,$$
$$y(0) - ay'(0) = 0, \quad y(l) + by'(l) = 0,$$

show that $\lambda = 0$ is an eigenvalue if and only if $a + b = -l$.

7. Use an energy argument to show that the eigenvalues for the problem

$$-(x^2 y')' = \lambda y, \quad 1 < x < e,$$
$$y(1) = y(e) = 0.$$

must be nonnegative. Find the eigenvalues and eigenfunctions.

8. Find eigenvalues and eigenfunctions for the problem

$$-y'' - 2by' = \lambda y, \quad 0 < x < 1, \ b > 0,$$
$$y(0) = y(1) = 0.$$

9. What, if anything, can be deduced about the sign of the eigenvalues for the problem

$$-y'' + xy = \lambda y, \quad 0 < x < 1,$$
$$y(0) = y(1) = 0.$$

What are the eigenvalues and eigenfunctions? (Use a computer algebra system to investigate this problem.)

10. Consider the regular problem

$$-y'' + q(x)y = \lambda y, \quad 0 < x < l,$$
$$y(0) = y(l) = 0,$$

where $q(x) > 0$ on $[0, l]$. Show that if λ and y are an eigenvalue and eigenfunction pair, then

$$\lambda = \frac{\int_0^l (y'^2 + qy^2)\, dx}{||y||^2}.$$

Is $\lambda > 0$? Can $y(x) = $ constant be an eigenfunction?

4.3 Integral Equations

4.3.1 Introduction

An integral equation is an equation where the unknown function occurs under an integral sign. Integral equations arise in the analysis of differential equations, and, in fact, many initial and boundary value problems for differential equations may be reformulated as integral equations, and vice-versa. But integral equations naturally arise in modeling as well. The two types of linear equations that we discuss are the **Fredholm equation**

$$\int_a^b k(x, y)u(y)\, dy - \lambda u(x) = f(x), \quad a \leq x \leq b, \tag{4.12}$$

and the **Volterra equation**

$$\int_a^x k(x, y)u(y)\, dy - \lambda u(x) = f(x), \quad a \leq x \leq b. \tag{4.13}$$

Here u is the unknown function, f is a given continuous function, and λ is a parameter. The given function k is called the **kernel**, and we assume that k is a continuous function on the square $a \leq x \leq b$, $a \leq y \leq b$. The interval $[a, b]$ is assumed to be finite. The difference between the Fredholm and Volterra equations is the variable upper limit of integration in the Volterra equation. By a solution of (4.12) or (4.13) we mean a continuous function $u = u(x)$ defined on the interval $[a, b]$ that, when substituted into the equation, reduces the equation to an identity on the interval. If $f \equiv 0$, then the equation is called **homogeneous**; otherwise it is **nonhomogeneous**. If $\lambda \equiv 0$, then the unknown appears only under the integral sign and the equation is said to be of the **first kind**; otherwise it is of the **second kind**. If $k(x, y) = k(y, x)$ then the kernel is said to be **symmetric**. Although (4.12) and (4.13) are similar, they require different methods of analysis. Most of the discussion deals with continuous solutions to integral equations; however, a more complete theory can be developed for square integrable functions.

We can streamline the discussion by introducing integral operator notation. For example, if we define the Fredholm operator K by

$$(Ku)(x) = \int_a^b k(x, y)u(y) \, dy, \tag{4.14}$$

then (4.12) can be written simply as

$$Ku - \lambda u = f, \tag{4.15}$$

where we have, as is common practice, suppressed the independent variable x. We think of K as operating on a function u to produce another function Ku. If the kernel k is continuous, then K maps continuous functions to continuous functions. Ultimately, we want to determine the solution (if any), which could be written symbolically in the form

$$u = (K - \lambda I)^{-1}f.$$

If this were true, then $(K - \lambda I)^{-1}$ would be the inverse of $K - \lambda I)$, and we would have succeeded in defining the inverse of an integral operator.

In the same way that we consider eigenvalue problems for matrices and for differential equations (Sturm–Liouville problems), we can consider eigenvalue problems for integral equations. A number λ is called an eigenvalue for the integral operator K if there exists a nontrivial solution u to the integral equation

$$Ku = \lambda u. \tag{4.16}$$

The set of eigenvalues form the **spectrum** of K, and the (geometric) **multiplicity** of an eigenvalue is the dimension of the function space spanned by its corresponding eigenfunctions. Much of the theory of integral equations deals with *solvability* questions for the Fredholm equation (4.15) (or the Volterra equation) and the *solution to the eigenvalue problem* (4.16).

To understand the mathematical issues in integral equations, we retreat to a familiar problem in linear algebra. There we encounter the algebraic eigenvalue problem: Given an $n \times n$ matrix A, determine numbers λ for which the matrix equation

$$A\mathbf{u} = \lambda\mathbf{u} \qquad (4.17)$$

has a nontrivial solution $\mathbf{u} \in \mathbb{R}^n$. The values of λ for which a solution $\mathbf{u}$ exists are the **eigenvalues**, and the corresponding solutions $\mathbf{u}$ are the associated **eigenvectors**. Eigenvalues may be real or complex numbers.

There are many reasons the algebraic eigenvalue problem is important, but we focus our thinking on a specific notion, namely how solving the eigenvalue problem (4.17) helps us to solve other linear systems, for example, the nonhomogeneous problem. We know from matrix theory that there are n eigenvalues, not necessarily distinct. Suppose, for our discussion, that the matrix A is a real, symmetric matrix ($A = A^{\mathrm{T}}$) and has n real, nonzero, distinct eigenvalues $\lambda_1, \ldots, \lambda_n$. Corresponding to each eigenvalue λ_i is an eigenvector $\mathbf{e}_i$, and the set of eigenvectors forms a linearly independent set. Moreover, because A is symmetric, the eigenvectors are mutually perpendicular, or $\mathbf{e}_i \cdot \mathbf{e}_j = 0$ for $i \neq j$. Dividing each vector by its length, we can always arrange for each eigenvectors to have unit length, that is, $\mathbf{e}_i \cdot \mathbf{e}_i = 1$. Therefore, the eigenvectors from an orthonormal basis for the space $\mathbb{R}^n$; this means that each vector $\mathbf{u}$ in $\mathbb{R}^n$ can be written uniquely as a linear combination of the $\mathbf{e}_i$, or

$$\mathbf{u} = \sum_{i=1}^{n} c_i \mathbf{e}_i. \qquad (4.18)$$

The c_i are the coordinates of $\mathbf{u}$ in this orthonormal basis and are given by $c_i = \mathbf{u} \cdot \mathbf{e}_i$, which is the projection of $\mathbf{u}$ onto the ith eigenvector.

The point we make is that knowledge of the eigenvalues and eigenvectors can be used to solve other algebraic problems. Take, for example, the nonhomogeneous linear system

$$A\mathbf{u} = \mu\mathbf{u} + \mathbf{f}, \qquad (4.19)$$

where $\mathbf{f}$ is a given vector, and μ is a given constant not equal to any of the eigenvalues of A. If a solution $\mathbf{u}$ exists, it must have the form (4.18) for some choice of the constants c_i, which are to be determined. Further, since $\mathbf{f}$ is a given vector, it has a known expansion in terms of the eigenvectors,

$$\mathbf{f} = \sum_{i=1}^{n} f_i \mathbf{e}_i,$$

where the coefficients $f_i = \mathbf{f} \cdot \mathbf{e}_i$ are known. Substituting these two expressions for $\mathbf{u}$ and $\mathbf{f}$ into (4.19) gives

$$A\left(\sum_{i=1}^{n} c_i \mathbf{e}_i\right) = \mu \sum_{i=1}^{n} c_i \mathbf{e}_i + \sum_{i=1}^{n} f_i \mathbf{e}_i.$$

Combining terms,

$$\sum_{i=1}^{n} c_i A e_i = \sum_{i=1}^{n} (\mu c_i + f_i) e_i.$$

But $A e_i = \lambda_i e_i$, because λ_i and e_i form an eigen-pair. Consequently,

$$\sum_{i=1}^{n} c_i \lambda_i e_i = \sum_{i=1}^{n} (\mu c_i + f_i) e_i.$$

Equating coefficients of the independent vectors e_i then gives

$$c_i \lambda_i = \mu c_i + f_i, \quad i = 1, \ldots, n.$$

Solving for the unknown coefficients c_i gives

$$c_i = \frac{f_i}{\lambda_i - \mu}, \quad i = 1, \ldots, n.$$

So the unique solution to (4.19) is

$$\mathbf{u} = \sum_{i=1}^{n} \frac{f_i}{\lambda_i - \mu} e_i, \quad \mu \neq \lambda_i.$$

The right side represents $(A - \mu I)^{-1} \mathbf{f}$, and so we have inverted the finite-dimensional operator $A - \mu I$.

If μ is an eigenvalue, say $\mu = \lambda_J$, then

$$c_i = \frac{f_i}{\lambda_i - \mu}, \quad i \neq J,$$

and $c_J \lambda_J = \mu c_J + f_J$. If $f_J \neq 0$, then (4.19) cannot have a solution; if $f_J = 0$, then c_J can be any arbitrary constant and (4.19) has infinitely many solutions of the form

$$\mathbf{u} = c_J e_J + \sum_{i=1, i \neq J}^{n} \frac{f_i}{\lambda_i - \mu} e_i.$$

In summary, knowledge of the eigen-structure of a matrix A makes the solution of other problems straightforward. This same idea, but based on expansions in terms of eigenfunctions, will be true for other operators, such as integral operators and differential operators. Knowledge of the eigen-structure makes it possible to invert such operators.

4.3.2 Volterra Equations

Volterra integral equations arise naturally in many settings and we begin with an example.

Example 4.6

(Inventory control problem) A shop manager determines that a percentage $k(t)$ of goods remains unsold at time t after she purchased the goods. At what rate should she purchase goods so that the stock remains constant? Typical of inventory problems, as well as population problems, we assume that this process is continuous. Let $u(t)$ be the rate (goods per unit time) at which goods are to be purchased, and let a be the amount of goods initially purchased. In the time interval $[\tau, \tau + \Delta\tau]$ she will buy an amount $u(\tau)\Delta\tau$, and at time t the portion remaining unsold is $k(t - \tau)u(\tau)\Delta\tau$. Therefore the amount of goods in the shop unsold at time t is

$$ak(t) + \int_0^t k(t - \tau)u(\tau)\,d\tau.$$

The first term represents the amount of the initial purchase left unsold at time t. Hence, this is the total amount of goods in the shop left unsold at time t, and so the shop manager should require that

$$ak(t) + \int_0^t k(t - \tau)u(\tau)\,d\tau = a.$$

So the solution to the shop manager's problem satisfies a nonhomogeneous Volterra integral equation. For example, if $k(t) = e^{-bt}$, then we can rearrange the problem as

$$a + \int_0^t e^{b\tau}u(\tau)\,d\tau = ae^{bt}.$$

Differentiating with respect to t gives the solution $u(t) = ab$.

Example 4.7

(Birthrates) In this example we set up an integral equation governing the number of newborns in a population where the mortality and fecundity rates are known. Let $u(a, t)$ be the population density of females of age a at time t. Thus, $u(a, t)\,da$ represents the number of females between the ages of a and $a + da$ at time t. Further, let $f(a)$ denote the fecundity rate of females at age a, measured in births per female per time. Then $f(a)u(a, t)\,da$ represents the rate of births at time t contributed by females between the ages a and $a + da$. Integrating over all ages gives the total number of birth per unit time at time t, or

$$B(t) = \int_0^\infty f(a)u(a,t)\,da.$$

There is no difficulty in using ∞ as the upper limit on the integral because the fecundity will be zero after some age when females cease reproduction. Now we separate ages into the two cases $a < t$ and $a > t$. For $a < t$, females of age a at time t were born at time $t - a$. So $u(a,t)\,da = B(t-a)s(a)\,da$, where $s(a)$ is the survivorship, or the fraction that survive from birth to age a. For $a > t$, females of age a were present at time $t = 0$ having age $a - t$. If $u_0(a)$ is the density of females at time zero, then $u(a,t)\,da = u_0(a-t)s(t)\,da$ for $a > t$. Consequently, the birthrate must satisfy the Volterra equation

$$B(t) = \int_0^t f(a)s(a)B(t-a)\,da + \int_t^\infty u_0(a-t)f(a)s(t)\,da.$$

The second integral is a known function of time, say $\phi(t)$; thus the birthrate B satisfies the Volterra equation

$$B(t) = \int_0^t f(a)s(a)B(t-a)\,da + \phi(t),$$

which is called the **renewal equation**. The renewal equation occurs often in applied mathematics, especially probability theory.

Example 4.8

Some integral equations, like the one obtained in the preceding examples, have kernels of a special type that permit a direct solution by Laplace transform methods. The Volterra integral equation

$$u(t) = f(t) + \int_0^t k(t-s)u(s)\,ds \qquad (4.20)$$

is said to be of *convolution type* because the integral is the convolution $k * u$ of k and u. We recall that the Laplace transform of a convolution is the product of the Laplace transforms , that is, $\mathcal{L}(k * u) = (\mathcal{L}k)(\mathcal{L}u)$. Therefore, taking the Laplace transform of both sides of (4.20) gives

$$\mathcal{L}u = \mathcal{L}f + (\mathcal{L}k)(\mathcal{L}u).$$

Solving for $\mathcal{L}u$ gives

$$\mathcal{L}u = \frac{\mathcal{L}f}{1 - \mathcal{L}k}.$$

Thus, the solution u is the inverse transform of the right side. For example, consider the integral equation

$$u(t) = t - \int_0^t (t - s)u(s)\,ds.$$

Here $f(t) = t$ and $k(t) = t$, both with transform $1/s^2$. Then, taking the Laplace transform of the equation gives

$$\mathcal{L}u = \frac{1}{s^2} - \frac{1}{s^2}\mathcal{L}u,$$

or

$$\mathcal{L}u = \frac{1}{1 + s^2}.$$

Therefore

$$u(t) = \mathcal{L}^{-1}\left(\frac{1}{1 + s^2}\right) = \sin t.$$

Initial value problems for ordinary differential equations can be reformulated as a Volterra integral equation. Consider

$$u' = f(t, u), \qquad u(t_0) = u_0, \tag{4.21}$$

and assume that f is a continuous function. If $u = u(t)$ is a solution of (4.21), then for all t we have

$$u'(t) = f(t, u(t)).$$

Replacing t by s and integrating from t_0 to t gives

$$\int_{t_0}^t u'(s)\,ds = \int_{t_0}^t f(s, u(s))\,ds,$$

or

$$u(t) = u_0 + \int_{t_0}^t f(s, u(s))\,ds, \tag{4.22}$$

where in the last step the fundamental theorem of calculus was applied. Equation (4.22) is a Volterra integral equation for u, and it is entirely equivalent to the initial value problem (4.21). To recover (4.21) from (4.22), we can differentiate (4.22) with respect to x. The initial condition is automatically contained in the integral form (4.22). One procedure for solving a Volterra equation, which sometimes works, is to recast it into an initial value problem and solve the differential problem.

To reformulate differential equations as integral equations, we can continually integrate the equation until all the derivatives are removed. The following lemma contains a formula for a repeated integral that is useful in carrying out this procedure for second-order equations.

Lemma 4.9

Let f be a continuous function for $x \geq a$. Then

$$\int_a^x \int_a^s f(y)\,dy\,ds = \int_a^x f(y)(x-y)\,dy$$

Proof

Let $F(s) = \int_a^s f(y)\,dy$. Then, using integration by parts,

$$\int_a^x \int_a^s f(y)\,dy\,ds = \int_a^x F(s)\,ds$$

$$= \int_a^x \frac{d(s)}{ds} F(s)\,ds$$

$$= xF(x) - aF(a) - \int_a^x sF'(s)\,ds$$

$$= x \int_a^x f(y)\,dy - \int_a^x yf(y)\,dy,$$

which is the result. $\qquad\qquad\square$

The next example shows how a second-order initial value problem can be turned into an integral equation.

Example 4.10

Consider the second-order initial value problem:

$$u'' + p(t)u' + q(t)u = f(t), \quad t > a; \qquad u(a) = u_0, u'(a) = u_1. \qquad (4.23)$$

Solving for u'' and integrating from a to t gives

$$u'(t) - u_1 = -\int_a^t p(s)u'(s)\,ds - \int_a^t (q(s)u(s) - f(s))\,ds.$$

If the first integral is integrated by parts the result is

$$u'(t) - u_1 = -p(t)u(t) - \int_a^t [(q(s) - p'(s))u(s) - f(s)]\,ds + p(a)u_0.$$

Integrating again gives

$$u(t) - u_0 = (p(a)u_0 + u_1)(t - a) - \int_a^t p(s)u(s)\,ds$$
$$- \int_a^t \int_a^r [(q(s) - p'(s))u(s) - f(s)]\,ds\,dr.$$

Lemma 4.9 finally implies

$$u(t) = (p(a)u_0 + u_1)(t - a) + u_0 - \int_a^t \{p(s) + (t - s)[q(s) - p'(s)]\}u(s)\,ds$$
$$+ \int_a^t (t - s)f(s)\,ds,$$

which is a Volterra equation is of the form

$$u(t) = \int_a^t k(t, s)u(s)\,ds + f(t).$$

The reverse process, transforming an integral equation into a differential equation, requires Leibniz rule for differentiating an integral with respect to a variable appearing in the upper limit of the integral and in the integrand (see Exercise 1).

It is common practice in applied science to apply iterative procedures to determine approximate solutions to various kinds of problems for which analytic solutions cannot be found. The integral equation (4.22), and therefore the initial value problem (4.21), can be solved approximately by fixed-point iteration. Beginning with an initial approximation $\phi_0(t)$ we generate a sequence of successive approximations $\phi_1(t)$, $\phi_2(t)$, . . . , via

$$\phi_{n+1}(t) = u_0 + \int_a^t f(s, \phi_n(s))\,ds, \quad n = 0, 1, 2, \ldots$$

This method is called **Picard's method**, and it can be shown under certain assumptions that the sequence $\phi_n(t)$ converges to the unique solution of (4.22).

For the linear Volterra equation of the form

$$u(t) = f(t) + \lambda \int_a^t k(t, s)u(s)\,ds,$$

a similar type of iteration can be defined. Let us write this equation in operator form as

$$u = f + \lambda K u, \tag{4.24}$$

where K is the Volterra integral operator

$$Ku(t) = \int_a^t k(t, s)u(s)\,ds. \tag{4.25}$$

By making a heuristic calculation we can guess what the solution representation is. First we write the integral equation as

$$(I - \lambda K)u = f.$$

Then, formally inverting the operator,

$$
\begin{aligned}
u &= (I - \lambda K)^{-1} f = \frac{1}{I - \lambda K} f \\
&= (I + \lambda K + \lambda^2 K^2 + \cdots + \lambda^j K^j + \cdots) f.
\end{aligned}
$$

This result is the content of the next theorem. Therefore, letting $u_0 = u_0(t)$ be given, we may define a sequence of successive approximations $u_n(t)$ by

$$
u_{n+1} = f + \lambda K u_n, \quad n = 0, 1, 2, \ldots \tag{4.26}
$$

We have the following result on convergence of this process.

Theorem 4.11

If u_0, f, k are continuous, then the sequence $u_n(t)$ defined by (4.26) converges uniformly to the unique solution of (4.24) on $[a, b]$ given by

$$
u(t) = f(t) + \sum_{j=1}^{\infty} \lambda^j K^j f(t), \tag{4.27}
$$

where K^j denotes the composition operator $K(K(\ldots (K) \ldots))$, j times.

The representation (4.27) of the solution to (4.24) is called the **Neumanm series** for u. (It is called the Born series in physics.) It represents the inverse operator $(I - \lambda K)^{-1}$. To prove this, note that after repeated iteration we can write

$$
u_{n+1} = f + \sum_{j=1}^{n} \lambda^j K^j f + \lambda^{n+1} K^{n+1} u_0. \tag{4.28}
$$

To examine the convergence of (4.28) we must look at the remainder term $\lambda^{n+1} K^{n+1} u_0$ as $n \to \infty$. Denoting $M = \max |k|$ and $C = \max |u_0|$, we have

$$
|K u_0(t)| \leq \int_a^t |k(t, s) u_0(s)| \, ds \leq (t - a) M C.
$$

Next

$$
|K^2 u_0(t)| \leq \int_a^t |k(t, s) K u_0(s)| \, dy \leq \int_a^t M(s - a) M C \, ds = C \frac{(t - a)^2 M^2}{2}.
$$

Continuing with an induction argument we obtain

$$
|K^{n+1} u_0(t)| \leq C \frac{(t - a)^{n+1} M^{n+1}}{(n + 1)!} \leq C \frac{(b - a)^{n+1} M^{n+1}}{(n + 1)!}.
$$

This bound is independent of t and it approaches zero as $n \to \infty$. Consequently $|\lambda^{n+1} K^{n+1} u_0| \to 0$ uniformly on $[a, b]$, and the unique solution of (4.24) is given by (4.27).

Examples are left to the exercises. We point out, however, one important consequence of the Theorem 4.11, namely that the equation $Ku = \mu u$ (set $f = 0$ and $\mu = 1/\lambda$), has only the trivial solution. Therefore, *there can be no eigenvalues for a Volterra operator* (4.25). This result is consistent with our inference that Volterra equations come from initial value problems, and linear initial value problems for differential equations have no eigenvalues. The conclusion is in sharp contrast to Fredholm equations, which arise from boundary value problems. We already know that boundary value problems (e.g., Sturm–Liouville problems), have a rich spectral structure, and we expect the same from Fredholm operators.

4.3.3 Fredholm Equations with Degenerate Kernels

Generally, integral equations are problems in infinite-dimensional spaces. However, a special class of Fredholm equations reduces, or degenerates, to a finite-dimensional problem. To this class belongs Fredholm integral equations whose kernel $k(x, y)$ is a finite sum of products of functions of x and y. These are essentially finite-dimensional problems, or problems leading to a linear system of algebraic equations.

Therefore, let us consider the integral equation of the form

$$Ku - \lambda u = f, \tag{4.29}$$

where λ is a parameter, f is a given continuous function, and K is the operator given by (4.14) with kernel

$$k(x, y) = \sum_{j=1}^{n} \alpha_j(x)\beta_j(y), \tag{4.30}$$

where the α_j and β_j are real-valued continuous functions. A kernel of this form is called **separable**, or **degenerate**. . Without any loss of generality, we can always assume that the α_j are independent, and the β_j are independent. When we write out equation (4.29), after substituting the kernel, we obtain

$$\sum_{j=1}^{n} \alpha_j(x) \int_a^b u(y)\beta_j(y)\, dy - \lambda u(x) = f(x).$$

The inner product notation is convenient for the remaining discussion. We notice that this last equation is just

$$\sum_{j=1}^{n} \alpha_j(x)c_j - \lambda u(x) = f(x), \tag{4.31}$$

where the unknown constants c_j are defined by the inner products

$$c_j = (u, \beta_j), \qquad j = 1, 2, \ldots \tag{4.32}$$

If we multiply (4.31) by $\beta_i(x)$ and integrate (this is the same as taking the inner product), we obtain

$$\sum_{j=1}^{n} (\beta_i, \alpha_j)c_j - \lambda c_i = (f, \beta_i), \qquad i = 1, 2, \ldots.$$

These represent n linear, algebraic equations in the unknowns $c_1, c_2, \ldots, c_n$. The system can be written concisely in matrix format by introducing the unknown column vector $\mathbf{c} = (c_1, \ldots, c_n)^{\mathrm{T}}$, the column vector $\mathbf{F} = ((f, \beta_1), \ldots, (f, \beta_n))^{\mathrm{T}}$, and the coefficient matrix A with entries (β_i, α_j). Thus we have reduced the solution of the integral equation (4.29) to the solution of the linear system

$$(A - \lambda I)\mathbf{c} = \mathbf{F}. \tag{4.33}$$

Determining $\mathbf{c}$ gives the c_j, and therefore (4.31) can be used (provided $\lambda \neq 0$) to determine the solution $u(x)$ of the separable integral equation (4.29).

First consider equation (4.29) for $\lambda \neq 0$. In this case

$$u(x) = \frac{1}{\lambda}\left(-f(x) + \sum_{j=1}^{n} \alpha_j(x)c_j\right), \tag{4.34}$$

where the c_j are obtained from (4.33). If $\det(A - \lambda I) \neq 0$, that is, if λ is not an eigenvalue of the matrix A, then (4.33) has a unique solution $\mathbf{c}$. If $\det(A - \lambda I) = 0$, then (4.33) either has no solution or infinitely many solutions, depending on whether $\mathbf{F}$ is in the range (column space) of the matrix $A - \lambda I$.[4] If $f = 0$, then (4.29) reduces to the eigenvalue problem $Ku = \lambda u$ for K; so the nonzero eigenvalues of the matrix A coincide with the eigenvalues of K. There are finitely many such eigenvalues (note that zero is not an eigenvalue of A), and the corresponding eigenfunctions $u(x)$ are defined by (4.31) with $f = 0$, where the c_j are the components of the corresponding eigenvectors of A. We can summarize our conclusions in the following theorem (such theorems are called **alternative theorems**).

---

[4] Recall from linear algebra that the range of $A - \lambda I$ is orthogonal to $\ker(A^{\mathrm{T}} - \lambda I)$, where ker denotes the kernel or nullspace. Thus, (4.33) has a solution if, and only if, $\mathbf{F} \perp \mathbf{v}$ for all $\mathbf{v}$ satisfying $(A^{\mathrm{T}} - \lambda I)\mathbf{v} = \mathbf{0}$.

Theorem 4.12

Consider the integral equation (4.29), where the kernel is given by (4.30) and $\lambda \neq 0$. Let A be the matrix $A \equiv ((\beta_i, \alpha_j))$. If λ is not an eigenvalue of A, then (4.29) has a unique solution given by (4.34); if λ is an eigenvalue of A, then (4.29) either has no solution or infinitely many solutions.

Next consider the case $\lambda = 0$. First, if $f = 0$, then (4.31) becomes

$$\sum_{j=1}^{n} c_j \alpha_j(x) = 0,$$

which implies $c_j = (u, \beta_j) = 0$ for all j, or u is orthogonal to all the β_j. There are infinitely many independent functions u that satisfy this condition. This is not obvious, and a complete discussion would take us beyond the scope of an elementary treatment. Suffice it to say that the n functions β_i span a finite-dimensional space, whereas u lies generally in an infinite dimensional space—for example, the square integrable functions. The space of functions orthogonal to the span of $\beta_1, \ldots, \beta_n$ is infinite dimensional. We refer to I. Stakgold (1998), for example, for a thorough treatment. Thus, there are infinitely many nontrivial solutions to the equation $Ku = 0$. This means, of course, that zero is an eigenvalue of K with *infinite multiplicity*; that is, the dimension of the eigenspace corresponding to $\lambda = 0$ is infinite.

If $\lambda = 0$ and $f \neq 0$, then (4.31) becomes

$$\sum_{j=1}^{n} c_j \alpha_j(x) = f(x).$$

Now it is clear that no solution can exist if f is not a linear combination of the α_j. If f is a linear combination of the α_j, say $f(x) = \sum_{j=1}^{n} f_j \alpha_j(x)$, then we must have $c_j = (u, \beta_j) = f_j$ for all j. Again, there are infinitely many solutions. We summarize the discussion in the following alternative theorem.

Theorem 4.13

Consider the equation $Ku = f$ with separable kernel (4.30). If f is not a linear combination of the α_j then there is no solution; if f is a linear combination of the α_j, then there are infinitely many solutions.

Finally, the preceding discussion implies the following theorem regarding the spectrum of a Fredholm operator with a separable kernel.

Theorem 4.14

The integral operator K with separable kernel (4.30) has finitely many nonzero eigenvalues of finite multiplicity, and zero is an eigenvalue of infinite multiplicity.

Example 4.15

Find the nonzero eigenvalues of the operator

$$Ku(x) = \int_0^1 (1 - 3xy)u(y)\,dy.$$

Here the kernel is degenerate with

$$\alpha_1(x) = 1, \alpha_2(x) = -3x, \beta_1(y) = 1, \beta_2(y) = y.$$

The matrix A is given by

$$A = \begin{pmatrix} 1 & -\frac{3}{2} \\ \frac{1}{2} & -1 \end{pmatrix},$$

and $\det(A - \lambda I) = \lambda^2 - \frac{1}{4} = 0$ gives the eigenvalues $\lambda = \pm\frac{1}{2}$. An eigenvector corresponding to $\lambda = \frac{1}{2}$ is $c = (3,1)^T$ and an eigenvector corresponding to $\lambda = -\frac{1}{2}$ is $(1,1)^T$. By formula (4.31), with $f = 0$, the eigenfunctions are

$$u(x) = 2(3\alpha_1(x)+\alpha_2(x)) = 6(1-x), \qquad u(x) = -2(\alpha_1(x)+\alpha_2(x)) = -2(1-3x).$$

But eigenfunctions are unique only up to a constant multiple, and therefore the eigenfunctions corresponding to $\lambda = \frac{1}{2}$ and $\lambda = -\frac{1}{2}$ are $u(x) = a(1-x)$ and $u(x) = b(1-3x)$, respectively, where a and b are arbitrary constants. Next consider the nonhomogeneous equation

$$\int_0^1 (1-3xy)u(y)\,dy - \lambda u(x) = f(x). \tag{4.35}$$

From the alternative theorem, if $\lambda \neq \pm\frac{1}{2}$, then (4.35) has a unique solution.

4.3.4 Symmetric Kernels

Many problems in the physical sciences lead naturally to a Fredholm equation with a continuous, symmetric kernel. First we consider the eigenvalue problem

$$Ku = \lambda u, \tag{4.36}$$

where K is the integral operator given by (4.14) and the kernel is real, symmetric, and continuous on the square $a \leq x, y \leq b$. We need the following lemma.

Lemma 4.16

If k is real, continuous, and $k(x, y) = k(y, x)$, then the integral operator K defined by (4.14) has the symmetry property

$$(Ku, v) = (u, Kv).$$

Proof

The proof is a straightforward calculation using the definition of inner product and interchanging order of integration. We have

$$
\begin{aligned}
(Ku, v) &= \int_a^b \left(\int_a^b k(x, y)u(y)\, dy \right) \overline{v(x)}\, dx \\
&= \int_a^b \int_a^b k(x, y)u(y)\overline{v(x)}\, dy\, dx \\
&= \int_a^b \int_a^b k(y, x)u(x)\overline{v(y)}\, dx\, dy \\
&= \int_a^b \int_a^b k(y, x)u(x)\overline{v(y)}\, dy\, dx \\
&= \int_a^b u(x) \overline{\left(\int_a^b k(x, y)v(y)\, dy \right)}\, dx \\
&= (u, Kv),
\end{aligned}
$$

which completes the proof. □

Integral operators that have the properties assumed in the lemma are called **symmetric operators**. Such operators have an extremely nice spectral structure. First we have the following theorem.

Theorem 4.17

Let K be the integral operator defined by (4.14) with a real, symmetric, continuous kernel. Then the eigenvalues, if they exist, are real, and eigenfunctions corresponding to distinct eigenvalues are orthogonal.

Proof

First observe that (Ku, u) is real for all u. This follows from

$$(Ku, u) = (u, Ku) = \overline{(Ku, u)}.$$

If a number equals its complex conjugate, then it must be real. Now let λ, u be an eigen-pair; that is, $Ku = \lambda u$. Then $(Ku, u) = (\lambda u, u) = \lambda(u, u)$. Since (Ku, u) and (u, u) are real, λ must be real. Finally, let $Ku = \lambda u$ and $Kv = \nu v$, where $\lambda \neq \nu$. Then $(u, v) = 0$ follows from

$$\lambda(u, v) = (Ku, v) = (u, Kv) = (u, \nu v) = \nu(u, v),$$

which completes the proof.   □

The question of existence of eigenvalues is more difficult and beyond the scope of this text. Although we shall state the basic result and use it in the sequel, we refer the reader to other sources for a proof (e.g., see Stakgold (1998)). We already know the spectral structure of Fredholm operators with degenerate kernels; there are finitely many nonzero eigenvalues of finite multiplicity, and zero is an eigenvalue of infinite multiplicity. Now we formulate a theorem for nonseparable kernels; the result is the **Hilbert[5]–Schmidt theorem.**

Theorem 4.18

Consider the integral operator K defined in (4.14) where the kernel k is real, continuous, symmetric, and not degenerate. Then K has infinitely many eigenvalues $\lambda_1, \lambda_2, \ldots$, each with finite multiplicity, and they can be ordered as

$$0 < \cdots \leq |\lambda_n| \leq \cdots \leq |\lambda_2| \leq |\lambda_1|,$$

with $\lim \lambda_n = 0$. Moreover, any square integrable function f can be expanded in terms of the set of orthonormal eigenfunctions $\phi_k(x)$ associated with the eigenvalues as

$$f(x) = \sum_{k=1}^{\infty} f_k \phi_k(x),$$

where the series converges to f in $L^2[a, b]$, and the coefficients are given by

$$f_k = (f, \phi_k), \qquad k = 1, 2, \ldots. \tag{4.37}$$

Having eigenfunctions opens the possibility of solving integral equations using eigenfunction expansions. Consider the integral equation

$$Ku - \mu u = f, \tag{4.38}$$

---

[5] David Hilbert (1862–1943) was one of the great mathematicians in the early 20th century. His formalist approach to problems became a key idea in studying abstract concepts in functional analysis. Hilbert spaces are now core structures in mathematics and science, especially quantum theory.

where the kernel of K is real, continuous, and symmetric, and f is a given continuous function. First we expand f and u in terms of the eigenfunctions of K as

$$f(x) = \sum_{k=1}^{\infty} f_k \phi_k(x), \qquad u(x) = \sum_{k=1}^{\infty} u_k \phi_k(x),$$

where the f_k are known and given by (4.37), and where the u_k are unknown and to be determined. Substituting these expansions into (4.38) gives

$$K \left(\sum_{k=1}^{\infty} u_k \phi_k \right) - \mu \sum_{k=1}^{\infty} u_k \phi_k = \sum_{k=1}^{\infty} f_k \phi_k.$$

But

$$K \left(\sum_{k=1}^{\infty} u_k \phi_k \right) = \sum_{k=1}^{\infty} u_k K \phi_k = \sum_{k=1}^{\infty} u_k \lambda_k \phi_k.$$

Therefore, equating the coefficients of ϕ_k yields

$$(\lambda_k - \mu) u_k = f_k, \quad k = 1, 2, 3, \dots. \tag{4.39}$$

Therefore, if μ is not an eigenvalue, we have the unique solution representation of (4.38) given by

$$u(x) = \sum_{k=1}^{\infty} \frac{f_k}{\lambda_k - \mu} \phi_k(x).$$

From the definition of the inner product, we can write this formula as

$$u(x) = \int_a^b \left(\sum_{k=1}^{\infty} \frac{\phi_k(x) \phi_k(y)}{\lambda_k - \mu} \right) f(y) \, dy.$$

The right side is a representation of $(K - \mu I)^{-1}$, the inverse operator of $K - \mu I$, acting upon f. (In this discussion we have ignored questions of convergence, but these issues can be made rigorous.)

If $\mu = \lambda_m$ for some fixed m, then in order to have a solution we must have $f_m = (f, \phi_m) = 0$, or f orthogonal to ϕ_m. In this case, then, u_m is arbitrary and we obtain infinitely many solutions of (4.38) given by

$$u(x) = c\phi_m(x) + \sum_{k=1, k \neq m}^{\infty} \frac{f_k}{\lambda_k - \mu} \phi_k(x),$$

where c is an arbitrary constant.

When there is a unique solution, Fredholm integral equations of the form (4.38) can be solved numerically by discretizing the domain and approximating the integral by a quadrature formula (e.g., the trapezoid rule). To this end, let $x_i = a + (i - 1)h$ for $i = 1, 2, \dots, N$, where $h = (b - a)/(N - 1)$ is the step size, and there are $N-1$ subintervals; note that $a = x_1$ and $b = x_N$. Then, evaluating the integral equation at the x_i gives

$$\int_a^b k(x_i, y)u(y)\, dy - \mu u(x_i) = f(x_i), \quad i = 1, 2, \ldots, N.$$

Now recall the trapezoid rule:

$$\int_a^b g(y)\, dy \approx \frac{b-a}{2(N-1)}\left(g(a) + 2\sum_{j=2}^{N-1} g(x_j) + g(b)\right).$$

Replacing the integral in the integral equation by its trapezoidal approximation, and introducing the notation u_i for the approximation to $u(x_i)$, we obtain

$$\frac{b-a}{2(N-1)}\left(k(x_i, a)u_1 + 2\sum_{j=2}^{N-1} k(x_i, x_j)u_j + k(x_i, b)u_N\right) - \mu u_i = f(x_i),$$

$$(4.40)$$

for $i = 1, 2, \ldots, N$. This is a system of N linear equations for the N unknowns $u_1, u_2, \ldots, u_N$, and it can be solved easily by matrix methods using a computer algebra system.

EXERCISES

1. Prove **Leibniz formula**

$$\frac{d}{dx}\int_{a(x)}^{b(x)} F(x, y)\, dy = \int_{a(x)}^{b(x)} F_x(x, y)\, dy + F(x, b(x))b'(x) - F(x, a(x))a'(x)$$

for differentiating an integral with respect to a parameter that appears in the integrand and in the limits of integration. Hint: Call the integral $I(x, a, b)$, where a and b depend on x, and differentiate, using the chain rule.

2. Let

$$A = \begin{pmatrix} 1 & 2 & 0 \\ 2 & 4 & 0 \\ 0 & 0 & 5 \end{pmatrix}.$$

a) Find the eigenvalues and an orthonormal set of eigenvectors.

b) In terms of the eigenvectors, solve $A\mathbf{x} - 2\mathbf{x} = (1, 4, 0)^T$, if possible.

c) In terms of the eigenvectors, solve $A\mathbf{x} - \mathbf{x} = (\frac{1}{2}, -1, 0)^T$, if possible.

d) In terms of the eigenvectors, solve $A\mathbf{x} - \mathbf{x} = (1, 4, 0)^T$, if possible.

3. Let

$$A = \begin{pmatrix} 2 & 2 \\ 1 & 3 \end{pmatrix},$$

and consider the system $(A - 4I)\mathbf{u} = \mathbf{b}$. Find conditions on the vector $\mathbf{b}$ for which this system has a solution.

4. Find the eigenvalues and orthonormal eigenfunctions associated with the following integral operators:

a) $Ku = \int_{-\pi}^{\pi} k(x+y)u(u)\, dy$ where $k(x) = \frac{k_0}{2} + \sum_{1}^{\infty} k_n \cos nx$, where k_n is a strictly decreasing sequence of positive numbers. Note that the set $\{1, \cos nx, \sin nx\}$ forms an orthogonal set on the interval.

b) $Ku = \int_{0}^{1} \min(x, y)u(y)\, dy$.

c) $Ku = \int_{0}^{\pi} k(x, y)u(y)\, dy$ where $k(x, y) = y(\pi - x)$ if $x > y$, and $k(x, y) = x(\pi - y)$ if $x < y$.

d) $Ku = \int_{0}^{x} \sin x \cos 2y\, u(y)\, dy$.

5. Find the expansion of the function

$$f(x) = \begin{cases} x, & 0 \le x < \pi/2, \\ \pi - x & \pi/2 \le x \le \pi, \end{cases}$$

in terms of the eigenfunctions of the operator K in Exercise 4(c).

6. Formulate the integral equation

$$u(t) = 1 + \int_{0}^{t} s \ln\left(\frac{s}{t}\right) u(s)\, ds$$

as an initial value problem.

7. Let K be the operator in Exercise 4(c). Solve the following problems, or state why there is no solution.

a) $Ku - 2u = 0$.

b) $Ku - \frac{\pi}{9}u = x(\pi - x)$.

c) $Ku - 2u = x(\pi - x)$.

d) $Ku - \frac{\pi}{9}u = \cos 2x$.

8. Find the first three nonzero terms of the Neumann series for the solution to the integral equation

$$u(t) = t + \mu \int_{0}^{t} (t - s)u(s)\, ds.$$

9. Solve the integral equation $u(t) = \int_0^t e^{t-s} u(s)\, ds$.

10. Reformulate the initial value problem

$$u'' - \lambda u = f(x), \quad x > 0; \quad u(0) = 1, \ u'(0) = 0$$

    as a Volterra integral equation.

11. Solve the integral equation $\int_0^t y u(y)\, dy - \lambda u(t) = f(t)$, $0 \le t \le 1$, where f is continuous and $\lambda \ne 0$. Hint: as an artifice, assume f' exists and then write the solution only in terms of f.

12. Consider the operator $Ku(x) = \int_{-1}^{1} k(x+y)u(u)\, dy$, where $k(x)$ is a continuous, even, periodic function with Fourier expansion

$$k(x) = \frac{k_0}{2} + \sum_{1}^{\infty} k_n \cos nx, \quad k_{n+1} < k_n.$$

    Find the eigenvalues and eigenfunctions of K.

13. Discuss the solvability of the following integral equations, and solve if possible.

    a) $u(x) = f(x) + \lambda \int_0^{1/2} u(y)\, dy$.

    b) $u(x) = b + \lambda \int_0^1 u(y)^2\, dy$.

    c) $u(x) = f(x) + \int_0^1 xy u(y)\, dy$.

14. Determine if the integral operator

$$Ku(x) = \int_0^{\pi} \sin x \sin 2y\, u(y)\, dy$$

    has eigenvalues.

15. Solve the integral equation

$$\int_0^1 k(x,u)u(y)\, dy - \lambda u(x) = x$$

    using eigenfunction expansions, where $k(x,y) = x(1-y)$ if $x < y$, and $k(x,y) = y(1-x)$ if $x > y$. Hint: Transform the problem into a Sturm–Liouville problem.

16. Replace $\sin xy$ by the first two terms in its power series expansion to obtain an approximate solution to

$$u(x) = x^2 + \int_0^1 \sin(xy)u(y)\, dy.$$

17. Investigate the solvability of

$$u(x) = \sin x + 3 \int_0^\pi (x + y)u(y)\, dy.$$

18. Solve the integral equation

$$\int_0^1 e^{x+y} u(y)\, dy - \lambda u(x) = f(x),$$

considering all cases.

19. Solve the nonlinear integral equation $u(x) = \lambda \int_0^1 yu(y)^2\, dy$.

20. Solve the integral equation

$$u(x) = f(x) + \int_0^1 (x\xi^2 + x^2\xi)u(\xi)\, d\xi,$$

considering all cases.

21. For the following integral equation, find a condition on μ for which the equation has a solution, and find the solution.

$$y(x) = \sin x + \mu \int_0^{\pi/2} \sin x \cos \xi\, y(\xi)d\xi.$$

22. For which value(s) of λ is there a unique solution to the integral equation

$$y(x) = \int_0^\pi (\sin \xi + \cos x)y(\xi)\, d\xi - \lambda y(x) = 1?$$

23. Reformulate the boundary value problem

$$u'' + \lambda u = 0, \quad 0 < x < l; \quad u(0) = u(l) = 0,$$

as a Fredholm integral equation.

24. Reformulate the initial value problem

$$u'' + u = t^2, \quad u(0) = 1, \quad u'(0) = 0,$$

as a Volterra integral equation of the form $u + Ku = f$, and find f and the kernel of K.

25. Find all solutions to the integral equation

$$y(x) = \mu \int_0^1 (x^4 + \xi^4)y(\xi)d\xi$$

which correspond to negative values of μ.

26. Find the eigenvalues and eigenfunctions of the integral operator

$$Ku(x) = \int_{-1}^{1} (1 - |x - y|)u(y)\, dy.$$

27. (This exercise requires perturbation methods from Chapter 2.) Consider the differential-integral equation

$$\varepsilon u' = u - u^2 - u \int_0^t u(s)ds, \quad u(0) = a < 1,$$

where ε is a small, positive parameter, which models a population $u = u(t)$ undergoing logistics growth and the cumulative effect of a toxin on the population. Find a uniformly valid approximation for $t > 0$.

28. Consider the differential–integral operator

$$Ku = -u'' + 4\pi^2 \int_0^1 u(s)\, ds, \quad u(0) = u(1) = 0.$$

Prove that eigenvalues of K, provided they exist, are positive. Find the eigenfunctions corresponding to the eigenvalue $\lambda = 4\pi^2$.

29. Use (4.40) to numerically solve the integral equation

$$\int_0^1 (1 - 3xy)u(y)\, dy - u(x) = x^3.$$

4.4 Green's Functions

What is a Green's function? Mathematically, it is the kernel of an integral operator that represents the inverse of a differential operator; physically, it is the response of a system when a unit point source is applied to the system. In this section we show how these two apparently different interpretations are actually the same.

4.4.1 Inverses of Differential Operators

To fix the notion we consider a regular Sturm–Liouville problem (SLP), which we write in the form

$$Au \equiv -(pu')' + qu = f, \quad a < x < b, \qquad (4.41)$$

$$B_1 u(a) \equiv \alpha_1 u(a) + \alpha_2 u'(a) = 0, \qquad (4.42)$$

$$B_2 u(b) \equiv \beta_1 u(b) + \beta_2 u'(b) = 0, \qquad (4.43)$$

where p, p', q, and f are continuous on $[a, b]$, and $p > 0$. So that the boundary conditions do not disappear, we assume that not both α_1 and α_2 are zero, and similarly for β_1 and β_2. For conciseness we represent this problem in operator notation

$$Lu = f, \tag{4.44}$$

where we consider the differential operator L as acting functions in $C^2[a, b]$ that satisfy the boundary conditions (4.42)–(4.43). Thus, L contains the boundary conditions in its definition, as well as the Sturm–Liouville operator A.

Recall the procedure in matrix theory when we have a *matrix* equation $L\mathbf{u} = \mathbf{f}$, where $\mathbf{u}$ and $\mathbf{f}$ are vectors and L is a square, invertible matrix. We immediately have the solution $\mathbf{u} = L^{-1}\mathbf{f}$, where L^{-1} is the inverse matrix. The inverse matrix exists if $\lambda = 0$ is not an eigenvalue of L, or when $\det L \neq 0$. We want to perform a similar calculation with the differential equation (4.44) and write the solution

$$u = L^{-1}f,$$

where L^{-1} is the inverse operator of L. Since L is a differential operator, we expect that the inverse operator to be an integral operator of the form

$$(L^{-1}f)(x) = \int_a^b g(x, \xi) f(\xi) \, d\xi, \tag{4.45}$$

with kernel g. Again drawing inferences from matrix theory, we expect the inverse to exist when $\lambda = 0$ is not an eigenvalue of L, that is, when there are no nontrivial solutions to the differential equation equation $Lu = 0$. If there are nontrivial solutions to $Lu = 0$, then L is not a one-to-one transformation ($u = 0$ is always a solution) and so L^{-1} does not exist.

If the inverse L^{-1} of the differential operator L exists, then the kernel function $g(x, \xi)$ in (4.45) is called the **Green's function** associated with L (recall that L contains in its definition the boundary conditions). This is the mathematical characterization of a Green's function. Physically, as we observe subsequently, the Green's function $g(x, \xi)$ is the solution to (4.44) when f is a unit point source acting at the point ξ; thus, it is the response of the system at x to a unit, point source at ξ. Our approach here will be to present the mathematical result first and then have the methodology unfold with a heuristic physical discussion. The following theorem summarizes our preceding comments and gives a precise representation of the Green's function.

Theorem 4.19

Consider the SLP (4.41)–(4.43), and assume that $\lambda = 0$ is not an eigenvalue of

L. Then L^{-1} exists and is given by (4.45) with

$$g(x,\xi) = \begin{cases} -\dfrac{u_1(x)u_2(\xi)}{p(\xi)W(\xi)}, & x < \xi \\[2mm] -\dfrac{u_1(\xi)u_2(x)}{p(\xi)W(\xi)}, & x > \xi \end{cases} \qquad (4.46)$$

Here, u_1 and u_2 are independent solutions of the homogeneous differential equation $Au = -(pu')' + qu = 0$ with $B_1 u_1(a) = 0$ and $B_2 u_2(b) = 0$, and $W = u_1 u_2' - u_1' u_2$ is the Wronskian of u_1 and u_2.

Before giving the proof, we make some remarks and give an example. Clearly, knowing the Green's function g allows us to immediately write down the unique solution to the inhomogeneous problem (4.41)–(4.43) as

$$u(x) = \int_a^b g(x,\xi)f(\xi)\,d\xi. \qquad (4.47)$$

Also, the Green's function can be expressed as a single equation in terms of the Heaviside step function $H(x)$ (where $H(x) = 0$ if $x < 0$, and $H(x) = 1$ if $x \geq 0$). Then

$$g(x,\xi) = -\frac{1}{p(\xi)W(\xi)}(H(x-\xi)u_1(\xi)u_2(x) + H(\xi-x)u_1(x)u_2(\xi)). \qquad (4.48)$$

Let us write down some of the basic properties of the Green's function. First it is clear that: (a) $g(x,\xi)$ satisfies the differential equation $Ag(x,\xi) = 0$ for $x \neq \xi$; (b) $g(x,\xi)$ satisfies the boundary conditions (4.42) and (4.43); (c) g is continuous on $[a,b]$ and, in particular, it is continuous at $x = \xi$; (d) however, g is not differentiable at $x = \xi$. It is easy to verify that there is a jump in the derivative at $x = \xi$ given by $g'(\xi^+,\xi) - g'(\xi^-,\xi) = -\frac{1}{p(\xi)}$. Here, prime denotes differentiation with respect to x. Thus g is a continuous curve with a corner at $x = \xi$.

Example 4.20

Consider the boundary value problem

$$-u'' = f(x), \quad 0 < x < 1; \quad u(0) = u(1) = 0. \qquad (4.49)$$

Here $Au = -u''$, and solutions to the homogeneous equation $Au = 0$ are linear functions $u(x) = ax + b$, where a and b are constants. Thus, we take $u_1(x) = ax$ and $u_2(x) = b(1-x)$, so that u_1 satisfies the left boundary condition and u_2 satisfies the right boundary condition. Also, $p = 1$ and the Wronskian is $W = -ab$. Therefore, the Green's function is given, according to (4.48), by

$$g(x,\xi) = \xi(1-x)H(x-\xi) + x(1-\xi)H(\xi-x).$$

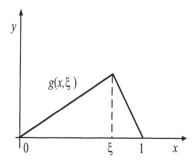

Figure 4.1   Green's function.

Consequently, the solution to (4.49) is given by

$$u(x) = \int_0^1 g(x, \xi) f(\xi) \, d\xi$$

The reader should verify that this is indeed the solution to (4.49). The Green's function is shown in Fig. 4.1.

Finally, the proof of the theorem is a straightforward calculation. First we write

$$v(x) = \int_a^b g(x, \xi) f(\xi) \, d\xi$$

$$= - \int_a^x \frac{u_1(\xi) u_2(x)}{p(\xi) W(\xi)} f(\xi) \, d\xi - \int_x^b \frac{u_1(x) u_2(\xi)}{p(\xi) W(\xi)} f(\xi) \, d\xi.$$

To prove the theorem, we must show that $Av = f$ and that the boundary conditions hold. The calculation of the derivatives of v can be accomplished by a direct application of the Leibniz rule, and we leave the details of this calculation to the reader.

4.4.2 Physical Interpretation

Next we investigate the physical interpretation of the Green's function (George Green, 1793–1841). Our discussion is heuristic and intuitive, but it is made precise in the next section. To fix the context we consider a steady-state heat flow problem. Consider a cylindrical bar of length L and cross-sectional area A, and let $u = u(x)$ denote the temperature at cross section x. Further, let $q = q(x)$ denote the energy flux across the face at x, measured in energy/(area

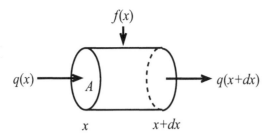

Figure 4.2 Small region.

· time), and let $f(x)$ be a given, distributed heat source over the length of the bar, measured in energy/(volume · time). By convention, the flux q is positive when the heat flow is to the right. Then, if x and $x + dx$ denote two arbitrary locations in the bar, in a steady-state mode the heat energy entering the cross section at x, minus the heat energy exiting the cross section at $x + dx$, plus the heat generated by the source is zero. See Fig. 4.2. That is,

$$Aq(x) - Aq(x + dx) + f(x)A\,dx = 0,$$

or, rearranging,

$$\frac{q(x + dx) - q(x)}{dx} = f(x).$$

These equations are merely a statement of conservation of energy. Now, taking the limit as $dx \to 0$, we obtain

$$q'(x) = f(x).$$

To obtain and equation for temperature, we assume Fourier's heat conduction law, which states that the flux is proportional to the negative temperature gradient, or

$$q(x) = -Ku'(x),$$

where K is the thermal conductivity of the bar, a physical constant. Then

$$-Ku''(x) = f(x),$$

which is the steady-state heat equation. For simplicity, we assume that the ends of a bar of unit length are held at zero degrees, and $K = 1$. Then the steady temperature distribution $u = u(x)$ along the length of the bar satisfies the boundary value problem

$$-u'' = f(x), \quad 0 < x < 1; \quad u(0) = u(1) = 0,$$

which is the same as (4.49). Now imagine that f is an idealized heat source of unit strength that acts only at a single point $x = \xi$ in $(0, 1)$, and we denote this source by $f \equiv \delta(x, \xi)$. Thus, it is assumed to have the properties

$$\delta(x, \xi) = 0, \quad x \neq \xi$$

and

$$\int_0^1 \delta(x, \xi)\, dx = 1.$$

A little reflection indicates that there is no ordinary function δ with these two properties; a function that is zero everywhere but one point must have zero integral. Nevertheless, this δ symbol has been used since the inception of quantum mechanics in the late 1920s (it was introduced by the mathematician-physicist P. Dirac); in spite of not having a rigorous definition of the symbol until the early 1950s, when the mathematician L. Schwartz gave a precise characterization of point sources, both physicists and engineers used the "δ function" with great success. In the next section we give a careful definition of the delta function, but for the present we continue with an intuitive discussion.

The differential equation for steady heat flow becomes, symbolically,

$$-u'' = \delta(x, \xi). \tag{4.50}$$

Thus, for $x \neq \xi$ we have $-u'' = 0$, which has solutions of the form $u = Ax + B$. To satisfy the boundary conditions we take

$$u = Ax, \quad x < \xi; \qquad u = B(1 - x), \quad x > \xi.$$

The reader should be reminded of the calculation we made in the last example. To determine the two constants A and B we use the fact that the temperature should be continuous at $x = \xi$, giving the relation $A\xi = B(1 - \xi)$. To obtain another condition on A and B we integrate the symbolic differential equation (4.50) over an interval $[\xi - \varepsilon, \xi + \varepsilon]$ containing ξ to get

$$-\int_{\xi-\varepsilon}^{\xi+\varepsilon} u''(x)\, dx = \int_{\xi-\varepsilon}^{\xi+\varepsilon} \delta(x, \xi)\, dx.$$

The right side is unity because of the unit heat source assumption; the fundamental theorem of calculus is then applied to the left side to obtain

$$-u'(\xi + \varepsilon) + u'(\xi - \varepsilon) = 1.$$

Taking the limit as $\varepsilon \to 0$ and multiplying by -1 yields

$$u'(\xi^+) - u'(\xi^-) = -1.$$

Observe that this condition is precisely the jump condition on the derivative at the point $x = \xi$ required of a Green's function. This last condition forces $-B - A = -1$. Therefore, solving the two equations for A and B simultaneously gives

$$A = (1 - \xi), \quad B = \xi.$$

Therefore the steady-state temperature in the bar, caused by a point source at $x = \xi$ is

$$u = (1 - \xi)x, \quad x < \xi; \quad u = \xi(1 - x), \quad x > \xi$$

or

$$u(x, \xi) = \xi(1 - x)H(x - \xi) + x(1 - \xi)H(\xi - x).$$

This is precisely the Green's function for the operator $L = -d^2/dx^2$ with boundary conditions $u(0) = u(1) = 0$ that we calculated in the last example. Consequently, this gives a physical interpretation for the Green's function; it is the response of a system (in this case the steady temperature response) to a point source (in this case a unit heat source). See Fig. 4.1.

This idea is fundamental in applied mathematics. The Green's function $g(x, \xi)$ is the solution to the *symbolic* boundary value problem

$$Ag(x, \xi) \equiv (-pg')' + qg = \delta(x, \xi), \tag{4.51}$$
$$B_1 g(a, \xi) = 0, \quad B_2 g(b, \xi) = 0, \tag{4.52}$$

where δ represents a point source at $x = \xi$ of unit strength. Furthermore, the solution (4.47) to the boundary value problem $Lu = f$ can therefore be regarded as a superposition of point sources of magnitude $f(\xi)$ over the entire interval $a < \xi < b$.

Example 4.21

Consider the differential operator $A = -d^2/dx^2$ on $0 < x < 1$ with the boundary conditions $u'(0) = u'(1) = 0$. In this case the Green's function does not exist since the equation $Lu = 0$ has nontrivial solutions (any constant function will satisfy the differential equation and the boundary conditions); stated differently, $\lambda = 0$ is an eigenvalue. Physically we can also see why the Green's function does not exist. This problem can be interpreted in the context of steady-state heat flow. The zero-flux boundary conditions imply that both ends of the bar are insulated, and so heat cannot escape from the bar. Thus it is impossible to impose a point source, which would inject heat energy at a constant, unit rate, and have the system respond with a time-independent temperature distribution; energy would build up in the bar, precluding a steady state.

In this heuristic discussion we noted that δ is not an ordinary function. Furthermore, the Green's function is not differentiable at $x = \xi$, so it remains to determine the meaning of applying a differential operator to g as in formulas (4.51)–(4.52). Our goal in the next section is to put these notions on firm ground.

Example 4.22

(Causal Green's function) The **causal Green's function** is the Green's function for an initial value problem. Consider the problem

$$Au \equiv -(pu')' + qu = f(t), \quad t > 0; \quad u(0) = u'(0) = 0 \qquad (4.53)$$

We assume p, p', and q are continuous for $t \geq 0$ and $p > 0$. The causal Green's function, also called the *impulse response function*, is the solution to (4.53) when f is a unit impulse applied at time τ, or, in terms of the delta function notation, when $f = \delta(t, \tau)$. Thus, symbolically, $Ag(t, \tau) = \delta(t, \tau)$, where g denotes the causal Green's function. We shall give a physical argument to determine g. Since the initial data is zero the response of the system is zero up until time τ; therefore, $g(t, \tau) = 0$ for $t < \tau$. For $t > \tau$ we require $Ag(t, \tau) = 0$, and we demand that g be continuous at $t = \tau$, or

$$g(\tau^+, \tau) = 0.$$

At $t = \tau$, the time when the impulse is given, we demand that g have a jump in its derivative of magnitude

$$g'(\tau^+, \tau) = -1/p(\tau).$$

This jump condition is derived in the same way that the jump condition is obtained in earlier. The continuity condition and the jump condition, along with the fact that g satisfies the homogeneous differential equation, are enough to determine $g(x, \tau)$ for $t > \tau$.

Now we return to the SLP problem (4.41)–(4.43) and ask what can be said if $\lambda = 0$ is an eigenvalue, that is, if the homogeneous problem has a nontrivial solution. In this case there may not be a solution, and if there is a solution, it is not unique. The following theorem summarizes the result.

Theorem 4.23

Consider the Sturm–Liouville problem (4.41)–(4.43), and assume there exists a nontrivial solution ϕ of the homogeneous problem $L\phi = 0$. Then (4.41)–

(4.43) has a solution if, and only if,

$$(\phi, f) \equiv \int_a^b \phi f \, dx = 0.$$

To prove necessity we assume a solution u exists. Then

$$(\phi, f) = (\phi, Au) = -\int_a^b \phi(pu')' \, dx + \int_a^b \phi q u \, dx.$$

The first integral may be integrated by parts twice to remove the derivatives from u and put them on ϕ. Performing this calculation gives, after collecting terms,

$$(\phi, f) = [p(u\phi' - \phi u')]_a^b + \int_a^b u(-(p\phi')' + q\phi) \, dx$$

$$= [p(u\phi' - \phi u')]_a^b + \int_a^b u A\phi \, dx$$

$$= [p(u\phi' - \phi u')]_a^b,$$

because $A\phi = 0$. Both ϕ and u satisfy the boundary conditions (4.42)–(4.43), and one can easily show that $[p(u\phi' - \phi u')]_a^b = 0$.

The proof of the "if" part of the theorem is less straightforward. We indicate the solution and leave the verification for the reader. Assume that $(\phi, f) = 0$ and let v be independent of ϕ and satisfy only the differential equation $Av = 0$, and not the boundary conditions. Now define

$$G(x, \xi) \equiv -\frac{1}{p(\xi)W(\xi)}(\phi(x)v(\xi)H(\xi - x) + \phi(\xi)v(x)H(x - \xi))$$

where H is the Heaviside function and $W = \phi v' - v\phi'$. Then

$$u(x) = c\phi(x) + \int_a^b G(x, \xi)f(\xi) \, d\xi$$

is a solution to $Lu = f$ for any constant c. (Note here that G is not the Green's function; as we have noted, the Green's function does not exist.)

4.4.3 Green's Function via Eigenfunctions

Suppose Green's function exists for the SLP (4.41)–(4.43). If we can solve the eigenvalue problem associated with the operator L, then we can find the Green's function. We know from Section 4.2 that the SLP problem

$$Lu = \lambda u, \quad a < x < b, \tag{4.54}$$

where L includes the boundary conditions (4.42)–(4.43), has infinitely many eigenvalues and corresponding orthonormal eigenfunctions λ_n and $\phi_n(x), n = 1, 2, \ldots$, respectively. Moreover, the eigenfunctions form a basis for the square-integrable functions on (a, b). Therefore we assume that the solution u of (4.41)–(4.43) is given in terms of the eigenfunctions as

$$u(x) = \sum_{n=1}^{\infty} c_n \phi_n(x), \qquad (4.55)$$

where the coefficients c_n are to be determined. Further, we write the given function f in terms of the eigenfunctions as

$$f(x) = \sum_{n=1}^{\infty} f_n \phi_n(x), \quad f_n = \int_a^b f(\xi) \phi_n(\xi)\, d\xi.$$

Substituting these expansions into (4.54) gives

$$L\left(\sum_{n=1}^{\infty} c_n \phi_n(x)\right) = \sum_{n=1}^{\infty} f_n \phi_n(x).$$

But

$$L\left(\sum_{n=1}^{\infty} c_n \phi_n(x)\right) = \sum_{n=1}^{\infty} c_n L(\phi_n(x))$$

$$= \sum_{n=1}^{\infty} c_n \lambda_n \phi_n(x)$$

$$= \sum_{n=1}^{\infty} f_n \phi_n(x).$$

Equating coefficients of the independent eigenfunctions gives

$$c_n = \frac{f_n}{\lambda_n} = \frac{1}{\lambda_n} \int_a^b f(\xi) \phi_n(\xi)\, d\xi.$$

Thus, from (4.55),

$$u(x) = \sum_{n=1}^{\infty} \frac{1}{\lambda_n} \left(\int_a^b f(\xi) \phi_n(\xi)\, d\xi\right) \phi_n(x).$$

Now we formally interchange the summation and integral to obtain

$$u(x) = \int_a^b \left(\sum_{n=1}^{\infty} \frac{\phi_n(x) \phi_n(\xi)}{\lambda_n}\right) f(\xi)\, d\xi.$$

Consequently, we have inverted the operator L and so we must have

$$g(x, \xi) = \sum_{n=1}^{\infty} \frac{\phi_n(x)\phi_n(\xi)}{\lambda_n},$$

which gives the Green's function in terms of the eigenvalues and eigenfunctions of the operator L. This expansion for g is called the **bilinear expansion**.

EXERCISES

1. Discuss the solvability of the boundary value problem

$$u'' + \pi^2 u = f(x), \quad 0 < x < 1; \quad u(0) = u(1) = 0.$$

2. Determine if there is a Green's function associated with the operator $Lu = u'' + 4u$, $0 < x < \pi$, with $u(0) = u(\pi) = 0$. Find the solution to the boundary value problem

$$u'' + 4u = f(x), \quad 0 < x < \pi; \quad u(0) = u(\pi) = 0.$$

3. Consider the boundary value problem

$$u'' - 2xu' = f(x), \quad 0 < x < 1; \quad u(0) = u'(1) = 0.$$

Find Green's function or explain why there isn't one.

4. Consider the boundary value problem

$$u'' + u' - 2u = f(x), \quad 0 < x < 1; \quad u(0) = u'(1) = 0.$$

Find Green's function or explain why there isn't one.

5. Use the method of Green's function to solve the problem

$$-(K(x)u')' = f(x), \quad 0 < x < 1; \quad u(0) = u(1) = 0; \quad K(x) > 0.$$

(Note: This differential equation is a steady-state heat equation in a bar with variable thermal conductivity $K(x)$.)

6. Consider a spring–mass system governed by the initial value problem

$$mu'' + ku = f(t), \quad t > 0; \quad u(0) = u'(0) = 0,$$

where $u = u(t)$ is the displacement from equilibrium, f is an applied force, and m and k are the mass and spring constants, respectively.

a) Show that the causal Green's function is

$$g(t, \tau) = \frac{1}{\sqrt{km}} \sin \sqrt{\frac{k}{m}} (t - \tau), \quad t > \tau.$$

b) Find the solution to the initial value problem and write it in the form

$$u(t) = \frac{1}{\sqrt{km}} \int_0^t \sin \sqrt{\frac{k}{m}} (t - \tau) f(\tau) \, d\tau.$$

7. By finding Green's function in two different ways, evaluate the sum

$$\sum_{n=1}^{\infty} \frac{\sin nx \sin n\xi}{n^2}, \quad 0 < x, \xi < \pi.$$

8. Find the inverse of the differential operator $Lu = -(x^2 u')'$ on $1 < x < e$ subject to $u(1) = u(e) = 0$.

4.5 Distributions

In the last section we showed, in an intuitive manner, that the Green's function satisfies a differential equation with a unit point source. Because the Green's function is not a smooth function, it leads us to the question of what it means to differentiate such a function. Also, we have not pinned down the properties of a point source (a delta function) in a precise way. If a point source is not a function, then what is it? The goal of this section is to come to answers to these questions. New notation, terminology, and concepts are required.

4.5.1 Test Functions

Let K be a set of real numbers. A real number c is said to be a **limit point** of set K if every open interval containing c, no matter how small, contains at least one point of K. If K is a set, then the **closure** of K, denoted by $\overline{K}$, is the set K along with all its limit points. A set is called **closed** if it contains all of its limit points. For example, the closure of the half-open interval $(a, b]$ is the closed interval $[a, b]$. A useful set in characterizing properties of a function ϕ is the set of points where it takes on nonzero values. The closure of this set is called the **support** of the function; precisely, we define supp $\phi \equiv \overline{\{x | \phi(x) \neq 0\}}$.

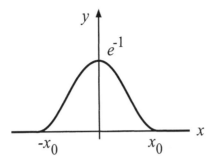

Figure 4.3 Plot of a test function.

So, the support of a function is the set of points where it is nonzero, along with the limit points of that set.

Now we introduce an important class of functions that plays a crucial role in our discussion. In the following, (a, b) denotes an open interval, which may be infinite. By the set $C_0^\infty(a, b)$ we mean the set of all continuous functions on (a, b) whose derivatives of all order exist and are continuous on (a, b), and which have their support contained in a closed, bounded subset of (a, b). These functions are called **test functions**. An example of a test function on $\mathbb{R}$ is the bell-shaped curve (see Fig. 4.3).

$$\phi(x) = \exp\left(-\frac{x_0^2}{x_0^2 - x^2}\right), |x| < x_0; \quad \phi(x) = 0, \quad |x| \geq x_0.$$

Therefore, a test function in a very smooth function on (a, b) with no corners or kinks in its graph or in any of its derivatives; its support must be entirely contained in (a, b) and, in fact, must lie in a closed, bounded interval strictly contained in (a, b); if a and b are finite, then a test function must be zero at a and b.

Finally, a function f is said to be **locally integrable** on (a, b) if

$$\int_c^d |f(x)| \, dx < \infty$$

for all subintervals $[c, d]$ of (a, b). Recall from calculus that an integrable function need not be differentiable, or even continuous.

As we know, a function with a corner (such as Green's function or the function $|x|$), does not have a derivative (tangent line) at that corner. However, it is possible to generalize the notion of derivative so that it makes sense to discuss the concept in a broader context. First we introduce the idea of a weak derivative. Let $u = u(x)$ be a continuously differentiable function on (a, b) and let $f = f(x)$ be its derivative. That is,

$$u' = f.$$

Multiplying both sides by a test function $\phi \in C_0^\infty(a, b)$ and integrating from a to b gives

$$\int_a^b u'\phi \, dx = \int_a^b f\phi \, dx.$$

Integrating the left side by parts and using the fact that $\phi(a) = \phi(b) = 0$ gives

$$-\int_a^b u\phi' \, dx = \int_a^b f\phi \, dx. \tag{4.56}$$

Now observe that equation (4.56) does not require u to be differentiable, just integrable. The integration by parts removed the derivative from u and put it on ϕ, which has derivatives of all order. We can take (4.56) as the basis of a definition for a derivative of an integrable function. If u and f are locally integrable functions, we say that f is the **weak derivative** of u on (a, b) if (4.56) holds for all $\phi \in C_0^\infty(a, b)$. Clearly, if u is continuously differentiable, then it has a weak derivative (the ordinary one); but a function can have a weak derivative without having an ordinary derivative. In fact, this idea can be extended so as not to require even local integrability of u or f.

Example 4.24

On the interval $(a, b) = (-1, 1)$ let $u(x) = |x|$ and $f(x) = H(x) - H(-x)$, where H is the Heaviside function. The function u is not differentiable at $x = 0$ in the usual sense. Yet $u' = f$ in the weak sense on $(-1, 1)$ because, from (4.56),

$$-\int_{-1}^1 |x|\phi'(x) \, dx = \int_{-1}^1 (H(x) - H(-x))\phi(x) \, dx, \quad \forall \phi \in C_0^\infty(-1, 1).$$

This is easily checked. The left side is

$$\int_{-1}^0 x\phi'(x) \, dx - \int_0^1 x\phi'(x) \, dx = -\int_{-1}^0 \phi(x) \, dx + \int_0^1 \phi(x) \, dx,$$

where we have integrated by parts and used the fact that ϕ vanishes at $x = -1$, 1. The right side is

$$\int_{-1}^0 -H(-x)\phi(x) \, dx + \int_0^1 H(x)\phi(x) \, dx = -\int_{-1}^0 \phi(x) \, dx + \int_0^1 \phi(x) \, dx,$$

and therefore (4.56) holds and $(d/dx)|x| = H(x) - H(-x)$ in a weak sense.

Example 4.25

Again take $(a, b) = (-1, 1)$ and let $u(x) = H(x)$. What is u' in the weak sense? Suppose $u' = f$ weakly. Then, by definition (4.56), we should have

$$-\int_{-1}^{1} H(x)\phi'(x)\,dx = \int_{-1}^{1} f(x)\phi(x)\,dx, \quad \forall \phi \in C_0^\infty(-1, 1).$$

The left side simplifies to

$$-\int_{0}^{1} H(x)\phi'(x)\,dx = -\int_{0}^{1} \phi'(x)\,dx = \phi(0).$$

So we require that

$$\int_{-1}^{1} f(x)\phi(x)\,dx = \phi(0), \quad \forall \phi \in C_0^\infty(-1, 1) \tag{4.57}$$

But, as the following calculation shows, there is no locally integrable function f for which (4.57) holds for all ϕ in $C_0^\infty(-1, 1)$. For, take the test function $\phi_a(x) = \exp(-a^2/(a^2 - x^2))$ for $|x| < a < 1$, and $\phi_a(x) = 0$ otherwise. Then

$$e^{-1} = \phi_a(0) = \int_{-a}^{a} f(x)\exp(-a^2/(a^2 - x^2))\,dx \le e^{-1}\int_{-a}^{a} |f(x)|\,dx.$$

The right side goes to zero as $a \to 0$ (because f is locally integrable), and so we obtain the contradiction $e^{-1} \le 0$. Therefore the weak derivative of the Heaviside function cannot be a locally integrable function. Intuitively, the derivative should be zero everywhere except at $x = 0$, where there is a unit jump in H; so the derivative should be localized at the point $x = 0$, similar to a point source. In fact, as we shall show, the derivative of the Heaviside function is the "delta function," in a sense to be defined.

4.5.2 Distributions

The last example showed that the concept of a weak derivative needs to be generalized further if we are to define the derivative of any locally integrable function. To accomplish this task we need to expand our idea of function and not think of the values of a function at points in its domain, but rather think of a function in terms of its action on other objects. For example, suppose $u = u(x)$ represents the distributed temperature at locations x in a bar. This is an idealized concept, because when we measure temperature at a point x, for example, we use a thermometer that has finite extent; it is not just localized at x. So, in fact, we are really measuring the average temperature $\bar{u}(x)$ of the points near x, or more precisely, a weighted average $\bar{u}(x) = \int u(\xi)\phi(x,\xi)d\xi$ over some interval containing x. We cannot measure the exact, idealized $u(x)$, just the average. The weight function $\phi(x,\xi)$ is characterized by properties of the thermometer at x. So sometimes it is better to think of a function in terms of its integrated values rather than its point values. In fact, often we have to give up our idea of requiring our functions to have values or local properties.

This idea leads us to define the notion of a *distribution,* or *generalized function.* A **distribution** is a mapping (transformation) that associates with each ϕ in $C_0^\infty(a,b)$ a real number. (Think of the ϕ as a thermometer.) We met the idea of mapping a function space to $\mathbb{R}$ in Chapter 3 on calculus of variations, where we considered functionals. A distribution is just a continuous linear functional on the set of test functions. If we denote the set of distributions by D', then $f \in D'$ implies $f \colon C_0^\infty(a,b) \to \mathbb{R}$. We denote the image of $\phi \in C_0^\infty(a,b)$ under f by (f,ϕ) rather than the usual function notation $f(\phi)$. Linearity implies $(f,\alpha\phi) = \alpha(f,\phi)$ and $(f,\phi_1 + \phi_2) = (f,\phi_1) + (f,\phi_2)$. The requirement that f is continuous is a slightly more technical concept, namely that of convergence in $C_0^\infty(a,b)$. We say that a sequence ϕ_n of test functions in $C_0^\infty(a,b)$ converges to zero in $C_0^\infty(a,b)$ (and we write $\phi_n \to 0$ in $C_0^\infty(a,b)$), if there is a single, closed bounded interval K in (a,b) containing the supports of all the ϕ_n, and on that interval K the sequence ϕ_n and the sequence of the derivatives $\phi_n^{(m)}$ converge uniformly to zero on K as $n \to \infty$. Then we say that f is **continuous** if $(f,\phi_n) \to 0$ for all sequences $\phi_n \to 0$ in $C_0^\infty(a,b)$.

Example 4.26

With each locally integrable function u on (a,b) there is a natural distribution u defined by

$$(u,\phi) \equiv \int_a^b u(x)\phi(x)\,dx, \quad \phi \in C_0^\infty(a,b). \tag{4.58}$$

Note that we speak of the distribution u and use the same symbol as for the function. One can verify that u defined by (4.58) satisfies the linearity and continuity properties for a distribution. Therefore, every locally integrable function is a distribution via (4.58).

Example 4.27

The distribution δ_ξ, $\xi \in (a,b)$, defined by

$$(\delta_\xi,\phi) = \phi(\xi), \quad \phi \in C_0^\infty(a,b), \tag{4.59}$$

is called the **Dirac** or **delta distribution** with pole at ξ. It is easily checked that δ_ξ is linear and continuous, and thus defines a distribution. The Dirac distribution δ_ξ acts by sifting out the value of ϕ at ξ. Note that we cannot represent (4.59) by an integral

$$(\delta_\xi, \phi) = \int_a^b \delta_\xi(x)\phi(x)\, dx = \phi(\xi),$$

because of the argument following equation (4.57); there we showed that there is no locally integrable function with this property. A distribution, like the Dirac distribution, that cannot be represented as (4.58) is called a **singular distribution**.

The customary notation used in distribution theory is confusing. Even when f is a singular distribution and no locally integrable function f exists for which $(f, \phi) = \int_a^b f(x)\phi(x)\, dx$, we still write (4.27), even though the integral makes no sense. The integral expression in (4.27) is only symbolic notation for the mathematically correct expression (f, ϕ). Further, even though it is impermissible to speak of the values of a distribution f at points x, we still on occasion write $f(x)$ to denote the distribution. For example, the Dirac distribution δ_0 with pole at zero is often written $\delta(x)$, and also δ_ξ is written $\delta_\xi(x)$ or $\delta(x, \xi)$.

Example 4.28

Consider $u(x) = H(x)$, the Heaviside function. We found earlier, in order for $u' = f$ to exist in a weak sense, it is necessary that

$$-\int_{-1}^1 H(x)\phi(x)\, dx = \int_{-1}^1 f(x)\phi(x)\, dx = \phi(0) \qquad (4.60)$$

for all test functions ϕ. There is no integrable function f that satisfies this equation. However, if f is the Dirac distribution $f = \delta_0$ with pole at zero, then according to (4.59), equation (4.60) is satisfied if we interpret the integral symbolically. Thus, in a *distributional* sense, $H' = \delta_0$, or $H'(x) = \delta(x)$.

Now let us take a more formal approach and underpin the issues we have raised. For conciseness, we use $D \equiv C_0^\infty(a, b)$ to denote the set of test functions on (a, b), and D' to denote the set of distributions (continuous linear functionals) on D. The interval (a, b) can be $(-\infty, \infty)$ or $(0, \infty)$. Equality of two distributions is defined by requiring the two distributions have the same action on all the test functions; we say that two distributions f_1 and f_2 are **equal** if $(f_1, \phi) = (f_2, \phi)$ for all $\phi \in D$, and we write $f_1 = f_2$ in D'.

It is possible to do algebra and calculus in the set D' of distributions just as we do with ordinary functions. It is easy to see that D' is a linear space under the definition $(f_1 + f_2, \phi) = (f_1, \phi) + (f_2, \phi)$ of addition, and $(c\, f, \phi) = c(f, \phi), c \in \mathbb{R}$, of scalar multiplication. To define rules for distributions we use the guiding principle that any general definition should be consistent with what is true for locally integrable functions, since locally integrable functions are distributions. For example, let $\alpha = \alpha(x)$ be an infinitely differentiable function (i.e., $\alpha \in C^\infty(a, b)$), and suppose we wish to define the multiplication αf, where $f \in D'$. If f is a locally integrable function, then

$$(\alpha f, \phi) = \int_a^b (\alpha f)\phi \, dx = \int_a^b f(\alpha\phi) \, dx = (f, \alpha\phi). \tag{4.61}$$

Observe that $\alpha\phi \in D$. Therefore we make the definition

$$(\alpha f, \phi) = (f, \alpha\phi)$$

for *any* $f \in D'$; we have then defined the distribution αf by its action on test functions through the action of f.

We can also define the derivative of a distribution f. Again, assume f and f' are locally integrable. Using the same argument that led to (4.56), namely integration by parts and the fact that test functions have support in the interval (a, b), we are led to

$$(f', \phi) = \int_a^b f'\phi \, dx = -\int_a^b f\phi' \, dx = -(f, \phi').$$

Therefore, if $f \in D'$ we define the derivative $f' \in D'$ by

$$(f', \phi) = -(f, \phi'), \quad \phi \in D.$$

f' is called the **distributional derivative** of f. In the same way (integrating by parts n times), we define the nth distributional derivative $f^{(n)}$ of a distribution f by

$$(f^{(n)}, \phi) = (-1)^n (f, \phi^{(n)}), \quad \phi \in D.$$

It is routine to verify that these definitions do indeed satisfy the requirements of a distribution (linearity and continuity). Thus, we have the interesting fact that a distribution has derivatives of all orders.

Example 4.29

Calculate the derivative of the Dirac distribution δ_ξ. By definition,

$$(\delta'_\xi, \phi) = -(\delta_\xi, \phi') = -\phi'(\xi).$$

Other properties of distributions are useful as well. For example, if $c \in \mathbb{R}$ and $f(x) \in D'(\mathbb{R})$, then we can define the translated distribution $f(x - c) \in D'(\mathbb{R})$ by

$$(f(x - c), \phi(x)) = (f(x), \phi(x + c)), \quad \phi \in D.$$

(see Exercise 5). For example, if $f(x) = \delta(x)$, the Dirac distribution with pole at zero, then $\delta(x - c)$ is defined by $(\delta(x - c), \phi(x)) = (\delta(x), \phi(x + c)) = \phi(c)$. But also, $(\delta_c(x), \phi(x)) = \phi(c)$. Therefore $\delta(x - c)$ has the same action on test functions as $\delta_c(x)$ and so

$$\delta_c(x) = \delta(x - c).$$

Thus we have yet another way to write the delta distribution with pole at c. Thus, the sifting property of the delta "function" is often written, as we have already done, symbolically as

$$\int_a^b \delta(x - \xi)\phi(x)\,dx = \phi(\xi).$$

In summary, a distribution is a generalization of the idea of a function. Rather than defined at points with local properties like functions are, distributions are global objects that are defined in terms of their integrated values, or their action on test functions.

4.5.3 Distribution Solutions to Differential Equations

We are now poised to define what is meant by a a weak solution and a distribution solution to a differential equation. Consider the differential operator L defined by

$$Lu \equiv \alpha u'' + \beta u' + \gamma u, \tag{4.62}$$

where α, β, and γ are C^∞ functions on (a, b). By a classical, or **genuine solution,** to the differential equation

$$Lu = f, \tag{4.63}$$

where f is continuous, we mean a twice, continuously differential function $u = u(x)$ that satisfies the differential equation identically for all x. In other words, a genuine solution has enough derivatives so that it makes sense to substitute it into the differential equation to check, pointwise, if it is indeed a solution. We can also interpret (4.63) in a distributional sense. That is, if u and f are distributions, then Lu is a distribution; if $Lu = f$, as distributions, then we say u is a **distribution solution** of (4.63). Recall, for $Lu = f$ as distributions, that we mean $(Lu, \phi) = (f, \phi)$ for all test functions $\phi \in D$. Clearly, if u is a classical solution, then it is a distribution solution, but the converse is not true.

There is still more terminology. A **fundamental solution** associated with L is a distribution solution to

$$Lu = \delta(x - \xi),$$

where the right side of (4.63) is a Dirac distribution. A fundamental solution must satisfy only a differential equation in a distributional sense, and it need not be unique.

T elaborate on these ideas further, let u be a distribution. Then Lu is a distribution and it is defined by the action

$$
\begin{aligned}
(Lu, \phi) &= (\alpha u'', \phi) + (\beta u', \phi) + (\gamma u, \phi) \\
&= (u'', \alpha\phi) + (u', \beta\phi) + (u, \gamma\phi) \\
&= (u, (\alpha\phi)'') - (u, (\beta\phi)') + (u, \gamma\phi) = (u, L^*\phi),
\end{aligned}
$$

where L^* is the **formal adjoint** operator defined by

$$
L^*\phi \equiv (\alpha\phi)'' - (\beta\phi)' + \gamma\phi.
$$

It follows that if u is a distribution solution to $Lu = f$, then

$$
(u, L^*\phi) = (f, \phi), \quad \forall \phi \in D. \tag{4.64}
$$

If both u and f are locally integrable, then (4.64) can be written

$$
\int_a^b u(x) L^*\phi(x)\, dx = \int_a^b f(x)\phi(x)\, dx, \quad \forall \phi \in C_0^\infty(a, b), \tag{4.65}
$$

and we say u is a **weak solution** of (4.63) if (4.65) holds. Note that (4.65) does not require u to have derivatives in the ordinary sense. A weak solution is a special case of a distribution solution; a weak solution is a function solution whose (weak) derivatives are functions, and (4.65) holds. Obviously, a classical solution is a weak solution, but the converse is not true.

In passing, we call an operator L **formally self-adjoint** if $L^* = L$. For example, the Sturm–Liouville operator $Au \equiv -(pu')' + qu$ is formally self-adjoint (see Exercise 9).

What we accomplished in the preceding discussion is significant. We replaced the usual question "Does this problem have any function solutions?" with the question "Does this problem have any distribution solutions or weak solutions?" Thus, we are not discarding physically meaningful solutions because of minor technicalities, for example, not having a derivative at some point. Now the symbolic differential equation for the Green's function,

$$
Ag(x, \xi) = \delta(x - \xi),
$$

as discussed in Section 4.4, is meaningful. It is to be interpreted as an equation for distributions and actually means

$$
(Ag(x, \xi), \phi) = (\delta(x - \xi), \phi), \quad \forall \phi \in D,
$$

or, equivalently,

$$
(g(x, \xi), A^*\phi) = \phi(\xi).
$$

Example 4.30

We verify that the function

$$u = g(x, \xi) = \xi(1 - x)H(x - \xi) + x(1 - \xi)H(\xi - x)$$

is the Green's function associated with the operator $L = -d^2/dx^2$ on $(0, 1)$ subject to the boundary conditions $u(0) = u(1) = 0$. That is, we show g satisfies the problem

$$-g''(x, \xi) = \delta(x - \xi), \tag{4.66}$$
$$g(0, \xi) = g(1, \xi) = 0, \tag{4.67}$$

in a distributional sense. We must show

$$(g(x, \xi), -\phi'') = \phi(\xi) \tag{4.68}$$

for all $\phi \in D(0, 1)$. To this end,

$$(g(x, \xi), -\phi'') = -\int_0^1 g(x, \xi)\phi''(x)\, dx$$
$$= -(1 - \xi)\int_0^\xi x\phi''(x)\, dx - \xi \int_\xi^1 (1 - x)\phi''(x)\, dx.$$

Now we integrate the two integrals by parts to get

$$\int_0^\xi x\phi''\, dx = \xi\phi'(\xi) - \phi(\xi),$$

and

$$\int_\xi^1 (1 - x)\phi''\, dx = (\xi - 1)\phi'(\xi) - \phi(\xi).$$

Upon substitution we obtain (4.68). So, interpreted properly, the Green's function is the solution to the boundary value problem (4.66)–(4.67) with a point source.

EXERCISES

1. Does $(u, \phi) = \phi(0)^2$ define a distribution in $D'(\mathbb{R})$?

2. Is the function $\phi(x) = x(1 - x)$ on $(0, 1)$ a test function? Why or why not?

3. Is $f(x) = 1/x$ locally integrable on $(0, 1)$?

4. Let $\phi \in D(\mathbb{R})$ be a test function. For which of the following does $\psi_n \to 0$ in $D(\mathbb{R})$?

   (a) $\psi_n(x) = \frac{1}{n}\phi(x)$.

   (b) $\psi_n(x) = \frac{1}{n}\phi\left(\frac{x}{n}\right)$.

   (c) $\psi_n(x) = \frac{1}{n}\phi(nx)$.

5. Prove the following statements.

   (a) $x\delta'(x) = -\delta(x)$.

   (b) $\alpha(x)\delta'(x) = -\alpha'(0)\delta(x) + \alpha(0)\delta'(x)$, where $\alpha \in C^\infty(\mathbb{R})$.

6. Show that $u(x, \xi) = \frac{1}{2}|x - \xi|$ is a fundamental solution for the operator $L = d^2/dx^2$ on $\mathbb{R}$.

7. Let $c \neq 0$ be a constant and $f(x)$ a distribution in $D'(\mathbb{R})$. Show that it is appropriate to define the distributions $f(x - c)$ and $f(cx)$ by

$$(f(x - c), \phi(x)) = (f(x), \phi(x + c)),$$

   and
$$(f(cx), \phi(x)) = \frac{1}{|c|}\left(f(x), \phi\left(\frac{x}{c}\right)\right).$$

8. Compute the distributional derivative of $H(x)\cos x$, where H is the Heaviside function. Does the derivative exist in a weak sense?

9. Show that the Sturm–Liouville operator $Au \equiv -(pu')' + qu$ is formally self-adjoint.

10. Find a fundamental solution associated with the operator L defined by $Lu = -x^2u'' - xu' + u$, $0 < x < 1$, such that $u(x, \xi) = x$ for $0 < x < \xi$.

11. In $D'(\mathbb{R})$ compute $(\frac{d}{dx} - \lambda)(H(x)e^{\lambda x})$.

REFERENCES AND NOTES

This chapter presented the core of classical applied mathematics—boundary value problems, Green's functions, integral equations, and distributions. There are many texts that specialize in one or more of these subjects. Comprehensive treatments, all more advanced than the present text, are given in Friedman (1956), Keener (2000), and Stakgold (1998).

Birkhoff, G. & Rota, G.-C. 1978. *Ordinary Differential Equations*, 2nd ed., John Wiley & Sons, New York.

Friedman, B. 1956. *Principles and Techniques of Applied Mathematics*, John Wiley & Sons, New York. (Reprinted by Dover Publications, 2001).

Keener, J. P. 2000. *Principles of Applied Mathematics*, revised edition, Perseus Press, Cambridge, MA.

Stakgold, I. 1998. *Green's Functions and Boundary Value Problems*, 2nd, ed., Wiley–Interscience, New York.

5

Discrete Models

From elementary courses we know that differential equations are often approximated by discrete, difference formulas; for example, the Euler and the Runge–Kutta numerical algorithms lead to difference equations. But, as well, discrete equations arise naturally without a continuous counterpart. The recent emphasis on quantitative methods in the biological sciences has brought discrete models to the forefront, not only in classical areas like population dynamics and epidemiology, but in newer applications in genomics arising from the accumulation of DNA sequence data, bioinformatics, and other phylogenetic processes. The digital revolution in electrical engineering has made discrete models and discrete transforms central in the development of advanced technological devices. Digital signal processing is a major areas of study. Discrete models are conceptually simpler that their differential equation counterparts. However, discrete models are less amenable to analytic solution techniques and their dynamics can be more complicated, often exhibiting cycling and chaotic behavior.

Another important endeavor is to understand how stochasticity, or randomness, enters and affects various systems. For example, there is always stochasticity in the environment that affects populations, noise in electrical circuits and devises that affect their response, and random forces on structures that affect their vibrations. We consider some aspects of randomness in this chapter as well.

5.1 One-Dimensional Models

5.1.1 Linear and Nonlinear Models

In differential equations the time t runs continuously, and so differential equations are often referred to as continuous time models. However, some processes are better formulated as **discrete time models** where time ticks off in discrete units $t = 0, 1, 2, 3, \ldots$ (say in days, months, or years, etc.). For example, if the money in a savings account is compounded monthly, then we need only compute the principal each month. In a fisheries model, the number of fish may be estimated once a year. Or, a wildlife conservationist may census a deer population in the spring and fall to estimate their numbers and make decisions on allowable harvesting rates. Data is usually collected at discrete times. In electrical engineering, continuous data is can be generated graphically on an oscilloscope, but digital data is equally common and is collected, for example, in a sampling process, in discrete time steps.

In a discrete time model the state function is a sequence x_t, i.e., x_0, x_1, x_2, x_3, ..., rather than a continuous function. The subscripts denote the time; e.g., in a weekly census of mosquitos grown in a laboratory, x_5 would denote the number of mosquitos at the fifth week. We graph the states, or sequence, x_t as a set of points (t, x_t) in a tx plane, often connecting them by straight line segments.

A one-stage, or first-order, **discrete time model**, the analog of a first-order differential equation, is an equation of the form

$$x_{t+1} = f(t, x_t), \quad t = 0, 1, 2, 3, \ldots, \tag{5.1}$$

where f is a given function. Such equations are called **difference equations** and recursion relations. Knowledge of the initial state x_0 allows us to compute the subsequent states recursively, in terms of the previously computed states. Thus, (5.1) is a deterministic update rule that tells us how to compute the next value in terms of the previous one. A sequence x_t that satisfies the model is a solution to the equation. If $f(t, x_t) = a_t x_t + b_t$, where a_t and b_t are given, fixed sequences, then the model is linear; otherwise, (5.1) is nonlinear. In the sequel we mostly examine the autonomous equation

$$x_{t+1} = f(x_t), \quad t = 0, 1, 2, 3, \ldots, \tag{5.2}$$

where the right side does not depend explicitly on t. Discrete models are also defined in terms of changes $\Delta x_t = x_{t+1} - x_t$ of the state x_t. Thus,

$$\Delta x_t = g(x_t)$$

defines a discrete model, where g is the given change. Finally, some models are also defined in terms of the relative change, or *per capita* change, $\Delta x_t / x_t$. Thus,

$$\frac{\Delta x_t}{x_t} = h(x_t),$$

where h is the given *per capita* change. Any form can be obtained easily from another by simple algebra.

An autonomous two-stage, or second-order, discrete time model has the form

$$x_{t+2} = f(x_{t+1}, x_t), \quad t = 0, 1, 2, 3, ...$$

Now a state depends upon two preceding states, and both x_0 and x_1 are required to start the process.

Difference equations are recursion formulas, and programs for computing the values x_t are easily composed on computer algebra systems and graphing calculators.

Example 5.1

The simplest discrete model, the **growth-decay** process, is linear and has the form

$$\begin{aligned} x_{t+1} &= x_t + rx_t \\ &= (1+r)x_t. \end{aligned} \tag{5.3}$$

For example, if x_t is the principal at month t in a savings account that earns 0.3% per month, then the principal at the $(t+1)$st month is $x_{t+1} = x_t + 0.003x_t$. We can perform successive iterations to obtain

$$\begin{aligned} x_1 &= (1+r)x_0, \\ x_2 &= (1+r)x_1 = (1+r)^2 x_0, \\ x_3 &= (1+r)x_2 = (1+r)^3 x_0, \end{aligned}$$

and so on. By induction, the solution to (5.3) is

$$x_n = (1+r)^t x_0.$$

If $r > 0$ then x_t grows geometrically and we have a *growth model*. If $-1 < r < 0$ then $1 + r$ is a proper fraction and x_t goes to zero geometrically; this is a *decay model*. If $-2 < r < -1$, then the factor $1 + r$ is negative and the solution x_t will oscillate between negative and positive values as it converges to zero. Finally, if $r < -2$ the solution oscillates without bound. It is also easily checked by direct substitution that the sequence

$$x_t = C(1+r)^t$$

is a solution to the equation for any value of C:

$$x_{t+1} = C(1+r)^{t+1} = (1+r)C(1+r)^t = (1+r)x_t.$$

If x_0 is fixed, then $C = x_0$. Discrete growth-decay models are commonplace in finance, in ecology, and in other areas. For example, in ecology we often write the model (5.3) as

$$\frac{\Delta x_t}{x_t} = r.$$

Then we can recognize r as the constant *per capita* growth rate. For populations in general, the constant r is given by $r = b - d + i - \epsilon$, where b, d, i, and ϵ are the birth, death, immigration, and emigration rates, respectively. When $r > 0$ the model (5.3) is called the **Malthus model** of population growth.

Example 5.2

The last example showed that the difference equation $x_{t+1} = \lambda x_t$ has general solution $x_t = C\lambda^t$, where C is any constant. Let us modify the equation and consider the linear model

$$x_{t+1} = \lambda x_t + p, \tag{5.4}$$

where p is a constant. We can consider this model to be, say, the monthly growth of principal x_t in a bank account where r $(\lambda = 1+r)$ is the monthly interest rate and p is a constant monthly addition to the account. In an ecological setting, λ may be the growth rate of a population and p a constant recruitment rate. We can solve this equation by recursion to find a formula for x_t. If x_0 is the initial value, then iteration yields

$$
\begin{aligned}
x_1 &= \lambda x_0 + p, \\
x_2 &= \lambda x_1 + p = \lambda^2 x_0 + \lambda p + p, \\
x_3 &= \lambda x_2 + b = \lambda^3 x_0 + \lambda^2 p + \lambda p + p, \\
&\cdots \\
x_t &= \lambda^t x_0 + \lambda^{t-1} p + \lambda^{t-2} p + \cdots + \lambda p + p.
\end{aligned}
$$

By the formula for the sum of a geometric sequence,

$$\lambda^{t-1} + \lambda^{t-2} + \cdots + \lambda + 1 = \frac{1 - \lambda^t}{1 - \lambda},$$

and consequently

$$x_t = \lambda^t x_0 + p\frac{1 - \lambda^t}{1 - \lambda}, \tag{5.5}$$

which is the solution to (5.4).

Example 5.3

(Logistics model) In the Malthus model of population growth,

$$\frac{\Delta x_t}{x_t} = r \qquad \text{or} \qquad x_{t+1} = (1+r)x_t,$$

the population x_t grows unboundedly if the *per capita* growth rate r, or the change in population per individual, is positive. Such a prediction cannot be accurate over a long time. If, for example, $r = b - d$, where b is the *per capita* birth rate and d is the *per capita* death rate, then

$$\frac{\Delta x_t}{x_t} = b - d.$$

We might expect that for early times, when the population is small, there are ample environmental resources to support a high birth rate, with the death rate being small. But for later times, as the population grows, there is a higher death rate as individuals compete for space and food (intraspecific competition). Thus, we should argue for a decreasing *per capita* growth rate r.as the population increases. The simplest such assumption is to take a linearly decreasing per capita rate, that is, $r(1 - x_t/K)$, where K is the *carrying capacity*, or the population where the growth rate is zero. See Fig. 5.1. Then

$$\frac{\Delta x_t}{x_t} = r\left(1 - \frac{x_t}{K}\right). \tag{5.6}$$

In the standard form

$$x_{t+1} = x_t\left(1 + r - \frac{r}{K}x_t\right), \tag{5.7}$$

which is a nonlinear. The discrete logistics model , quadratic in the population, is the simplest nonlinear model that we can develop. It is tempting to identify $b = r$ as a constant birth rate and $d = \frac{r}{K}x_t$ as the death rate depending upon the population. But an alternative is to take $b = r - \frac{r}{2K}x_t$ and $d = \frac{r}{2K}x_t$, with both depending upon population. A plot of the population is shown in Fig. 5.2 with $r = 0.5$, $K = 100$, and $x_0 = 20$. It shows a steady increase up to the carrying capacity, where it levels off. The accompanying MATLAB *m-file* produces the sequence in Fig. 5.2 for $t = 1, ..., 20$.

```
function Logistics
r=0.5; K=100;
x=15; xhistory=x;
for t=1:50;
    x=x.*(1+r-r*x/K);
    xhistory=[xhistory, x];
end
```

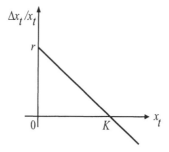

Figure 5.1 *Per capita* growth rate for the logistics model.

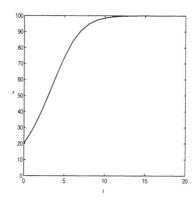

Figure 5.2 Logistics population growth.

plot(xhistory)

In the next section we observe that this is not the whole story; different values of the parameters can lead to interesting, unusual, and complex behavior.

Example 5.4

(The Ricker model). The **Ricker model** is a nonlinear, discrete time ecological model for the yearly population x_t of a fish stock where adult fish cannibalize the young. The dynamics is

$$x_{t+1} = bx_t e^{-cx_t},$$

where $b > 1$ and c is positive. In a heuristic manner, one can think of the model in the following way. In year t there are x_t adult fish, and they would normally give rise to bx_t adult fish the next year, where b is the number of fish produced per adult. However, if adults eat the younger fish, only a fraction of

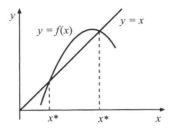

Figure 5.3 Graphical method for finding equilibrium as the intersection points of the graphs of $y = x$ and $y = f(x)$.

those will survive to the next year to be adults. We assume the probability of a fish surviving cannibalism is e^{-cx_t}, which decreases as the number of adults increases. Thus, $bx_t e^{-cx_t}$ is the number of adults the next year. One cannot "solve" this model to obtain a formula for the fish stock x_t, so we must be satisfied to plot its solution using iteration, as was done for the logistics model in the last example.

5.1.2 Equilibria, Stability, and Chaos

An important question for discrete models, as it is for continuous models, is whether the state of the system approaches an equilibrium as time gets large (i.e., as t approaches infinity). Equilibrium solutions are constant solutions, or constant sequences. We say $x_t = x^*$ is an **equilibrium solution** of $x_{t+1} = f(x_t)$ if

$$x^* = f(x^*) \tag{5.8}$$

Equivalently, x^* is a value that makes the change Δx_t zero. Graphically, we can find equilibria x^* as the intersection points of the graph of $y = f(x)$ and $y = x$ in an xy plane. See Fig. 5.3.

Example 5.5

(Logistics model) Setting $\Delta x_t = 0$ in the logistics model gives $rx^* \left(1 - \frac{x^*}{K}\right) = 0$ or, $x^* = 0$ and $x^* = K$. These two equilibria represent extinction and the population carrying capacity, respectively. Graphically, we can plot $y = x$ vs. $y = f(x) = x\left(1 + r - \frac{r}{K}x\right)$; the equilibria are at the intersections points of the two curves. See Fig. 5.4.

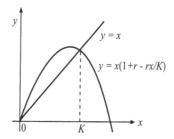

Figure 5.4 Graphical method for finding equilibrium 0 and K for the logistics model.

Example 5.6

(Ricker model) An equilibrium state for the Ricker model must satisfy

$$x^* = bx^* e^{-cx^*},$$

or

$$x^*(1 - be^{-cx^*}) = 0.$$

Therefore one equilibrium state is $x^* = 0$, which corresponds to extinction. Setting the other factor equal to zero gives

$$be^{-cx^*} = 1$$

or

$$x^* = \frac{\ln b}{c}.$$

If $b > 1$ we obtain a positive, viable equilibrium population.

If there is an equilibrium solution x^* to a discrete model (5.8), we always ask about its permanence, or stability. For example, suppose the system is in equilibrium and we perturb it by a small amount (natural perturbations are present in all physical and biological systems). Does the system return to that state or does it do something else (e.g., go to another equilibrium state, or blow up)? We say an equilibrium state is **locally asymptotically stable** if small perturbations decay and the system returns to the equilibrium state. If small perturbations of the equilibrium do not cause the system to deviate too far from the equilibrium, we say the equilibrium is stable. If a perturbation grows, then we say the equilibrium state is **unstable**. In the next paragraph we use a familiar argument to determine the stability of an equilibrium population.

Let x^* be an equilibrium state for (5.8), and assume y_0 represents a small deviation from x^* at $t = 0$. Then this perturbation will be propagated in time, having value y_t at time t. Does y_t decay, or does it grow? The dynamics of the state $x^* + y_t$ (the equilibrium state plus the deviation) must still satisfy the dynamical equation. Substituting $x_t = x^* + y_t$ into (5.8) gives

$$x^* + y_{t+1} = f(x^* + y_t).$$

We can simplify this equation using the assumption that the deviations y_t are small. We can expand the right side in a Taylor series centered about the value x^* to obtain

$$x^* + y_{t+1} = f(x^*) + f'(x^*)y_t + \frac{1}{2!}f''(x^*)y_t^2 + \frac{1}{3!}f''(x^*)y_t^3 + \cdots.$$

Because the deviations are small we can discard the higher powers of y_t and retain only the linear term. Moreover, $x^* = f(x^*)$ because x^* is an equilibrium. Therefore, small perturbations are governed by the linearized perturbation equation, or **linearization**

$$y_{t+1} = f'(x^*)y_t.$$

This difference equation is the growth-decay model (note that $f'(x^*)$ is a constant). For conciseness, let $\lambda = f'(x^*)$. Then the difference equation has solution

$$y_t = y_0 \lambda^t.$$

If $|\lambda| < 1$ then the perturbations y_t decay to zero and x^* is locally asymptotically stable; if $|\lambda| > 1$ then the perturbations y_t grow and x^* is unstable. If $\lambda = 1$ then the linearization gives no information about stability and further calculations are required. The stability indicator λ is called the **eigenvalue**. Therefore, if the absolute value of the slope of the tangent line to $f(x)$ at the intersection with $y = x$ is less than 1, then the equilibrium is asymptotically stable; if the absolute value of the slope is greater than 1, the equilibrium is unstable. In other words, we get stability if f is not too steep at the intersection point.

Example 5.7

Consider the Ricker model

$$x_{t+1} = bx_t e^{-cx_t}.$$

where $c > 0$ and $b > 1$. Let us check the equilibrium

$$x^* = \frac{\ln b}{c}.$$

for stability. Here $f(x) = bxe^{-cx}$ and we must calculate the eigenvalue, or the derivative of f evaluated at equilibrium. To this end $f'(x) = (-cx + 1)be^{-cx}$; evaluating at x^* gives

$$\lambda = f'(x^*) = (-cx^* + 1)be^{-cx^*} = 1 - \ln b.$$

We get asymptotic stability when

$$-1 < 1 - \ln b < 1,$$

or

$$1 < b < e^2 \approx 7.39.$$

If $b < e^2$, then the perturbations decay and the equilibrium is asymptotically stable. Let us follow up on this by performing some simulations with different values of b and observe the effect. Fix $c = 0.001$. Figure 5.5 shows the results for $b = 6.5\,9, 13, 18$. The equilibrium is at $x^* = 1000 \ln b$. For $b = 6.5$, within the stability range, the solution does indeed represent an asymptotically stable, decaying oscillation. For $b = 9$, however, a periodic, 2-cycle appears, alternating between high and low population values. When b is increased further, there is some value of b where suddenly the 2-cycle bifurcates into a 4-cycle. The plot shows the periodic 4-cycle when $b = 13$. This period doubling continues as b increases further, giving 8-cycles, 16-cycles, and so on. It can be shown that these cycles are stable limit cycles. But there is a critical value of b where the patterns disappear, and there are seemingly random fluctuations in the population ($b = 18$ shows this case). When b exceeds this critical value these non-periodic fluctuations become highly sensitive to initial conditions, a behavior known as **chaos**. The frames in Fig. 5.6 show the solution for conditions $x_0 = 99$ and $x_0 = 101$ with $b = 18$. We observe much different solutions. In the chaotic regime, the solution is deterministic (rather than stochastic), but highly unstable with respect to initial data.

Chaotic behavior is common in discrete models, even in one dimension. Such behavior does not appear in differential equations until dimension 3, other than in forced systems.

Next we illustrate a graphical procedure to determine the stability of an equilibrium solution of $x_{t+1} = f(x_t)$. The discussion refers to Fig. 5.7. We first sketch the curves $y = x$ and $y = f(x)$ on a set of axes, as in Fig. 5.4. We then mark the beginning value x_0 on the x axis. To find x_1 we move vertically to the graph of $f(x)$, because $x_1 = f(x_0)$, and mark x_1 on the y axis. To find x_2 we reflect x_1 back to the x axis through the line $y = x$. Now we have x_1 on the x axis. We repeat the process by moving vertically to the curve $f(x)$, which gives x_2. We reflect it back to the x axis, and so on. This

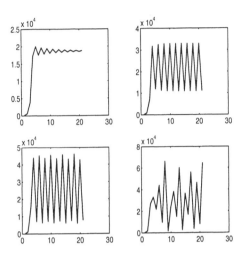

Figure 5.5 Population dynamics for the Ricker model for $b = 6.5, 9, 13, 18$, showing a decaying oscillation, 2-cycle, 4-cycle, and chaos, as b increases.

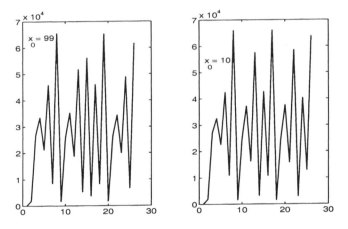

Figure 5.6 Plots of discrete solutions of the Ricker equation in the case the initial populations are close, in the chaos regime.

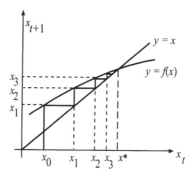

Figure 5.7 Cobweb diagram for a stable equilibrium.

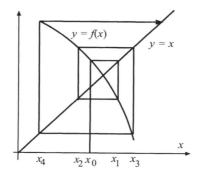

Figure 5.8 Cobweb diagram for an unstable equilibrium.

sequence of moves is accomplished by starting at x_0, going vertically to $f(x)$, going horizontally to the diagonal, vertically to the curve, horizontally to the diagonal, alternating back and forth. The plot of these vertical and horizontal moves that give the segments between the curve $f(x)$ and the diagonal is called the **cobweb diagram**. If the equilibrium is asymptotically stable, the cobweb will converge to the equilibrium represented by the intersection point. If the equilibrium is unstable, the cobweb will diverge from the intersection point. Figure 5.8 shows a divergent cobweb.

EXERCISES

1. Solve the following linear difference equations and plot their solutions:

   a) $x_{t+1} = \frac{1}{2}x_t$, $x_0 = 100$.

   b) $x_{t+1} = -1.3x_t$, $x_0 = 25$.

c) $x_{t+1} = -0.2x_t + 0.7$, $x_0 = 50$.

d) $x_{t+1} = 4x_t + 1$, $x_0 = 1$.

2. Write each of the models in Exercise 1 in terms of the *per capita* rate $\Delta x_t/x_t$.

3. If you borrow $15,000 to purchase a new car at a monthly rate of 0.9% and you pay off your loan in 4 years, what is your monthly payment?

4. Use iteration to show that the solution to the difference equation $x_{t+1} = \lambda x_t + p_t$ can be written

$$x_t = x_0\lambda^t + \sum_{k=0}^{t-1} p_k \lambda^{t-k-1}.$$

5. The next three exercises investigate second-order discrete models that are the analogs of the differential equation $u'' = pu' - qu$. Consider a second-order linear discrete model of the form

$$x_{t+2} = -px_{t+1} - qx_t$$

Assume that the solution sequence takes the form $x_t = r^t$ for some r to be determined. Show that r satisfies the quadratic equation (called the **characteristic equation**)

$$r^2 + pr + q = 0.$$

In the case that there are two distinct real roots r_1 and r_2 we have two distinct solutions r_1^t and r_2^t. Show that for any two constants A and B, the discrete function $x_t = Ar_1^t + Br_2^t$ is a solution to the difference equation. This expression is called the general solution, and all solutions have this form. Use this method to find the general solution of

$$x_{t+2} = x_{t+1} + 6x_t.$$

Find a specific solution that satisfies the conditions $x_0 = 1$ and $x_1 = 5$.

6. Referring to Exercise 5, show that if the characteristic equation has two equal real roots ($r = r_1 = r_2$), then $x_n = r^n$ and $x_n = nr^n$ are both solutions to the equation. The general solution is $x_t = Ar^t + Btr^t = r^t(A + Bt)$. Find the general solution of the discrete model

$$x_{t+2} = 4x_{t+1} - 4x_t.$$

Find a specific solution that satisfies the conditions $x_0 = 1$ and $x_1 = 0$.

7. Investigate the case when the roots of the characteristic equation in Exercise 5 are complex conjugates, that is, $r = a \pm bi$. Find two real solutions. Hint: use the exponential representation of the complex number $a + bi$:

$$a + bi = \rho e^{i\theta}.$$

What is the general solution to the discrete model

$$x_{t+2} = x_{t+1} - x_t?$$

Find and plot the solution if the initial states are $x_0 = 1$ and $x_1 = \frac{1}{2}$.

8. Find the general solution to the **Beverton–Holt** model equation

$$x_{t+1} = \frac{a x_t}{b + x_t}$$

by making a substution $w_t = \frac{1}{x_t}$.

9. Draw the cobweb diagram for the model

$$x_{t+1} = \frac{8 x_t}{1 + 2 x_t}$$

when $x_0 = 0.5$.

10. A density-limited population model is

$$x_{t+1} = \frac{a x_t}{(1 + b x_t)^c},$$

where a, b, and c are positive.

a) Use scaling to reduce the number of parameters in the model two.

b) Using the scaled equation, find the equilibria and determine conditions for stability for each.

c) Plot, in two-dimensional parameter space, the region of stability in the case there is a positive equilibrium.

11. The *Gilpin–Ayala* population model is given by

$$\Delta P_t = r P_t \left(1 - \left(\frac{P_t}{K} \right)^{\theta} \right),$$

where r, K, and θ are positive constants. (a) Find the equilibria. (b) Find a formula for the *per capita* population growth. (c) Write the model in the form $P_{t+1} = f(P_t)$ and determine the stability of the equilibria. (d) Sketch the region in the $r\theta$ plane that gives values for which the nonzero equilibrium is stable.

5.2 Systems of Difference Equations

5.2.1 Linear Models

A linear model is a system of difference equations of the form

$$\mathbf{x}_{t+1} = A\mathbf{x}_t, \quad t = 0, 1, 2, ..., \tag{5.9}$$

where $\mathbf{x}_t$ is an n-vector of quantities at time t and A is a constant $n \times n$ matrix. Such models arise naturally in Markov[1] processes (discussed below), where A is a Markov matrix with nonnegative entries and whose columns sum to 1. Linear models occurs in other areas, for example, for stage-structured models in biology where A is a Leslie matrix. By iteration, if $\mathbf{x}_0$ is an initial vector, then

$$\mathbf{x}_t = A^t \mathbf{x}_0 \tag{5.10}$$

is the solution to (5.9). We can get a better representation of the solution in terms of the eigenvalues and eigenvectors of A. Suppose that A has n distinct, real eigenvalues λ_k with associated independent eigenvectors $\mathbf{v}_k$, $k = 1, 2, ..., n$. Then the eigenvectors form a basis for $\mathbb{R}^n$ and we can express the initial vector

$$\mathbf{x}_0 = c_1\mathbf{v}_1 + \cdots + c_n\mathbf{v}_n,$$

where the c_k are the coordinates of $\mathbf{x}_0$ in the basis. Then, using linearity,

$$\mathbf{x}_t = A^t\mathbf{x}_0 = c_1A^t\mathbf{v}_1 + \cdots + c_nA^t\mathbf{v}_n = c_1\lambda_1^t\mathbf{v}_1 + \cdots + c_n\lambda_n^t\mathbf{v}_n, \tag{5.11}$$

which is the solution.

We recall that we can find the power A^t by diagonalizing the matrix A. Let D be the diagonal matrix with the eigenvalues on the diagonal, and let Q be the matrix whose columns are the associated eigenvectors. Then, from linear algebra, $Q^{-1}AQ = D$ (we say the matrix Q diagonalizes A). Thus $A = QDQ^{-1}$ and $A^t = (QDQ^{-1})^t = QD^tQ^{-1}$, where D^t is a diagonal matrix with the entries λ_i^t on the diagonal.

An important case occurs when there is a real, strictly **dominant eigenvalue**, say λ_1. This means $|\lambda_1| > |\lambda_2| \geq |\lambda_3| \geq \cdots \geq |\lambda_n|$ Then

$$
\begin{aligned}
\mathbf{x}_t &= c_1\lambda_1^t\mathbf{v}_1 + c_2\lambda_2^t\mathbf{v}_2 + \cdots + c_n\lambda_n^t\mathbf{v}_n \\
&= \lambda_1^t\left[c_1\mathbf{v}_1 + c_2\left(\frac{\lambda_2}{\lambda_1}\right)^t\mathbf{v}_2 + \cdots + c_n\left(\frac{\lambda_n}{\lambda_1}\right)^t\mathbf{v}_n\right].
\end{aligned}
\tag{5.12}
$$

---

[1] A. Markov (1856–1922) was a Russian mathematician who made significant contributions to analysis, number theory, approximation theory, and probability.

We have $|\lambda_i/\lambda_1| < 1$ for $i = 2, 3, ..., n$, and therefore all the terms in parentheses decay to zero as $t \to \infty$. Consequently, for large t the solution to the linear model behaves approximately as

$$\mathbf{x}_t \approx c_1 \lambda_1^t \mathbf{v}_1.$$

Asymptotically, if $\lambda_1 > 0$, the solution exhibits geometric growth or decay, depending upon the magnitude of λ_1. In this case, the dominant eigenvalue λ_1 is called the **growth rate**, and the vector $c_1 \mathbf{v}_1$ is the **asymptotic solution**, or in the context of population models, the stable age structure.

If some of the eigenvalues of A are complex, but the set of eigenvalues is still distinct, the argument above can be repeated. Using Euler's formula, $e^{i\theta} = \cos\theta + i\sin\theta$, every complex term in solution (5.11) can be written in terms of real and imaginary parts, each of which is a real solution. For example, if $\mathbf{v} = \mathbf{u} + i\mathbf{w}$ and $\lambda = \alpha + i\beta = re^{i\theta}$ is a complex eigen-pair, where r is the modulus of λ and θ is its argument, then

$$
\begin{aligned}
\mathbf{v}\lambda^t &= (\mathbf{u} + i\mathbf{w})(\alpha + i\beta)^t \\
&= r^t(\mathbf{u} + i\mathbf{w})(\cos\theta t + i\sin\theta t) \\
&= r^t(\mathbf{u}\cos\theta t - \mathbf{w}\sin\theta t) + ir^t(\mathbf{u}\sin\theta t + \mathbf{w}\cos\theta t).
\end{aligned}
$$

The real and imaginary parts of this vector are two linearly independent, real solutions. The complex conjugate eigen-pair $\mathbf{v} = \mathbf{u} - i\mathbf{w}$ and $\lambda = \alpha - i\beta$ gives the same two independent solutions. In summary, the two complex terms in (5.11) associated with complex conjugate eigen-pairs may be replaced by the two real, independent vector solutions. Therefore, we may always write the solution as a real solution. Complex eigenvalues clearly lead to oscillations in the system. The analysis of the case of a repeated eigenvalue of A is left to the Exercises.

Notice that (5.12) also implies that the solution to the linear model (5.9) satisfies the condition $\mathbf{x}_t \to \mathbf{0}$ if, and only if, all the eigenvalues have modulus less that one, or $|\lambda_i| < 1$. In this case we $\mathbf{x} = \mathbf{0}$ say is an **asymptotically stable** equilibrium. If there is an eigenvalue with $|\lambda| > 1$, then the equilibrium $\mathbf{x} = \mathbf{0}$ is **unstable**.

Previously, when we discussed population models, we had in mind tracking one or more species and we lumped all the individuals of a given species into one quantity, without regard to development stage, size, or age. However, some animals, such as insects, have well-defined life stages, such as egg, larva, pupa, and adult. We are interested in developing dynamics of how those animals might go through its life stages, keeping track of the population of each stage.

Example 5.8

Consider a hypothetical animal that has two life stages, juvenile and adult. At time t (say, given in weeks) we let J_t denote the number of juveniles and A_t denote the number of adults. Over each time step of one week we assume that juveniles are born, some die, and some graduate to adulthood. For adults, some die and some are recruited from the graduating juvenile stage. In symbols, for the juvenile population we may write

$$J_{t+1} = J_t - mJ_t - gJ_t + fA_t. \tag{5.13}$$

In words, the number of juveniles at the next week (J_{t+1}) equals the number at the last week (J_t), minus the number that died (mJ_t), minus the number that became adults (gJ_t), plus the number of births (fA_t). The constant m is the weekly juvenile mortality rate, g the weekly graduation rate, and f the average, weekly fecundity (fertility) of adults. Because adults reproduce, we assume the number of juveniles produced each week is proportional to the number of adults present. For example, if each adult on the average gives rise to three juveniles each week, then $f = 3$. Then, 100 adults would give rise to 300 juveniles per week. In the same way, for the adult population,

$$A_{t+1} = A_t - \mu A_t + gJ_t, \tag{5.14}$$

where μ is the adult weekly mortality rate. The last term represents those juveniles that became adults. Equations (5.13)–(5.14) represent a system of two coupled, first-order difference equations. They provide a way to take information from time t and project it forward to the next time $t + 1$. Therefore, if we know the initial numbers in each stage, we can recursively calculate the numbers at time t. Of course, doing this by hand is tedious, so we use software. A MATLAB m-file produces the time series plots shown in Fig. 5.9.

```
function stagepopulation
clear
m=0.5; g=0.5; f=2; mu=0.9; timesteps=35;
J=80; Jhistory=J; A=30; Ahistory=A;
for t=1:timesteps;
x=(1-m-g)*J+f*A;
y=g*J+(1-mu)*A;
J=x; A=y;
Ahistory=[Ahistory J];
Jhistory=[Jhistory A];
end
time=0:timesteps;
```

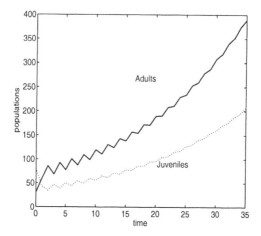

Figure 5.9 Population of adults (solid) and juveniles (dashed).

plot(time,Jhistory), hold on
plot(time,Ahistory), hold off

The stage-based population model in the last example can be formulated in matrix form as

$$\begin{pmatrix} J_{t+1} \\ A_{t+1} \end{pmatrix} = \begin{pmatrix} 1 - m - g & f \\ g & 1 - \mu \end{pmatrix} \begin{pmatrix} J_t \\ A_t \end{pmatrix},$$

and can be formulated in vector notation as

$$\mathbf{x}_{t+1} = P\mathbf{x}_t, \tag{5.15}$$

where

$$\mathbf{x}_t = \begin{pmatrix} J_t \\ A_t \end{pmatrix}, \quad P = \begin{pmatrix} 1 - m - g & f \\ g & 1 - \mu \end{pmatrix}.$$

The matrix P is called the **population projection matrix**, or **Leslie matrix**[2] (after P. H. Leslie, a biologist who developed the ideas in the 1940s). By iteration we see immediately that the solution to (5.15) is

$$\mathbf{x}_t = P^t\mathbf{x}_0,$$

where $\mathbf{x}_0$ is a vector containing the initial stage populations.

---

[2] Actually, the Leslie matrix was introduced to study age-structured models. Later, L. P. Lefkovitch in 1965 generalized the models to stage-structured population models. We use the term Leslie matrix for any population projection matrix.

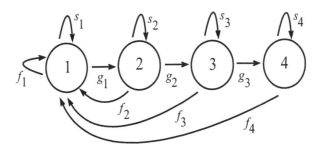

Figure 5.10 Leslie diagram.

Example 5.9

In the preceding example let us choose $g = m = 0.5$, $f = 2$, and $\mu = 0.9$. Then the model is

$$\mathbf{x}_{t+1} = \begin{pmatrix} 0 & 2 \\ 0.5 & 0.1 \end{pmatrix} \mathbf{x}_t.$$

The eigen-pairs are

$$\lambda_1 = 1.051, \quad \mathbf{v}_1 = \begin{pmatrix} 0.885 \\ 0.465 \end{pmatrix}; \qquad \lambda_2 = -0.951, \quad \mathbf{v}_2 = \begin{pmatrix} 0.903 \\ -0.429 \end{pmatrix}.$$

The general solution is

$$\mathbf{x}(t) = c_1 (1.051)^t \begin{pmatrix} 0.885 \\ 0.465 \end{pmatrix} + c_2 (-0.951)^t \begin{pmatrix} 0.903 \\ -0.429 \end{pmatrix}.$$

The dominant eigenvalue is $\lambda_1 = 1.051$ and the first term produces growth of about 5%. The second term will give a decaying oscillation. This is our observation in Fig. 5.9. Over a long time, the juvenile-adult population will approach a stable age structure defined by the eigenvector $(0.885, 0.465)^T$, which, when normalized to make the sum of the entries one, gives $(0.656, 0.344)^T$. Therefore, in the long run, 65.6% of the population will be juveniles, and 34.4% of the population will be adults. In yet different words, there will be about 1.9 juveniles for every adult.

In general, $\mathbf{x}_t$ may be an n-vector whose entries represent the populations of each of n stages at time t. If $n = 4$, for example, $\mathbf{x}_t = (x_{1t}, x_{2t}, x_{3t}, x_{4t})^T$, and the Leslie matrix (generally, we use the letter A) has the form

$$A = \begin{pmatrix} s_1 + f_1 & f_2 & f_3 & f_4 \\ g_1 & s_2 & 0 & 0 \\ 0 & g_2 & s_3 & 0 \\ 0 & 0 & g_3 & s_4 \end{pmatrix}, \tag{5.16}$$

where f_i is the fecundity of the ith stage, s_i is survival rate (one minus the mortality rate) of the ith stage, and g_i is the fraction of the ith stage that pass to the jth stage during the fixed time step. The **Leslie diagram** in Fig. 5.10 shows pictorially how the rates and the stages are connected. In still a more general case, the entries in the Leslie matrix may depend upon t, or they may depend upon the populations themselves, making the model nonlinear.

When there is a positive, dominant eigenvalue, say λ_1, of the Leslie matrix A, we have $x_t \sim c_1 v_1 \lambda_1^t$ for large t. Therefore λ_1 is the long term growth rate and $c_1 v_1$ represents the stable stage structure. Because c_1 is a constant, we can just use v_1 to determine the proportion, or percentage, of each stage to the whole population. For example, if $v_1 = (1, 2, 3)^T$ is an eigenvector associated with a dominant eigenvalue, then the vector $(\frac{1}{6}, \frac{1}{3}\frac{1}{2})$ shows the fraction of each stage in the stable age structure.

We can ask if there is a dominant eigenvalue for a given model. A simple version of the **Perron–Frobenius** theorem states that if A is a non-negative matrix (all of its entries are nonnegative) and if for some positive integer k the matrix A^k has all positive entries, then A has a simple, positive dominant eigenvalue with a positive eigenvector. There are other theorems that guarantee the same conclusion under less strenuous hypotheses.

One interesting question in a stage-structured population model is to determine how values in the Leslie matrix, which represent fecundities and survivorship probabilities, most affect the dominant eigenvalue λ, which is the long term growth rate. This is particularly important in making management decisions about ecosystems. We can measure the sensitivity of λ to changes in an entry a_{ij} by calculating the derivative $\partial \lambda / \partial a_{ij}$. (Note that the dominant eigenvalue is a function of the entries in A.) The matrix of these values,

$$S = \left(\frac{\partial \lambda}{\partial a_{ij}} \right),$$

is called the **sensitivity matrix**.

Example 5.10

For a 2×2 matrix

$$A = \begin{pmatrix} a & b \\ c & d \end{pmatrix}$$

with non-negative entries the dominant eigenvalue is easily found from the roots of the characteristic equation,

$$\lambda = \frac{1}{2} \left(\operatorname{tr} A + \sqrt{D} \right), \quad D = (\operatorname{tr} A)^2 - 4 \det A.$$

Then the sensitivity matrix is

$$
S = \begin{pmatrix} \frac{\partial \lambda}{\partial a} & \frac{\partial \lambda}{\partial b} \\ \frac{\partial \lambda}{\partial c} & \frac{\partial \lambda}{\partial d} \end{pmatrix} = \begin{pmatrix} \frac{1}{2}\left(1 + \frac{a-d}{D}\right) & \frac{c}{\Delta} \\ \frac{b}{\Delta} & \frac{1}{2}\left(1 - \frac{a-d}{D}\right) \end{pmatrix}.
$$

For example, consider the two-stage (juveniles and adults) model

$$
x_{t+1} = 1.7 y_t, \quad y_{t+1} = 0.6 x_t + 0.1 y_t
$$

with matrix

$$
A = \begin{pmatrix} 0 & 1.7 \\ 0.6 & 0.1 \end{pmatrix}.
$$

The value 1.7 is the fecundity of adults, and 0.1 is their survival rate; 0.6 is the fraction of juveniles that survive and pass to the adult stage. The dominant eigenvalue is $\lambda = 1.06$, which means that the long time growth rate of the population is 6% per season. The sensitivity matrix is

$$
S = \begin{pmatrix} 0.475 & 0.297 \\ 0.841 & 0.525 \end{pmatrix}.
$$

Therefore, we see that the $2, 1$ entry, the survivorship of juveniles affects the dominant eigenvalue most. We can estimate the change $\Delta \lambda$ in the long term growth rate when the survivorship is changed by Δc with the linear approximation

$$
\Delta \lambda = \frac{\partial \lambda}{\partial c} \Delta c = 0.841 \cdot \Delta c.
$$

Now we proceed in general. Consider the n-dimensional model

$$
\mathbf{x}_{t+1} = A \mathbf{x}_t
$$

with $\lambda, \mathbf{v}$ the dominant eigen-pair. Then

$$
A \mathbf{v} = \lambda \mathbf{v}.
$$

We know that λ is also an eigenvalue of A^T having some eigenvector $\mathbf{w}$. Therefore $A^T \mathbf{w} = \lambda \mathbf{w}$, or, taking the transpose,

$$
\mathbf{w}^T A = \lambda \mathbf{w}^T.
$$

Therefore $\mathbf{w}$ is a **left eigenvector** of A.

To calculate the change in the dominant eigenvalue with respect to the change of an entry in the matrix, we change the ij-entry of A by a small amount Δa_{ij}, while holding the remaining entries fixed. Let ΔA be the matrix with Δa_{ij} in the ij-position and zeros elsewhere. Then the perturbed matrix $A + \Delta A$ has dominant eigenvalue $\lambda + \Delta \lambda$ and eigenvector $\mathbf{v} + \Delta \mathbf{v}$, or

$$
(A + \Delta A)(\mathbf{v} + \Delta \mathbf{v}) = (\lambda + \Delta \lambda)(\mathbf{v} + \Delta \mathbf{v}).
$$

Multiplying out and discarding the higher order terms gives, to leading order,

$$A(\Delta \mathbf{v}) + (\Delta A)\mathbf{v} = \lambda(\Delta \mathbf{v}) + (\Delta \lambda)\mathbf{v}.$$

Multiplying by $\mathbf{w}^{\mathrm{T}}$ on the left gives

$$\mathbf{w}^{\mathrm{T}}(\Delta A)\mathbf{v} = (\Delta \lambda)\mathbf{w}^{\mathrm{T}}\mathbf{v},$$

or

$$\Delta \lambda = \frac{\mathbf{w}^{\mathrm{T}}(\Delta A)\mathbf{v}}{\mathbf{w}^{\mathrm{T}}\mathbf{v}} = \frac{w_i v_j \Delta a_{ij}}{\mathbf{w}^{\mathrm{T}}\mathbf{v}}.$$

In the limit we have

$$S = \left(\frac{\partial \lambda}{\partial a_{ij}} \right) = \frac{1}{\mathbf{w}^{\mathrm{T}}\mathbf{v}}(w_i v_j) = \frac{1}{\mathbf{w}^{\mathrm{T}}\mathbf{v}}\mathbf{w}\mathbf{v}^{\mathrm{T}}, \tag{5.17}$$

which is a simple representation of the sensitivity matrix. As a final note, it is always possible to normalize $\mathbf{v}$ and $\mathbf{w}$ so that $\mathbf{w}^{\mathrm{T}}\mathbf{v} = 1$.

MATLAB is useful to calculate the sensitivity matrix S. The following command line statements calculate S for the model in the last example.

```
A=[0  1.7; 0.6  0.1]
[V,D]=eig(A)
v=V(:,2)
[W,d]=eig(A')
w=W(:,2)
S=(1/(w'*v))*(w*v')
```

Unfortunately, the sensitivity matrix may give information that is hard to interpret when entries in A have large differences. For example, fecundities are usually high compared to survivorships. Therefore asking how the dominant eigenvalue changes with respect to a unit change in an entry is not always practical. It is better to compute the *relative* changes, or the elasticity. We define the **elasticity matrix** E by

$$E = \left(\frac{a_{ij}}{\lambda} \frac{\partial \lambda}{\partial a_{ij}} \right) \simeq \left(\frac{\Delta \lambda / \lambda}{\Delta a_{ij} / a_{ij}} \right).$$

In MATLAB this is simply E = A.*S/max(eig(A)). When an entry of E is large, we say the entry is *elastic*, and when it is small, we say it is *inelastic*. These terms are familiar terms in economics with regard to the elasticity of demand for a product with respect to price changes; there,

$$\text{elasticity} = \frac{\% \text{ change in demand}}{\% \text{ change in price}} \simeq \frac{p}{D} \frac{dD}{dp},$$

where the demand D is a function of price p.

For reference or study, the book by Caswell (2001) has a thorough discussion of matrix models in the biological sciences.

EXERCISES

1. Find the solution to the discrete model

$$X_{t+1} = X_t + 4Y_t$$
$$Y_{t+1} = \frac{1}{2}X_t,$$

if $X_0 = 6$ and $Y_0 = 0$.

2. For each of the following Leslie matrices find the dominant eigenvalue (if any), the corresponding eigenvector, and the stable age structure. Interpret each entry in context of a 3-stage population model.

a) $A = \begin{pmatrix} 0 & 1 & 5 \\ 0.3 & 0 & 0 \\ 0 & .5 & 0 \end{pmatrix}$.

b) $A = \begin{pmatrix} 0 & 1 & 1 \\ \frac{2}{3} & 0 & 0 \\ 0 & \frac{1}{3} & 0 \end{pmatrix}$.

3. Describe the dynamics of the linear model

$$\mathbf{x}_{t+1} = \begin{pmatrix} 0 & 0 & f \\ a & 0 & 0 \\ 0 & b & 0 \end{pmatrix} \mathbf{x}_t,$$

where a, b, and f are positive constants. Discuss the two cases $abf > 1$ and $abf < 1$.

4. Consider the linear model

$$\mathbf{x}_{t+1} = \begin{pmatrix} ac & bc \\ s(1-a) & 0 \end{pmatrix} \mathbf{x}_t,$$

where s, b, $c > 0$ and $0 < a < 1$.

a) Determine conditions (if any) for which there are infinitely many equilibria.

b) In the case there is a single equilibrium, determine conditions on the parameters that ensure all solutions converge to zero as $t \to \infty$.

5. Prove that the quadratic equation $\lambda^2 - p\lambda + q = 0$ has roots satisfying $|\lambda| < 1$ if, and only if,

$$|p| < 1 + q < 2.$$

The condition is called the **Jury condition**.

6. For many bird species the fecundity and survivorship of adults is independent of the age of the adult bird. So, we can think of the population as composed of two classes, juveniles and adults.

   a) Set up a general Leslie model to describe the dynamics of the female bird population, assuming juveniles do not reproduce and that the juvenile stage lasts only one year. Adult females may live more than one year.

   b) Find the eigenvalues and eigenvectors of the Leslie matrix for this model.

   c) Identify the growth rate and dominant eigenvalue, and determine the long time fraction of the total population that are juveniles. Examine all cases.

7. A population is divided into three age classes, ages 0,1,and 2. During each time period 20% of the females of age 0 and 70% of the females of age 1 survive until the end of the following breeding season. Females of age 1 have an average fecundity of 3.2 offspring per female and the average fecundity of age 2 females is 1.7. No female lives beyond three years.

   a) Set up the Leslie matrix.

   b) If initially the population consists of 2000 females of age 0, 800 females of age 1, and 200 females of age 2, find the age distribution after three years.

   c) Determine the asymptotic behavior of the population.

8. Show that if $\lambda = 0$ is an eigenvalue of a matrix A, then A^{-1} does not exist.

9. Consider a linear model $x_{t+1} = Ax_t$. Under what condition(s) on A does the system have a single equilibrium?

10. Consider the stage-structured model

$$x_{t+1} = \begin{pmatrix} \frac{1}{2} & 1 & \frac{3}{4} \\ \frac{2}{3} & 0 & 0 \\ 0 & \frac{1}{3} & 0 \end{pmatrix} x_t$$

   a) Find the eigenvalues and eigenvectors, and identify the growth rate and stable age structure.

   b) Generate the sensitivity matrix S and determine which entry in A most strongly affects the growth rate.

   c) Compute the elasticity matrix E. Do the entries of E change your conclusion in part (b)?

11. A 3-stage model of coyote populations (pups, yearlings, and adults) has a Leslie matrix given by

$$\begin{pmatrix} 0.11 & 0.15 & 0.15 \\ 0.3 & 0 & 0 \\ 0 & 0.6 & 0.6 \end{pmatrix}.$$

Repeat parts (a) and (b) of Exercise 10.

12. A female population with three stages, juveniles J, youth Y, and adults A have the following vital statistics. In a time step (one season), no animal remains in the same stage; but, on the average, during each time step 50% of the juveniles become youth, and 60% of the youth survive to become adults. On the average, each youth produces 0.8 offspring (juveniles) per time step, and each adult produces 2.2 offspring per time step. Juveniles do not reproduce.

   a) Draw a Leslie diagram with arrows indicating how the stages interact.

   b) Write down the discrete model and identify the population projection matrix L.

   c) If $J_0 = 25$, $Y_0 = 50$, and $A_0 = 25$, draw plots (on the same set of axes) of the populations of each stage for the first 30 seasons.

   d) What is the the growth rate of the total population? Over a long time, what percentage of the population is in each stage? Explain your observations in part (c) based on the eigenvalues and eigenvectors of the population projection matrix L. What are the absolute values of the eigenvalues?

   e) Under the same initial conditions, plot the natural logarithms of the populations versus time? Explain the difference observed from part (d).

   f) Compute the sensitivity matrix S.

   g) The dominant eigenvalue is most sensitive to a change in what nonzero entry in L?

   h) Compute the elasticity of the dominant eigenvalue to adult fecundity. Interpret your result in words.

   i) If measures are taken to increase the survivorship of youths by 10% , what would be the approximate population growth rate?

   j) How much would the fecundity of adults have to decrease to drive the population to extinction?

5.2.2 Nonlinear Interactions

Next we set up and analyze two-dimensional nonlinear discrete systems

$$x_{t+1} = f(x_t, y_t), \tag{5.18}$$
$$y_{t+1} = g(x_t, y_t), \tag{5.19}$$

where f and g are given functions, and $t = 0, 1, 2, \ldots$. As in the case of differential equations, we can visualize a solution geometrically as time series, where the sequences x_t, and y_t are plotted against t, or as a sequence of points (x_t, y_t) in the xy plane, or phase plane. We often connect the points of the sequence to make the plots continuous.

The importance of equilibria again arises in understanding the behavior of nonlinear systems. An equilibrium solution is a constant solution $y_t = x^*$, $y_t = y^*$. It plots as a single point, sometimes called a **critical point**, in the phase plane; as in the case of differential equations, the behavior of the system is determined by its local behavior near the critical points. Therefore, stability, or permanence of an equilibrium, becomes an issue.

An equilibrium solution satisfies the equations

$$x^* = f(x^*, y^*), \quad y^* = g(x^*, y^*).$$

To analyze the orbital behavior near the equilibrium we consider small perturbations u_t and v_t from x^* and y^*, respectively, or

$$x_t = x^* + u_t, \quad y_t = y^* + v_t.$$

Substituting into the system (5.18)–(5.19) and linearizing leads, in the usual way, to the linearized perturbation equations

$$\mathbf{u}_{t+1} = J\mathbf{u}_t, \tag{5.20}$$

where $\mathbf{u}_t = (u_t, v_t)^{\mathrm{T}}$ and $J = J(x^*, y^*)$ is the **Jacobian matrix**

$$J = \begin{pmatrix} f_x(x^*, y^*) & f_y(x^*, y^*) \\ g_x(x^*, y^*) & g_y(x^*, y^*) \end{pmatrix}. \tag{5.21}$$

Let λ_1 and λ_2 denote the eigenvalues of J. From the section on linear systems, we have the following stability result:

- If $|\lambda_1| < 1$ and $|\lambda_2| < 1$, then the perturbations decay and (x^*, y^*) is locally asymptotically stable.

- If either $|\lambda_1| > 1$ or $|\lambda_2| > 1$, then some of the perturbations will grow and (x^*, y^*) is unstable.

This result translates into the following general condition for asymptotic stability in terms of the characteristic equation

$$\det(J - \lambda I) = \lambda^2 - (\operatorname{tr} J)\lambda + \det J = 0$$

for the Jacobian matrix. Necessary and sufficient conditions for local asymptotic stability, $|\lambda_1| < 1$ and $|\lambda_2| < 1$, are

$$|\operatorname{tr} J| < 1 + \det J, \quad \det J < 1. \tag{5.22}$$

Conditions (5.22) are called the **Jury conditions**. A proof is requested in the Exercises.

To plot orbits in the phase plane it is useful to find the nullclines where x_t or y_t does not change. Therefore the nullclines are defined by $\Delta x_t = 0$, $\Delta y_t = 0$, or the loci

$$f(x, y) = x, \quad g(x, y) = y,$$

respectively. The x- and y-nullclines intersect at the critical points. The behavior of the orbits near the critical points is similar to that encountered in differential equations: nodal structure, saddle structure, and rotational structure (spirals and ellipses).

Example 5.11

(**Nicholson–Bailey model**) Many insect predators are parasitoids. A parasitoid finds prey, called hosts, and lays its eggs inside or on the victim. The eggs hatch into larvae, which then consume the host. The host then dies and the larvae metamorphose and emerge as adults. The population dynamics is based upon the probability of the host avoiding detection and has the general form

$$
\begin{aligned}
x_{t+1} &= r x_t F(y_t), \\
y_{t+1} &= c y_t \left(1 - F(y_t)\right),
\end{aligned}
$$

where x_t is the host population and y_t is the prey population; F is the escapement function, or probability of avoiding detection in one season, r is the geometric growth factor, and c is the number of new parasitoids per host discovered. Think of F as the fraction of hosts that survive to the next time step. The dynamics is discrete because both host and parasitoid usually have life cycles synchronized with seasons. The model was developed in the 1930s by A. J. Nicholson and V. A. Bailey, and it is fundamental in population ecology. There are various forms for the escapement function F. The usual model is to base it upon a **Poisson random variable** (see Section 5.3). A discrete random variable X is Poisson distributed if

$$\Pr(X = k) = \frac{\lambda^k}{k!}e^{-\lambda}, \quad k = 0, 1, 2, \ldots,$$

where λ is a positive constant. A Poisson distribution describes the occurrence of discrete, random events, such as encounters between a predator and its prey, and X is the probability of k encounters in a given time period, say over the lifespan of the prey. The constant λ is the average number of encounters during the period. In terms of hosts and parasitoids, the average encounter rate for a given host should be proportional to the number of predators searching, or $\lambda = ay_t$, during that period. So, the probability of no encounters is the zeroth-order term $(k = 0)$, or

$$F(y_t) = e^{-ay_t}.$$

Therefore

$$\begin{aligned}
x_{t+1} &= rx_t e^{-ay_t}, & (5.23) \\
y_{t+1} &= cy_t \left(1 - e^{-ay_t}\right), & (5.24)
\end{aligned}$$

The equilibria are found from the equations

$$x = rxe^{-ay}, \quad y = cy\left(1 - e^{-ay}\right),$$

which give two equilibria

$$x^* = y^* = 0; \quad x^* = \frac{r\ln r}{(r-1)ac}, \quad y^* = \frac{\ln r}{a}.$$

Note that $r > 1$ must be an implicit assumption to the model. To determine stability we must compute the Jacobian matrix (5.21). This tedious calculation is left to the reader. We obtain

$$J(x^*, y^*) = \begin{pmatrix} re^{-ay^*} & -ax^* e^{-ay^*} \\ c(1 - e^{-ay^*}) & acx^* e^{-ay^*} \end{pmatrix}.$$

At extinction,

$$J(0, 0) = \begin{pmatrix} r & 0 \\ 0 & 0 \end{pmatrix},$$

which has eigenvalues $r > 1$ and 0. Hence, the extinct population is unstable. The nonzero equilibrium has Jacobian matrix

$$J(x^*, y^*) = \begin{pmatrix} 1 & -\frac{\ln r}{c(r-1)} \\ \frac{c(r-1)}{r} & \frac{\ln r}{r-1} \end{pmatrix}.$$

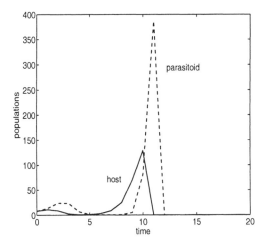

Figure 5.11 Simulations of the Nicholson–Bailey model. The dashed curve is the parasitoid abundance and the solid curve is the host abundance.

To determine stability we apply the Jury conditions (5.22). The trace and determinant are

$$\text{tr}\, J = 1 + \frac{\ln r}{r-1}, \quad \det J = \frac{r \ln r}{r-1}.$$

We show that $\det J > 1$, and so the equilibrium is unstable. Equivalently, we show $g(r) = r - 1 - r \ln r < 0$ for $r > 1$. But $g(1) = 0$ and $g'(r) = -\ln r < 0$. Therefore we must have $g(r) < 0$.

The results of the Nicholson–Bailey model are perplexing. There are no stable equilibria, yet in nature hosts and parasitoids endure, co-evolving together. Does this mean it is an invalid model? To get a sense of the dynamics we simulate the solution with $x_0 = 10$, $y_0 = 10$, taking $a = 1$, $c = 1$, and $r = 2$ Fig. 5.11 shows the unstable oscillations that result. If time is continued, the oscillations continue and become greater. There have been explanations that attempt to resolve this issue, but more to the point, the Nicholson–Bailey model has become the basis of many other models with modifications that attempt to stabilize the dynamics. For example, the **Beddington model**

$$
\begin{aligned}
x_{t+1} &= e^{b(1-x_t/K)} x_t e^{-a y_t}, \\
y_{t+1} &= c y_t \left(1 - e^{-a y_t}\right),
\end{aligned}
$$

where the growth rate r is replaced by a saturating density dependent rate $e^{b(1-x_t/K)}$, has a positive stable equilibrium for a large set of parameter values. Texts in mathematical ecology investigate some of these models in detail.

Example 5.12

(Epidemics) Consider a disease in a population of fixed size N, and let S_t, I_t, and R_t be the susceptible, infective, and removed classes at time t, respectively. See Section 1.4.3. The susceptibles are those who can catch the disease but are currently not infected, the infectives are those who are infected with the disease and are contagious, and the removed individuals have either recovered permanently, are naturally immune, or have died. We have $N = S_t + I_t + R_t$. With homogeneous mixing, the dynamics is given by

$$\begin{aligned}
S_{t+1} &= S_t - \alpha S_t I_t, \\
I_{t+1} &= I_t + \alpha S_t I_t - \gamma I_t, \\
R_{t+1} &= R_t + \gamma I_t,
\end{aligned}$$
(5.25)

where the individuals in a given class at time $t+1$ equals the number at time t, plus or minus those added or subtracted during that time step, The constant α is the **transmission coefficient** and γ is the **removal rate**. This model is the **discrete SIR model**. To interpret the constants we can argue as follows. If S_0 and I_0 are the number susceptible and infected initially, then $\alpha S_0 I_0$ measures the number that become infected at the outset; thus αS_0 is the number that become infected by a contact with a single infected individual during the initial time step. Therefore, α represents the probability that one susceptible person becomes infected. Looked at in a different way, α^{-1} is the average time that it takes for an individual to get infected. The removal rate can be interpreted similarly; γ^{-1} is the average duration of the infectious period. To understand how the disease progresses we observe that

$$\begin{aligned}
\Delta I_t &> 0 \quad \text{if} \quad S_t > \frac{\gamma}{\alpha}; \\
\Delta I_t &< 0 \quad \text{if} \quad S_t < \frac{\gamma}{\alpha}; \\
\Delta I_t &= 0 \quad \text{if} \quad S_t = \frac{\gamma}{\alpha}.
\end{aligned}$$

Because $\Delta S_t < 0$, we know S_t is always decreasing. Therefore, if $S_0 < \frac{\gamma}{\alpha}$ then $S_t < \frac{\gamma}{\alpha}$ for all t, and the disease decreases. On the other hand, if $S_0 > \frac{\gamma}{\alpha}$ then $S_t > \frac{\gamma}{\alpha}$ for all t, and the disease increases in the population and an epidemic occurs. If we define $R = \frac{\alpha}{\gamma} S_0$, then an epidemic occurs if $R > 1$. Epidemiologists refer to R as the **basic reproduction number** of the infection. Notice that the epidemic peaks when $S_t = \frac{\gamma}{\alpha}$, where ΔI_t changes sign. Simulations of the SIR equations (5.25) are investigated in the Exercises. Both time series plots and orbits in the SI phase plane illustrate the dynamics. Fig. 5.12 shows a typical orbit in the phase plane when there is an epidemic. The initial point is on the line $S_0 + I_0 = N$, and the orbit increases up to a maximum, which lies

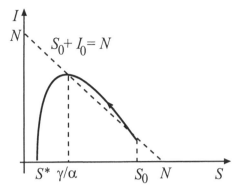

Figure 5.12 Orbit in the SIR model.

on the vertical line $S_t = \frac{\gamma}{\alpha}$. Then the disease dies out, leaving a number S^* of infectives who never get the disease.

The Exercises will show the variety and scope of nonlinear problems in science.

EXERCISES

1. Consider the linear model

$$\Delta P_t = P_t - 2Q_t, \quad \Delta Q_t = -P_t.$$

(a) Find the equilibria. (b) Draw the nullclines and the arrows indicating the direction of the "flow" the on the nullclines and in the regions between the nullclines. (c) On the line $Q_t = P_t$ in the phase plane, draw the arrows. (d) Does the equilibrium look stable or unstable, or can you tell? (e) Construct a rough phase plane diagram that indicates the solution paths. (f) Find the equations of these curves by finding the solution P_t and Q_t in terms of t.

2. Consider the following model connecting the cinnabar moth and ragweed. The life cycle of the moth is as follows. It lives for one year, lays it eggs on the plant and dies. The eggs hatch the next spring. The number of eggs is proportional to the plant biomass the preceding year, or

$$E_{t+1} = aB_t.$$

The plant biomass the next year depends upon the biomass the current year and the number of eggs according to

$$B_{t+1} = ke^{-cE_t/B_t}.$$

a) Explain why this last equation is a reasonable model (plot biomass vs eggs and biomass vs biomass).

b) What are the units of k, c, and a?

c) Rewrite the model equations in terms of the "re-scaled" variables $X_t = B_t/k$ and $Y_t = E_t/ka$. Call $b = ac$. What are the units of X_t, Y_t, and b?

d) Find the equilibrium and determine a condition on b for which the equilibrium is stable.

e) If the system is perturbed from equilibrium, describe how it returns to equilibrium in the stable case.

3. Examine the nonlinear model

$$\Delta x_t = r x_t \left(1 - \frac{x_t}{K}\right) - s x_t y_t,$$
$$\Delta y_t = -d y_t + \epsilon x_t y_t$$

with respect to equilibria and their stability. Here, all the parameters are positive.

4. Consider the nonlinear fisheries model

$$\begin{pmatrix} x_{t+1} \\ y_{t+1} \end{pmatrix} = \begin{pmatrix} ame^{-\beta x_t} & \frac{3}{2}mb \\ ae^{-\beta x_t} & 0 \end{pmatrix} \begin{pmatrix} x_t \\ y_t \end{pmatrix},$$

where a, m, b, $\beta > 0$.

a) Find the equilibria (use the notation $R = am\left(1 + \frac{3b}{2}\right)$.) Find a condition on R that guarantees the nontrivial equilibrium lies in the first quadrant.

b) For a positive equilibrium, determine conditions on R, a, and b under which the equilibrium is asymptotically stable.

c) For a fixed a, plot the stability region in the b, $\ln R$ parameter plane and identify the stability boundary.

5. Analyze the Nicholson–Bailey model if the escapement function is $F(y) = \frac{1}{[1+(a/k)y]^k}$.

6. Suppose one infected individual is introduced into a population of 500 individuals, all susceptible. The likelihood that an individual becomes infected is 0.1%, and an infective is contagious for an average of 10 days.

a) Compute the basic reproduction number of the disease and determine if an epidemic occurs.

b) Plot the time series S_t, I_t, and R_t over a 60-day period. At what time does the disease reach a maximum? How many do not get the disease? Over a long time, how many do not get the disease?

c) Plot the orbit in an SI phase plane.

7. Modify the SIR model to include vital dynamics. That is, assume susceptibles are born at a constant rate m, and at the same time individuals in each class die at a per capita rate m. Working in the SI phase plane, investigate the model thoroughly, finding equilibria (if any) and their stability, and sketching the orbits. Explain the long term dynamics of the disease.

8. A nonlinear juvenile-adult model where the fecundity depends upon the adult population is given by

$$
\begin{aligned}
J_{t+1} &= cbA_t e^{-A_t} J_t + be^{-A_t} A_t, \\
A_{t+1} &= \tau J_t - \sigma A_t.
\end{aligned}
$$

Here, $b, c > 0$ and $0 < \tau, \sigma < 1$. Find the equilibria and examine their stability.

5.3 Stochastic Models

Many dynamical processes are not deterministic, but rather contain elements of uncertainty. This is particularly the case in the life sciences, especially in ecology, which provide a context for the discussion in much of this section. We show how randomness arises and how it can be introduced into discrete dynamical models.

It is assumed the reader has studied elementary probability. Some of the concepts and definitions are presented in the first subsection, and the notation is established.

5.3.1 Elementary Probability

A **random variable** (often denoted RV) is a variable whose numerical value (a real number) is determined by the outcome of a random experiment, and it is characterized by a rule that dictates what the probabilities of various outcomes are. The outcome may be the number in a population, the charge on a capacitor, the length of a random phone call, the total time it takes to log onto the internet, the IQ of an individual, the time it takes a forager to find a food item, and so on. Typically we use capital letters to denote random variables and lowercase letters to denote real variables or specific values of random variables. To know the random variable X means to know the *probability* that X takes a value less than or equal to a number b, or symbolically,

$$\Pr(X \leq b).$$

There are two types of random variables, discrete and continuous. A discrete RV takes on discrete values, either finitely or infinity many, and a continuous RV takes on a continuum of values.

A **continuous random variable** X is characterized by its **probability density function** (pdf) denoted by $f_X(x)$. Pdfs are also called probability distributions. We assume the pdf exists for all real x. The subscript X on f reminds us that it is associated with the random variable X. The pdf has the properties:

1. $f_X(x) \geq 0$ for all x.

2. $\int_{-\infty}^{\infty} f_X(x)\, dx = 1$.

3. $\Pr(X \leq x) = \int_{-\infty}^{x} f_X(y)\, dy$.

Condition (1) says that the pdf is nonnegative; condition (2) requires that the total area under the graph of the pdf is unity, and condition (3) dictates the value of the probability of X taking a value smaller than x . The values of the random variable itself are usually not of interest. It is the density function that contains all the information about the random variable X and allows us to compute probabilities. The pdf need not be a continuous function. If there is no confusion about what the random variable is, we drop the subscript on $f_X(x)$ and just write $f(x)$. The following are important continuous distributions. (Other distributions are introduced in the Exercises.) As an exercise, the reader should plot the pdf in each case.

− The most common, and most important, pdf is the **normal** density

$$f(x) = \frac{1}{\sqrt{2\pi\sigma^2}} e^{-(x-\mu)^2/2\sigma^2}. \tag{5.26}$$

Its graph is the standard bell-shaped curve. The number μ is the average, or mean, of the density, and σ is the standard variation, which measures the width of the curve at its inflection points, and hence the spread of the distribution (see below). If a random variable X has the normal density we say that X is normally distributed and we write $X \sim \mathrm{N}(\mu, \sigma)$. The distribution $\mathrm{N}(0, 1)$ is called the **standard normal.**

- The **uniform** probability density on an interval $[c, d]$ is the step function

$$f(x) = \frac{1}{d - c} 1_{[c,d]}(x) = \begin{cases} \frac{1}{d-c}, & c \le x \le d \\ 0, & \text{otherwise} \end{cases}$$

The uniform density models situations were all values in an interval are equally probable.

- The **exponential** probability density with parameter λ is defined by

$$f_X(x) = \begin{cases} \lambda e^{-\lambda x}, & x \ge 0 \\ 0, & x < 0 \end{cases}.$$

The exponential distribution models, for example, the length X of a phone call, or the time X of survival of a cancer patient after treatment.

The function $F_X(x)$, which measures the probability of X taking on a value less than x, is called the **cumulative distribution function (cdf)**; this should not be confused with the probability distribution, or pdf. Thus

$$F_X(x) = \Pr(X \le x) = \int_{-\infty}^{x} f_X(x)\, dx.$$

The cdf has the properties

$$F_X(x) \ge 0, \quad F_X(-\infty) = 0, \quad F_X(+\infty) = 1,$$

and

$$F_X(x) \quad \text{is a nondecreasing function of } x.$$

The cdf is continuous from the right. That is, for each real number a,

$$\lim_{x \to a^+} F_X(x) = F_X(a).$$

If the pdf $f_X(x)$ is continuous at x, then the fundamental theorem of calculus guarantees that

$$f_X(x) = \frac{dF_X(x)}{dx}.$$

Therefore, the pdf is the derivative of the cdf and the cdf is the integral of the pdf. Note that

$$\Pr(x < X \le x+h) = \Pr(X \le x+h) - \Pr(X \le x) = \int_{x}^{x+h} f_X(y)\, dy \simeq f_X(x)h,$$

or

$$f_X(x) \simeq \frac{1}{h} \Pr(x < X \le x + h).$$

This equation explains what the pdf $f_X(x)$ actually is—it is the probability per unit length, and hence the name "probability density". The probability of the random variable taking a value between x and $x + h$ is approximately the area under the density curve between x and $x + h$. Also observe that

$$\Pr(a < X \le b) = F_X(b) - F_X(a).$$

In summary, the probability law characterizing a random variable can be defined by the pdf or the cdf; both carry all the probabilistic and statistical information about X.

Usually we are uninterested in the entire distribution. For example, when shopping for a new car we do not necessarily want to see the entire set of gas mileage data for all cars of the type we wish to buy; we are content to know the average gas mileage. Or, we may want to know the average height of a person in a population, or the average IQ of an entering class of students. A measure of the average, or central tendency, of a random variable X is its **mean**, or **expected value**,

$$\mu = \mathrm{E}(X) = \int_{-\infty}^{\infty} x f_X(x)\, dx.$$

The expected value is the coarsest probabilistic information about X. Other common measures of the central tendency are the mode and the median. A measure of how the possible values of the random variable X are spread about its mean is called the **variance,** which is defined by

$$\mathrm{Var}(X) = \mathrm{E}((X - \mu)^2) = \int_{-\infty}^{\infty} (x - \mu)^2 f_X(x)\, dx.$$

The **standard deviation** σ is the square root of the variance, or

$$\sigma = \sqrt{\mathrm{Var}(X)}.$$

Example 5.13

The normal distribution $X \sim \mathrm{N}(\mu, \sigma)$ has expected value $\mathrm{E}(X) = \mu$ and $\mathrm{Var}(X) = \sigma^2$. Expected values and variances for other random variables are contained in the Exercises.

If X is a random variable then we can form combinations of X to create new random variables. That is, we can define $Y = g(X)$, where g is a given real-valued function. This equation defines a transformation of a random variable. Let us assume that g is a strictly increasing function whose first derivative is continuous on all of the real line, and let us denote the inverse g^{-1} of g by φ. That is $g^{-1} = \varphi$ Thus $y = g(x)$ if, and only if, $x = \varphi(y)$, or $g(\varphi(y)) = y$ and

$\varphi(g(x)) = x$. We can determine the pdf for Y in terms of the pdf for X by calculating

$$\begin{aligned}
\Pr(a < Y \le b) &= \Pr(a < g(X) \le b) \\
&= \Pr(\varphi(a) < X \le \varphi(a)) \\
&= \int_{\varphi(a)}^{\varphi(b)} f_X(x)\,dx.
\end{aligned}$$

In this last integral we can perform a change variables via $x = \varphi(y)$, $dx = \varphi'(y)\,dy$. Then

$$\Pr(a < Y \le b) = \int_a^b f_X(\varphi(y))\varphi'(y)\,dy.$$

By definition of the probability law, therefore, the pdf for Y is

$$f_Y(y) = f_X(\varphi(y))\varphi'(y). \tag{5.27}$$

If the function g is not strictly increasing, then the density can be found by breaking up the problem into intervals over which g is monotone increasing or monotone decreasing. As an aside, we recall from calculus that the relationship between the derivatives of φ and g is $\varphi'(y) = 1/g'(x)$.

There is an interesting corollary of this result. Nearly the same calculation can be performed to obtain

$$\Pr(c < X \le d) = \Pr(c < \varphi(Y) \le b) = \Pr(g(c) < Y \le g(d)).$$

In terms of integrals,

$$\int_c^d f_X(x)\,dx = \int_{g(c)}^{g(d)} f_Y(y)\,dy.$$

In words, "probability is conserved" by a strictly increasing, continuously differentiable transformation of the random variable.

Next we inquire about the expected value of $Y = g(X)$. We have

$$\begin{aligned}
\mathrm{E}(g(X)) &= \mathrm{E}(Y) = \int_{-\infty}^{\infty} y f_Y(y)\,dy \\
&= \int_{-\infty}^{\infty} g(x) f_Y(g(x)) g'(x)\,dx \\
&= \int_{-\infty}^{\infty} g(x) f_X(\varphi(g(x))) \varphi'(g(x)) g'(x)\,dx \\
&= \int_{-\infty}^{\infty} g(x) f_X(x)\,dx,
\end{aligned}$$

where we have used (5.27) and the fact that g and φ are inverses. Using these facts, we can easily show

$$E(aX + b) = aE(X) + b, \quad a, b \in \mathbb{R}, \tag{5.28}$$

and

$$\text{Var}(X) = E(X^2) - E(X)^2. \tag{5.29}$$

One of the key results in probability theory is the **central limit theorem**. If Y_1, Y_2,... is a sequence of independent and identically distributed random variables with finite mean μ and variance σ^2, then the limiting distribution of the random variable

$$S = \frac{\sum_{i=1}^{n} \frac{Y_i}{n} - \mu}{\sigma/\sqrt{n}}$$

is the standard normal $N(0, 1)$.

The preceding definitions and results go over to **discrete random variables** in an obvious way. Let X be a discrete RV that assumes the values x_1, x_2,...,x_k, ...The analog of the pdf is the discrete **probability mass function**

$$p_k = \text{Pr}(X = x_k), \quad k = 1, 2, 3, ...,$$

which gives the probabilities of X taking the various values. Clearly $\sum_k p_k = 1$. The **expected value** of X is the weighted average

$$\mu = E(X) = \sum_k x_k p_k,$$

and more generally

$$E(g(X)) = \sum_k g(x_k) p_k.$$

The **variance** is

$$\text{Var}(X) = E((X - \mu)^2),$$

and relations (5.28) and (5.29) still hold. Sometimes the **coefficient of variation** CV, defined by $CV = \sigma/\mu$, is used as measure of variation; for example, $CV = 0.40$ carries the rough interpretation that on average X varies 40% from its mean value. Some important discrete random variables are presented below, and the reader should plot the pmf in each case for several values of the parameters.

– Some experiments have only two outcomes, success S and failure F. If N independent trials of a given experiment are performed, we can let the random variable X be the number of successes. If $\text{Pr}(S) = p$ and $\text{Pr}(F) = 1 - p$, then the probability mass function is

$$\text{Pr}(X = k) = \binom{N}{k} p^k (1 - p)^{N-k},$$

which is the probability of obtaining k successes in the N trials. This pmf is the **binomial distribution**. We write $X \sim \text{bin}(N, p)$ to say X is binomial distributed.

– Again consider successive independent trials where an event A occurs with probability p. Let X be the number of trials until A occurs the first time. Then

$$\Pr(X = k) = p(1 - p)^k, \quad k = 1, 2, 3, \ldots$$

is called the **geometric distribution** with parameter p.

– Again in the binomial setting, let X be the number of trials required until the rth success occurs. Then

$$\Pr(X = k) = \binom{k - 1}{r - 1} p^r (1 - p)^{k - r}, \quad k = r, r + 1, \ldots$$

is the **negative binomial distribution**.

– The **Poisson distribution** with parameter $\lambda > 0$ is

$$\Pr(X = k) = \frac{\lambda^k e^{-\lambda}}{k!}, \quad k = 0, 1, 2, \ldots.$$

The Poisson RV models the probability of unpredictable, rare events over a short time scale, where λ is the average occurrence of the event.

Finally, we review some ideas in **conditional probability**. If A and B are two events with $\Pr B > 0$, then the probability that A occurs, given that B has occurred, is

$$\Pr(A|B) = \frac{\Pr(A \cap B)}{\Pr B}.$$

A key result, the total probability theorem, is used in the sequel. It states that if A is composed of mutually disjoint events $A_1, \ldots, A_n$, that is, $A = \cup_{k=1}^{n} A_k$ with $A_i \cap A_j \neq \phi$, then

$$\Pr A = \sum_{k=1}^{n} P(A|A_k) \Pr A_k.$$

Finally, we say any set of events $A_1, \ldots, A_n$ is **independent** if

$$\Pr(A_{i_1} \cap A_{i_2} \cap \cdots \cap A_{i_r}) = \Pr(A_{i_1}) \Pr(A_{i_2}) \cdots \Pr(A_{i_r})$$

for all sequences of integers $i_1 < i_2 < \cdots < i_k$, $k = 2, 3, \ldots$. A set of events is **dependent** if they are not independent.

5.3.2 Stochastic Processes

A **stochastic process** (abbreviated **SP**) is an indexed family of random variables X_t (or $X(t)$), where t belongs to some index set T. That is, for each assigned $t \in T$, the quantity X_t is a random variable. Usually the index set is the time interval $T = [0, \infty)$ or the discrete set of integers $T = \{0, 1, 2, 3, ...\}$. When the latter is the index set, the SP is often called a random sequence. When the index set is the finite set $T = \{1, 2, 3, ..., n\}$, then the SP is a random vector. There are actually four types of real, stochastic processes X_t that can be considered. Two characterizations are based on whether the index set is discrete or is continuous (an interval), and two more are defined by whether X_t itself takes on discrete values or a continuum of values. As an example of a discrete process, let X_t denote the result of a roll of a single die at times $t \in T = \{0, 1, 2, 3, ...\}$. At each instant of time t, X_t is a random variable that takes on one of the values 1,2,3,4,5,6, each with equal probability (one-sixth). A typical **realization** of the process is $1, 1, 3, 6, 1, 4, 5, 2, 5, ...$ In this case of a die, the X_t are statistically independent, which is generally not the case for stochastic processes. In fact, statistical dependence is usually what makes stochastic processes interesting and stochastic processes are frequently characterized by their memory; that is, they are defined by how a random variable X_t depends on earlier random variables X_s, $s < t$. An example of a continuous SP is a Poisson process. For this process $T = [0, \infty)$ and X_t is discrete; X_t counts the number of times a certain event occurs from time $t = 0$ to time t. The event could be an accident at an intersection, the number of customers arriving at a bank, the number of breakdowns of machines in a plant, or discoveries of prey items by a predator.

Generally, to define a SP we must specify the index set T, the space where the X_t lie, and relations between the X_t for different t (for example, joint distributions, correlations, or other dependencies). In fact, the probability law for a SP is a specification of all the joint distributions; but simplifications are often introduced to make the process tractable. The types of SPs studied most often in applied probability theory include Markov processes, Poisson processes, Gaussian processes, Wiener processes, and martingales—all of these are defined by certain dependencies, or memory. Therefore, definitions of these processes depend on the notion of conditional probability.

Because a stochastic process X_t is defined as a random variable for each assigned time t, there is a probability density function $f(x, t)$ for X_t. Observe that the density has *both* spatial and temporal dependence. From the density function we can calculate the expected value and variance at each time t via

$$\mu(t) = \mathrm{E}(X_t) = \int_{-\infty}^{\infty} x f(x, t)\, dx$$

and

$$\sigma^2(t) = \text{Var}(X_t) = \int_{-\infty}^{\infty} (x - \mu(t))^2 f(x, t) \, dx.$$

Similar definitions hold for discrete random variables.

EXERCISES

1. Calculate the mean, standard deviation, and cumulative distribution function of a uniform random variable X with density $\frac{1}{d-c} 1_{[c,d]}(x)$.

2. Verify that the standard deviation of a normal random variable with density $N(0, \sigma)$ is, in fact, σ.

3. Show that the expected value of the exponential distribution with parameter λ is $E(X) = \frac{1}{\lambda}$, and the variance is $\text{Var}(X) = \frac{1}{\lambda^2}$. Show that the the cumulative distribution function is

$$F(x) = 1 - e^{-\lambda x}, \quad x > 0.$$

4. If X is a continuous RV, find the pdf of the random variable $Y = X^2$.

5. The famous Scottish physicist J. C. Maxwell showed that molecular collisions in a gas result in a speed distribution where the fraction of molecules in the speed range $[v, v + dv]$ is given by the **Maxwell–Boltzmann** distribution

$$f(v) = \sqrt{\frac{m}{2\pi kT}} v^2 e^{-mv^2/2kT}, \quad v \geq 0.$$

where n is the total number of particles, m is the molecular mass of each particle, T is the temperature, and k is Boltzmann's constant ($k = 1.38 \times 10^{-23}$ joules/deg K). Sketch a generic graph of the distribution and compute the average speed of a molecule in a gas.

6. Verify formulas (5.28) and (5.29), and show that

$$\text{Var}(a + bX) = b^2 \text{Var}(X), \quad a, b \in \mathbb{R}.$$

7. If $X \sim N(\mu, \sigma)$, show that $Z = \frac{X-\mu}{\sigma} \sim N(0, 1)$.

8. Find the mean and variance of a Poisson random variable.

9. A Cauchy random variable has density defined by

$$f(x) = \frac{a}{\pi} \frac{1}{a^2 + x^2}.$$

Compute the cumulative distribution function and draw a graph. What is $E(X)$ and $\text{Var}(X)$?

10. On the average there are about 15 cyclones formed off the U.S. Pacific coast each year. In a given year, what is the probability of there being at most 6 cyclones?

11. A random variable X is **gamma distributed** with parameters a and λ if its pdf is

$$f(x) = \frac{\lambda(\lambda x)^{a-1}e^{-\lambda x}}{\Gamma(a)}, \quad x \geq 0,$$

and $f(x) = 0$ for $x < 0$, where

$$\Gamma(a) = \int_0^\infty z^{a-1}e^{-z}dz$$

is the gamma function. Sketch the pdf for $a = 3.5$ and $\lambda = 2$.

12. Let X be a normal random variable with mean μ and variance σ^2. The random variable Y defined by $Y = e^X$ is said to be a **log-normal** random variable. Show that the pdf of Y is given by

$$f_Y(y) = \frac{1}{y\sqrt{2\pi\sigma^2}}e^{-(\ln y - \mu)^2/2\sigma^2},$$

and plot the pdf for various values of σ. What is the mean and variance of Y? (The log-normal density is often associated with phenomena that involve multiplicative effects of a large number of independent random events.)

13. Let $X \sim N(0,1)$. Find the pdf of the random variable X^2. (This pdf is called the **chi-squared** distribution.)

14. Let X be a discrete RV with $\Pr(X = x_j) = p_j$, $j = 0, 1, 2, \dots$. The **probability generating function** is defined by

$$G(t) = E(t^X) = \sum_{j=0}^\infty p_j t^j.$$

a) Show that $E(X) = G'(1)$ and $\text{Var}(X) = G''(1) + G'(1) - G'(1)^2$.

b) Use the probability generating function to find the mean and variance of a binomial random variable $X \sim \text{bin}(N, p)$.

15. The **probability generating function** for a continuous RV X is defined by

$$G(t) = E(t^X) = \int_{-\infty}^\infty t^x f(x)\,dx.$$

a) Show that $E(X) = G'(1)$ and $\text{Var}(X) = G''(1) + G'(1) - G'(1)^2$.

b) Use the probability generating function to find the mean and variance of an exponential distribution X with parameter λ.

16. Let X be a discrete RV with $\Pr(X = x_j) = p_j$, $j = 0, 1, 2, \ldots$. The **moment generating function** is defined by

$$M(t) = \mathrm{E}(e^{tX}) = \sum_{j=0}^{\infty} p_j e^{jt}.$$

a) Show that $\mathrm{E}(X) = M'(0)$ and $\mathrm{Var}(X) = M''(0) - M'(0)^2$.

b) Use the moment generating function to find the mean and variance of a binomial random variable $X \sim \mathrm{bin}(N, p)$.

17. The **moment generating function** for a continuous RV X is defined by

$$G(t) = \mathrm{E}(e^{tX}) = \int_{-\infty}^{\infty} e^{tx} f(x)\, dx.$$

a) Show that $\mathrm{E}(X) = M'(0)$ and $\mathrm{Var}(X) = M''(0) - M'(0)^2$.

b) Use the moment generating function to find the mean and variance of an exponential distribution X with parameter λ.

18. Let X be a continuous RV. Prove Chebyshev's inequality:

$$\Pr(|X - \mu| \geq k\sigma) \leq \frac{1}{k^2},$$

where μ and σ are the mean and standard variation, and $k > 0$.

19. Let X be a continuous RV with mean μ and standard variation σ. If g is a sufficiently smooth function, show that $\mathrm{Var}(g(X)) \cong g'(\mu)\sigma^2$.

20. If X is geometrically distributed, show that $\mathrm{E}(X) = \frac{1}{p}$ and $\mathrm{Var}(X) = \frac{1-p}{p^2}$.

21. Show that if X is Poisson distributed with parameter λ, then $\mathrm{E}(X) = \lambda$ and $\mathrm{Var}(X) = \lambda$.

22. Consider a set of N objects of which r of them are of a special type. Choosing n objects, without replacement, show that the probability of getting exactly k of the special objects is

$$\Pr(X = k) = \frac{\binom{r}{k}\binom{N-r}{n-k}}{\binom{N}{n}}, \quad k = 0, 1, \ldots, n.$$

This distribution is called the **hypergeometric distribution**. Show that $\mathrm{E}(X) = \frac{nr}{N}$.

5.3.3 Environmental and Demographic Models

There are myriad reasons for random variations in population growth and death rates. Ecologists often categorize these random effects as **environmental stochasticity** and **demographic stochasticity**. The former includes random weather patterns and other external sources where all individuals are affected equally and there is no individual variation; the latter involves natural variability in behavior, growth, vital rates, and other genetic factors that occur in all populations, even though the environment may be constant. Demographic stochasticity is commonly modeled by a Gaussian (normal) process, but environmental variations can take many forms. For example, an environment may suddenly have a catastrophic event, such a flood, or a bonanza year where the conditions for reproduction are unusually high.

In engineering one can identify similar processes, those that have external noise, akin to environmental stochasticity, and those that have internal or system noise, which is like demographic stochasticity.

There are two approaches to creating stochastic discrete models. One is to add, in one way or another, stochasticity to a known deterministic model. The other is to construct the probabilistic model directly. In the sequel we illustrate these constructions.

To fix the idea of environmental stochasticity and adding stochasticity to a known deterministic model, consider the Ricker population model

$$x_{t+1} = bx_t e^{-cx_t},$$

where b is the average yearly birthrate. Suppose, on the average, once every 25 years there is a weather event that lowers the birth rate that year by 30%. We create a fixed, uniformly distributed random process Z_t having range $[0, 1]$. At each time step we modify the birthrate according to the rule: *if* $Z_t < 0.04$ *then* adjust the birthrate to $0.7b$; *else* make no adjustment. In this manner, the birthrate becomes a random variable B_t and we can write the dynamics as

$$X_{t+1} = B_t X_t e^{-cX_t}, \tag{5.30}$$

where X_t, the population, is now a random process. It is interesting to simulate this process and the Exercises will guide the reader through this activity.

A word about notation—on the right side of (5.30) we mean the specific value $X_t = x_t$ of the random variable, or population, calculated at the time step t; some authors prefer to write the stochastic model (5.30) as $X_{t+1} = B_t x_t e^{-cx_t}$.

To model demographic changes we might assume the birthrate is normally distributed. This assumption follows from the observation that birthrates do not vary much from the average, with large deviations rare. Then we have

$$X_{t+1} = G_t X_t e^{-cX_t},$$

where for each time t we choose the birthrate from the distribution $G_t \sim N(b, \sigma)$, a normal distribution with mean b and standard deviation σ (or variance σ^2). Computer algebra systems, and well as calculators, have commands that generate different types of random variables (e.g., in MATLAB, *rand* generates a uniform random variable on $[0, 1]$, and *randn* generates a normal random variable $N(0, 1)$.

Generally, a discrete model has the form

$$x_{t+1} = f(x_t, r), \tag{5.31}$$

where r is a parameter. (The analysis can clearly be extended to vector functions and sets of parameters.) In the two examples above stochasticity was added to the model by redefining the parameter r to be random process R_t, giving a stochastic process X_t satisfying

$$X_{t+1} = f(X_t, R_t).$$

The values of R_t are usually chosen from a fixed distribution with mean μ and variance σ^2. For example, we may choose $R_t = \mu + \sigma W_t$, where $W_t \sim N(0, 1)$. Alternately, the values of R_t may be chosen themselves to satisfy a linear autoregressive process, for example, $R_t = aR_{t-1} + b + W_t$, where $W_t \sim N(0, 1)$, and a and b are fixed constants. How stochasticity is put into a model depends upon what one seeks and is therefore strictly an issue of modeling. Different stochastic mechanisms produce different patterns of variability.

Sometimes stochasticity is introduced in a special way to simplify the analysis. Models of the form

$$Y_{t+1} = g(Y_t) + E_t,$$

where the noise E_t is additive, are called **nonlinear autoregressive (NLAR)** processes; such processes, which have constant variance, are straightforward to analyze and apply data fitting techniques. For example, environmental stochasticity can be introduced in (5.31) by adding noise on a logarithm scale, creating an NLAR process. Taking the logarithm of both sides of (5.31) and then adding randomness gives

$$\ln X_{t+1} = \ln f(X_t, r) + E_t, \tag{5.32}$$

where $E_t \sim N(0, \sigma)$. This is an NLAR process for the transformed variable $Y_t = \ln X_t$. Therefore, the model

$$X_{t+1} = f(X_t, r)e^{E_t}$$

is often studied as an environmental stochasticity model. Alternately, demographic stochasticity is introduced in the discrete model (5.31) by adding stochasticity on a square root scale, producing

$$X_{t+1} = ([\sqrt{f(X_t, r)} + E_t]^+)^2.$$

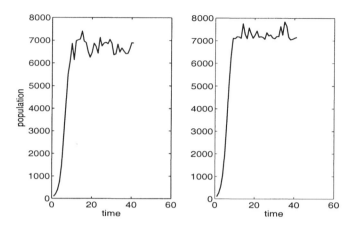

Figure 5.13 Realizations of the stochastic Ricker model for environmental and demographic stochasticity. The parameters are $b = 2$, $c = 0.001$, with $X_0 = 40$. For environmental stochasticity $E_t \sim N(0, 0.05)$, and for demographic stochasticity, $E_t \sim N(0, 0.7)$.

Again, this is an NLAR model for the transformed process $Y_t = \sqrt{X_t}$. The notation $[q]^+$ means the positive part of the quantity q.

Example 5.14

Consider the Ricker model where $f(x) = bxe^{-cx}$. Figure 5.13 shows realizations of two different random processes, one with environmental stochasticity and one with demographic stochasticity.

In many models the stochasticity enters only through the initial condition. These situations are modeled by the process

$$X_{t+1} = f(X_t, r), \quad X_0 = C,$$

where C is a fixed random variable. The solution to this random equation can be obtained by solving the deterministic problem, if possible; in this case, the random initial condition is evolved deterministically. The solution has the form

$$X_t = G(t, C, r),$$

which is a transformation of a random variable C (treating t and r as parameters) and therefore the statistical properties of X_t can, in principle, be determined exactly. Simulations, or realizations of these processes, are smooth curves with random initial conditions.

Later in this chapter we examine stochastic models that are obtained directly from probability considerations.

Example 5.15

Consider the random difference equation

$$
\begin{aligned}
X_{t+1} &= aX_t, \quad t = 0, 1, 2, \ldots, \\
X(0) &= X_0,
\end{aligned}
$$

where a is a positive constant and X_0 is a random variable. Treating the equation deterministically, we obtain

$$
X_t = X_0 a^t.
$$

Then

$$
\begin{aligned}
\mathrm{E}(X_t) &= \mathrm{E}(X_0 a^t) = a^t \mathrm{E}(X_0), \\
\mathrm{Var}(X_t) &= \mathrm{Var}(X_0 a^t) = a^{2t} \mathrm{Var}(X_0).
\end{aligned}
$$

For example, if $0 < a < 1$, then the expected value of X_t goes to zero as $t \to \infty$, and the variance goes to zero as well; this means the spread of the distribution gets narrower and narrower. In fact, we can determine the pdf for X_t as follows. Letting $f_{X_0}(x)$ denote the pdf of X_0, we have

$$
\begin{aligned}
\Pr(X_t \leq x) &= \Pr(X_0 a^t \leq x) = \Pr(X_0 \leq xa^{-t}) \\
&= \int_{-\infty}^{xa^{-t}} f_{X_0}(y)\, dy = \int_{-\infty}^{x} f_{X_0}(za^{-t}) a^{-t}\, dz.
\end{aligned}
$$

Therefore the pdf of the process is

$$
f_{X_t}(x, t) = f_{X_0}(xa^{-t}) a^{-t}.
$$

For a specific case let X_0 be uniformly distributed on the interval $[0, b]$. That is,

$$
f(x) = \frac{1}{b} 1_{[0,b]}(x).
$$

Then

$$
f_{X_t}(x, t) = \frac{a^{-t}}{b} 1_{[0,b]}(xa^{-t}) = \frac{a^{-t}}{b} 1_{[0,ba^t]}(x),
$$

which is uniform. As an exercise, the reader should plot this density for $t = 1, 2, 3$ for $a > 1$ and $0 < a < 1$. This method is successful because this simple discrete model can be solved exactly.

EXERCISES

1. A deterministic model for the yearly population of Florida sandhill cranes is
$$x_{t+1} = (1 + b - d)x_t,$$
where $b = 0.5$ is the birthrate and $d = 0.1$ is the deathrate. Initially the population is 100. (Adapted from Mooney & Swift, 1999.)

a) Find a formula for the population after t years and plot the population for the first 5 years.

b) On average, a flood occurs once every 25 years, lowering the birthrate 40% and raising the deathrate 25%. Set up a stochastic model and perform 20 simulations over a 5-year period. Plot the simulations on the same set of axes and compare to the exact solution. Draw a frequency histogram for the number of ending populations for the ranges (bins) 200–250, 250–300, and so on, continuing in steps of 50.

c) Assume the birthrates and deathrates are normal random variables, $N(0.5, 0.03)$ and $N(0.1, 0.08)$, respectively. Perform 20 simulations of the population over a 5-year period and plot them on the same set of axes. On a separate plot, for each year draw side-by-side box plots indicating the population quartiles. Would you say the population in year 5 is normally distributed?

2. A "patch" has area a, perimeter s, and a strip (band) of width w inside the boundary of a from which animals disperse. Only those in the strip disperse. Let u_t be the number of animals in a at any time t. The *per capita* growth rate of all the animals in a is r. The rate that animals disperse from the strip is proportional to the fraction of the animals the strip, with proportionality constant is ϵ, which is the emigration rate for those in the strip.

a) Argue that the dynamics is ruled by
$$u_{t+1} = ru_t - \epsilon\left(\frac{ws}{a}u_t\right).$$

b) Determine conditions on the parameters r, w, s, ϵ, and a under which the population is growing.

c) Why do you expect $s = k\sqrt{a}$ for some constant k? Write down the model in the case that the region is a circle.

d) Suppose at each time step the emigration rate is chosen from a normal distribution with mean 0.35 and standard deviation 0.5. With a growth rate of 1.06 in a circle of radius 100 m, with width $w = 10$ m, simulate the dynamics of the process. Take $u_0 = 100$. Sketch three simulations (realizations) on the same set of axes. Does the population grow or become extinct?

e) In a given time step, suppose the animals disperse with $\epsilon = 0.25$ or $\epsilon = 0.45$, each with equal probability. Simulate the dynamics of the process and sketch three realizations.

3. Consider the population growth law $x_{t+1} = r_t x_t$, where the growth rate r_t is a fixed positive sequence and x_0 is given.

a) Let x_T be the population at time T. Which quantity is a better measure of an *average* growth rate over the time interval $t = 0$ to $t = T$, the arithmetic mean $r_A = \frac{1}{T}(r_0 + r_1 + \cdots + r_{T-1})$, or the geometric mean $r_G = (r_0 r_1 \cdots r_{T-1})^{1/T}$?

b) Consider the stochastic model process $X_{t+1} = R_t X_t$, where $R_t = 0.86$ or $R_t = 1.16$, each with probability one-half. Does the population eventually grow, die out, or remain the same? If $X_0 = 100$ is fixed, what is the expected value of the population after 500 generations? Run some simulations to confirm your answer.

4. A stochastic model is defined by

$$X_{t+1} = R_t X_t,$$

where the initial population $X_0 = x_0$ is fixed, and the growth rate R_t at each time is chosen from a fixed distribution with mean μ and variance σ^2. Use the central limit theorem to prove that X_t approaches a log-normal random variable for large t. What is the mean and variance of X_t for large t?

5. Suppose you are playing a casino game where on each play you win or lose one dollar with equal probability. Beginning with $100, plot several realizations of a process where you play 200 times. Perform the same task if the casino has the probability 0.52 of winning. In the latter case, perform several simulations and estimate the average length of a game.

6. In a generation, suppose each animal in a population produces two offspring with probability $\frac{1}{4}$, one offspring with probability $\frac{1}{2}$, and no offspring with probability $\frac{1}{4}$. Assume an animal itself does not survive over the generation. Illustrate five realizations of the population history over 200 generations when the initial population is 8, 16, 32, 64, and 128. Do your results say anything about extinction of populations?

7. Suppose μ is the probability that each individual in a population will die during a given year. If X_t is the population at year t, then the probability there will be n individuals alive the next year can be modeled by a binomial random variable,

$$\Pr(X_{t+1} = n | X_t = x_t) = \binom{x_t}{n}(1 - \mu)^n \mu^{x_t - n}.$$

In other words, the population at the next time step is the probability of having n successes out of x_t trials, where $1 - \mu$ is the probability of success (living).

a) At time $t = 0$ assume a population has 50 individuals, each having mortality probability $\mu = 0.7$. Use a computer algebra system to perform 10 simulations of the population dynamics over 7 years and plot the results.

b) For the general model, find the expected value $E(X_{t+1})$ and the variance $\text{Var}(X_{t+1})$.

8. Consider a random process X_t governed by the discrete model

$$X_{t+1} = \frac{2X_t}{1 + X_t},$$

where X_0 is a random variable with density $f_{X_0}(x)$. Find a formula for the density $f(x, t)$ of X_t.

9. Experimental data for the yearly population of Serengeti wildebeest can be fit by the Beverton–Holt model

$$X_{t+1} = \frac{(1 - h)rKX_t}{K + (r - 1)X_t},$$

where K is the carrying capacity, r is a growth factor, and h is the annual harvesting rate The carrying capacity is a function of rainfall and is modeled by $K = 20748R$, where R is the annual rainfall. Actual measured rainfalls over a 25-year period are given in the vector

$$\mathbf{R} = (100, 36, 100, 104, 167, 107, 165, 71, 91, 77, 134, 192, 235,$$
$$159, 211, 257, 204, 300, 187, 84, 99, 163, 97, 228, 208).$$

Using $X_0 = 250,000$ and $r = 1.1323$, simulate the population dynamics over 50 years for different harvesting rates, $h = 0.05, 0.075, 0.1, 0.125, 0.15$, and each year randomly draw the annual rainfall from the vector $\mathbf{R}$. Run enough simulations to make some conclusions about how the fraction of populations that collapsed depends upon the harvesting rate. Assume the threshold value of 150,000 wildebeest as representing a collapsed herd.

10. Develop a demographic stochastic model based upon the discrete model

$$x_{t+1} = \frac{bx_t}{1 + cx_t}.$$

Take $c = 1$ and $b = 6$. With $x_0 = 200$, run simulations of the model with noise $E_t \sim N(0, \sigma)$ for several different values of σ, and comment upon the results.

5.4 Probability-Based Models

5.4.1 Markov Processes

Stochastic processes X_t are characterized by the dependencies among the random variables $X_0, X_1, X_2, \ldots$. A Markov process is a special stochastic where only the current value of the random variable is relevant for future predictions. The past history, or how the present emerged from the past, is irrelevant. Many processes in economics, biology, games of chance, engineering, and other areas are of this type—for example, stock prices, a board position in the game of monopoly, the population of a plant species, and so on.

We illustrate the Markov model by considering a specific problem in phylogenetics.

We may recall from elementary biology that the double helix DNA molecule is the genetic material that contains information on coding for and synthesizing proteins. DNA is constructed from four bases, the purines A and G, and the pyrimidines C and T. In the DNA molecule, across the rungs of the double helix molecular ladder, A and T are paired and G and C are paired. Therefore, to represent a DNA molecule mathematically all we need consider is a single sequence, e.g., $ATCCGTAGATGG$, along one side of the ladder. As time evolves, DNA must be copied from generation to generation. During this process there are natural mutations, and our goal is to set up a simple model for determining how these mutations evolve to change an ancestral DNA sequence. Such models are important, for example, in determining phylogenetic distances between different species and to understand the evolution of the genome in organisms, both plant and animal. Phylogentics helps identify related species and to construct evolutionary trees; understanding how viruses evolve, for example, aids in developing strategies for disease prevention and treatment.

The simplest mutations in copying a DNA sequence are called base *substitutions*—a base at a given location is simply replaced by another one at that same location over one generation. There are other types of mutations, but they are not considered in our introductory discussion. If a purine is replaced by a purine or a pyrimidine with a pyrimidine, then the substitution is a *transition*; if the base class is interchanged, then the substitution is a *transversion*.

To fix the idea, consider mutations of a sequence over three generations:

$$S_0 \quad : \quad ATC\mathbf{C}GT\mathbf{A}GATGG$$

$$S_1 \quad : \quad ATC\mathbf{C}AT\mathbf{A}GTTGG$$

$$S_2 \quad : \quad ATC\mathbf{C}GT\mathbf{A}GTTGG$$

The changes are highlighted. Comparing the first two generations, in the fifth position there is a transition and in the ninth position there is a transversion.

Consider an arbitrary sequence. We let P_A, P_G, P_C, and P_T denote the probabilities that A, G, C, and T appear in at a given location. In the sequence S_0, for example, $P_A = 3/12$. Next we fix an arbitrary location at some generation t. During the next generation, $t+1$, the base at that location may stay the same, or it may mutate to another base. We let the sixteen numbers $P(X|Y)$, where X and Y represent the four bases, denote the conditional probabilities that the location has base X at time $t+1$, given that it was at Y at time t. We can arrange these numbers in a *transition matrix*

$$M = \begin{pmatrix} P(A,A) & P(A,G) & P(A,C) & P(A,T) \\ P(G,A) & P(G,G) & P(G,C) & P(G,T) \\ P(C,A) & P(C,G) & P(C,C) & P(C,T) \\ P(T,A) & P(T,G) & P(T,C) & P(T,T) \end{pmatrix}. \tag{5.33}$$

Notice the order of the entries. Columns and rows are labeled in the order, A, G, C, T, and the column denotes the state of the ancestral base, and the row is the state of the descendent base. In the transition matrix M, the entries are nonnegative and the columns sum to one because one of the transitions must occur. Such a matrix is called a *Markov matrix*. We assume that, over many generations of equal time steps, the same transition matrix characterizes the transitions at each time step. Now, if $\mathbf{p}_t = (P_A, P_G, P_C, P_T)^T$ denotes the *probability vector* at time t, the model

$$\mathbf{p}_{t+1} = M\mathbf{p}_t, \tag{5.34}$$

$t = 0, 1, 2, \ldots$, is an example of a *discrete* **Markov process**. It dictates how the probabilities of bases occurring in a generation change over time. Therefore, if $\mathbf{p}_0$ is given, then $\mathbf{p}_t$ can be calculated by iteration.

We recognize (5.34) as a linear, Leslie-type model that we examined in an earlier section. The solution is

$$\mathbf{p}_t = M^t \mathbf{p}_0,$$

where the entries in the power matrix M^t denote the probabilities of being in a given state (base) after t time steps, given the initial state. For example, for the transition matrix (5.33) for a DNA sequence, the $4, 2$ element in M^t would be the probability that the base is T after t time steps, given that it was G at time $t = 0$.

In general, a discrete Markov process, or Markov chain, can be defines as follows. Consider a sequence of random variables $X_0, X_1, X_2, ...$, where each random variable is defined on a finite or countably infinite state space, which we denote by $S = \{0, 1, 2, 3, ...\}$. The stochastic process X_t is **Markov,** if for any $t = 0, 1, 2, 3, ...$, it has the property

$$\Pr(X_{t+1} = a_{t+1} | X_t = a_t, ..., X_0 = a_0) = \Pr(X_{t+1} = a_{t+1} | X_t = a_t), \quad a_t \in S.$$

In other words, the future behavior depends only upon the present state and not on the previous history. The **transition matrix** M for the process is the matrix whose entries are $m_{ij} = \Pr(X_{t+1} = i | X_t = j)$. Thus, m_{ij} is the probability of moving from state j at time t to state i at time $t + 1$. Notice that M may be an infinite matrix if the state space is infinite. Clearly, by definition, the entries in each column of M are non-negative and sum to one. We sometimes graphically show the states and elements of the transition matrix on a Leslie type diagram, as in Fig. 5.10. The entries of M^t are denoted by $m_{ij}^{(t)}$ and represent the probabilities of moving from the jth state to the ith state in t time steps. A vector $\mathbf{p}$ with nonnegative entries that sum to one is called a **probability vector**. There are many examples of Markov processes in all areas of science, engineering, and economics.

Example 5.16

(Jukes–Cantor model) In the **Jukes–Cantor** model of base substitution,

$$M = \begin{pmatrix} 1-a & \frac{a}{3} & \frac{a}{3} & \frac{a}{3} \\ \frac{a}{3} & 1-a & \frac{a}{3} & \frac{a}{3} \\ \frac{a}{3} & \frac{a}{3} & 1-a & \frac{a}{3} \\ \frac{a}{3} & \frac{a}{3} & \frac{a}{3} & 1-a \end{pmatrix},$$

where $1 - a$ is the probability that there is no change in a base, while $a/3$ is the probability of change to any other base. The number a is called the *mutation rate per generation, or the number of substitutions per site per time step;* values of a range from 10^{-9} for some plants, 10^{-8} for mammals, to 10^{-2} for viruses, which mutate quickly. We are assuming a is small constant and does not depend upon time or upon location in the sequence.

It is straightforward to verify that $\lambda = 1$ is an eigenvalue of M with eigenvector $(1, 1, 1, 1)^{\mathrm{T}}$, and $\lambda = 1 - \frac{4}{3}a$ is an eigenvalue of multiplicity 3 with independent eigenvectors $(1, 1, -1, -1)^{\mathrm{T}}, (1, -1, 1, -1)^{\mathrm{T}}, (1, -1, -1, 1)^{\mathrm{T}}$. Then, if Q is the matrix whose columns are the eigenvectors, then Q diagonalizes M, which means $M = QDQ^{-1}$, where D is the diagonal matrix

$$
D = \begin{pmatrix}
1 & 0 & 0 & 0 \\
0 & 1 - \frac{4}{3}a & 0 & 0 \\
0 & 0 & 1 - \frac{4}{3}a & 0 \\
0 & 0 & 0 & 1 - \frac{4}{3}a
\end{pmatrix}.
$$

Therefore

$$
M^t = QD^tQ^{-1} = Q \begin{pmatrix}
1 & 0 & 0 & 0 \\
0 & \left(1 - \frac{4}{3}a\right)^t & 0 & 0 \\
0 & 0 & \left(1 - \frac{4}{3}a\right)^t & 0 \\
0 & 0 & 0 & \left(1 - \frac{4}{3}a\right)^t
\end{pmatrix} Q^{-1}.
$$

These matrices can be multiplied out to obtain M^t. One can check that the diagonal entries of M^t are $q(t) = \frac{1}{4} + \frac{3}{4}\left(1 - \frac{4}{3}a\right)^t$, and the off-diagonal entries are all given by $\frac{1}{4} - \frac{1}{4}\left(1 - \frac{4}{3}a\right)^t$. Therefore $q(t)$ is the expected fraction of sites in the sequence that are observed to remain unchanged after t time steps. It follows that $1 - q(t)$ is the fraction of sites expected to differ from the original ancestral sequence after t time steps.

It is not hard to prove the following important theorem (a proof for a two-dimensional matrix is requested in Exercise 1).

Theorem 5.17

Let $M = (m_{ij})$ be an $n \times n$ Markov matrix having nonnegative entries m_{ij} and column sums equal to one. Then $\lambda = 1$ is an eigenvalue of M and the corresponding eigenvectors have nonnegative entries; the remaining eigenvalues λ_i satisfy $|\lambda_i| \leq 1$. Moreover, if the entries of M are strictly positive, then $\lambda = 1$ is a dominant eigenvalue with a single eigenvector.

If the state space is finite, then over a long time the system may approach a stationary probability distribution defined by a probability vector $\mathbf{p}$ that is an eigenvector of $\lambda = 1$. That is, $M\mathbf{p} = \mathbf{p}$. Depending upon the geometric multiplicity of $\lambda = 1$, the stationary distribution may or may not be unique. A stationary distribution $\mathbf{p}$ is said to be **limiting** if $M^t\mathbf{p}_0 \to \mathbf{p}$ for every initial probability vector $\mathbf{p}_0$. There are general conditions on a Markov matrix that guarantee a unique limiting stationary distribution (e.g., see Allen, 2003).

There are many interesting questions about processes that satisfy the Markov conditions. For example: Which states can lead to some other state?

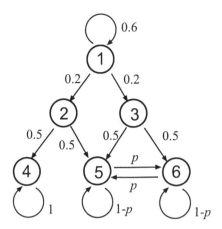

Figure 5.14 Six-state process.

Are there states that are visited infinitely often, or are there states in which that trap the process? Are there states where the process will always end up? Figure 5.14 shows a Leslie type diagram for a six-state system, and it illustrates different types of states. The mutually communicating states $\{5,6\}$ and the state $\{4\}$ are trapping states, and regardless of the initial probability vector, the process eventually will migrate to one of these states. States $\{1\}$, $\{2\}$, and $\{3\}$ are transient. Note that 1 communicates with 2, but 2 does not communicate with 1. There is a technical language that precisely characterizes the states of a process, and we refer to Allen (2003) for a full discussion. Other questions regarding the probabilities and expected times for moving from one state to another are discussed in the next section.

EXERCISES

1. Let M be a 2×2 transition matrix for a two-state Markov process. Prove Theorem 16 in this case.

2. For the Markov matrix

$$
M = \begin{pmatrix} p & 0 & q \\ q & p & 0 \\ 0 & q & p \end{pmatrix},
$$

   with $p, q > 0$, find the unique stationary probability distribution. Find a 3×3 Markov matrix with a non-unique stationary probability distribution.

3. Find the transition matrix for the process shown in Fig. 5.14.

4. A regional rent-a-car company has offices located in Lincoln, Omaha, and ten other small cities. Their records show that each week, 55% of the cars rented in Lincoln returned to Lincoln, 35% are returned to Omaha, and the remaining cars are returned elsewhere. Of those rented in Omaha, 75% are returned to Omaha, 5% to Lincoln, and 20% elsewhere. Of the cars rented in the ten other cities, 25% are returned Omaha and 5% to Lincoln. Draw a Leslie type diagram illustrating the states and the transition probabilities. Over the long run, if conditions remain the same, what percentage of the cars in their fleet will end up in Lincoln?

5. Consider the matrix
$$M = \begin{pmatrix} 0 & 1 \\ 1 & 0 \end{pmatrix}.$$
Show that there is a stationary probability distribution, but it is not limiting.

6. At each discrete instant of time, measured in minutes, a molecule is either inside a cell membrane, or outside. If it is inside the cell, it has probability 0.2 of passing outside, and if it is outside it has probability 0.1 of entering the cell.

   a) If it is inside the cell initially, what is the probability that the molecule is inside the cell at time t? Over the long run, what is the probability that it is inside the cell?

   b) If there are 100 molecules inside the cell initially, what is the probability that there are exactly 80 of the molecules remaining in the cell after 1 minute? What is the probability that there are 10 or fewer molecules in the cell after 20 minutes? If N_t is the RV representing the number of molecules in the cell at time t, plot the probability mass function for N_{20}.

7. In a small ecosystem there are two food patches X and Y, and foraging animals move week to week, from patch to patch, with the following probabilities: $P(X, X) = 0.6$, $P(X, Y) = 0.7$, $P(Y, X) = 0.4$, $P(Y, Y) = 0.3$. Write down the transition matrix.

   a) What is the limiting distribution of animals in the two patches?

   b) If an animal is patch X at time $t = 5$ weeks, what is the probability that it is in patch Y at $t = 10$ weeks?

8. In the Jukes–Cantor model of base substitution a given base changed to another base with equal probability. A generalization of this model is the two-parameter **Kimura model** where it is assumed that transitions and transversions have different probabilities, b and c, respectively.

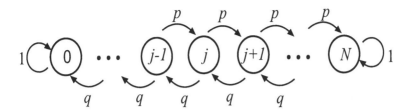

Figure 5.15 Graph for an random walk with absorbing end states.

a) Write down the Markov matrix M for the Kimura model.

b) Find a limiting stationary distribution.

5.4.2 Random Walks

A random walk is a special type of Markov process where the transitions can only occur between adjacent states. We begin the discussion with a classic example where there are $N + 1$ states $\{0, 1, 2, ..., N\}$, and if the system is in state j, then it moves to $j + 1$ with probability p and to $j - 1$ with probability $q = 1 - p$. Figuratively, we think of an individual walking on a lattice of points, which are the states. At each instant the individual tosses a loaded coin with $\Pr(\text{heads}) = p$ and $\Pr(\text{tails}) = 1 - p$. If heads comes up he moves to the right, and if tails shows he moves to the left—thus the terminology **random walk**. We assume the boundary states 0 and N are **absorbing**, that is, $m_{00} = m_{NN} = 1$. In other words, if the individual reaches one of these states, then he can never get out. A boundary state that is not absorbing is called a reflecting boundary. Fig. 5-15 shows the graph. With a little imagination the reader can understand why random walks arise in many areas, including gambling (the states represent a player's fortune and p is the probability of winning on a given play), diffusion (the states are geometrical positions and a particle or an animal moves to the left or right with given probabilities), or population models (the states are population levels and there are probabilities of births and deaths in any time period).

We address only one important question in the context of a particle executing a random walk. If a particle is located at position j at time, then what is the probability it will reach position 0 before it reaches N? Let

$$u_j = \text{the probability that the particle will}$$
$$\text{reach 0 before } N \text{ when it starts at } j,$$

for $1 \leq j \leq N - 1$. To obtain an analytic expression for this probability, we argue as follows. Beginning at j, there are two ways to proceed. In the first step the particle can jump to $j + 1$ with probability p and then have probability u_{j+1} of reaching 0 before N; or, it can jump to $j - 1$ with probability q and then have probability u_{j-1} of reaching 0 before N. According to the law of total probability,

$$u_j = pu_{j+1} + qu_{j-1}, \quad j = 1, ..., N - 1. \tag{5.35}$$

Also, at the boundaries,

$$u_0 = 1, \quad u_N = 0. \tag{5.36}$$

Equation (5.35) is a second-order difference equation that can be solved by the techniques introduced in Exercises 5 and 6 of Section 5.1. Solutions have the form $u_j = r^j$. Substituting this expression into (5.35) we obtain the characteristic equation

$$pr^2 - r + q = 0.$$

In the case $p \neq q$ the roots are $r = 1$, $r = q/p$, and in the case $p = q = \frac{1}{2}$, the roots are $1, 1$. Therefore, the general solution to (5.35) is

$$u_j = c_1 + c_2 \left(\frac{q}{p}\right)^j, \quad p \neq q, \tag{5.37}$$

$$u_j = c_1 + c_2 j, \quad p = q. \tag{5.38}$$

The arbitrary constants may be found from the boundary conditions (5.36). We obtain

$$u_j = \frac{\left(\frac{q}{p}\right)^j - \left(\frac{q}{p}\right)^N}{1 - \left(\frac{q}{p}\right)^N}, \quad p \neq q,$$

$$u_j = \frac{N - j}{N}, \quad p = q.$$

By exactly the same argument we can calculate the probability

$$v_j = \text{the probability that the particle will}$$
$$\text{reach } N \text{ before 0 when it starts at } j.$$

The boundary conditions are now $v_0 = 0$, $v_N = 1$. We get

$$v_j = \frac{1 - \left(\frac{q}{p}\right)^j}{1 - \left(\frac{q}{p}\right)^N}, \quad p \neq q,$$

$$v_j = \frac{j}{N}, \quad p = q.$$

Clearly we have $u_j + v_j = 1$. Therefore, the probability that the particle hits a boundary is 1.

We may not be surprised by the last conclusion. But we can ask what is the expected time to hit a boundary. Let T_j be the RV

$$T_j = \text{the first time a particle arrives at}$$
$$0 \text{ or } N \text{ when it starts at } j.$$

Further, let $\tau_j = E(T_j)$. Then

$$\tau_j = p(1 + \tau_{j+1}) + q(1 + \tau_{j-1}).$$

The reasoning underlying this difference relation is if the particle begins at j, it may move to $j+1$ or $j-1$ with probability p and q, respectively. If it moves to $j+1$ the expected time it hits a boundary $1 + \tau_{j+1}$, counting the jump just taken. If it moves to $j-1$ the expected time is $1 + \tau_{j-1}$. Therefore, rearranging,

$$p\tau_{j+1} - \tau_j + q\tau_{j-1} = -1, \tag{5.39}$$

which is a second-order nonhomogeneous difference equation. The boundary conditions are $\tau_0 = \tau_N = 0$. The homogeneous equation is the same as (5.35), and we get the same solution (5.37)–(5.38). To find particular solution we use the method of undetermined coefficients, as in differential equations, and we find particular solutions $\frac{j}{q-p}$ and $-j^2$ in the cases $p \neq q$ and $p = q$, respectively. In summary, forming the general solution of (5.39) (composed of the sum of the solution to the homogeneous equation and a particular solution) and evaluating the constants using the boundary conditions, we find the expected time for absorption at a boundary is

$$\tau_j = \frac{1}{q-p}\left[j - N\left(\frac{1 - \left(\frac{q}{p}\right)^j}{1 - \left(\frac{q}{p}\right)^N} \right)\right], \quad p \neq q,$$

$$\tau_j = j(N - j), \quad p = q,$$

where j is the initial position. The exercises will request calculations and plots of these times.

Next we describe a combinatorial approach to a random walk problem. Consider a migrating animal, or perhaps a dispersing plant species, that moves in a linear habitat such as a river, or along the bank of a river. As before we zone the habitat into a lattice of sites $..., -2, -1, 0, 1, 2, ...$. For this example we assume the habitat is long and has no boundaries. The animal's location at time t ($t = 0, 1, 2, 3, ...$) is the random variable X_t. Starting at $X_0 = 0$, at each time step the animal moves to the right with probability p or to the left with probability $q = 1 - p$. We are interested in calculating the probability mass function, or distribution, of X_t. In other words, what is the probability of the animal being at location k after t time steps? A specific example will lead us to the correct reasoning.

Example 5.18

Consider the probability of being at location $k = 2$ after $t = 10$ time steps. To accomplish this, the animal must move to the right $r = 6$ steps and to the left $l = 4$ steps. A possible path is $RLRRLRLLRR$. The probability that this single path will occur is clearly $p^6 q^4$. How many such paths can be formed with $r = 6$ and $l = 4$? It is clearly "10 choose 6", or $\binom{10}{6} = 210$ paths. Therefore the probability of arriving at $k = 2$ after 10 steps is $\binom{10}{6} p^6 q^4$.

Therefore, we can argue in general that if there are t steps terminating at k, then $t = r + l$ and $k = r - l$

$$\Pr(X_t = k) = \binom{t}{r} p^r q^{t-r}, \quad r = \frac{t+k}{2}.$$

This holds for $t \geq |k|$. If $t < |k|$, then it is impossible to reach k and $\Pr(X_t = k) = 0$. In summary, we have shown that X_t is binomially distributed.

Example 5.19

Assuming $p = q = \frac{1}{2}$, after $2n$ time steps what is the probability that the animal will be back at the origin? We have

$$\Pr(X_{2n} = 0) = \binom{2n}{n} \left(\frac{1}{2}\right)^n \left(\frac{1}{2}\right)^n = \frac{(2n)!}{n!^2 2^{2n}}.$$

With some work we can calculate the right side, but if n is large it will be cumbersome. Fortunately, there is an excellent approximation for $n!$ for large n, and it works well even for n values as low as 6. **Stirling's approximation** is

$$n! \sim \sqrt{2\pi} n^{n+\frac{1}{2}} e^{-n}.$$

Using this formula, we find

$$\Pr(X_{2n} = 0) \sim \frac{1}{\sqrt{n\pi}}.$$

For large t, there is a relationship between the binomial distribution and the normal distribution. Although it is out of our scope to prove it, we present the key result. If t is large and neither p nor q are too close to zero, the binomial $X \sim \text{bin}(t, p)$ can be closely approximated with the standardized normal random variable given by

$$Z = \frac{X - tp}{\sqrt{tpq}}.$$

The approximation is good if tp and tq are greater than 5. In symbols,

$$\lim_{t \to \infty} \Pr\left(a < \frac{X - tp}{\sqrt{tpq}} \leq b\right) = \frac{1}{\sqrt{2\pi}} \int_a^b e^{-s^2} ds.$$

This approximation enables us to easily calculate probabilities that an animal executing a random walk on the integers will lie in some given interval after many steps.

EXERCISES

1. Consider the random walk on the lattice $\{0, 1, 2, ..., 100\}$. If $p = 0.45$, find the expected time τ_j for absorption at a boundary, $j = 0, 1, 2, ..., 100$. Plot τ_j vs. j.

2. A gambler is playing a casino game, and on each play he wins $1 with probability $p = 0.46$, and he loses $1 with probability $q = 0.54$. He starts with $50 and decides to play the game until he goes broke, or until he wins $100. What is the probability of going broke before winning $100? How long should he expect to play? What if the odds are even?

3. Referring to Exercise 2, when $p = 0.46$, simulate three games and plot realizations on the same set of axes.

4. Compute both 6! and 10! exactly and by Stirling's approximation.

5. An animal, starting at the origin, is executing a random walk on the integers with $p = 0.6$. After 40 steps what is the probability $\Pr(X_{40} \leq 10)$? If $p = 0.5$, show that after a million steps the animal will almost certainly lie within 4000 units of the origin. If $p = 0.5$, what is the probability that its position after 50 time steps lies in the interval $5 \leq k \leq 10$?

6. Write a computer program that simulates a random walk on the integers, with $X_0 = 0$ and $p = 0.5$. Sketch three realizations on the same set of axes.

5.4.3 The Poisson Process

Consider a stochastic process taking place in continuous time in which events of a certain type can occur at various instants. For example: phone calls arriving at a relay station, discoveries of food items by a forager, alpha particles emitted from a radioactive source, or customers arriving at bank. These processes can be modeled by a **Poisson process**, defined by the conditions:

$$\Pr(\text{one event in } dt) = \lambda dt + \text{o}(dt),$$
$$\Pr(\text{no event in } dt) = 1 - \lambda dt + \text{o}(dt),$$
$$\Pr(\text{two or more events in } dt) = \text{o}(dt),$$

for small intervals of time dt. The parameter λ is the **rate parameter** of the process. Now, let

$$X_t = \text{ the number of events in } [0, t],$$

and denote the distribution function by

$$p_n(t) = \Pr(X_t = n).$$

We can calculate the probabilities recursively in the way we now describe. We have

$$
\begin{aligned}
p_0(t + dt) &= \Pr(\text{no events in } [0, t + dt]) \\
&= \Pr(\text{no events in } [0, t]) \cdot \Pr(\text{no events in } [t, t + dt]) \\
&= p_0(t)(1 - \lambda dt + \text{o}(dt)).
\end{aligned}
$$

Therefore

$$\frac{p_0(t + dt) - p_0(t)}{dt} = -\lambda p_0(t) + \frac{\text{o}(dt)}{dt}.$$

Taking the limit as $dt \to 0$ gives

$$p_0'(t) = -\lambda p_0(t),$$

with $p_0(0) = 1$. Therefore the probability that no event occurs up to time t is an exponentially decreasing function

$$p_0(t) = e^{-\lambda t}.$$

We use this same method to compute $p_1(t)$. Observe

$$
\begin{aligned}
p_1(t + dt) &= \Pr(\text{one event in } [0, t + dt]) \\
&= \Pr(\text{no events in } [0, t]) \cdot \Pr(\text{one event in } dt) \\
&\quad + \Pr(\text{one event in } [0, t]) \cdot \Pr(\text{no events in } dt) \\
&= p_0(t)\lambda dt + p_1(t)(1 - \lambda dt) + \text{o}(dt)).
\end{aligned}
$$

Therefore,

$$p_1'(t) = -\lambda p_1(t) + \lambda p_0(t),$$

with $p_1(0) = 0$.

This method can be extended indefinitely to the other distributions, and we can obtain the system of differential equations

$$p'_n(t) = -\lambda p_n(t) + \lambda p_{n-1}(t), \quad n = 0, 1, 2, ..., \tag{5.40}$$

with $p_n(0) = 0$. These are called the **Kolmogorov equations** [3] for the process. They can be solved recursively to get

$$p_n(t) = \frac{(\lambda t)^n}{n!} e^{-\lambda t}, \tag{5.41}$$

for $n = 0, 1, 2, ...$ From this it easily follows that

$$p_n(t) = \frac{\lambda t}{n} p_{n-1}(t).$$

Beginning with $p_0(t)$, one can easily calculate the remaining distributions.

The probability distribution (5.41) is the Poisson distribution with parameter λt.

To simulate a Poisson process we need to know the inter-event period, or the time between events. But this, too, is a random variable T. Without loss of generality, we can assume that T is the waiting time from 0 to the first event. To determine how T is distributed, we compute the cumulative distribution function $F_T(t)$. We have

$$
\begin{aligned}
F_T(t) &= \Pr(T \le t) = 1 - \Pr(T > t) \\
&= 1 - \Pr(\text{no events in } [0, t]) \\
&= 1 - p_0(t) = 1 - e^{-\lambda t}.
\end{aligned}
$$

The pdf is the derivative, or

$$f_T(t) = \lambda e^{-\lambda t}, \quad t \ge 0,$$

which is the exponential distribution. Therefore, the inter-event time is exponentially distributed and

$$E(T) = \frac{1}{\lambda}, \quad \text{Var}(T) = \frac{1}{\lambda^2}.$$

This means the average waiting time is $1/\lambda$. An important assumption underlying a Poisson process is that it is memoryless. That is, occurrence of an event at time $t = t_0$ has no influence on the future occurrence of events. Symbolically, $\Pr(T > t + t_0 | T > t_0) = \Pr(T > t)$, which is easily obtained from the definition of conditional probability.

---

[3] A. Kolmogorov (1903–1987) was instrumental in developing the modern, axiomatic theory of probability.

In applications to predation events in ecology, the Poisson model describes a search for randomly distributed prey, where success does not deplete the number of prey. This may not be valid in some situations, for example, when the prey are clumped or depleted.

Now we return to the question of simulation of a Poisson process. The key idea is to note that if T is Poisson distributed, then $T = -\frac{\ln U}{\lambda}$, where U is a uniformly distributed random variable on $[0, 1]$. In MATLAB, for example, a sequence of event times $t(k)$ can be calculated recursively in a loop $t(k + 1) = t(k) - \frac{\ln(\text{rand})}{\lambda}$, where $t(0) = 0$.

Example 5.20

Assume a simple birth model where no individual dies, and each has probability of giving birth in a small interval dt to a single offspring with probability $b\,dt$. Then, if n is the population at time t, the population at time $t + dt$ is about $n + bn\,dt + o(dt)$, where $bn\,dt$ is the probability of n individuals giving birth to a single individual (one event) in the interval dt. We can model the population with a Poisson process with rate parameter $\lambda = \lambda_n = bn$, now depending on n. The random variable X_t represents the number of individuals in the population at time t. That is, $p_n(t) = \Pr(X_t = n)$. The Kolmogorov equations (5.40) become

$$p_n'(t) = -bn p_n(t) + b(n - 1)p_{n-1}(t), \quad n = 0, 1, 2, \dots. \tag{5.42}$$

It is shown in Exercise 2 that X_t is negatively binomial distributed.

In this chapter we have only scratched the surface of a large body of theory and applications of stochastic processes. In particular, we have not discussed continuous random walks, where each time step is infinitesimally small. These processes, called Wiener processes or Brownian motion, are the basis of diffusion theory and play a large role in the development of stochastic differential equations. The references at the end of the chapter may be consulted for readable treatments of these important concepts.

EXERCISES

1. Show that if T is Poisson distributed with parameter λ, then $T = -\frac{\ln U}{\lambda}$, where U is a uniformly distributed random variable on $[0, 1]$. Use a computer algebra system to simulate a realization of a Poisson process with $\lambda = 1$.

2. Suppose a population is of fixed size N at $t = 0$ and undergoes a pure birth process with Kolmogorov equations (5.42).

a) Show, by direct substitution, that

$$p_n(t) = \binom{n-1}{N-1} e^{-Nbt}(1 - e^{-bt})^{n-N}, \quad n = N, N+1, N+2, \dots$$

satisfies the Kolmogorov equations.

b) Perform and sketch three simulations of the birth process when $N = b = 1$ over the interval $[0, 5]$. (Use Exercise 1 with $\lambda = bn$.)

c) Plot the densities $p_n(t)$ for $n = 0, 1, 2, \dots$ at times $t = 0, 1, 2, 3$. What does the graph of $p_n(t)$ vs. t look like for a fixed n?

d) Show that the mean and variance of X_t are Ne^{bt} and $Ne^{bt}(e^{bt} - 1)$, respectively. (Look up the negative binomial distribution.)

REFERENCES AND NOTES

Discrete models have come to play an increasingly important role in biological dynamics, and there are many recent books with excellent chapters discussing those models. These books include Allman & Rhodes (2004), Edelstein–Keshet (1988), Kot (2001), Mooney & Swift (1999), and Renshaw (1991). The classical article by May (1976) is required reading. The text by Allen (2003) focuses on stochastic aspects of discrete models, and Caswell (2001) is the standard reference on matrix models. Difference equations, from a mathematical viewpoint, are treated in Elaydi (2003). Of the many outstanding texts on probability and random processes, the recent book by Olofsson (2005) is a general account and can be recommended. The undergraduate probability text by Chung (1979) is highly recommended. Regarding stochastic differential equations and Brownian motion we refer to Soong (1973), Chapter 8 of Allen (2003), and the article by Higham (2001).

Allen, L. J. S. 2003. *An Introduction to Stochastic Processes with Applications to Biology*, Pearson Education, Upper Saddle River, NJ.

Allman, E. S. & Rhodes, J. A. 2004. *Mathematical Models in Biology*, Cambridge University Press, Cambridge.

Caswell, H. 2001. *Matrix Population Models*, 2nd ed., Sinauer Associates, Sunderland, MA.

Chung, K. L. 1979. *Elementary Probability Theory with Stochastic Processes*, 3rd ed., Springer–Verlag, New York.

Edelstein-Keshet, L. 1988. *Mathematical Models in Biology*, McGraw-Hill, New York (reprinted by SIAM, Philadelphia, 2003).

Elaydi, S. 2003. *An Introduction to Difference Equations*, 3rd ed., Springer-Verlag, New York.

Higham, D. J. 2001. An algorithmic introduction to numerical simulation of stochastic differential equations, *SIAM Review* **43**(3), 525–546.

Kot, M. 2001. *Elements of Mathematical Ecology*, Cambridge University Press, Cambridge.

May, R. M. 1976. Simple mathematical models with very complicated dynamics, *Nature* **261**, 459–467.

Mooney, D. & Swift, R. 1999. *A Course in Mathematical Modeling*, Mathematical Association of America, Washington, DC.

Olofsson, P. 2005. *Probability, Statistics, and Stochastic Processes*, Wiley–Interscience, New York.

Renshaw, E. 1991. *Modelling Biological Populations in Space and Time*, Cambridge University Press, Cambridge.

Soong, T. T. 1973. *Random Differential Equations in Science and Engineering*, Academic Press, New York.

6
Partial Differential Equations

Partial differential equations is one of the most fundamental areas in applied analysis, and it is hard to imagine any area of science and engineering where the impact of this subject is not felt. Partial differential equation models, for example, form the basis of electrodynamics (Maxwell's equations), quantum mechanics (the Schrödinger equation), continuum mechanics (equations governing fluid, solid, and gas dynamics; acoustics), reaction–diffusion systems in biological and chemical sciences, only to mention a few. The goal in this and the next chapter are to introduce some important models and key techniques that expose the solution structure of such equations.

6.1 Basic Concepts

Ordinary differential equations describe how a system evolves in time. Yet many systems, perhaps most, evolve in both space and time, giving rise to two or more independent variables. For example, the temperature u in a slender, laterally insulated, metal bar of length l, subject to prescribed temperatures at its ends, depends upon the location x in the bar and upon the time t. That is, $u = u(x,t)$. Later we show, using conservation of energy, that the temperature u must satisfy a partial differential equation

$$u_t(x,t) - k u_{xx}(x,t) = 0,$$

where k is a numerical constant characterizing the thermal conductivity properties of the material composing the bar; the equation holds for $t > 0$ and $0 < x < l$. It is called the **heat equation**, and it is the basic equation of heat transfer.

Consider another example from population dynamics. Imagine, for example, toxic bacteria distributed throughout a canal of length L. If we ignore spatial variation then we may model the total population $u = u(t)$ of bacteria by the usual logistics equation

$$\frac{du(t)}{dt} = ru(t)\left(1 - \frac{u(t)}{K}\right),$$

where r is the growth rate and K is the carrying capacity. Here, the population depends only on time t and we have an ordinary differential equation. However, in a nonhomogeneous environment the parameters r and K may depend upon geographical location x. Even if the environment is homogeneous, the bacteria may be distributed non-uniformly in the canal, and they may move from high population densities to low densities (this is called diffusion), giving spatial dependence. If we wish to consider how the spatial distribution of bacteria changes in time, then we must include a spatial independent variable and write $u = u(x, t)$, where now u is interpreted as a population density, or population per unit length; that is, $u(x, t)\Delta x$ represents approximately the number of bacteria between x and $x + \Delta x$. If we include diffusion of the bacteria then the model modifies to

$$u_t(x, t) = Du_{xx}(x, t) + ru(x, t)\left(1 - \frac{u(x, t)}{K}\right),$$

where D is the diffusion constant. This is **Fisher's equation**,[1] one of the fundamental equations in structured population dynamics.

It is common, as in ordinary differential equations, to drop the independent variables in the notation and just write, for example, the heat equation as $u_t - ku_{xx} = 0$, the dependence of u on x and t being understood. In the next section we derive these model equations.

In general, a second-order partial differential equation in two independent variables is an equation of the form

$$G(x, t, u, u_x, u_t, u_{xx}, u_{xt}, u_{tt}) = 0, \tag{6.1}$$

where (x, t) lies in some domain D in $\mathbb{R}^2$. By a *solution* we mean a twice continuously differentiable function $u = u(x, t)$ on D, which when substituted

---

[1] R. A. Fisher (1890–1962), who made significant contributions to statistics and to evolutionary biology and genetics, formulated the equation in 1937 as a model for the spread of an advantageous gene throughout a population.

into (6.1), reduces it to an identity for (x, t) in D. We assume that u is twice continuously differentiable so that it makes sense to calculate the second-order derivatives and substitute them into (6.1). A solution of (6.1) may be represented graphically as a smooth surface in three dimensional xtu space lying above the domain D. We regard x as a position or spatial coordinate and t as time. The domain D in $\mathbb{R}^2$ where the problem is defined is referred to as a space-time domain and problems that include time as an independent variable are called **evolution problems**. Such problems model how how a system evolves in time, particularly those that govern diffusion processes and wave propagation. When two spatial coordinates, say x and y, are the independent variables we refer to the problem as an **equilibrium** or **steady-state** problem; time is not involved. A large portion of the literature on theoretical partial differential equations deals with questions of regularity of solutions. That is, given an equation and auxiliary conditions, how smooth are the solutions? We do not address these questions here, and we refer to an advanced text (see References and Notes).

A partial differential equation of type (6.1) has infinitely many solutions. Just as the general solution of an ordinary differential equation depends on arbitrary constants, the general solution of a partial differential equation depends on arbitrary functions.

Example 6.1

Consider the simple partial differential equation

$$u_{tx} = tx.$$

Integrating with respect to x gives

$$u_t = \tfrac{1}{2}tx^2 + f(t),$$

where f is an arbitrary function. Integrating with respect to t gives the general solution

$$u = \tfrac{1}{4}t^2x^2 + g(t) + h(x),$$

where h is an arbitrary function and $g(t) = \int f(t)dt$ is also arbitrary. Thus the general solution depends on two arbitrary functions; any choice of g and h will yield a solution.

For ordinary differential equations, initial or boundary conditions fix values for the arbitrary constants of integration and thus often pick out a unique solution. Similarly, partial differential equations do not live in a vacuum. They are usually accompanied by initial or boundary conditions that select out one of its many solutions. A condition given at $t = 0$ along some segment of the x axis is called an initial condition. A condition given along any other curve in the xt plane is called a boundary condition. Initial or boundary conditions involve specifying values of u, its derivatives, or combinations of both along the given curves in the xt plane.

Example 6.2

Heat flow in a bar of length l is governed by the partial differential equation

$$u_t - k u_{xx} = 0, \quad t > 0, \quad 0 < x < l,$$

where k is a physical constant and $u = u(x, t)$ is the temperature in the bar at location x at time t. An auxiliary condition of the form

$$u(x, 0) = f(x), \quad 0 < x < l,$$

is an **initial condition** because it is given at $t = 0$. We regard $f(x)$ as the initial temperature distribution in the bar. Conditions of the form

$$u(0, t) = h(t), \quad u(l, t) = g(t), \quad t > 0,$$

are **boundary conditions**, and $h(t)$ and $g(t)$ represent specified temperatures imposed at the boundaries $x = 0$ and $x = l$ for $t > 0$, respectively. These functions are represented in Fig. 6.1. Graphically, the surface $u = u(x, t)$ representing the solution has f, g, and h as its boundaries. We can visualize the solution by sketching the solution surface or by plotting several time snapshots, or spatial profiles space frozen in time. Figure 6.1 shows a graph of one profile $u(x, t_0)$ for some fixed t_0. It is a cross section of the solution surface. Often a sequence of profiles $u(x, t_1), u(x, t_2), \ldots$, for $t_1 < t_2 < \ldots$, are plotted on the same set of xu axes to indicate how the temperature profiles change in time.

The general solution of a partial differential equation is usually difficult or impossible to find. Therefore, for partial differential equations, we seldom solve a boundary value problem by determining the general solution and then finding the arbitrary functions from the initial and boundary data. This is in sharp contrast to ordinary differential equations where the general solution is found and the arbitrary constants are evaluated from the initial or boundary conditions. Generalizations of the partial differential equation (6.1) can be made in various directions—higher-order derivatives, several independent variables, and several unknown functions (governed by several equations). Later we observe that, there are fundamentally three types of partial differential equations, those that govern diffusion processes, those that govern wave propagation, and those that model equilibrium phenomena. These types are termed **parabolic**, **hyperbolic**, and **elliptic** equations, respectively. Mixed types also occur.

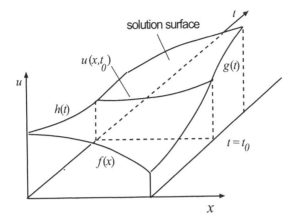

Figure 6.1 A solution surface $u = u(x,t)$ and a time snapshot $u = u(x,t_1)$ at time $t = t_1$. The functions h and g define the boundary conditions, and f specifies the initial condition.

6.1.1 Linearity and Superposition

The separation of partial differential equations into the classes of linear and nonlinear equations is a significant one. Linear equations have a linear algebraic structure to their solution set; that is, the sum of two solutions of a linear homogeneous equation is again a solution, as is a constant multiple of a solution. These facts, termed the **superposition principle**, often aid in constructing solutions. Superposition for linear equations often allows one to construct solutions that can meet diverse boundary or initial requirements. This observation is the basis of the Fourier method, or method of eigenfunction expansions, for linear equations. Linear equations are also susceptible to transform methods for finding solutions, for example, Laplace transforms and Fourier transforms. On the other hand, none of these principles or methods are valid for nonlinear equations. In summary, there is a profound difference between these two classes of problems [see Logan (1994), for example, for an introductory treatment of nonlinear equations].

To formulate these concepts more precisely, we can regard the partial differential equation (6.1) as defining a differential operator L acting on the unknown function $u(x,t)$ and we write (6.1) as

$$Lu(x,t) = f(x,t), \quad (x,t) \in D.$$

Or, suppressing the independent variables,

$$Lu = f, \quad (x, t) \in D. \tag{6.2}$$

In (6.2) all terms involving u are put on the left in the term Lu, and f is a known function. If $f = 0$ on D, then (6.2) is **homogeneous**; if f is not identically zero, then (6.2) is **nonhomogeneous**. The heat equation $u_t - ku_{xx} = 0$ can be written $Lu = 0$, where L is the partial differential operator $\partial/\partial t - k\partial^2/\partial x^2$, and it is clearly homogeneous. The partial differential equation $uu_t + 2txu - \sin tx = 0$ can be written $Lu = \sin tx$, where L is the differential operator defined by $Lu = uu_t + 2txu$. This equation is nonhomogeneous. The definition of linearity depends on the operator L in (6.2). We say that (6.2) is a **linear** equation if L has the properties

(i) $L(u + w) = Lu + Lw$,

(ii) $L(cu) = cLu$,

where u and w are functions and c is a constant. If (6.2) is not linear, then it is **nonlinear** .

Example 6.3

The heat equation is linear because

$$L(u + w) = (u + w)_t - k(u + w)_{xx}$$
$$= u_t + w_t - ku_{xx} - kw_{xx}$$
$$= Lu + Lw,$$

and

$$L(cu) = (cu)_t - k(cu)_{xx}$$
$$= cu_t - cku_{xx}$$
$$= cLu.$$

Example 6.4

The differential equation $uu_t + 2txu - \sin tx = 0$ is nonlinear because

$$L(u + w) = (u + w)(u + w)_t + 2tx(u + w)$$
$$= uu_t + wu_t + ww_t + uw_t + 2txu + 2txw,$$

yet

$$Lu + Lw = uu_t + 2txu + ww_t + 2txw.$$

Note that the nonhomogeneous term $\sin tx$ does not affect linearity or nonlinearity.

It is clear that $Lu = f$ is linear if Lu is first-degree in u and its derivatives; that is, no products involving u and its derivatives occur. Hence the most general linear equation of second order is of the form

$$au_{tt} + bu_{xt} + u_{xx} + du_t + eu_x + gu = f(x,t), \quad (x,y) \in D, \qquad (6.3)$$

where the functions a, b, c, d, e, g, and f are given continuous functions on D. If any of the coefficients $a, \ldots, g$, or f depend on u, we say that the equation is **quasi-linear**. Referring to an earlier remark, (6.3) is hyperbolic (wave-like), parabolic (diffusion-like), or elliptic (equilibrium type) on a domain D if the discriminant $b(x,t)^2 - 4a(x,t)c(x,t)$ is positive, zero, or negative, respectively, on that domain. For example, the heat equation $u_t - ku_{xx} = 0$ has $b^2 - 4ac = 0$ and is parabolic on all of $\mathbb{R}^2$. Two other well-known equations of mathematical physics, the wave equation $u_{tt} - c^2 u_{xx} = 0$ and Laplace's equation $u_{xx} + u_{yy} = 0$, are hyperbolic and elliptic, respectively.

If $Lu = 0$ is a linear homogeneous equation and u_1 and u_2 are two solutions, then it obviously follows that $u_1 + u_2$ is a solution because $L(u_1 + u_2) = Lu_1 + Lu_2 = 0 + 0 = 0$. Also cu_1 is a solution because $L(cu_1) = cLu_1 = c \cdot 0 = 0$. A simple induction argument shows that if $u_1, \ldots, u_n$ are solutions of $Lu = 0$ and $c_1, \ldots, c_n$ are constants, then the finite sum, or linear combination, $c_1 u_1 + \cdots + c_n u_n$ is also a solution; this is the superposition principle for linear equations. If certain convergence properties hold, the superposition principle can be extended to infinite sums $c_1 u_1 + c_2 u_2 + \cdots$.

Another form of a superposition principle is a continuous version of the one just cited. In this case let $u(x, t; \alpha)$ be a family of solutions on D, where α is a real number ranging over some interval I. That is, suppose $u(x, t; \xi)$ is a solution of $Lu = 0$ for *each* value of $\xi \in I$. Then *formally*[2] we may superimpose these solutions by forming

$$u(x,t) = \int_I c(\xi) u(x, t; \xi) \, d\xi,$$

where $c(\alpha)$ is a function representing a continuum of coefficients, the analog of $c_1, c_2, \ldots, c_n$. If we can write

$$Lu = L \int_I c(\xi) u(x, t; \xi) \, d\xi$$

$$= \int_I c(\xi) Lu(x, t; \xi) \, d\xi$$

$$= \int_I c(\xi) \cdot 0 \, d\xi = 0,$$

---

[2] Generally, a *formal* calculation in mathematics is one that lacks complete rigor but that can usually be justified under certain circumstances.

then $u(x,t)$ is also a solution. This sequence of steps, and hence the superposition principle, depends on the validity of pulling the differential operator L inside the integral sign. For such a principle to hold, these formal steps would need careful analytical verification.

Example 6.5

Consider the heat equation

$$u_t - ku_{xx} = 0, \quad t > 0, \quad x \in \mathbb{R}. \tag{6.4}$$

It is straightforward to verify that

$$u(x,t;\xi) = \frac{1}{\sqrt{4\pi kt}} e^{-(x-\xi)^2/4kt}, \quad t > 0, \quad x \in \mathbb{R},$$

is a solution of (6.4) for any $\xi \in I = \mathbb{R}$. This solution is called the **fundamental solution** of the heat equation. We can formally superimpose these solutions to obtain

$$u(x,t) = \int_{-\infty}^{\infty} c(\xi) \frac{1}{\sqrt{4\pi kt}} e^{-(x-\xi)^2/4kt} \, d\xi,$$

where $c(\xi)$ is some function representing the coefficients. It can be shown that if $c(\xi)$ is continuous and bounded, then differentiation under the integral sign can be justified and $u(x,t)$ is therefore a solution. The fundamental solution is recognized as a normal probability density function (bell-shaped curve) with mean ξ and variance $2kt$. Therefore, as time increases the spatial profiles, or temperature distributions, spread. We should not be surprised that a probability density is related to solutions to the heat equation because heat transfer involves the random collisions of the particles making up the medium. We discuss the connection between diffusion and randomness in a subsequent section.

Example 6.6

This example shows how we can use superposition to construct solutions that satisfy auxiliary conditions. The pure initial value problem for the heat equation,

$$\begin{aligned} u_t - ku_{xx} &= 0, \quad t > 0, \quad x \in \mathbb{R}, \\ u(x,0) &= f(x), \quad x \in \mathbb{R}, \end{aligned}$$

is called the Cauchy problem. If f is a bounded continuous function, then its solution is given by the integral representation

$$u(x,t) = \int_{-\infty}^{\infty} f(\xi) \frac{1}{\sqrt{4\pi kt}} e^{-(x-\xi)^2/4kt} \, d\xi.$$

We derive this formula by the Fourier transform method in Section 6.5.2. But we can observe formally that the solution is valid by the superposition principle and the fact that $u(x, 0) = f(x)$. This latter fact comes from making the change of variables $r = (x - \xi)/\sqrt{4kt}$ and rewriting the solution as

$$u(x, t) = \frac{1}{\sqrt{\pi}} \int_{-\infty}^{\infty} f(x - r\sqrt{4kt})e^{-r^2} \, dr.$$

Then, taking the limit as $t \to 0^+$ and formally pulling the limit under the integral gives $u(x, 0) = \frac{1}{\sqrt{\pi}} \int_{-\infty}^{\infty} f(x)e^{-r^2} \, dr = f(x)\frac{1}{\sqrt{\pi}} \int_{-\infty}^{\infty} e^{-r^2} \, dr = f(x)$.

EXERCISES

1. Sketch the solution surface and temperature time profiles of the fundamental solution of the heat equation

$$u(x, t; 0) = \frac{1}{\sqrt{4\pi kt}} e^{-x^2/4kt}$$

   for different times t. What is the behavior of the profiles as t approaches zero? Comment on the differences one would observe in the profiles for large k and for small k.

2. Find the general solution of the following partial differential equations in terms of arbitrary functions.

   a) $u_{xx} + u = 6y$, where $u = u(x, y)$.

   b) $tu_{xx} - 4u_x = 0$, where $u = u(x, t)$. (Hint: Let $v = u_x$.)

   c) $u_{xt} + (1/x)u_t = t/x^2$, where $u = u(x, t)$.

   d) $u_{tx} + u_x = 1$, where $u = u(x, t)$.

   e) $uu_t = x - t$, where $u = u(x, t)$

3. Find a formula for the solution to the

$$u_{xt} = f(x, t), \quad x, t > 0,$$

   that satisfies the auxiliary conditions $u(x, 0) = g(x)$, $x > 0$ and $u(0, t) = h(t)$, $t > 0$, where f, g, and h are given, well-behaved functions with $g(0) = h(0)$, $g'(0) = h'(0)$.

4. Find the general solution of the first-order linear equation

$$u_t + cu_x = f(x, t),$$

   by changing variables to the new spatial coordinate $z = x - ct$. (Here, f is a given function and c is a constant.)

5. Introduce polar coordinates $x = r\cos\theta$ and $y = r\sin\theta$ to show that the general solution $u = u(x, y)$ of the equation

$$yu_x - xu_y = 0$$

   is $u = \psi(x^2 + y^2)$, where ψ is an arbitrary function.

6. Determine regions in the plane where the equation $yu_{xx} - 2u_{xy} + xu_{yy} = 0$ is hyperbolic, elliptic, or parabolic.

7. Show that the equation $u_t - (K(x)u_x)_x = f(t, x)$, where K and f are given functions, is linear.

8. Show that a parabolic equation $\alpha u_{tt} + \beta u_{tx} + \gamma u_{xx} = F(x, t, u, u_x, u_t)$, where α, β, γ are constant, can be reduced to an equation of the form $U_{\eta\eta} = G(\xi, \eta, U, U_\xi, U_\eta)$ for $U = U(\xi, \eta)$ by a linear transformation of the variables, $\xi = ax + bt$, $\eta = cx + dt$.

9. Find all solutions of the heat equation $u_t = ku_{xx}$ of the form $u(x, t) = U(z)$, where $z = x/\sqrt{kt}$.

6.2 Conservation Laws

Many fundamental equations in the natural and physical sciences are obtained from **conservation laws**. Conservation laws are balance laws, or equations that express the fact that some quantity is balanced throughout a process. In thermodynamics, for example, the first law states that the change in internal energy in a given system is equal to, or is balanced by, the total heat added to the system plus the work done on the system. Thus the first law of thermodynamics is an energy balance law, or conservation law. As another example, consider a fluid flowing in some region of space that consists of chemical species undergoing chemical reaction. For a given chemical species, the time rate of change of the total amount of that chemical in the region must equal the rate at which the chemical flows into the region, minus the rate at which it flows out, plus the rate at which the species is created, or consumed, by the chemical reactions. This is a verbal statement of a conservation law for the amount of the given chemical species. Similar balance or conservation laws occur in all branches of science. In population ecology, for example, the rate of change of a given animal population in a certain region must equal the birthrate, minus the deathrate, plus the migration rate into or out of the region.

Figure 6.2 Cylindrical tube of cross-sectional area A showing a cross section at x and a finite section I: $a \leq x \leq b$.

6.2.1 One Dimension

Mathematically, conservation laws translate into differential equations, which are then regarded as the *governing equations* or *equations of motion* of the process. These equations dictate how the process evolves in time. First we formulate the basic one-dimensional conservation law, out of which will come some of the basic models and concepts.

Let us consider some quantity (mass, animals, energy, momentum, or whatever) distributed along a tube of cross-sectional area A (see Fig. 6.2), and let $u = u(x, t)$ denote its density, or concentration, measured in the amount of the quantity per unit volume. An implicit assumption is that there is enough of the quantity distributed in space that it makes sense to talk about its density. (If there were only a few items, then a probability model would be more appropriate.) By definition, u varies in only one spatial direction, the axial direction x. Therefore, by assumption, u is constant in any cross section of the tube, which is the one-dimensional assumption. Next consider an *arbitrary* segment of the tube denoted by the interval $I = [a, b]$. The total amount of the quantity u inside I at time t is

$$\text{Total amount of the quantity in } I = \int_a^b u(x, t) A \, dx.$$

Assume there is motion of the quantity in the tube in the axial direction. We define the **flux** of u at x at time t to be the scalar function $J(x, t)$; that is, $J(x, t)$ is the amount of the quantity u flowing through the cross section at x at time t, per unit area, per unit time. Thus the dimensions of J are $[J] =$ amount/(area·time). Notice that flux is density times velocity. By convention we take J to be positive if the flow at x is in the positive x direction, and J is negative at x if the flow is in the negative x direction. Therefore, at time t the *net* rate that the quantity is flowing into the interval I is the rate it is flowing in at $x = a$ minus the rate it is flowing out at $x = b$. That is,

$$\text{Net rate of quantity flowing into } I = AJ(a, t) - AJ(b, t).$$

Finally, the quantity u may be created or destroyed inside I by some external or internal source (e.g., by a chemical reaction if u were a species concentration, or by birth or death if u were a population density, or by a heat source or sink if u were energy density—a heat source could be resistive heating if the tube were a wire through which a current is flowing). We denote this *source function*, which is a local function acting at each x, by $f(x, t, u)$ with dimensions $[f]$ =amount $/(\text{vol·time})$. Consequently, f is the rate that u is created (or destroyed) at x at time t, per unit volume. Note that the source function f may depend on u itself, as well as space and time. If f is positive, we say that it is a *source*, and if f is negative, we say that it is a *sink*. Given f, we may calculate the total rate that u is created in I by integration. We have

$$\text{Rate that quantity is produced in } I \text{ by sources} = \int_a^b f(x, t, u(x, t)) A \, dx.$$

The fundamental conservation law may now be formulated for the quantity u: For any interval I, we have

Rate of change of total amount of the quantity in I

$\quad =$ Net rate of quantity flowing into I

$\quad +$ Rate that quantity is produced by sources in I.

In terms of the expressions we introduced above, after canceling the constant cross-sectional area A, we have

$$\frac{d}{dt} \int_a^b u(x, t) \, dx = J(a, t) - J(b, t) + \int_a^b f(x, t, u) \, dx. \qquad (6.5)$$

In summary, (6.5) states that the rate that u changes in I must equal the net rate at which u flows into I plus the rate that u is produced in I by sources. Equation (6.5) is called a **conservation law in integral form**, and it holds even if u, J, or f are not smooth (continuously differentiable) functions. The latter remark is important when we consider in subsequent chapters physical processes giving rise to shock waves, or discontinuous solutions.

If some restrictions are place on the triad u, J, and f, (6.5) may be transformed into a single partial differential equation. Two results from elementary integration theory are required to make this transformation: (i) the fundamental theorem of calculus, and (ii) the result on differentiating an integral with respect to a parameter in the integrand. Precisely,

(i) $\int_a^b J_x(x, t) \, dx = J(b, t) - J(a, t)$.

(ii) $\frac{d}{dt} \int_a^b u(x, t) \, dx = \int_a^b u_t(x, t) \, dx$.

These results are valid if J and u are continuously differentiable functions on $\mathbb{R}^2$. Of course, (i) and (ii) remain correct under less stringent conditions, but the assumption of smoothness is all that is required in the subsequent discussion. Therefore, assuming smoothness of u and J, as well as continuity of f, equations (i) and (ii) imply that the conservation law (6.5) may be written

$$\int_a^b [u_t(x,t) + J_x(x,t) - f(x,t,u)]\, dx = 0, \quad \text{for all intervals } I = [a,b]. \quad (6.6)$$

Because the integrand is a continuous function of x, and because (6.6) holds for all intervals of integration I, it follows that the integrand must vanish identically; that is,

$$u_t + J_x = f(x,t,u), \quad x \in \mathbb{R}, \quad t > 0. \quad (6.7)$$

Equation (6.7) is a partial differential equation relating the density $u = u(x,t)$ and the flux $J = J(x,t)$. Both are regarded as unknowns, whereas the form of the source f is given. Equation (6.7) is called a **conservation law in differential form**, in contrast to the integral form (6.6). The J_x term is called the *flux term* because it arises from the movement, or transport, of u through the cross section at x. The source term f is often called a *reaction term* (especially in chemical contexts) or a *growth* or *interaction* term (in biological contexts). Finally, we have defined the flux J as a function of x and t, but this dependence on space and time may occur through dependence on u or its derivatives. For example, a physical assumption may require us to posit $J(x,t) = J(x,t,u(x,t))$, where the flux is dependent upon u itself.

If there are several quantities with densities u_1, u_2,..., then we can write down a conservation law for each. The fluxes and source terms may involve several of the densities, making the resulting system of conservation laws coupled.

6.2.2 Several Dimensions

It is straightforward to formulate conservation laws in several spatial dimensions. For review, and to set the notation, we introduce some basic definitions. In $\mathbb{R}^n$, points are denoted by $\mathbf{x} = (x_1, ..., x_n)$, and $d\mathbf{x} = dx_1 \cdots dx_n$ represents a volume element. In $\mathbb{R}^2$ we often use (x,y), and in $\mathbb{R}^3$ we use (x,y,z). If $u = u(\mathbf{x},t)$ is a scalar function, then its integral over a suitable region (volume) Ω is

$$\int_\Omega u(\mathbf{x},t)\, d\mathbf{x}.$$

In $\mathbb{R}^2$ we have $d\mathbf{x} = dxdy$, and in $\mathbb{R}^3$ we have $d\mathbf{x} = dxdydz$. The boundary of Ω is denoted by $\partial\Omega$. We always assume that the domain Ω is has a smooth boundary $\partial\Omega$, or piecewise smooth boundary composed of finitely many smooth components. We say Ω is **open** if it includes none of its boundary, and we denote $\overline{\Omega} = \Omega \cup \partial\Omega$ as the **closure** of Ω. This assumption will cover the cases of spheres, rectangles, parallelopideds, and so on, which are the domains important in most science and engineering problems. We refer to these type of domains as *well-behaved*. The set of continuous functions on a domain Ω is denoted by $C(\Omega)$. Surface integrals are denoted by

$$\int_{\partial\Omega} u(\mathbf{x}, t)\, dA.$$

Nearly always, surface integrals are flux integrals (discussed below). Partial derivatives are represented by subscripts, or $\frac{\partial u}{\partial x_k} = u_{x_k}$. The **gradient** of u is the vector

$$\nabla u = (u_{x_n}, ..., u_{x_n}).$$

The gradient points in the direction of the maximum increase of the function. The **directional derivative** of u in the direction of a unit vector $\mathbf{n} = (n_1, ..., n_n)$ is denoted by

$$\frac{du}{dn} = \mathbf{n} \cdot \nabla u.$$

It measures the change in u in the direction $\mathbf{n}$. If $J(\mathbf{x}, t) = (J_1(\mathbf{x}, t), ..., J_n(\mathbf{x}, t))$ is a vector field, then its **divergence** is the scalar function

$$\nabla \cdot \mathbf{J} = \frac{\partial J_1}{\partial x_1} + \cdots + \frac{\partial J_n}{\partial x_n}.$$

Two common, alternate notations for the divergence and gradient are div $\mathbf{J}$ and grad u, respectively. The **flux** of a vector field $J(\mathbf{x}, t)$ through an oriented surface element $\mathbf{n}dA$, where $\mathbf{n}$ is an oriented unit normal vector (see Fig. 6.3 for the picture in $\mathbb{R}^3$), is defined to be

$$\text{flux} = \mathbf{J}(\mathbf{x}, t) \cdot \mathbf{n}\, dA.$$

There are two useful differential identities that are straightforward to prove. The first is that the divergence of the gradient is the **Laplacian** $\Delta u = \frac{\partial^2 u}{\partial x_1^2} + \cdots + \frac{\partial^2 u}{\partial x_n^2}$, or

$$\Delta u = \nabla \cdot \nabla u.$$

Another common notation for the Laplacian Δ is ∇^2. The second is the product rule

$$\nabla \cdot (u\mathbf{J}) = u\nabla \cdot \mathbf{J} + \nabla u \cdot \mathbf{J}.$$

Two simple integral theorems in $\mathbb{R}^n$ are analogs of the one-dimensional results used in the preceding section.

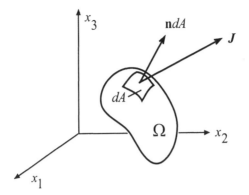

Figure 6.3 A surface with surface element $\mathbf{n}dA$, oriented by its unit normal vector $\mathbf{n}$, along with the flux vector ϕ.

(i) Let f be a continuous function on a well-behaved domain Ω with

$$\int_D f(\mathbf{x})\, d\mathbf{x} = 0.$$

for every subdomain $D \subset \Omega$. Then $f(\mathbf{x}) = 0$ for all $x \in \Omega$.

(ii) Let $u = u(\mathbf{x}, t)$ and $u_t(\mathbf{x}, t)$ be continuous on a well-behaved domain $x \in \Omega$, $t \in I$, where I is an open interval. Then

$$\frac{d}{dt} \int_\Omega u(\mathbf{x}, t)\, d\mathbf{x} = \int_\Omega u_t(\mathbf{x}, t)\, d\mathbf{x}.$$

The most important integral identity in multivariable calculus in the **divergence theorem**, which is an n-dimensional version of the fundamental theorem of calculus. It states that the integral of a partial derivative over a volume is a boundary, or surface, integral. It has been proved for very general domains Ω, but we only require it for well-behaved domains. A natural mathematical question concerns conditions on the functions for which the divergence theorem holds. Generally, we assume that all functions are continuous on the closed domain consisting of Ω and its boundary $\partial\Omega$, and have as many continuous partial derivatives in Ω that occur in the formula. We do not formulate the most general conditions. Let Ω be an open, well-behaved domain. Then, by $u \in C^1(\overline{\Omega})$ we mean that u is continuous on Ω, and its first partial derivatives are continuous on Ω and can be extended continuously to the boundary of Ω. By $u \in C^2(\Omega)$ to mean that u has continuous second partials in Ω. Many of our theorems will require $u \in C^1(\overline{\Omega}) \cap C^2(\Omega)$, that is, u is twice continuously differentiable in the open domain Ω and its first derivatives can be extended to the closure.

Theorem 6.7

(Divergence theorem) Let Ω be a well-behaved, open, bounded region in $\mathbb{R}^n$ and let $u \in C^1(\overline{\Omega}) \cap C^2(\Omega)$. Then

$$\int_\Omega u_{x_k} d\mathbf{x} = \int_{\partial\Omega} u n_k dA,$$

where n_k is the kth component of the outward unit normal vector $\mathbf{n}$ to the bounding surface.

If J is a vector field, we can apply this result to each component and add to obtain the standard vector version of the theorem:

$$\int_\Omega \nabla \cdot \mathbf{J} dx = \int_{\partial\Omega} \mathbf{J} \cdot \mathbf{n} dA.$$

The divergence theorem leads to fundamental integral identities. We list these important results.

(iii) **Integration by parts formula**

$$\int_\Omega w u_{x_k} d\mathbf{x} = -\int_\Omega u w_{x_k} d\mathbf{x} + \int_{\partial\Omega} u w n_k \, dA.$$

(iv) **Green's first identity**

$$\int_\Omega (u\Delta w + \nabla u \cdot \nabla w) dx = \int_{\partial\Omega} u \frac{dw}{dn} \, dA.$$

(v) **Green's second identity**

$$\int_\Omega u\Delta w \, dx = \int_\Omega w\Delta u \, dx + \int_{\partial\Omega} \left(u \frac{dw}{dn} - w \frac{du}{dn} \right) dA.$$

It is impossible to overestimate the utility of integration by parts formulae in the subject of partial differential equations. Formula (v) is just an integration by parts formula for the Laplacian operator. All of these identities apply for suitable functions u and w over well-behaved, bounded domains, as in the statement of the divergence theorem. The proofs are requested in the Exercises.

With this preparation we are ready to write down the conservation law in several variables. Let Ω be a region in $\mathbb{R}^n$ and let $u = u(\mathbf{x}, t)$ be the density, or concentration, of some quantity distributed throughout Ω. Let $J(\mathbf{x}, t)$ be the flux vector for that quantity, where $\mathbf{J} \cdot \mathbf{n} dA$ measures the rate that the quantity crosses an oriented surface element in the medium. In this domain Ω let V be an arbitrary, well-behaved bounded region (for example, a sphere). Then the total amount of the quantity in V is given by the volume integral

$$\text{Total amount in } V = \int_V u(\mathbf{x}, t) \, d\mathbf{x}.$$

The time rate of change of the total amount of the quantity in V must be balanced by the rate that the quantity is produced in V by sources, plus the net rate that the quantity flows through the boundary of V. We let $f(\mathbf{x}, t, u)$ denote the source term, so that the rate that the quantity is produced in V is given by

$$\int_V f(\mathbf{x}, t, u) \, d\mathbf{x}.$$

The net *outward* flux of the quantity u through the boundary ∂V is given by the surface integral

$$\int_{\partial V} \mathbf{J} \cdot \mathbf{n} \, dA,$$

where $\mathbf{n} = \mathbf{n}(\mathbf{x})$ is the outward unit normal. Therefore, the conservation or balance law for u is given by

$$\frac{d}{dt} \int_V u \, d\mathbf{x} = -\int_{\partial V} \mathbf{J} \cdot \mathbf{n} \, dA + \int_V f(\mathbf{x}, t, u) \, d\mathbf{x}. \tag{6.8}$$

The minus sign on the flux term occurs because outward flux decreases the rate that u changes in V. This integral law is the higher-dimensional analog of (6.6).

The integral form of the conservation law (6.8) can be reformulated as a local condition, that is, a partial differential equation, provided that u and J are sufficiently smooth functions. In this case the surface integral can be rewritten as a volume integral over V using the divergence theorem, and the time derivative can be brought under the integral sign by (ii). Hence,

$$\int_V u_t \, d\mathbf{x} = \int_V \nabla \cdot \mathbf{J} \, d\mathbf{x} + \int_V f(\mathbf{x}, t, u) \, d\mathbf{x}.$$

From (i), the arbitrariness of V implies the differential form of the balance law throughout the domain Ω:

$$u_t + \nabla \cdot \mathbf{J} = f(\mathbf{x}, t, u), \quad x \in \Omega, \quad t > 0. \tag{6.9}$$

Equation (6.9) is the higher-dimensional version of equation (6.7), the **local conservation law** in one dimension.

6.2.3 Constitutive Relations

Because the conservation equation (6.7) [or (6.9)] is a single partial differential equation for two unknown quantities (the density u and the flux J), intuition indicates that another equation is required to have a well determined system. This additional equation is often an equation that is based on an assumption about the physical properties of the medium, which, in turn, is based on empirical reasoning or experimental observation. Equations expressing these assumptions are called **constitutive relations** or equations of state. Thus constitutive equations are on a different level from the basic conservation law; the latter is a fundamental law of nature connecting the density u to the flux J, whereas a constitutive relation is often an approximate equation whose origin lies in empirics. In the next few paragraphs we examine some possibilities for modeling the flux, or motion of the quantity under examination.

Example 6.8

(Diffusion). In many physical problems it is observed that the amount of the substance that moves through a cross section at x at time t is proportional to its density gradient u_x, or $J(x,t) \propto u_x(x,t)$. If $u_x > 0$, then $J < 0$ (the substance moves to the left), and if $u_x < 0$, then $J > 0$ (the substance moves to the right). Figure 6.4 illustrates the situation. The substance moves down the density gradient, from regions of high concentration to regions of low concentrations. We say the motion is "down the gradient." By the second law of thermodynamics, heat behaves in this manner; heat flows from hotter regions to colder regions, and the steeper the temperature distribution curve, the more rapid the flow of heat. This movement is closely connected to the random motion of molecules caused by collisions of the particles. If u represents a concentration of organisms, one may observe that they move from high densities to low densities with a rate proportional to the density gradient. Therefore, we assume the basic constitutive law

$$J(x,t) = -Du_x(x,t), \tag{6.10}$$

which is known as **Fick's law**. It models diffusion. The positive proportionality constant D is called the **diffusion constant** and has dimensions $[D] = \text{length}^2/\text{time}$. Fick's law accurately describes the behavior of many physical and biological systems. This assumption reduces the conservation law (6.7) to a single second-order linear partial differential equation for the unknown density $u = u(x,t)$ given by

$$u_t = Du_{xx} + f. \tag{6.11}$$

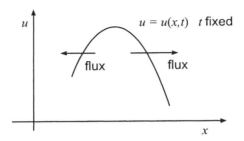

Figure 6.4 Time snapshot of the concentration $u(x, t)$ at a fixed time t. The arrows indicate the direction of the flow, from higher concentrations to lower concentrations. The flow is said to be *down the gradient*. The minus occurs because positive flux requires a negative derivative.

In several dimensions Fick's law is

$$\mathbf{J}(\mathbf{x}, t) = -D\nabla u, \tag{6.12}$$

giving

$$u_t = D\Delta u + f. \tag{6.13}$$

Equations (6.11) and (6.13) are called **reaction–diffusion equations**, and they govern processes where the flux, called a diffusive flux, is specified by Fick's law. If the source term vanishes, they are called diffusion equations. The diffusion constant D defines a characteristic time, or time scale, τ for a diffusion process. If L is a length scale (e.g., the length of the container), the quantity

$$\tau = \frac{L^2}{D}$$

is the only constant in the process with dimensions of time, and τ gives a measure of the time required for discernible changes in concentration to occur. The diffusion constant may depend upon $\mathbf{x}$ in a nonhomogeneous medium, and well as the density u itself. These cases are mentioned below.

Example 6.9

(Classical heat equation) In the case where u is an energy density (i.e., a quantity with dimensions of energy per unit volume), the diffusion equation describes the transport of energy in the medium. In a homogeneous medium with density ρ and specific heat C, the energy density is

$$u(\mathbf{x}, t) = C\rho T(\mathbf{x}, t),$$

where T is the temperature. The dimensions are $[C]$ =energy/(mass·degree) and $[\rho]$ =mass/volume. For example, the units of energy could be joules or calories. The conservation law may be written

$$C\rho T_t + \nabla \cdot \mathbf{J} = f,$$

where f is a heat source. In heat conduction Fick's law has the form

$$\mathbf{J} = -K\nabla T, \tag{6.14}$$

where K is the thermal conductivity, measured in energy/(mass·time·degree). In the context of heat conduction, Fick's law is called **Fourier's law** of heat conduction. It follows that the temperature T satisfies the equation

$$C\rho T_t - K\nabla \cdot \nabla T = f,$$

or

$$T_t = k\Delta T + \frac{1}{C\rho}f, \quad k = \frac{K}{C\rho}. \tag{6.15}$$

Equation (6.15) is called the **heat equation** with heat source f, and the constant k, which plays the role of the diffusion constant, is called the **diffusivity**, measured as length-squared per unit time. Thus the diffusivity k in heat flow problems is the analog of the diffusion constant, and the heat equation is just the diffusion equation. In this discussion we have assumed that the physical parameters C, K, and ρ of the medium are constant; however, these quantities could depend on the temperature T, which is an origin of nonlinearity in the problem. If these parameters depend on location $\mathbf{x}$, then the medium is nonhomogeneous and the conservation law in this case becomes

$$T_t = \frac{1}{C(\mathbf{x})\rho(\mathbf{x})}\nabla \cdot (K(\mathbf{x})\nabla T) + \frac{f}{C(\mathbf{x})\rho(\mathbf{x})}.$$

The variable conductivity cannot be pulled outside the divergence. Notice that if the system is in a steady-state where the temperature is not changing in time and the source is independent of t, then $T_t = 0$ and the temperature depends on $\mathbf{x}$ alone, $T = T(\mathbf{x})$. Then the last equation becomes

$$-\nabla \cdot (K(\mathbf{x})\nabla T) = f(\mathbf{x}, T).$$

This is the steady-state heat equation (it is addressed in detail in later sections).

Example 6.10

(Advection) **Advection**, or convection or drift, occurs when the bulk motion of the medium carries the substance under investigation along with it at a given velocity. For example, wind can carry propagules, water can carry chemicals or animals, air motion can transport heat energy, and so on. The advective flux, which measures the amount of the substance crossing a section, is given by the velocity times the concentration, or

$$\mathbf{J}(\mathbf{x}, t) = \mathbf{c}u(\mathbf{x}, t),$$

where $\mathbf{c} = \mathbf{c}(\mathbf{x}, t)$ is the velocity of the medium. Then, combined with the conservation law, we obtain the reaction-advection equation

$$u_t + \nabla \cdot (\mathbf{c}u) = f.$$

If the advection speed $\mathbf{c}$ is a constant vector, this equation reduces to

$$u_t + \mathbf{c} \cdot \nabla u = f.$$

These types of equations give rise to wave propagation, rather than diffusion, and are examined in Chapter 7.

Example 6.11

(Advection–diffusion) When both advection and diffusion occur, the flux is composed of two parts and the model becomes

$$u_t = \nabla \cdot (D(x)\nabla u) - \nabla \cdot (\mathbf{c}(\mathbf{x})u) + f.$$

Here we have assumed D and $\mathbf{c}$ are spatially dependent. In one dimension, with constant diffusion,

$$u_t = Du_{xx} - cu_x + f.$$

This is the **reaction–advection–diffusion** equation. If D or $\mathbf{c}$ depend on u itself, then the equation is a nonlinear partial differential equation.

Example 6.12

(Growth) If $u = u(x, t)$ is a population density that grows logistically and Fick's law models its diffusion, then the population law in one dimension is

$$u_t = Du_{xx} + ru\left(1 - \frac{u}{K}\right),$$

which is Fisher's equation (see Section 6.1). This important equation is a prototype of nonlinear reaction-diffusion equations. Other types of growth rates are also of great interest in population ecology. For example, the reaction term $f(u) = ru$ models growth or decay, and the Allee growth rate $f(u) = ru(a - u)(u - b)$, $0 < a < b$, models growth where the rate is negative for small populations.

6.2.4 Probability and Diffusion

We now show a connection between diffusion and randomness. For example, we think of heat flow, governed by the diffusion equation, is caused by random molecular collisions which cause the energy to disperse from high-energy regions to low-energy regions. This idea is based upon the kinetic theory of gases, and the quantities involved are statistical averages of the behavior of large numbers of particles. For animal populations, diffusion is a different type of mechanism, but the constitutive law is the same as for heat flow (Fick's law). If there are only a few particles or a few animals, then it is not sensible to define a density and we cannot apply the direct arguments above to obtain a partial differential equation. In this case we must rely on probability theory and make deductions about the spread of the quantity of interest. We fix the idea by examining animal populations. As we shall observe, information about how an organism moves on a short time scale can be used to determine where it is likely to be at any time. We translate the short time behavior into a probability density function (pdf) $u(x,t)$ for the random variable $X = X(t)$ giving the location of the organism at time t. The analysis leads to a partial differential equation for $u(x,t)$.

Suppose an organism is located at the origin $x = 0$ at time $t = 0$. The movement rule is a simple random walk in one dimension. At each time step the organism jumps a distance h either to the left or right with equal probability. As time advances in equal discrete steps, the organism moves on the lattice of points $0, \pm h, \pm 2h, \pm 3h, \ldots$ on the x axis. We let $u(x,t)h$ be the probability that the organism is located in the interval between x and $x+h$ at time t, given that it was at $x = 0$ at $t = 0$. Rather than calculating the probability density function directly, we show that it must satisfy a special partial differential equation, the Fokker–Planck equation. This result follows from the *master equation*

$$u(x, t + \tau) = \frac{1}{2}u(x - h, \tau) + \frac{1}{2}u(x + h, \tau),$$

which states that the probability that the organism is in the interval $[x, x + h]$ at time $t + \tau$ equals the probability that it was either in $[x - h, x]$ at time t and jumped to the right, or it was in the interval $[x + h, x + 2h]$ at time t and it jumped to the left (we have canceled a h factor in the last equation). Expanding each term in the master equation in a Taylor series, we get

$$u(x, t) + p_t(x, t)\tau + \frac{1}{2}u_{tt}(x, t)\tau^2 + \cdots$$

$$= \frac{1}{2}\left(u(x, t) - u_x(x, t)h + \frac{1}{2}u_{xx}(x, t)h^2 + \cdots \right)$$

$$+ \frac{1}{2}\left(u(x, t) + u_x(x, t)h + \frac{1}{2}u_{xx}(x, t)h^2 + \cdots \right),$$

where the "dots" denote higher order terms in the increments h and τ. Simplifying, we obtain

$$u_t + \frac{\tau}{2}u_{tt}(x,t) = \frac{h^2}{2\tau}u_{xx} + \cdots.$$

Now we pass from a discrete random walk to a continuous random walk. Taking the limit as $h, \tau \to 0$ in such a way that the ratio of h^2 to 2τ remains constant, or $h^2/2\tau = D$, we get

$$u_t = Du_{xx},$$

which is the one-dimensional diffusion equation. If the probabilities of moving to the left or right in the random walk are not equal, then a *drift* term arises in the preceding argument and the probability density function satisfies the advection-diffusion equation (see the Exercises). These equations for the probability densities in the random walk models are special cases of what is known as the **Fokker–Planck equation**. In passing from a discrete random walk to a continuous one, we have masked a lot detailed analysis; suffice it to say that this process can be put on a firm mathematical foundation using the theory of stochastic processes.

We have derived the diffusion equation in two ways—from a conservation law and from a probability model. Previously, we defined the **fundamental solution** to the one-dimensional diffusion equation as

$$u(x,t) = \frac{1}{\sqrt{4\pi Dt}}e^{-x^2/4Dt}, \quad t > 0.$$

It will be derived later using Fourier transforms. This solution is the probability density function for the random variable $X = X(t)$, which is the location of an individual performing a continuous random walk beginning at the origin. That is,

$$\Pr(a < X(t) \le b) = \int_a^b u(x,t)\,dx.$$

It is straightforward to show that $\int_{-\infty}^{\infty} u(x,t)dx = 1$. We recognize the distribution u as a normal density with variance $2Dt$. Therefore, the probability density profiles spread as time increases, and the location of the individual becomes less certain. As $t \to 0^+$ we have $u(x,t) \to 0$ if $x \ne 0$, and $u(0,t) \to +\infty$. In fact, we have $u(x,0) = \delta_0(x)$, in a distributional sense (fundamental solution).

The **fundamental solution** to the diffusion equation $u_t = D\Delta u$ in $\mathbb{R}^n$ is

$$u(\mathbf{x},t) = \frac{1}{(4\pi Dt)^{n/2}}e^{-|\mathbf{x}|^2/4Dt}, \quad \mathbf{x} \in \mathbb{R}^n, \ t > 0.$$

This solution, which one can verify, can be derived using Fourier transforms (Section 5.5). The function $u(\mathbf{x} - \xi, t)$ describes the evolution of organisms or heat, for example, from a point release of unit magnitude at a location ξ at time $t = 0$. Thus, the initial condition is a delta distribution $\delta_\xi(\mathbf{x})$. So it is the solution to a problem with a unit point source. An example from population ecology illustrates the utility of this solution.

Example 6.13

(Invasion of species) In the plane, consider a diffusing animal population whose growth rate is λu. The model is the growth–diffusion equation

$$u_t = D\Delta u + \lambda u, \quad \mathbf{x} \in \mathbb{R}^2, \ t > 0.$$

A change of dependent variables to $v(\mathbf{x}, t) = u(\mathbf{x}, t)e^{-\lambda t}$ transforms the growth–diffusion equation into the diffusion equation for v,

$$v_t = D\Delta v, \quad \mathbf{x} \in \mathbb{R}^2, \ t > 0.$$

The fundamental solution is, using $R = |\mathbf{x}| = \sqrt{x^2 + y^2}$,

$$v(R, t) = \frac{1}{4\pi Dt}e^{-R^2/4Dt}, \quad R > 0, \ t > 0.$$

Therefore, a point release of u_0 animals at the origin will result in the radial population density

$$u(R, t) = \frac{u_0}{4\pi Dt}e^{\lambda t - R^2/4Dt}, \quad R > 0, \ t > 0.$$

We ask how fast the animals invade the region. Imagine that the population is detectable at a given radius when it reaches a critical threshold density u_c; that is,

$$\frac{u_0}{4\pi Dt}e^{\lambda t - R^2/4Dt} = u_c.$$

This leads to gives

$$\frac{R}{t} = \sqrt{4\lambda D - \frac{4D}{t}\ln\left(\frac{4\pi u_c}{u_0}t\right)}.$$

For large times t the limit is

$$\frac{R}{t} = \sqrt{4\lambda D}.$$

Therefore the invasion speed approaches a constant determined by the growth rate λ and the diffusion constant D. This result has been confimed by various observations, one of the most famous being the muskrat invasion in Europe and Asia beginning in 1905 by an accidental release in Bohemia.

6.2.5 Boundary Conditions

In the context of heat conduction, we now discuss the types of auxiliary conditions that lead to a well-defined mathematical problem that is physically meaningful. The heat conduction equation is the canonical example of a parabolic partial differential equation. Basically, evolution problems require an initial condition, and if spatial boundaries are present, then boundary conditions are needed. We want to formulate problems with conditions that make the mathematical problem **well-posed**. This means that a solution must exist, the solution must be unique, and the solution should be stable in the sense that it should depend continuously on initial and boundary data.

Consider a bar of length l and constant cross-sectional area A. The temperature $u = u(x,t)$ (we use u as the dependent variable, rather than T) must satisfy the one-dimensional heat equation (see (6.15))

$$u_t - ku_{xx} = 0, \quad 0 < x < l, \quad t > 0, \tag{6.16}$$

where we have assumed no heat sources.

An auxiliary condition on the temperature u at time $t = 0$ is called an **initial condition** and is of the form

$$u(x,0) = f(x), \quad 0 < x < l, \tag{6.17}$$

where $f(x)$ is the given initial temperature distribution. Conditions prescribed at $x = 0$ and $x = l$ are **boundary conditions**. If the temperature u at the ends of the bar are prescribed, then the boundary conditions take the form

$$u(0,t) = g(t), \quad u(l,t) = h(t), \quad t > 0, \tag{6.18}$$

where g and h are prescribed functions. A boundary condition of this type is called a **Dirichlet condition** . Other boundary conditions are possible. It is easy to imagine physically that one end of the bar, say at $x = 0$, is insulated so that no heat can pass through. By Fourier's heat law this means that the flux at $x = 0$ is zero or

$$u_x(0,t) = 0, \quad t > 0. \tag{6.19}$$

Condition (6.19) is called an **insulated boundary condition**. Or, one may prescribe the flux at an end as a given function of t, for example,

$$-Ku_x(0,t) = g(t), \quad t > 0, \tag{6.20}$$

which is called a **Neumann condition**. If the conductivity K depends on u, then condition (6.20) is a nonlinear boundary condition. Finally, a boundary condition may be specified by Newton's law of cooling,

$$-Ku_x(0,t) = \alpha(u(0,t) - \psi(t)), \quad t > 0,$$

which requires the flux be proportional to the difference between the temperature at the end and the temperature ψ of the outside environment. This is a **Robin condition**, or **radiation condition**.

Remark 6.14

For the advection–diffusion equation

$$u_t = Du_{xx} - cu_x,$$

the flux is $J(x,t) = -Du_x + cu$; so specifying the flux at a boundary amounts to specifying J, and not just the diffusive flux $-Du_x$.

If an initial condition is given and a boundary condition is prescribed at one end (say at $x = 0$) of a very long rod, then the problem can be considered on an semi-infinite domain $x \geq 0$ on which the heat conduction equation holds. For example the initial boundary value problem

$$
\begin{aligned}
u_t - ku_{xx} &= 0, & & t > 0, \quad 0 < x < \infty, \\
u(x,0) &= 0, & & 0 < x < \infty, \\
u(0,t) &= g(t), & & t > 0,
\end{aligned}
$$

models heat flow in an infinite medium $x \geq 0$ of constant diffusivity k, initially at zero degrees, subject to the maintenance of the end at $x = 0$ at $g(t)$ degrees. Here we expect the problem to model heat conduction in a long bar for short enough time so that any condition at the far end would not affect the temperature distribution in the portion of the rod that we are interested in studying. As a practical example, heat flow in an infinite medium arises from the study of the underground temperature variations, given the temperature changes at ground level on the earth ($x = 0$). It seems clear that problems on the infinite interval $-\infty < x < \infty$ are also relevant in special physical situations where the medium is very long in both directions. In this case only an initial condition can be prescribed. To restrict the class of solutions to those that are physically meaningful, a condition is sometimes prescribed at infinity; for example, u is bounded at infinity or $\lim_{x \to +\infty} u(x,t) = 0$, $t > 0$.

In summary, by an *initial boundary value problem* for the heat conduction equation we mean the problem of solving the heat equation (6.16) subject to the initial initial condition (6.17) and some type of boundary conditions imposed at the endpoints. It may be clear from the physical context which auxiliary conditions should be prescribed to obtain a unique solution; in some cases it is not clear. A large body of mathematical literature is devoted to proving the existence and uniqueness of solutions to various kinds of partial differential equations subject to sundry auxiliary data.

An example of such a uniqueness theorem is the following:

Theorem 6.15

The initial boundary value problem

$$u_t - ku_{xx} = 0, \ 0 < x < l, \ 0 < t < T,$$
$$u(x, 0) = f(x), \ 0 < x < l,$$
$$u(0, t) = g(t), \ u(l, t) = h(t), \ 0 < t < T,$$

where $f \in C[0, l]$ and $g, h \in C[0, T]$, has a solution $u(x, t)$ on the rectangle $R: 0 \leq x \leq l, \ 0 \leq t \leq T$, then the solution is unique.

We prove this theorem to show an example of the **energy argument**. By way of contradiction, assume solutions are not unique and there are two distinct solutions $u_1(x, t)$ and $u_2(x, t)$. Then their difference $w(x, t) \equiv u_1(x, t) - u_2(x, t)$ must satisfy the boundary value problem

$$w_t - kw_{xx} = 0, \ 0 < x < l, \ 0 < t < T,$$
$$w(x, 0) = 0, \ 0 < x < l,$$
$$w(0, t) = w(l, t) = 0, \ 0 < t < T.$$

If we show $w(x, t) \equiv 0$ on R, then $u_1(x, t) = u_2(x, t)$ on R, which is a contradiction. To this end define the **energy integral**

$$E(t) = \int_0^l w^2(x, t) \, dx.$$

Taking the time derivative

$$E'(t) = \int_0^l 2ww_t \, dx = 2k \int_0^l ww_{xx} \, dx = 2kww_x \big|_0^l - 2k \int_0^l w_x^2 \, dx,$$

where the last equality was obtained using integration by parts. Because the boundary term vanishes,

$$E'(t) = -2k \int_0^l w_x^2(x, t) dx \leq 0.$$

Thus $E(t)$ is nonincreasing, which along with the facts that $E(t) \geq 0$ and $E(0) = 0$, implies $E(t) = 0$. Therefore $w(x, t) = 0$ identically in $0 \leq x \leq l$, $0 \leq t \leq T$, since w is continuous in both its arguments. Existence of a solution is established later by actually exhibiting the solution.

The energy method is an important technique in proving uniqueness and obtaining estimates of solutions in partial differential equations in one and in several dimensions. In addition to existence and uniqueness questions the notion of continuous dependence of the solution on the initial or boundary data is important. From a physical viewpoint it is reasonable that small changes in the initial or boundary temperatures should not lead to large changes in the overall temperature distribution. A mathematical model should reflect this stability in that small changes in the auxiliary data should lead to only small changes in the solution. Stated differently, the solution should be stable under small perturbations of the initial or boundary data. (We remark, parenthetically, that although stability seems desirable, many important physical processes are unstable, and so **ill-posed** problems are often studied well.)

Example 6.16

(**Hadamard's**[3] **example**) Consider the partial differential equation

$$u_{tt} + u_{xx} = 0, \quad t > 0, \quad x \in \mathbb{R} \tag{6.21}$$

subject to the initial conditions

$$u(x, 0) = 0, \quad u_t(x, 0) = 0, \quad x \in \mathbb{R}. \tag{6.22}$$

Note that (6.21) is elliptic and not parabolic like the heat equation. The solution of this pure initial value problem is clearly the zero solution $u(x, t) \equiv 0$ for $t \geq 0$, $x \in \mathbb{R}$. Now let us change (6.22) to

$$u(x, 0) = 0, \quad u_t(x, 0) = 10^{-n} \sin 10^n x, \tag{6.23}$$

which represents a very small change in the initial data, provided n is large. As one can check, the solution of (6.21) subject to this new boundary condition is

$$u(x, t) = 10^{-2n} \sin(10^n x) \sinh(10^n t). \tag{6.24}$$

For large values of t, the function $\sinh(10^n t)$ behaves like $\exp(10^n t)$. Therefore the solution grows exponentially with t. Therefore, for the initial value problem (6.21)–(6.22), an arbitrarily small change in the initial data leads to an arbitrarily large solution, and the problem is not well posed. Initial value problems for elliptic equations do no lead to well-posed problems. In numerical calculations stability is essential. For example, suppose a numerical scheme is devised to propagate the initial and boundary conditions. Those conditions can never be represented exactly in the computer, as small errors will exist because of roundoff or truncation of the data. If the problem itself is unstable, then these small errors in the data may be propagated in the numerical scheme in such a way that the calculation becomes meaningless.

---

[3] Jacques Hadamard (1865–1963) was a French mathematician who contributed to analysis, number theory, and partial differential equations.

EXERCISES

1. Solve the nonlinear Cauchy problem

$$
\begin{aligned}
u_t &= ku_{xx} + u_x^2, \quad t > 0, \quad x \in \mathbb{R}, \\
u(x, 0) &= f(x), \quad x \in \mathbb{R},
\end{aligned}
$$

by making the transformation $w = e^u$.

2. Using the energy method to prove that a solution to the initial boundary value problem

$$
\begin{aligned}
& u_t - ku_{xx} = 0, \ 0 < x < l, \ 0 < t < T, \\
& u(x, 0) = f(x), \ 0 < x < l, \\
& u_x(0, t) = 0, \ u(l, t) = h(t), \ 0 < t < T,
\end{aligned}
$$

must be unique.

3. Use the energy method to prove uniqueness for the problem

$$
\begin{aligned}
u_t &= \Delta u, \quad \mathbf{x} \in \Omega, \ t > 0, \\
u(\mathbf{x}, 0) &= f(\mathbf{x}), \quad \mathbf{x} \in \Omega, \\
u(\mathbf{x}, t) &= g(\mathbf{x}), \quad \mathbf{x} \in \partial\Omega, \ t > 0,
\end{aligned}
$$

4. A homogeneous (constant ρ, C, and K) metal rod has cross-sectional area $A(x)$, $0 < x < l$, and there is only a small variation of $A(x)$ with x, so that the assumption of constant temperature in any cross section remains valid. There are no sources and the flux is given by $-Ku_x(x, t)$. From a conservation law obtain a partial differential equation for the temperature $u(x, t)$ that reflects the area variation of the bar.

5. In the absence of sources, derive the diffusion equation for radial motion in the plane,

$$
u_t = \frac{D}{r}(ru_r)_r,
$$

from first principles. That is, take an arbitrary domain between circles $r = a$ and $r = b$ and apply a conservation law for the density $u = u(r, t)$, assuming the flux is $J(r, t) = -Du_r$. Assume no sources.

6. A fluid, having density ρ, specific heat C, and conductivity K, flows at a constant velocity V in a cylindrical tube of length L and radius R. The temperature at position x is $T = T(x, t)$, and diffusion of heat is ignored. As it flows, heat is lost through the lateral side at a rate jointly proportional to the area and to the difference between the temperature T_e of the external environment and the temperature $T(x, t)$ of the fluid (Newton's

law of cooling). Derive a partial differential differential equation model
for the temperature $T(x, t)$. Find the general solution of the equation by
transforming to a moving coordinate system $z = x - Vt$, $\tau = t$.

7. Show that the growth–diffusion equation $u_t = D\Delta u + ru$ can be transformed into a pure diffusion equation via the transformation $v = ue^{-rt}$.

8. Repeat the random walk argument in Section 5.2.4 in the case that the
probabilities of moving to the right and to the left are p and $1 - p$, respectively, and show in appropriate limits that one obtains the advection-
diffusion equation $p_t = -cp_x + Dp_{xx}$.

9. Show that the nonhomogeneous problem

$$u_t - Du_{xx} = f(x),\ 0 < x < l,\ t > 0,$$
$$u(0, t) = g(t),\ u(l, t) = h(t),\ t > 0,$$
$$u(x, 0) = u_0(x),\ 0 < x < l.$$

can be transformed into a problem with homogeneous boundary conditions
by subtracting from u a linear function of x that satisfies the boundary
conditions for all time.

10. Consider the following initial boundary value problem for the diffusion
equation.

$$u_t - Du_{xx} = f(x),\ 0 < x < l,\ t > 0,$$
$$u_x(0, t) = A,\ u_x(l, t) = B,\ t > 0,$$
$$u(x, 0) = u_0(x),\ 0 < x < l.$$

Show that if the solution is independent of time, i.e. $u = u(x)$, then
$DA - DB = \int_0^l f(x)\, dx$. Give a physically meaningful interpretation of
this condition.

11. Solve the initial, boundary value problem

$$u_t - ku_{xx} = 0,\ x > 0,\ t > 0$$
$$u(0, t) = 1,\ u(\infty, t) = 0,\ t > 0,$$
$$u(x, 0) = 0,\ x > 0,$$

for the heat equation by assuming a solution of the form $u(x, t) = U(z)$
where $z = x/\sqrt{kt}$. Write the solution in terms of the error function erf.

12. Green's theorem in the plane is usually stated in calculus texts as

$$\int\int_R (Q_x(x, y) - P_y(x, y))\, dxdy = \int_C P(x, y)\, dx + Q(x, y)\, dy,$$

where P and Q are smooth functions in a well-behaved domain R in the xy plane with boundary C. The line integral is taken counterclockwise. Verify that Green's theorem is really the divergence theorem in two dimensions.

13. Show that the nonlinear initial boundary value problem

$$u_t = u_{xx} - u^3,\ 0 < x < l,\ t > 0,$$
$$u(0,t) = 0,\ u(l,t) = 0,\ t > 0,$$
$$u(x,0) = 0,\ 0 < x < l,$$

has only the trivial solution.

14. Let $u = u(\mathbf{x}, t)$ be a positive solution to the equation

$$u_t = \Delta u + f(\mathbf{x}),\quad \mathbf{x} \in \Omega,\ t > 0,$$

where Ω is a well-behaved, bounded domain, and let $s = \ln u$ be the local **entropy** of u. Let $S = \int_\Omega s d\mathbf{x}$ denote the total entropy in Ω, let $J = -\int_{\partial\Omega} \nabla s \cdot \mathbf{n} dA$ be the net entropy flux through the boundary, and let $Q = \int_\Omega f u^{-1} d\mathbf{x}$ be the entropy change due to the source f in Ω. Prove that

$$\frac{dS}{dt} + J \geq Q.$$

15. Two homogeneous rods of length L_1 and L_2 have the same density and specific heats, but their thermal conductivities are K_1 and K_2, which differ. The two rods are welded together with a weld joint that puts them in perfect contact. The composite rod is laterally insulated and the temperature of the left end is held at 0 degrees, while the right end is held at τ degrees. Assume the system is allowed to come to a steady state where the temperature distribution in the system depends only upon x.

   a) Discuss why the temperature and the flux should be continuous at the joint.

   b) Determine the steady-state temperature distribution in the composite rod.

6.3 Equilibrium Equations

6.3.1 Laplace's Equation

Laplace's equation, the prototype of elliptic equations, is one of the basic equations of mathematics. Elliptic equations model equilibrium, or time-independent, phenomena. One way to motivate equilibrium states is to imagine that these states arise from a long-time, or asymptotic, limit of transient states, which are.solutions to evolution problems. For example, suppose that conditions on the boundary $\partial\Omega$ are fixed and do not depend on time, and sources, if present, are time-independent. Then, after a long time we expect that the effects of the initial condition in the region Ω will decay away, giving an equilibrium state $u = u(\mathbf{x})$ that satisfies a steady-state type equation.

Consider the initial boundary value problem for the diffusion equation,

$$
\begin{aligned}
u_t - D\Delta u &= f(\mathbf{x}), & \mathbf{x} \in \Omega,\ t > 0, \\
u(\mathbf{x}, t) &= g(\mathbf{x}), & \mathbf{x} \in \partial\Omega,\ t > 0, \\
u(\mathbf{x}, 0) &= u_0(\mathbf{x}), & \mathbf{x} \in \Omega,\ t > 0.
\end{aligned}
$$

Over the long run we may expect that the effects of the the initial data to be small and an equilibrium solution $u = u(x)$ to emerge that satisfies the equilibrium problem

$$
\begin{aligned}
-D\Delta u &= f(\mathbf{x}), & \mathbf{x} \in \Omega, & \qquad (6.25) \\
u(\mathbf{x}) &= g(\mathbf{x}), & \mathbf{x} \in \partial\Omega. & \qquad (6.26)
\end{aligned}
$$

Equation (6.25) is **Poisson's** [4] **equation**.

If there are no sources, then the equation (6.25) reduces to

$$
\Delta u = 0, \quad \mathbf{x} \in \Omega, \qquad (6.27)
$$

which is **Laplace's equation**. This partial differential equation is possibly the most analyzed equation in analysis. It also arises in many other natural settings. Readers who have studied complex analysis may recall that the two-dimensional Laplace equation is satisfied by both the real and imaginary parts of a holomorphic (analytic) function $F(z) = u(x, y) + iv(x, y)$ on a domain Ω in the complex plane; that is, $\Delta u = 0$ (or $u_{xx} + u_{yy} = 0$) and $\Delta v = 0$ in Ω. Thus Laplace's equation plays an important role in complex analysis; conversely, complex analysis can be the basis of an approach for studying Laplace's equation in two dimensions.These equations also arise naturally in potential theory. For example, if $\mathbf{E}$ is a static electric field in Ω induced by charges lying outside Ω, then $\nabla \times \mathbf{E} = 0$. If $V(\mathbf{x})$ is a potential function, which means $\mathbf{E} = -\nabla V$, then it follows that the potential function V satisfies Laplace's equation. Laplace's equation is also important in fluid mechanics and many other areas of application. Solutions to Laplace's equation are called *harmonic* functions. In one dimension Laplace's equation is just $-u''(x) = 0$, which has linear solutions

---

[4] S. Poisson (1781–1840), a French mathematician and scientist, made great contributions to physics, mathematics, and probability. He ranks as one of the most prolific and gifted scientists of all time.

$u(x) = ax + b$. So, steady-state temperature profiles in a rod are linear. The constants a and b are determined from boundary conditions.

What kinds of auxiliary conditions are appropriate for Laplace's equation, or equation (6.25)? From the preceding remarks we expect to impose only boundary conditions along $\partial\Omega$, and not initial conditions (in Hadamard's example we observed that the initial value problem for Laplace's equation in two dimensions was not well-posed). Therefore, we impose only boundary conditions, such as the Dirichlet condition (6.26), the Neuman condition

$$\frac{du}{dn} = g(\mathbf{x}), \quad \mathbf{x} \in \partial\Omega, \tag{6.28}$$

which specifies the flux across the boundary, or a Robin condition

$$\frac{du}{dn} + a(\mathbf{x})u = g(\mathbf{x}), \quad \mathbf{x} \in \partial\Omega. \tag{6.29}$$

In some applications in three dimensions, it is convenient to work in either cylindrical or spherical coordinates. These coordinates are defined by the equations

$$x = r\cos\theta$$
$$y = r\sin\theta \quad \text{(cylindrical coordinates } r, \theta, z)$$
$$z = z$$

and

$$x = r\sin\phi\cos\theta$$
$$y = r\sin\phi\sin\theta \quad \text{(spherical coordinates } r, \phi, \theta)$$
$$z = r\cos\phi$$

where θ is the polar angle and ϕ is the azimuthal angle. An application of the chain rule permits us to write the Laplacian in cylindrical and spherical coordinates as

$$\Delta u = u_{rr} + \frac{1}{r}u_r + \frac{1}{r^2}u_{\theta\theta} + u_{zz} \quad \text{(cylindrical)}$$

and

$$\Delta u = \frac{1}{r^2}\frac{\partial}{\partial r}(r^2 u_r) + \frac{1}{r^2\sin\phi}\frac{\partial}{\partial\phi}(\sin\phi\, u_\phi) + \frac{1}{r^2\sin^2\phi}u_{\theta\theta} \quad \text{(spherical)}$$

The details of these calculations are left as an exercise that should be done once in everyone's life; but not twice. In two-dimensional with circular symmetry, polar coordinates are appropriate.

Example 6.17

Find all radial solutions to Laplace's equation in two dimensions. We write Laplace's equation in polar coordinates as

$$u_{rr} + \frac{1}{r}u_r = 0, \quad u = u(r).$$

Notice that the partial derivatives are actually ordinary derivatives. We can immediately write

$$v_r + \frac{1}{r}v = 0, \quad v = u_r(r).$$

This equation is separable and can be integrated to get

$$v(r) = \frac{a}{r},$$

where a is a constant. Then

$$u(r) = a \ln r + b,$$

where b is another arbitrary constant. For example, equilibrium temperatures between two concentric circles, with consant values on the boundaries, vary logarithmically. This is in contrast to the linear distribution in a bar, or linear geometry.

6.3.2 Basic Properties

The identities in the previous subsection permit us to obtain in an easy manner some basic properties of equilibrium problems. For example, we can prove that solutions to the Dirichlet problem are unique.

Theorem 6.18

Let h be a continuous function on $\partial\Omega$ and f be continuous on $\overline{\Omega}$. If the Dirichlet problem

$$\Delta u = f, \quad \mathbf{x} \in \Omega; \quad u = h, \quad \mathbf{x} \in \partial\Omega$$

has a solution $u \in C^1(\overline{\Omega}) \cap C^2(\Omega)$, then it is unique.

To prove this assume that there are two such solutions, u_1 and u_2. Then

$$\Delta u_1 = 0, \quad \mathbf{x} \in \Omega, \ u_1 = h, \ \mathbf{x} \in \partial\Omega; \quad \Delta u_2 = 0, \quad \mathbf{x} \in \Omega, \ u_2 = h, \ \mathbf{x} \in \partial\Omega.$$

Therefore the difference $w \equiv u_1 - u_2$ must satisfy the homogeneous problem

$$\Delta w = 0, \quad \mathbf{x} \in \Omega; \ w = 0, \ \mathbf{x} \in \partial\Omega.$$

Now, in Green's identity (9), take $u = w$ to obtain

$$\int_\Omega \nabla w \cdot \nabla w \, d\mathbf{x} = 0$$

Hence $\nabla w = 0$ and so $w = const.$ in Ω; since w is continuous in $\overline{\Omega}$ and zero on the boundary we must have $u_1 - u_2 = w = 0$ in $\overline{\Omega}$, or $u_1 = u_2$.

In a similar way we have the following theorem:

Theorem 6.19

Let g be continuous on $\partial\Omega$ and f be continuous on $\overline{\Omega}$. If the Neumann problem

$$\Delta u = f, \quad \mathbf{x} \in \Omega; \quad \frac{du}{dn} = g, \quad \mathbf{x} \in \partial\Omega,$$

has a solution $u \in C^1(\overline{\Omega}) \cap C^2(\Omega)$, then necessarily

$$\int_\Omega f \, dx = \int_{\partial\Omega} g \, dS \tag{6.30}$$

The proof follows immediately from setting $w = 1$ in Green's identity (10). Physically, (11) means that, in a steady state, the net flux (of heat, for example) through the boundary should be balanced by the total amount of heat being created in the region by the sources.

Additional applications of the Green's identities are found in the Exercises.

Another important property of (nonconstant) solutions to Laplace's equation is that they attain their maximum and minimum values on the boundary of the domain, and not the interior. This result is called the **maximum principle**, and we refer to the references for a proof.

Theorem 6.20

(Maximum principle) Let Ω be an open, bounded, well-behaved domain in $\mathbb{R}^n$. If $\Delta u = 0$ in Ω and u is continuous in $\overline{\Omega}$, then the maximum and minimum values of u are attained on $\partial\Omega$.

EXERCISES

1. Start with the product rule for derivatives and derive the integration by parts formula.

2. Use the divergence theorem to derive Green's identities.

3. Consider the initial boundary value problem for the n-dimensional heat equation given by

$$u_t - k\Delta u = \rho(\mathbf{x}, t), \quad \mathbf{x} \in \Omega, \ t > 0,$$
$$u(\mathbf{x}, t) = g(\mathbf{x}, t), \quad \mathbf{x} \in \partial\Omega, \ t > 0,$$
$$u(\mathbf{x}, 0) = f(\mathbf{x}), \quad \mathbf{x} \in \Omega,$$

where ρ, g, and f are continuous. Use an energy argument to prove that solutions are unique under reasonable smoothness conditions on u.

4. Let $q > 0$ and continuous in $\overline{\Omega} \subset \mathbb{R}^n$, f be continuous in $\overline{\Omega}$, and g be continuous on $\partial\Omega$. Prove that the Neumann problem

$$-\Delta u + q(\mathbf{x})u = f(\mathbf{x}), \quad \mathbf{x} \in \Omega,$$
$$\frac{du}{dn} = g(\mathbf{x}), \quad \mathbf{x} \in \partial\Omega$$

can have at most one solution in $C^1(\overline{\Omega}) \cap C^2(\Omega)$.

5. A radially symmetric solution to Laplace's equation $\Delta u = 0$ in $\mathbb{R}^3$ is a solution of the form

$$u = u(r), \quad r = \sqrt{x_1^2 + x_2^2 + x_3^2},$$

where u just depends only on the distance from the origin. Find all radially symmetric solutions in $\mathbb{R}^3$.

6. Prove that if the Dirichlet problem

$$-\Delta u = \lambda u, \quad \mathbf{x} \in \Omega,$$
$$u = 0, \quad \mathbf{x} \in \partial\Omega,$$

has a nontrivial solution, then the constant λ must be positive.

7. Let $\mathbf{f}$ be a smooth vector field on $\mathbb{R}^n$ with the property that

$$|\mathbf{f}(\mathbf{x})| \leq \frac{1}{|\mathbf{x}|^3 + 1}, \quad \mathbf{x} \in \mathbb{R}^n.$$

Prove that $\int_{\mathbb{R}^n} \nabla \cdot \mathbf{f} \, d\mathbf{x} = 0$.

8. Problems in the calculus of variations (Chapter 3) can be generalized to multiple integral problems. Consider the problem of minimizing the functional

$$J(u) = \int_\Omega L(\mathbf{x}, u, \nabla u) \, d\mathbf{x},$$

over all $u \in C^2(\Omega)$ with $u(\mathbf{x}) = f(\mathbf{x})$, $\mathbf{x} \in \partial\Omega$, where f is a given function on the boundary; Ω is a bounded well-behaved region in $\mathbb{R}^n$.

a) Show that the first variation is

$$\delta J(u,h) = \int_\Omega [L_u h + L_{\nabla u} \cdot \nabla h] \, d\mathbf{x}$$

$$= \int_\Omega (L_u - \nabla \cdot L_{\nabla u}) h \, d\mathbf{x} - \int_{\partial \Omega} h L_{\nabla u} \cdot \mathbf{n} \, dA,$$

where $L_{\nabla u} = (L_{u_{x_1}}, ..., L_{u_{x_n}})$, and where $h \in C^2(\Omega)$ with $h(\mathbf{x}) = 0$, $\mathbf{x} \in \partial \Omega$.

b) Show that a necessary condition for u to minimize J is that u must satisfy the Euler equation

$$L_u - \nabla \cdot L_{\nabla u} = 0, \quad \mathbf{x} \in \Omega.$$

c) If u is not fixed on the boundary $\partial \Omega$, find the natural boundary condition.

9. Find the Euler equation for the following functionals.

a) $J(u) = \int_\Omega (\nabla u \cdot \nabla u) \, d\mathbf{x}$.

b) $J(u) = \int_\Omega (x^2 u_x^2 + y^2 u_y^2) d\mathbf{x}$, where $u = u(x,y)$, $d\mathbf{x} = dxdy$.

c) $J(u) = \frac{1}{2} \int_0^l \int_\Omega (u_t^2 - c^2 \nabla u \cdot \nabla u - mu^2) dxdt$ where $u = u(\mathbf{x},t)$, and c and m are constants. (Hint: Treat time t as an additional independent variable and work in $\mathbb{R}^{n+1}$.)

10. The biomass density u (mass per unit volume) of zooplankton in a very deep lake varies as a function of depth x and time t. Zooplankton diffuse vertically with diffusion constant D, and buoyancy effects cause them to migrate toward the surface at a constant speed of ag, where g is the acceleration due to gravity and a is a positive constant. At any time, in any vertical tube of unit cross-sectional area, the total biomass of zooplankton is a constant U.

a) From first principles derive a partial differential equation model for the biomass density of zooplankton.

b) Find the *steady-state* biomass density as a function of depth. (Formulate any reasonable boundary conditions, or other auxiliary conditions, that are needed to solve this problem.)

11. Consider the initial boundary value problem

$$u_t = u_{xx} - u^3, \quad x \in (0,1), \ t > 0,$$
$$u(x,0) = 0, \quad x \in (0,1),$$
$$u(0,t) = u(1,t) = 0, \quad t > 0.$$

Use an energy method to show that the solution must be identically zero.

6.4 Eigenfunction Expansions

In Chapter 4 we noted the importance of the eigenvalue problem for transformations on $\mathbb{R}^n$ (matrices), for Sturm–Liouville operators, and for integral operators. The same, or course, is true for partial differential operators. Basically, these problems have the form

$$Lu = \lambda u, \quad \mathbf{x} \in \Omega, \tag{6.31}$$
$$B(u) = 0, \quad \mathbf{x} \in \partial\Omega, \tag{6.32}$$

where L is a linear partial differential operator and B is a linear operator on the boundary. Specification of the operator includes both L and B. If there is a value of λ for which (6.31)–(6.32) has a nontrivial solution $u(\mathbf{x})$, then λ is called an eigenvalue and $u(\mathbf{x})$ is a corresponding eigenfunction. Any constant multiple of an eigenfunction is also an eigenfunction with respect to the same eigenvalue. The set of all eigenfunctions form an natural orthogonal basis for the function spaces involved in solving boundary value problems.

6.4.1 Spectrum of the Laplacian

Consider

$$-\Delta u = \lambda u, \quad \mathbf{x} \in \Omega, \tag{6.33}$$
$$u = 0, \quad \mathbf{x} \in \partial\Omega. \tag{6.34}$$

The technique we use to find the **spectrum** is called separation of variables. It is widely used method for solving linear boundary value problems on bounded domains with boundaries along the coordinate direction. Here the method is illustrated on the Laplacian operator, and in the next section we show how it equally applies to evolution problems. The negative sign on the Laplacian will give positive eigenvalues.

Example 6.21

Take $\Omega = [0, \pi] \times [0, \pi] \subset \mathbb{R}^2$. Then the partial differential equation is $-(u_{xx} + u_{yy}) = \lambda u$ for $u = u(x, y)$. The Dirichlet boundary conditions become $u(0, y) = u(\pi, y) = 0$ for $0 \le y \le \pi$, and $u(x, 0) = u(x, \pi) = 0$ for $0 \le x \le \pi$. We assume a separable solution of the form $u(x,y) = X(x)Y(y)$ for some functions X and Y to be determined. Substituting we get

$$-X''(x)Y(y) - X(x)Y''(y) = \lambda X(x)Y(y),$$

which can be rewritten as

$$-\frac{X''(x)}{X(x)} = \frac{Y''(y) + \lambda Y(y)}{Y(y)},$$

with a function of x on one side and a function of y on the other. Now comes a crucial observation. A function of x can equal a function of y for all x and y only if both are equal to a constant. That is,

$$-\frac{X''(x)}{X(x)} = \frac{Y''(y) + \lambda Y(y)}{Y(y)} = \mu$$

for some constant μ, called the *separation constant*. Therefore the partial differential equation separates into two ordinary differential equations for the two spatial variables:

$$-X''(x) = \mu X(x), \quad -Y''(y) = (\lambda - \mu)Y(y).$$

Next we substitute the assumed form of u into the boundary conditions to obtain $X(0) = X(\pi) = 0$, and $Y(0) = Y(\pi) = 0$. Consequently we have obtained two boundary value problems of Sturm–Liouville type,

$$\begin{aligned} -X''(x) &= \mu X(x), \quad X(0) = X(\pi) = 0, \\ -Y''(y) &= (\lambda - \mu)Y(y), \quad Y(0) = Y(\pi) = 0. \end{aligned}$$

The first problem was solved in Chapter 4 and we have eigen-pairs

$$X_n(x) = \sin nx, \quad \mu_n = n^2, \quad n = 1, 2, 3, \dots.$$

Then the Y problem becomes

$$-Y''(y) = (\lambda - n^2)Y(y), \quad Y(0) = Y(\pi) = 0.$$

This problem is again of Sturm-Liouville type and will have nontrivial solutions $Y_k(y) = \sin ky$ when $\lambda - n^2 = k^2$, for $k = 1, 2, 3, \dots$ Therefore, double indexing λ, we get eigenvalues and eigenfunctions for the negative Laplacian with Dirichlet boundary conditions as

$$u_{n,k}(x,y) = \sin nx \sin ky, \quad \lambda_{n,k} = n^2 + k^2, \quad n, k = 1, 2, 3, \dots$$

Observe that there are infinitely many positive, real eigenvalues, whose limit is infinity, and the corresponding eigenfunctions are orthogonal, that is,

$$\int_\Omega (\sin nx \sin ky)(\sin mx \sin ly)dA = 0 \quad \text{if} \quad n \neq m \text{ or } k \neq l.$$

This example illustrates what to expect from the Dirichlet problem (6.33)–(6.34). In fact, we have, in general, the following properties.

1. The eigenvalues are real.

2. There are infinitely many eigenvalues that can be ordered as $0 < \lambda_1 \leq \lambda_2 \leq \lambda_3 \leq \cdots$ with $\lambda_n \to +\infty$ as $n \to \infty$.

3. Eigenfunctions corresponding to distinct eigenvalues are orthogonal in the inner product $(u, v) = \int_\Omega u(\mathbf{x})v(\mathbf{x})d\mathbf{x}$.

4. The set the eigenfunctions $u_n(\mathbf{x})$ is complete in the sense that any square-integrable function $f(\mathbf{x})$ on Ω can be uniquely represented in its generalized Fourier series

$$f(\mathbf{x}) = \sum_{n=1}^{\infty} c_n u_n(\mathbf{x}), \quad c_n = \frac{(f, u_n)}{||u_n||^2},$$

where the c_n are the Fourier coefficients, and the norm is $||u_n|| = \sqrt{(u_n, u_n)}$. Convergence of the series is in the $L^2(\Omega)$ sense, meaning

$$\int_\Omega \left(f(\mathbf{x}) - \sum_{n=1}^{N} c_n u_n(\mathbf{x}) \right)^2 d\mathbf{x} \to 0 \quad \text{as} \quad N \to \infty.$$

The results are exactly the same when the Dirichlet boundary condition (6.34) is replaced by a Neumann boundary condition $\frac{du}{dn} = 0$, with the exception that $\lambda = 0$ is also an eigenvalue, with eigenfunction $u_0(x) = \text{constant}$. When a Robin boundary condition $\frac{du}{dn} + a(\mathbf{x})u = 0$ is imposed, again there is a zero eigenvalue with constant eigenfunction provided $a(x) \geq 0$. If $a(x)$ fails to be nonnegative, then there may also be negative eigenvalues.

Remark 6.22

The previous results extend to the eigenvalue problem

$$-\nabla \cdot (p\nabla u) + qu = \lambda w u, \quad \mathbf{x} \in \Omega,$$

Here, $w = w(\mathbf{x}) > 0$, $p = p(\mathbf{x}) > 0$, $q = q(\mathbf{x})$, p, q and w are continuous on $\overline{\Omega}$, and p has continuous first partial derivatives on Ω. The boundary conditions may be Dirichlet ($u = 0$ on $\partial\Omega$) Neumann ($\frac{du}{dn} = 0$ on $\partial\Omega$), or Robin ($\frac{du}{dn} + a(\mathbf{x})u = 0$ on $\partial\Omega$, with $a(x) \geq 0$). In this case the eigenfuctions are orthogonal $u_n(x)$ with respect to the inner product

$$(u, v)_w = \int_\Omega u(\mathbf{x})v(\mathbf{x})w(\mathbf{x}) \, d\mathbf{x},$$

and $||u||_w = \sqrt{(u,u)_w}$. The Fourier series takes the form

$$f(\mathbf{x}) = \sum_n^\infty c_n u_n(\mathbf{x}), \quad c_n = \frac{(f, u_n)_w}{||u_n||_w^2},$$

where convergence is

$$\int_\Omega \left(f(\mathbf{x}) - \sum_{n=1}^N c_n u_n(\mathbf{x}) \right)^2 w(\mathbf{x})\, d\mathbf{x} \to 0 \quad \text{as} \quad N \to \infty.$$

Just as for one-dimensional Sturm–Liouville problems and for integral equations, the completeness of the set of eigenfunctions allows us to solve the nonhomogeneous problem.

Theorem 6.23

(Fredholm alternative) Consider the boundary value problem

$$-\nabla \cdot (p\nabla u) + qu = \mu w u + f(\mathbf{x}), \quad \mathbf{x} \in \Omega,$$

with homogeneous Dirichlet, Neumann, or Robin boundary conditions, where the coefficient functions satisfy the conditions in Remark 6.22, and f is a given function. Then

(a) If μ is not an eigenvalue of the corresponding homogeneous problem ($f = 0$), then there is a unique solution for all functions f with $\int_\Omega f(\mathbf{x}) w(\mathbf{x})^{-2} d\mathbf{x} < \infty$.

(b) If μ is an eigenvalue of the homogenous problem, then there is no solution or infinitely many solutions, depending upon the function f.

The proof of this theorem is straightforward and follows exactly the format of the proofs in Chapter 4, namely, to expand u and f in eigenfunction expansions and solve for the coefficients of u.

6.4.2 Evolution Problems

The method of separation of variables is, in general, a way of constructing solutions to partial differential equations in the form of an eigenfuction expansion.

Let $u = u(\mathbf{x}, t)$ satisfy the initial boundary value problem

$$\begin{aligned}
u_t &= k\Delta u, & \mathbf{x} \in \Omega, \ t > 0, \\
u(\mathbf{x}, t) &= 0, & \mathbf{x} \in \partial\Omega, \ t > 0, \\
u(\mathbf{x}, 0) &= f(\mathbf{x}) & \mathbf{x} \in \Omega.
\end{aligned}$$

on a well-behaved bounded domain Ω. The strategy is simple. We seek functions of the form $u(\mathbf{x}, t) = F(\mathbf{x})T(t)$ which satisfy the partial differential equation and the Dirichlet boundary condition, but not necessarily the initial condition. We find that there are many such solutions, and because the problem is linear, any linear combination of them will satisfy the equation and boundary condition. We then form the linear combination and select the constants so that the sum also satisfies the initial condition.

To this end substitute $u(\mathbf{x}, t) = F(\mathbf{x})T(t)$ into the partial differential equation to obtain

$$F(\mathbf{x})T'(t) = kT(t)\Delta F(\mathbf{x}),$$

which we write as

$$\frac{T'(t)}{kT(t)} = \frac{\Delta F(\mathbf{x})}{F(\mathbf{x})} = -\lambda,$$

for some λ. The crucial part of the argument is to note that λ is a constant because it cannot be both a function of only t and only x. When this occurs, we say the differential equation is separable, and λ is called the separation constant. Consequently, we have

$$-\Delta F = \lambda F, \quad T' = -\lambda kT. \tag{6.35}$$

The partial differential equation separated into a spatial part and a temporal part. The sign we put on λ does not matter because it all comes out correctly in the end. But, note that the choice of a negative sign gives the standard negative Laplacian examined previously and makes the temporal part decay, which is expected from a diffusion process. Next, the boundary condition implies $u(\mathbf{x}, t) = F(\mathbf{x})T(t) = 0$ for $x \in \partial\Omega$, or

$$F(\mathbf{x}) = 0, \quad \mathbf{x} \in \partial\Omega. \tag{6.36}$$

(A choice of $T(t) = 0$ would give the uninteresting trivial solution.)

The first equation in (6.35) along with the Dirichlet boundary condition (6.36) is the eigenvalue problem for the negative Laplacian that we studied in the previous section. The eigen-pairs are λ_n, $F_n(\mathbf{x})$, $n = 1.2, 3, \ldots$. The eigenvalues are positive and the eigenfunctions are orthogonal. In the context of an evolution problem, the spatial eigenfunctions are called the *normal modes*. With these eigenvalues, the temporal equation $T' = -\lambda_n kT$ has solutions $T_n(t) = e^{-\lambda_n kt}$. Therefore we have constructed solutions $u_n(\mathbf{x}, t) = F_n(\mathbf{x})e^{-\lambda_n kt}$, $n = 1, 2, 3, \ldots$, that satisfy the partial differential equation and the Dirichlet boundary condition. Next we form the linear combination

$$u(\mathbf{x}, t) = \sum_{n=1}^{\infty} a_n F_n(\mathbf{x})e^{-\lambda_n kt}.$$

The initial condition now forces

$$f(\mathbf{x}) = \sum_{n=1}^{\infty} a_n F_n(\mathbf{x}),$$

which is the **generalized Fourier series** for f. The coefficients must be the **Fourier coefficients**

$$a_n = \frac{1}{(F_n, F_n)} \int_{\Omega} f(\mathbf{x}) F_n(\mathbf{x}) \, d\mathbf{x}, \quad n = 1, 2, 3, \dots$$

Recall that $(F_n, F_n) = ||F_n||^2 = \int_{\Omega} F_n(\mathbf{x})^2 d\mathbf{x}$.

Therefore we have derived a formal solution to the initial boundary value problem in the form of an infinite series. The validity of such a series representation must be proved by checking that it does indeed satisfy the problem. This requires substitution into the partial differential equation and boundary conditions, so conditions that guarantee term by term differentiation must be verified.

Example 6.24

In one spatial dimension, consider the initial boundary value problem for the heat equation:

$$u_t = k u_{xx}, \quad 0 < x < l, \ t > 0, \tag{6.37}$$

$$u(0, t) = 0, \quad u(l, t) = 0, \ t > 0, \tag{6.38}$$

$$u(x, 0) = f(x), \quad 0 < x < l. \tag{6.39}$$

This models heat flow in a bar of length l with the ends held at constant zero temperature. The function f defines the initial temperature distribution. Substituting $u(x, t) = F(x)T(t)$ into the partial differential equation and boundary conditions leads to the equation $T'(t) = -\lambda k T(t)$, and the eigenvalue problem

$$-F'' = \lambda F, \quad 0 < x < 1, \quad F(0) = F(l) = 0, \tag{6.40}$$

which is the Sturm-Liouville problem discussed in Chapter 4. The eigenvalues and eigenfunctions are

$$\lambda_n = \frac{n^2 \pi^2}{l^2}, \quad F_n(x) = \sin \frac{n \pi x}{l}, \quad n = 1, 2, \dots$$

The solution to the time equation is $T_n(t) = e^{-n^2 \pi^2 kt/l^2}$. We form the series

$$u(x, t) = \sum_{n=1}^{\infty} a_n e^{-n^2 \pi^2 kt/l^2} \sin \frac{n \pi x}{l},$$

where the a_n are to be determined. But the initial condition implies

$$f(x) = \sum_{n=1}^{\infty} a_n \sin \frac{n\pi x}{l},$$

which is the generalized Fourier series for f. The a_n are the Fourier coefficients

$$a_n = \frac{1}{(F_n, F_n)} \int_0^l f(\xi) \sin \frac{n\pi \xi}{l} \, d\xi, \quad n = 1, 2, 3, \ldots$$

We have $(F_n, F_n) = \int_0^l \sin^2 \frac{n\pi x}{l} \, dx = \frac{l}{2}$. Therefore the solution to (6.37)–(6.39) is given by the infinite series

$$u(x, t) = \frac{2}{l} \sum_{n=1}^{\infty} \left(\int_0^l f(\xi) \sin \frac{n\pi \xi}{l} \, d\xi \right) e^{-n^2 \pi^2 t / l^2} \sin \frac{n\pi x}{l}.$$

By rearranging the terms and formally switching the order of summation and integration, we can write the solution in a particularly nice form as

$$u(x, t) = \int_0^l G(x, \xi, t) f(\xi) \, d\xi.$$

where

$$G(x, \xi, t) = \frac{2}{l} \sum_{n=1}^{\infty} e^{-n^2 \pi^2 t / l^2} \sin \frac{n\pi \xi}{l} \sin \frac{n\pi x}{l}.$$

This example shows how to proceed when there is one spatial variable. In higher dimensions the eigenvalue problem will be higher-dimensional, and it can be solved by separating the spatial variables.

Remark 6.25

The eigenfunction expansion method performs equally well on any evolution equation of the form

$$u_t = \frac{1}{w} \nabla \cdot (p \nabla u) - qu, \quad \mathbf{x} \in \Omega, \ t > 0,$$

with initial condition $u(\mathbf{x}, 0) = f(x)$, $\mathbf{x} \in \Omega$. Here, $w = w(\mathbf{x}) > 0$, $p = p(\mathbf{x}) > 0$, $q = q(\mathbf{x})$, p, q and w are continuous on $\overline{\Omega}$, and p has continuous first partial derivatives on Ω. All of these equations are separable. The boundary conditions may be Dirichlet ($u = 0$ on $\partial\Omega$) Neumann ($\frac{du}{dn} = 0$ on $\partial\Omega$), or Robin ($\frac{du}{dn} + a(x)u = 0$ on $\partial\Omega$, with $a \geq 0$). The domain Ω must be bounded, and the boundary conditions have to be given on $x = $ constant planes. In polar, cylindrical, and spherical coordinates the boundary conditions must be given on constant coordinate curves and surfaces. The associated eigenvalue problem will have the form

$$-\nabla \cdot (p\nabla F) + qF = \lambda w F, \quad \mathbf{x} \in \Omega,$$

with either Dirichlet, Neumann, or Robin boundary conditions.

Remark 6.26

The eigenfunction method is also applicable to equations of the form

$$u_{tt} = \frac{1}{w}\nabla \cdot (p\nabla u) - qu, \quad \mathbf{x} \in \Omega, \ t > 0,$$

with the same caveats and limitations as in the last Remark. These are wave equations and have second-order time derivatives. When variables are separated, the temporal equation will be second-order, and in this case two initial conditions, $u(\mathbf{x}, 0) = f(\mathbf{x})$ and $u_t(\mathbf{x}, 0) = g(\mathbf{x})$, $\mathbf{x} \in \Omega$, are required. Wave equations are discussed in the next chapter.

Remark 6.27

Nonhomogenous problems having a partial differential equation of the form

$$u_t = \frac{1}{w}\nabla \cdot (p\nabla u) - qu + f(\mathbf{x}), \quad \mathbf{x} \in \Omega, \ t > 0,$$

can be solved by assuming a solution of the form

$$u(\mathbf{x}, t) = \sum_{n=1}^{\infty} g_n(t) F_n(\mathbf{x}),$$

where the $F_n(x)$ are the eigenfunctions of the nonhomogeneous spatial problem. Upon substitution into the nonhomogeneous equation, the unknown coefficients $g_n(t)$ can be determined.

EXERCISES

1. Use the eigenfunction expansion method to solve the following problems:

   a)

$$u_t = ku_{xx}, \quad 0 < x < l, \ t > 0,$$
$$u_x(0, t) = u_x(l, t) = 0, \ t > 0,$$
$$u(x, 0) = f(x), \ 0 < x < l.$$

b)

$$u_{tt} = c^2 u_{xx} - a^2 u, \quad 0 < x < l, \ t > 0,$$
$$u(0,t) = u(l,t) = 0, \quad t > 0,$$
$$u(x,0) = f(x), \ u_t(x,0) = 0, \ 0 < x < l.$$

c)

$$u_{xx} + u_{yy} = 0, \quad 0 < x < \pi, \ y > 0,$$
$$u(0,y) = u(\pi, y) = 0, \quad y > 0,$$
$$u(x,0) = x(\pi - x), \quad 0 < x < \pi; \ u \text{bounded}$$

2. Organisms of density $u(x,t)$ are distributed in a patch of length l. They diffuse with diffusion constant D, and their growth rate is ru. At the ends of the patch a zero density is maintained, and the initial distribution is $u(x,0) = f(x)$. Because there is competition between growth in the interior of the patch and escape from the boundaries, it is interesting to know whether the population increases or collapses. Show that the condition on the patch size, $l < \pi \sqrt{D/r}$, ensures death of the population.

3. Consider the partial differential operator

$$Lu = -\nabla \cdot (p\nabla u) - qu, \quad \mathbf{x} \in \Omega,$$

where $p = p(\mathbf{x}) > 0$, $q = q(\mathbf{x})$, p and q are continuous on $\overline{\Omega}$, and p has continuous first partial derivatives on $\overline{\Omega}$.

a) Prove the integration by parts formula

$$\int_\Omega vLu\,dx = \int_\Omega uLv\,dx + \int_{\partial\Omega} p\left(u\frac{dv}{dn} - v\frac{du}{dn}\right)dA.$$

b) Consider the eigenvalue problem $Lu = \lambda u$, $\mathbf{x} \in \Omega$ with a Dirichlet boundary condition $u = 0$, $\mathbf{x} \in \partial\Omega$. Prove that the eigenvalues are positive and that distinct eigenvalues have corresponding orthogonal eigenfunctions.

4. Give a proof of the Fredholm alternative theorem for the problem

$$-\Delta u + qu = \mu wu + f(\mathbf{x}), \quad \mathbf{x} \in \Omega,$$

with a homogeneous Dirichlet boundary condition.

6.5 Integral Transforms

The eigenfunction expansion method is, for the most part, applicable to problems on bounded spatial domains. Transform methods, on the other hand, are usually applied to problems on infinite or semi-infinite spatial domains. In this section we introduce two fundamental integral transforms, the Laplace transform and the Fourier transform. Basically, the idea is to transform a partial differential equation into an ordinary differential equation in a transform domain, solve the ordinary differential equation, and then return to the original domain by an inverse transform.

6.5.1 Laplace Transforms

Laplace transforms are encountered in elementary differential equations courses as a technique for solving linear ordinary differential equations with constant coefficients. They are particularly useful for nonhomogeneous problems where the source term is piecewise continuous or is an impulse function (a delta function). We assume the reader is familiar with these elementary methods. Here we show that they easily extend to partial differential equations.

First, we recall some ideas for functions of a single variable. If $u = u(t)$ is a piecewise continuous function on $t \geq 0$ that does not grow too fast (for example, assume that u is of *exponential order*, i.e., $|u(t)| \leq c \exp(at)$ for t sufficiently large, where $a, c > 0$), then the **Laplace transform** of u is defined by

$$(\mathcal{L}u)(s) \equiv U(s) = \int_0^\infty u(t)e^{-st}\, dt \qquad (6.41)$$

We often write this as $\mathcal{L}u = U$. The Laplace transform is an example of an integral transform; it takes a given function $u(t)$ in the time domain and maps it to a new function $U(s)$ in the so-called transform domain. U and s are called the transform variables. The transform is linear in that $\mathcal{L}(c_1 u + c_2 v) = c_1 \mathcal{L}u + c_2 \mathcal{L}v$, where c_1 and c_2 are constants. If the transform $U(s)$ is known, then $u(t)$ is called the **inverse transform** of $U(s)$ and we write $(\mathcal{L}^{-1}U)(t) = u(t)$, or just $\mathcal{L}^{-1}U = u$. Pairs of Laplace transforms and their inverses have been tabulated in extensive tables (see Table 6.1 for a brief list), and they are available on most computer algebra systems.

The importance of the Laplace transform, like other transforms, is that it changes derivative operations in the time domain to multiplication operations in the transform domain. In fact, we have the important operational formulas

$$(\mathcal{L}u')(s) = sU(s) - u(0), \qquad (6.42)$$
$$(\mathcal{L}u'')(s) = s^2 U(s) - su(0) - u'(0). \qquad (6.43)$$

Table 6.1 Laplace Transforms

$u(t)$	$U(s)$
1	$s^{-1}, \quad s > 0$
e^{at}	$\frac{1}{s-a}, \quad s > a$
t^n, n a positive integer	$\frac{n!}{s^{n+1}}, \quad s > 0$
$\sin at$ and $\cos at$	$\frac{a}{s^2+a^2}$ and $\frac{s}{s^2+a^2}, \quad s > 0$
$\sinh at$ and $\cosh at$	$\frac{a}{s^2-a^2}$ and $\frac{s}{s^2-a^2}, \quad s > |a|$
$e^{at}\sin bt$	$\frac{b}{(s-a)^2+b^2}, \quad s > a$
$e^{at}\cos bt$	$\frac{s-a}{(s-a)^2+b^2}, \quad s > a$
$t^n \exp(at)$	$\frac{n!}{(s-a)^{n+1}}, \quad s > a$
$H(t-a)$	$s^{-1}\exp(-as), \quad s > 0$
$\delta(t-a)$	$\exp(-as)$
$H(t-a)f(t-a)$	$F(s)\exp(-as)$
$f(t)e^{-at}$	$F(s+a)$
$H(t-a)f(t)$	$e^{-as}\mathcal{L}(f(t+a))$
$\operatorname{erf}\sqrt{t}$	$s^{-1}(1+s)^{-1/2}, \quad s > 0$
$\frac{1}{\sqrt{t}}\exp\left(\frac{-a^2}{4t}\right)$	$\sqrt{\pi/s}\exp(-a\sqrt{s}), \quad (s > 0)$
$1 - \operatorname{erf}\left(\frac{a}{2\sqrt{t}}\right)$	$s^{-1}\exp(-a\sqrt{s}), \quad s > 0$
$\frac{a}{2t^{3/2}}\exp\left(\frac{-a^2}{4t}\right)$	$\sqrt{\pi}\exp(-a\sqrt{s}), \quad s > 0$
$u^{(n)}(t)$	$s^n U(s) - s^{n-1}u(0) - s^{n-2}u'(0) - \cdots - u^{(n-1)}(0)$
$\int_0^t u(\tau)v(t-\tau)\,d\tau$	$U(s)\,V(s)$

Formulae (6.42) and (6.43) are readily proved using the definition (6.41) and integration by parts. The strategy for solving differential equations is simple. Taking the Laplace transform of an ordinary differential equation for $u(t)$ results in an algebraic equation for $U(s)$ in the transformed domain. Solve the algebraic equation for $U(s)$ and then recover $u(t)$ by inversion.

In general, to determine $u(t)$ from knowledge of its transform $U(s)$ would take us into the realm of complex variables (which is not a prerequisite for this book). However, we can indicate the general formula for the inverse transform. The **inversion formula** is

$$u(t) = (\mathcal{L}^{-1}U)(t) = \frac{1}{2\pi i}\int_{a-i\infty}^{a+i\infty} U(s)e^{st}\,ds.$$

The integral in this formula is a complex contour integral taken over the infinite vertical line, called a Bromwich path, in the complex plane from $a - i\infty$ to $a + i\infty$. The number a is any real number for which the resulting Bromwich path lies to the right of any singularities (poles, essential singular points, or branch points and cuts) of the function $U(s)$. Calculating inverse transforms using the Bromwich integral usually involves a difficult contour integration in the complex plane and using the residue theorem. In this text we only use a table or a computer algebra system.

A result that is useful in calculations is the **convolution theorem**. Often, solving a differential equation results in having to invert the product of two transforms in the transform domain. The convolution theorem tells what the inverse transformation is.

Theorem 6.28

(Convolution theorem) Let u and v be piecewise continuous on $t \geq 0$ and of exponential order. Then

$$\mathcal{L}(u * v)(s) = U(s)V(s),$$

where

$$(u * v)(t) \equiv \int_0^t u(t - y)v(y)\, dy$$

is the **convolution** of u and v, and $U = \mathcal{L}u, V = \mathcal{L}v$. Furthermore,

$$\mathcal{L}^{-1}(UV)(s) = (u * v)(t).$$

Therefore, the Laplace transform is additive, but not multiplicative. That is, the Laplace transform of a product is not the product of the Laplace transforms. The convolution theorem tells what to take the transform of in order to get a product of Laplace transforms, namely the convolution.

The same strategy applies to partial differential equations where the unknown is a function of two variables, for example, $u = u(x, t)$. Now we transform on t, as before, with the variable x being a parameter unaffected by the transform. In particular, we define the Laplace transform of $u(x, t)$ by

$$(\mathcal{L}u)(x, s) \equiv U(x, s) = \int_0^\infty u(x, t)e^{-st}\, dt.$$

Then time derivatives transform as in (6.42) and (6.43); for example

$$
\begin{aligned}
(\mathcal{L}u_t)(x, s) &= sU(x, s) - u(x, 0), \\
(\mathcal{L}u_{tt})(x, s) &= s^2 U(x, s) - su(x, 0) - u_t(x, 0).
\end{aligned}
$$

On the other hand, spatial derivatives are left unaffected, for example

$$(\mathcal{L}u_x)(x,s) = \int_0^\infty \frac{\partial}{\partial x} u(x,t)e^{-st}\, dt = \frac{\partial}{\partial x} \int_0^\infty u(x,t)e^{-st}\, dt = U_x(x,s),$$

and $(\mathcal{L}u_{xx})(x,s) = U_{xx}(x,s)$ On taking Laplace transforms, a partial differential equation in x and t is reduced to an ordinary differential equation in x with s as a parameter; all the t derivatives are turned into multiplication in the transform domain.

Example 6.29

Let $u = u(x,t)$ denote the concentration of a chemical contaminant, and let $x > 0$ be a semi-infinite region that initially contains no contaminant. For times $t > 0$ we impose a constant, unit concentration on the boundary $x = 0$, and we ask how the contaminant diffuses into the region. Assuming a unit diffusion constant (the problem can be rescaled to make the diffusion constant unity), the mathematical model is

$$u_t - u_{xx} = 0, \; x > 0, \; t > 0,$$
$$u(x,0) = 0, \; x > 0,$$
$$u(0,t) = 1, \; t > 0, \; u(x,t) \text{ bounded.}$$

Taking Laplace transforms of both sides of the equation yields

$$sU(x,s) - U_{xx}(x,s) = 0.$$

This is an ordinary differential equation with x as the independent variable, and the solution is

$$U(x,s) = a(s)e^{-\sqrt{s}x} + b(s)e^{\sqrt{s}x}.$$

Because we require bounded solutions we set $b(s) = 0$. Then

$$U(x,s) = a(s)e^{-\sqrt{s}x}.$$

Next we take Laplace transforms of the boundary condition to get $U(0,s) = 1/s$, where we have used $\mathcal{L}(1) = 1/s$. Therefore $a(s) = 1/s$ and the solution in the transform domain is

$$U(x,s) = \frac{1}{s}e^{-\sqrt{s}x}.$$

Consulting Table 5.1, we find that the solution is

$$u(x,t) = 1 - \mathrm{erf}\left(\frac{x}{\sqrt{4t}}\right).$$

Figure 6.5 shows several time snapshots of the concentration of the contaminant as it diffuses into the medium.

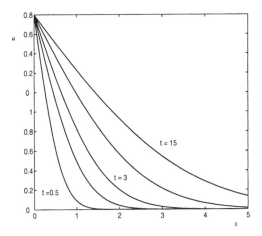

Figure 6.5 Concentration profiles for different times t.

The Laplace transform of a distribution can be defined as well, and we may formally carry out transformations of delta functions and solve equations with point source functions. See Table 6.1 for the relevant entries. The theoretical basis for these calculations is contained in Section 6.7.2.

6.5.2 Fourier Transforms

It is impossible to overestimate the important role that Fourier analysis played in the evolution of mathematical analysis over the last two centuries. Much of this development was motivated theoretical issues associated with problems in engineering and the sciences. Now, Fourier analysis is a monument of applied mathematics, forming a base of signal processing, imaging, probability theory, and differential equations, only to mention a few such areas.

From a partial differential equations viewpoint, the Fourier transform is integral operator with properties similar to the Laplace transform in that derivatives are turned into multiplication operations in the transform domain. Thus the Fourier transform, like the Laplace transform, is useful as a computational tool in solving differential equations. The **Fourier transform** of a function in one dimension $u = u(x)$, $x \in \mathbb{R}$, is defined by the equation

$$(\mathcal{F}u)(\xi) \equiv \hat{u}(\xi) = \int_{-\infty}^{\infty} u(x)e^{i\xi x}\, dx.$$

The variable ξ is called the frequency (or wave number) variable. There is one important remark about notation. There is no standard convention on how to define the Fourier transform; some put a factor of $1/2\pi$ or $1/\sqrt{2\pi}$ in front of the integral, and some have a negative power in the exponential or even a factor of 2π. Therefore one needs to be aware of these differences when consulting other sources because the formulas change depending upon the convention. Another issue is to decide what set of functions on which to define the transform. Certainly, if u is integrable on $\mathbb{R}$, that is, $\int_{-\infty}^{\infty} |u|\, dx < \infty$, then $|\hat{u}(\xi)| = \left| \int_{-\infty}^{\infty} u(x) e^{i\xi x}\, dx \right| \leq \int_{-\infty}^{\infty} |u|\, dx < \infty$, and $\hat{u}$ exists. But, as the example below shows, if u is integrable then $\hat{u}$ is not necessarily integrable; this causes problems, therefore, with inversion of the transform. Consequently, in the theory of Fourier transforms one often takes the domain to be a much smaller class of functions, for example, the square-integrable functions $L^2(\mathbb{R})$, or the **Schwartz class** of functions $\mathcal{S}$. Schwartz class consists of very smooth functions that decay, along with all their derivatives, very rapidly at infinity. Specifically, $u \in \mathcal{S}$ if $u \in C^\infty(\mathbb{R})$ and u and all its derivatives $u^{(k)}$ have the property $\lim_{|x| \to \infty} (x^p u^{(k)})(x) = 0$, for any positive integer p. This assumption makes the theory go through in a highly aesthetic way (see, for example, Strichartz (1994)).

Example 6.30

Compute the Fourier transform of the characteristic function

$$u(x) = \chi_{[-a,a]}(x) = \begin{array}{ll} 1, & |x| \leq a, \\ 0, & |x| > a. \end{array}$$

We have

$$\hat{u}(\xi) = \int_{-\infty}^{\infty} \chi_{[-a,a]}(x) e^{i\xi x}\, dx. = \int_{-a}^{a} e^{i\xi x}\, dx = \frac{2 \sin a\xi}{\xi}.$$

Notice that u has its support, or nonzero values, in a bounded interval, and it is integrable. Its Fourier transform is in $C^\infty(\mathbb{R})$, yet it is not integrable.

In general, one can show that if u is integrable, then $\hat{u}$ is continuous, bounded, and $\lim_{|\xi| \to \infty} \hat{u}(\xi) = 0$. If $u \in \mathcal{S}$, then $\hat{u} \in \mathcal{S}$.

In this text our goal is not a thorough analysis of transforms; rather, we want to study the mechanics and applications of Fourier transforms. Our strategy is to perform the calculations formally, making what smoothness assumptions that are required to do the calculations. Once a result, say, a solution formula, is obtained, we could then make assumptions that are needed for its validity. For example, we may require a high degree of smoothness and rapid decay at infinity to apply the transform method and invert it, but at the end we may find that the solution is valid under much less strenous conditions.

A basic property of the one-dimensional Fourier transform is

$$(\mathcal{F}u^{(k)})(\xi) = (-i\xi)^k \hat{u}(\xi), \tag{6.44}$$

confirming our comment that derivatives are transformed to multiplication by a factor of $(-i\xi)^k$. This operational formula is easily proved using integration by parts (like for the Laplace transform), assuming u and its derivatives are continuous and integrable. For functions of two variables, say $u = u(x, t)$, the variable t acts as a parameter and we define the Fourier transform by

$$(\mathcal{F}\, u)(\xi, t) \equiv \hat{u}(\xi, t) = \int_{-\infty}^{\infty} u(x, t)e^{i\xi x}\, dx.$$

Then, under Fourier transformation, x derivatives turn into multiplication and t derivatives remain unaffected; for example,

$$(\mathcal{F}\, u_x)(\xi, t) = (-i\xi)\hat{u}(\xi, t),$$
$$(\mathcal{F}\, u_{xx})(\xi, t) = (-i\xi)^2 \hat{u}(\xi, t),$$
$$(\mathcal{F}\, u_t)(\xi, t) = \hat{u}_t(\xi, t).$$

Solving a differential equation for u first involves transforming the problem into the transform domain and then solving for $\hat{u}$. Then one is faced with the inversion problem, or the problem of determining the u for which $\mathcal{F}\, u = \hat{u}$. One of the nice properties of the Fourier transform is the simple form of the **inversion formula**, or **inverse transform**:

$$(\mathcal{F}^{-1}\hat{u})(x) \equiv u(x) = \frac{1}{2\pi} \int_{-\infty}^{\infty} \hat{u}(\xi)e^{-i\xi x}\, d\xi. \tag{6.45}$$

This result, stated precisely, is the content of the **Fourier integral theorem**. For example, if $u \in S$, then $\hat{u} \in S$, and conversely, and (6.45) is valid. Under the weaker condition that both u and $\hat{u}$ are integrable, then (6.45) holds at all points where u is continuous. The reader should note the similarity of (6.45) to a Fourier series expansion of a function, but where the integral replaces the sum, the values $\hat{u}(\xi)$ represent a continuum of Fourier coefficients, and the eigenfunctions are $e^{-i\xi x}$.

Some Fourier transforms can be calculated directly, as in the preceding example; many require complex contour integration. A table of Fourier transforms appears at the end of this section (Table 6.2). In the next example we calculate the transform of Gaussian function $u(x) = e^{-ax^2}, a > 0$, using a differential equation technique.

Example 6.31

We want to calculate $\hat{u}$ where

$$\hat{u}(\xi) = \int_{-\infty}^{\infty} e^{-ax^2} \, e^{i\xi x} \, dx.$$

Differentiating with respect to ξ and then integrating by parts gives

$$\hat{u}'(\xi) = i \int_{-\infty}^{\infty} x e^{-ax^2} e^{i\xi x} \, dx$$

$$= \frac{-i}{2a} \int_{-\infty}^{\infty} e^{-ax^2} \, e^{i\xi x} \, dx$$

$$= \frac{-\xi}{2a} \hat{u}(\xi).$$

Therefore we have a differential equation $\frac{d\hat{u}}{d\xi} = (-\xi)/(2a)\hat{u}$ for $\hat{u}$. Separating variables and integrating gives the general solution

$$\hat{u}(\xi) = C e^{-\xi^2/4a}.$$

The constant C can be determined by noticing

$$\hat{u}(0) = \int_{-\infty}^{\infty} e^{-ax^2} \, dx = \sqrt{\frac{\pi}{a}}.$$

Consequently we have

$$\mathcal{F}\left(e^{-ax^2}\right) = \sqrt{\frac{\pi}{a}} \, e^{-\xi^2/4a}. \qquad (6.46)$$

So, the Fourier transform of a Gaussian function is a Gaussian; conversely, the inverse transform of a Gaussian is a Gaussian. Observe that a Gaussian with large spread (variance) gets transformed to a Gaussian with a narrow spread, and conversely.

A convolution relation also holds for Fourier transforms. If u and v are absolutely integrable, then we define the **convolution** of u and v by

$$(u * v)(x) = \int_{-\infty}^{\infty} u(x - y)v(y) \, dy.$$

Then we have the following theorem.

Theorem 6.32

(Convolution theorem) If u and v are in $\mathcal{S}$, then

$$\mathcal{F}\,(u * v)(\xi) = \hat{u}(\xi)\hat{v}(\xi).$$

Then

$$(u * v)(x) = \mathcal{F}^{-1}(\hat{u}(\xi)\hat{v}(\xi)).$$

This formula, which can also be stated under less strenuous conditions, asserts that the inverse transform of a product is a convolution. As with Laplace transforms, this is a useful relationship in solving differential equations because we frequently end up with a product of transforms in the frequency domain. The convolution theorem follows from formally interchanging the order of integration in the following string of equalities.

$$\mathcal{F}(u * v)(\xi) = \int_{-\infty}^{\infty} \left(\int_{-\infty}^{\infty} u(x-y)v(y)\,dy \right) e^{i\xi x}\,dx$$

$$= \int_{-\infty}^{\infty} \int_{-\infty}^{\infty} u(x-y)v(y)e^{i\xi x}\,dx\,dy$$

$$= \int_{-\infty}^{\infty} \int_{-\infty}^{\infty} u(r)v(y)e^{i\xi r}\,e^{i\xi y}\,dr\,dy$$

$$= \int_{-\infty}^{\infty} u(r)e^{i\xi r}\,dr \int_{-\infty}^{\infty} v(y)e^{i\xi y}\,dy$$

$$= \hat{u}(\xi)\hat{v}(\xi).$$

Example 6.33

Consider the ordinary differential equation

$$u'' - u = f(x), \quad x \in \mathbb{R}.$$

Assume that u and its derivatives, and f, are sufficiently smooth and decay rapidly at infinity. Proceeding formally, we take the transform of both sides to obtain

$$(-i\xi)^2\hat{u} - \hat{u} = \hat{f},$$

Hence

$$\hat{u}(\xi) = -\frac{1}{1+\xi^2}\hat{f}(\xi).$$

In the transform domain the solution is a product of transforms, and so we apply the convolution theorem. From Table 6.2

$$\mathcal{F}\left(\tfrac{1}{2}e^{-|x|}\right) = \frac{1}{1+\xi^2}.$$

Therefore

$$u(x) = -\tfrac{1}{2}e^{-|x|} * f(x) = -\tfrac{1}{2}\int_{-\infty}^{\infty} e^{-|x-y|}f(y)\,dy.$$

We can show, in fact, that this is a solution if f is continuous and integrable.

Example 6.34

Use Fourier transforms to solve the pure initial value problem for the diffusion equation:

$$u_t - ku_{xx} = 0, \quad x \in \mathbb{R}, t > 0; \quad u(x,0) = f(x), x \in \mathbb{R}. \tag{6.47}$$

We assume the functions u and f are in $\mathcal{S}$. Taking Fourier transforms of the equation gives

$$\hat{u}_t = -\xi^2 k\hat{u},$$

which is an ordinary differential equation in t for $\hat{u}(\xi,t)$, with ξ as a parameter. Its solution is

$$\hat{u}(\xi,t) = C(\xi)e^{-\xi^2 kt}.$$

The initial condition gives $\hat{u}(\xi,0) = \hat{f}(\xi)$, and so $C(\xi) = \hat{f}(\xi)$. Therefore

$$\hat{u}(\xi,t) = e^{-\xi^2 kt}\hat{f}(\xi).$$

Replacing a by $1/4kt$ in formula (6.46) gives

$$\mathcal{F}\left(\frac{1}{\sqrt{4\pi kt}}e^{-x^2/4kt}\right) = e^{-\xi^2 kt}.$$

By the convolution theorem we have

$$u(x,t) = \int_{-\infty}^{\infty} \frac{1}{\sqrt{4\pi kt}}e^{-(x-y)^2/4kt}f(y)\,dy. \tag{6.48}$$

This solution was derived under the assumption that u and f are in the Schwartz class of functions. But, now that we have it, we can show that it is a solution under milder restrictions on f. For example, one can prove that (6.48) is a solution to (5.5.7) if f is a continuous, bounded function on $\mathbb{R}$. If f is only piecewise continuous, then u remains a solution for $t > 0$, but at point of discontinuity of f, say x_0, we have $\lim_{t\to 0+} u(x_0,t) = \tfrac{1}{2}(f(x_0^-) + f(x_0^+))$.

Example 6.35

Next we work a problem involving Laplace's equation in the upper half-plane. Consider

$$\begin{aligned} u_{xx} + u_{yy} &= 0, \quad x \in \mathbb{R},\, y > 0, \\ u(x,0) &= f(x), \quad x \in \mathbb{R}. \end{aligned}$$

We append the condition that the solution u stay bounded as $y \to \infty$. This example is similar to the last example. Taking the transform (on x, with y as a parameter) of the equation we obtain

$$\hat{u}_{yy} - \xi^2 \hat{u} = 0,$$

which has general solution

$$\hat{u}(\xi, y) = a(\xi) e^{-\xi y} + b(\xi) e^{\xi y}.$$

The first term will grow exponentially if $\xi < 0$, and the second term will grow if $\xi > 0$. To maintain boundedness for all ξ we can take

$$\hat{u}(\xi, y) = c(\xi) e^{-|\xi| y}.$$

Applying the Fourier transform to the boundary condition yields $a(\xi) = \hat{f}(\xi)$. Therefore the solution in the transform domain is

$$\hat{u}(\xi, y) = e^{-|\xi| y} \hat{f}(\xi).$$

By the convolution theorem we obtain

$$u(x, y) = \frac{y}{\pi} \frac{1}{x^2 + y^2} * f = \frac{y}{\pi} \int_{-\infty}^{\infty} \frac{f(\eta) \, d\eta}{(x - \eta)^2 + y^2}.$$

After the fact, let us motivate how consideration of the Fourier transform arises naturally from examining the initial value problem (6.47) for the heat equation using a separation of variables technique and superposition. If we assume a solution of the form $u(x, t) = T(t) Y(x)$, then substitution gives

$$\frac{T'(t)}{kT(t)} = \frac{Y''(x)}{Y(x)} \equiv -\xi^2,$$

where the separation constant $-\xi^2$ has been chosen to be negative, in anticipation of having a decaying solution. A solution to the time equation is $T = e^{-\xi^2 kt}$, while a solution of the Y equation is $Y = e^{-i\xi x}$. These functions of x represent a continuum of bounded eigenfunctions. Therefore we obtain a one-parameter family of solutions (one for each choice of $\xi \in \mathbb{R}$) of the heat equation of the form

$$u(x, t; \xi) = e^{-i\xi x} e^{-\xi^2 kt}.$$

We superimpose these solutions and define

$$u(x, t) = \int_{-\infty}^{\infty} c(\xi) e^{-i\xi x} e^{-\xi^2 kt} \, d\xi,$$

Table 6.2 Fourier Transforms

$u(x)$	$\widehat{u}(\xi)$
$\delta_a(x)$	$e^{ia\xi}$
e^{-ax^2}	$\sqrt{\frac{\pi}{a}}e^{-\xi^2/4a}$
$H(x)$	$\pi\delta(\xi) - \frac{i}{\xi}$
$\delta(x - x_0)$	$e^{ix_0\xi}$
$e^{-a|x|}$	$\frac{2a}{a^2+\xi^2}$
$H(a - |x|)$	$\frac{2\sin(a\xi}{\xi}$
$\delta(t - a)$	$e^{ia\xi}$
$u^{(n)}(x)$	$(-i\xi)^n\widehat{u}(\xi)$
$u * v$	$F(\xi)\,G(\xi)$

where $c(\xi)$ represents a continuum of coefficients. By superposition, u satisfies the heat equation provided c is nice enough. The initial condition implies

$$u(x,0) = f(x) = \int_{-\infty}^{\infty} c(\xi)e^{-i\xi x}\,d\xi,$$

which, up to a constant multiple $\frac{1}{2\pi}$, is the inverse Fourier transform of $c(\xi)$. So, in this manner, the transform integral arises naturally. This being the case, the Fourier coefficients $c(\xi)$ are given by $c(\xi) = \frac{1}{2\pi}\hat{f}(\xi)$, and

$$u(x,t) = \frac{1}{2\pi}\int_{-\infty}^{\infty}\hat{f}(\xi)e^{-i\xi x}e^{-\xi^2 kt}\,d\xi.$$

This integral can be calculated to obtain the solution (6.48) of the Cauchy problem. We again notice that the solution resembles a continuous Fourier expansion.

As in the case of the Laplace transform, the Fourier transform can be extended to distributions (see Table 6.2).

We end this section with some comments on eigenvalue problems on infinite domains. For example, consider the eigenvalue problem

$$-\Delta u = \lambda u, \quad \mathbf{x} \in \mathbb{R}^n.$$

On a bounded domain there were accompanying boundary conditions (Dirichlet, Neumann, Robin) on u. Now we must decide what kind of boundary conditions to impose at infinity. For example, do we assume u is square-integrable, bounded, or goes to zero as $|\mathbf{x}| \to \infty$? The one-dimensional problem reveals the issues. Consider

$$-u'' = \lambda u, \quad x \in \mathbb{R}.$$

The general solution is: $u(x) = ax + b$ if $\lambda = 0$, $u(x) = ae^{-\sqrt{-\lambda}x} + be^{\sqrt{-\lambda}x}$ if $\lambda < 0$, and $u(x) = ae^{-i\sqrt{\lambda}x} + be^{i\sqrt{\lambda}x}$, if $\lambda > 0$. None of these solutions is square-integrable on $\mathbb{R}$ or goes to zero at $x = \pm\infty$; so in these cases there would be no eigenvalues. However, if we impose a boundedness condition on all of $\mathbb{R}$, then any $\lambda \geq 0$ gives a nontrivial solution $u(x, \lambda)$. In this case we say λ belongs to the **continuous spectrum** of the operator $-\frac{d^2}{dx^2}$, and the associated eigenfunctions are $e^{-i\sqrt{\lambda}x}$, $e^{i\sqrt{\lambda}x}$, $\lambda \geq 0$. These are sometimes given the unfortunate misnomers *improper eigenvalues* and *improper eigenfunctions*. We can write this set of eigenfunctions as e^{-ikx}, where $-\infty < k < \infty$. Then, if f is a given function, we can think of representing it as an as eigenfunction expansion, but this time not as a sum (as for the case of discrete eigenvalues), but as an integral

$$f(x) = \frac{1}{2\pi} \int_{-\infty}^{\infty} c(k)e^{-ikx}dk.$$

But this means that the continuum $c(k)$ of Fourier coefficients is given by the Fourier transform of f. In a distributional sense, the eigenfunctions e^{-ikx} are orthogonal. Therefore, the entire theory of eigenfunctions expansions seems to remain valid for operators with continuous spectrum. Indeed, this is true and there is a close connection to eigenfunction expansions and transform theory.

These ideas have their origin in quantum theory. The eigenvalue problem for the Schrödinger operator

$$-\Delta u + V(\mathbf{x})u = \lambda u, \quad \mathbf{x} \in \mathbb{R}^n,$$

where V is the potential function, can have both eigenvalues (with square-integrable eigenfunctions, called bound states) and continuous spectrum (with bounded eigenfunctions, called unbound states). The eigenfunction expansion then involves both Fourier series (sums) and Fourier integrals. These topics are beyond our scope and we refer the reader to Keener (2000) or Stakgold (1998).

EXERCISES

1. Use a transform method to solve the diffusion problem

$$
\begin{aligned}
u_t &= Du_{xx}, \quad 0 < x < l, \ t > 0, \\
u(0,t) &= u(l,t) = 1, \quad t > 0, \\
u(x,0) &= 1 + \sin\left(\frac{\pi x}{l}\right), \quad 0 < x < l.
\end{aligned}
$$

2. A toxic chemical diffuses into a semi-infinite domain $x \geq 0$ from its boundary $x = 0$, where the concentration is maintained at $g(t)$. The model is

$$u_t = Du_{xx}, \quad x > 0, \ t > 0,$$
$$u(x, 0) = 0, \quad x > 0,$$
$$u(0, t) = g(t), \quad t > 0.$$

Determine the concentration $u = u(x, t)$ and write the solution in the form of $u(x, t) = \int_0^t K(x, t - \tau)g(\tau)d\tau$, identifying the kernel K.

3. The small, transverse deflections from equilibrium of an elastic string under the influence of gravity satisfy the forced wave equation

$$u_{tt} = c^2 u_{xx} - g,$$

where c^2 is a constant and g is the acceleration due to gravity; $u(x, t)$ is the actual deflection at location x at time t. (Wave equations are derived in Chapter 7.) At $t = 0$ a semi-infinite string ($x \geq 0$), supported underneath, lies motionless along the x axis. Suddenly the support is removed and the string falls under the force of gravity with the deflection at $x = 0$ is held fixed at the origin; the initial velocity $u_t(x, 0)$ of the string is zero. Determine the displacement of the string and sketch several time snapshots of the solution.

4. Show that the inverse Fourier transform of $e^{-a|\xi|}$, $a > 0$, is

$$\frac{a}{\pi} \frac{1}{x^2 + a^2}.$$

5. Verify the following properties of the Fourier transform.

   a) $(\mathcal{F}u)(\xi) = 2\pi(\mathcal{F}^{-1}u)(-\xi)$.

   b) $\mathcal{F}(e^{iax}u)(\xi) = \hat{u}(\xi + a)$.

   c) $\mathcal{F}(u(x + a)) = e^{-ia\xi}\hat{u}(\xi)$.

6. Find the Fourier transform of the following functions.

   a) $u(x) = H(x)e^{-ax}$, where H is the heaviside function.

   b) $u(x) = xe^{-ax^2}$.

7. Use Fourier transforms to find the solution to the initial value problem for the advection–diffusion equation

$$u_t - cu_x - u_{xx} = 0, \quad x \in \mathbb{R}, \quad t > 0;$$
$$u(x, 0) = f(x), \quad x \in \mathbb{R}.$$

8. Solve the Cauchy problem for the nonhomogeneous heat equation:

$$u_t = u_{xx} + F(x,t), \quad x \in \mathbb{R}, \ t > 0; \quad u(x,0) = 0, \ x \in \mathbb{R}.$$

9. The Fourier transform and its inverse in $\mathbb{R}^n$ are given by

$$\hat{u}(\xi) \quad = \quad \int_{\mathbb{R}^n} u(\mathbf{x}) e^{i\xi \cdot \mathbf{x}} \, d\mathbf{x}, \quad \mathbf{x}, \xi \in \mathbb{R}^n,$$

$$u(\mathbf{x}) \quad = \quad \frac{1}{(2\pi)^n} \int_{\mathbb{R}^n} \hat{u}(\mathbf{x}) e^{-i\xi \cdot \mathbf{x}} \, d\mathbf{x}.$$

a) Show that the Fourier transform of the Laplacian is $\mathcal{F}(\Delta u) = -|\xi|^2 \hat{u}(\xi)$.

b) Find the Fourier transform of the n-dimensional Gaussian $u(\mathbf{x}) = e^{-a|\mathbf{x}|^2}$.

c) Solve the Cauchy problem for the diffusion equation in $\mathbb{R}^n$.

$$u_t = D\Delta u, \quad \mathbf{x} \in \mathbb{R}^n, \quad u(\mathbf{x},0) = f(\mathbf{x}), \quad \mathbf{x} \in \mathbb{R}^n.$$

10. Prove the **Plancherel relation** $\int_{-\infty}^{\infty} |u(x)|^2 dx = \frac{1}{2\pi} \int_{-\infty}^{\infty} |\hat{u}(\xi)|^2 d\xi$.

11. Use the Plancherel relation to evaluate the integral

$$\int_{-\infty}^{\infty} \frac{dx}{(1+x^2)^2}.$$

12. Find a formula for the solution to the initial value problem

$$u_t = iu_{xx}, \quad x \in \mathbb{R}, \ t > 0; \quad u(x,0) = e^{-x^2}, \ x \in \mathbb{R}.$$

13. Solve the initial boundary value problem

$$u_t = Du_{xx}, \quad x > 0, \ t > 0,$$
$$u(x,0) = f(x), \quad x > 0,$$
$$u(0,t) = 0, \quad t > 0,$$

by extending f to the entire real line as an odd function and then using Fourier transforms.

14. Solve the following initial value problem for the heat equation.

$$u_t = ku_{xx}, \quad x \in \mathbb{R}, \ t > 0,$$
$$u(x,0) = u_0, \ |x| \le a; \quad u(x,0) = 0, \quad |x| > a,$$

where u_0 is a constant. Write your solution in terms of the erf function. For large times t, show that u decays like $\frac{1}{\sqrt{t}}$, or more specifically, $u \sim u_0 a / \sqrt{\pi k t}$.

15. Show that the solution to Laplace's equation in the upper half-plane with a Neumann boundary condition,

$$
\begin{aligned}
u_{xx} + u_{yy} &= 0, \quad x \in \mathbb{R}, \; y > 0, \\
u_y(x, 0) &= f(x), \quad x \in \mathbb{R},
\end{aligned}
$$

is given by

$$
u(x, y) = \frac{1}{2\pi} \int_{-\infty}^{\infty} \ln[(x - \eta)^2 + y^2] f(\eta) d\eta.
$$

(Hint: let $v(x, y) = u_y(x, y)$.)

6.6 Stability of Solutions

6.6.1 Reaction–Diffusion Equations

We already noted the broad occurrence of diffusion problems in biological systems. In this section we investigate another aspect of such problems, namely the persistence of equilibrium states in systems governed by **reaction–diffusion** systems. At issue is the stability of those states: if a system is in an equilibrium state and it is slightly perturbed, does the system return to that state, or does it evolve to a completely different state?

Underpinned by the seminal work of Alan Turing[5] in 1952 on the chemical basis of morphogenesis, it has been shown in extensive research that diffusion induced instabilities can give rise to spatial patterns in all sorts of biological systems. Reaction-diffusion models have been used to explain the evolution of form and structure in developmental biology (morphogenesis), tumor growth, ecological spatial patterns, aggregation in slime molds, patterns on animal coats, and many other observed processes in molecular and cellular biology.

In this section we introduce the basic idea of stability in reaction–diffusion models and we observe that such systems can have instabilities that lead to density variations and patterns.

A similar theory of local stability of equilibrium solutions can be developed for partial differential equations. In the interval $0 < x < L$ consider the reaction–diffusion equation

$$
u_t = Du_{xx} + f(u), \tag{6.49}
$$

with no-flux boundary conditions $u_x(0, t) = u_x(\pi, t) = 0$. Let $u(x, t) = u_e$ be a constant equilibrium solution (so that $f(u_e) = 0$). To fix the idea let

---

[5] A. Turing was a British mathematician who was instrumental in breaking the German codes during World War II.

$a = f'(u_e) > 0$. Note that the equilibrium solution satisfies the partial differential equation and the boundary conditions. Next, let $U(x,t)$ be a small perturbation from the equilibrium solution, or $u(x,t) = u_e + U(x,t)$. The *perturbation equation* is determined by substituting this expression into (6.49) and the boundary conditions:

$$U_t = DU_{xx} + f(u_e + U), \quad U_x(0,t) = U_x(L,t) = 0.$$

To linearize this equation we expand the right side in a Taylor series and discard the nonlinear terms:

$$f(u_e + U) = f(u_e) + f'(u_e)U + \cdots = f'(u_e)U + \cdots.$$

Then we obtain the *linearized perturbation equation*

$$U_t = DU_{xx} + aU, \quad a = f'(u_e) > 0, \tag{6.50}$$

subject to boundary conditions

$$U_x(0,t) = U_x(L,t) = 0. \tag{6.51}$$

This problem, which is on a bounded interval, can be solved by the separation of variables method. We assume $U = g(x)h(t)$ and then substitute into (6.50) and boundary conditions (6.51) to obtain

$$gh' = Dg''h + agh,$$

or

$$\frac{h'}{h} = \frac{Dg''}{g} + a = \lambda.$$

Then $h = Ce^{\lambda t}$ and

$$g'' + \left(\frac{a - \lambda}{D}\right)g = 0, \quad g'(0) = g'(L) = 0.$$

This equation cannot have nontrivial exponential solutions that satisfy the boundary conditions; therefore $a - \lambda \geq 0$. In the case $\lambda = a$ we obtain a constant solution, and in the case $a - \lambda > 0$ we obtain

$$g(x) = A \sin \sqrt{\frac{a - \lambda}{D}} x + B \cos \sqrt{\frac{a - \lambda}{D}} x.$$

Applying $g'(0) = 0$ forces $A = 0$; then $g'(L) = 0$ implies

$$\sin \sqrt{\frac{a - \lambda}{D}} L = 0,$$

or

$$\sqrt{\frac{a - \lambda}{D}} L = n\pi, \quad n = 1, 2, \ldots$$

Therefore, incorporating the case $\lambda = a$, we obtain eigenvalues

$$\lambda = \lambda_n = a - \frac{Dn^2\pi^2}{L^2}, \quad n = 0, 1, 2, \dots \tag{6.52}$$

The modal solutions are therefore

$$U_n(x, t) = e^{\lambda_n t} \cos \frac{n\pi}{L} x, \quad n = 0, 1, 2, \dots \tag{6.53}$$

What conclusions can we draw from this calculation? Because the general solution $U(x, t)$ to the boundary value problem (6.50)–(6.51) is a linear combination of the modal solutions (6.53),

$$U(x, t) = c_0 e^{at} + \sum_{n=1}^{\infty} c_n e^{\lambda_n t} \cos \frac{n\pi}{L} x,$$

it will decay if all of the modes decay, and it will grow if one of the modes grow. The constants c_n are determined by the initial perturbation. The spatial part of the modal solutions, or the cosine, remains bounded. The amplitude factor $e^{\lambda_n t}$, and therefore the eigenvalues λ_n, will determine growth or decay. Let us examine the modes. In the case $n = 0$ the eigenvalue is $\lambda_n = a$ and the modal solution is $c_0 e^{at}$, which grows exponentially. In fact, any mode satisfying $a \geq \frac{Dn^2\pi^2}{L^2}$ is unstable. Therefore, instabilities are likely to occur for modes of low frequency (n small), or in systems of large size L or low diffusion properties D. Oppositely, small systems with large diffusion constants are stabilizing, as are the high frequency modes and geometrically small regions. In the most general case the initial perturbation will nearly always contain all modes and the steady state will be locally unstable.

Above we examined the stability of a constant solution, but the idea applies to nonconstant equilibrium solutions as well.

6.6.2 Pattern Formation

The mechanism causing instabilities in the preceding problem causes patterns to form. We apply this idea to a problem in cell aggregation. A slime mold population is a collection of unicellular amoeboid cells that feed on bacteria in the soil. When the food supply is plentiful, the bacteria are generally distributed uniformly throughout the soil. But, as the food supply becomes depleted and starvation begins, the amoeba start to secrete a chemical (cyclic AMP) that acts as an attractant to the other amoeba and aggregation sites form. The rest of the story is even more interesting as the aggregation sites evolve into slugs that ultimately develop into sporangiophores consisting of a stalk and head containing new spores. The spores are released and the process begins anew. We are interested here in only the first part of this complicated problem, the onset of aggregation. We work in one spatial dimension and imagine the amoeba are distributed throughout a tube of length L.

Let $a = a(x, t)$ and $c = c(x, t)$ denote the density and concentration of the cellular amoeba and cAMP, respectively. The fundamental conservation laws are (see Section 6.2)

$$a_t = -J_x^{(a)}, \quad c_t = -J_x^{(c)} + F,$$

where $J^{(a)}$ and $J^{(c)}$ are the fluxes of the amoeba and the chemical, respectively. There are no source terms in the amoeba equation because we will not consider birth and death processes on the time scale of the analysis. The source term in the chemical equation consists of two parts: production of the chemical by the amoeba and degradation of the chemical in the soil. We assume the chemical is produced by the amoeba at a rate proportional to the density of the amoeba, and the chemical degrades at a rate proportional to its concentration; that is,

$$F = fa - kc.$$

The motion of the chemical is caused by diffusion only, and we assume Fick's law:

$$J^{(c)} = -\delta c_x,$$

where δ is the diffusion constant. The amoeba are also assumed to randomly diffuse via Fick's law, but there is another flux source for the amoeba, namely attraction to the chemical. We assume this attraction is "up the chemical gradient", toward high concentrations of c. Additionally, the rate of diffusion should depend on the amoeba population because that will increase the magnitude of the chemical concentration released. This type of flow, induced by chemical gradients, is called **chemotaxis**. Therefore we assume

$$\begin{aligned} J^{(a)} &= \text{random flux} + \text{chemotatic flux} \\ &= -\mu a_x + \nu a c_x, \end{aligned}$$

where μ is the amoeba motility and ν is the strength of the chemotaxis, both assumed to be positive constants. Note that the random flux, having the form of Fick's law, has a negative sign because flow is "down the gradient" (from high to low densities), and the chemotatic flux has a positive sign because that term induces flow "up the chemical gradient" (from low to high concentrations). Putting all these equations all together gives a system of reaction–diffusion equations,

$$a_t = \mu a_{xx} - \nu(ac_x)_x, \quad c_t = \delta c_{xx} + fa - kc. \tag{6.54}$$

Both a and c satisfy no-flux boundary conditions (i.e., $a_x = c_x = 0$ at $x = 0$, L,), which means there is no escape from the medium.

Notice that there will be an equilibrium solution $a = \bar{a}$, $c = \bar{c}$ to (6.54) provided

$$f\bar{a} = k\bar{c}.$$

That is, the production of the chemical equals its degradation. This equilibrium state represents the spatially uniform state in the soil before aggregation begins. To determine the local stability of this uniform state we let

$$a = \bar{a} + u(x, t), \quad c = \bar{c} + v(x, t),$$

where u and v are small perturbations. Substituting these quantities into (6.54) gives, after simplification, the perturbation equations

$$u_t = \mu u_{xx} - \nu((\bar{a} + u)v_x)_x, \quad v_t = \delta v_{xx} + fu - kv.$$

These equations are nonlinear because of the uv_x term in the amoeba equation. If we discard the nonlinear terms on the assumption that the product of small terms is even smaller, then we obtain the linearized perturbation equations

$$u_t = \mu u_{xx} - \nu\bar{a}v_{xx}, \quad v_t = \delta v_{xx} + fu - kv. \tag{6.55}$$

Easily one can see that the perturbations satisfy Neumann, no-flux boundary conditions.

Motivated by our knowledge of the form of solutions to linear equations, we assume there are modal solutions of the form

$$u(x, t) = c_1 e^{\sigma t} \cos rx, \quad v(x, t) = c_2 e^{\sigma t} \cos rx, \tag{6.56}$$

where r and σ are to be determined, and c_1 and c_2 are some constants. Notice the form of these assumed solutions. The spatial part is bounded and periodic with wave number, or frequency, r, and have wavelength $2\pi/r$. The temporal part is exponential with growth factor σ, which may be a real or complex number. If σ is negative or has negative real part, then the perturbation will decay and the equilibrium state will return (the uniform state is stable); if σ is positive or has positive real part, then the perturbations will grow and the equilibrium will be unstable. Let us ask why solutions of (6.55) should be of this form (6.56) without going through the entire separation of variables argument. Equations (6.55) are linear with constant coefficients, and both equations contain both unknowns u and v and their derivatives. If we are to have a solution, then all the terms must match up in one way or another in order to cancel. So both u and v must essentially have the same form. Because there is a first derivative in t, the time factor must be exponential so it will cancel, and the spatial part must be a sine or a cosine because of the appearance second spatial derivative. We anticipate the cosine function because of the no-flux boundary conditions, as in Chapter 4. If we substitute (6.56) into (6.55) we obtain the two algebraic equations

$$(\sigma + \mu r^2)c_1 - \nu \bar{a} r^2 c_2 = 0, \quad -fc_1 + (\sigma + k + \delta r^2)c_2 = 0,$$

which relate all the parameters. We may regard these as two linear, homogeneous equations for the constants c_1 and c_2. If we want a nontrivial solution for c_1 and c_2, then matrix theory dictates that the determinant of the coefficient matrix must be zero. That is,

$$(\sigma + \mu r^2)(\sigma + k + \delta r^2) - f\nu \bar{a} r^2 = 0,$$

which is an equation relating the temporal growth factor σ, the wave number r, and the other constants in the problem. Expanded out, this equation is a quadratic in σ,

$$\sigma^2 + \gamma_1 \sigma + \gamma_2 = 0,$$

where

$$\gamma_1 = r^2(\mu + \gamma) + k > 0, \quad \gamma_2 = r^2[\mu(\delta r^2 + k) - f\nu \bar{a}].$$

The roots of the quadratic are

$$\sigma = \frac{1}{2}\left(-\gamma_1 \pm \sqrt{\gamma_1^2 - 4\gamma_2}\right).$$

Clearly one of the roots (taking the minus sign) is always negative or has negative real part. The other root can have positive or negative real part, depending upon the value of the discriminant $\gamma_1^2 - 4\gamma_2$. We are interested in determining if there are parameter choices that lead to an instability; so we ask when σ positive. Hence, γ_2 must be negative, or

$$\mu(\delta r^2 + k) < f\nu \bar{a}.$$

If this inequality holds, there is an unstable mode and perturbations will grow. We analyze this further.

The number r is the wave number of the perturbations. Applying the no-flux boundary conditions forces

$$r = \frac{n\pi}{L}, \quad n = 0, 1, 2, 3, \ldots$$

For each value of n we obtain a wave number $r = r_n$ and a corresponding growth factor σ_n. The nth mode will therefore grow and lead to local instability when

$$\mu\left(\delta \frac{n^2\pi^2}{L^2} + k\right) < f\nu \bar{a}. \tag{6.57}$$

We can now ask what factors destabilize the uniform, equilibrium state in the amoeba–cAMP system and therefore promote aggregation. That is, when is (

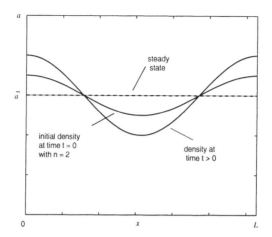

Figure 6.6 Plot showing the growing amoeba density at time t when the uniform state is unstable to local perturbations in the mode $n = 2$. This instability gives rise to the aggregation sites in the regions where the solution has greater amplitude.

6.57) likely to hold? We can list the factors that may make the left side of the inequality (6.57) smaller than the right side: low motility μ of the bacteria; low degradation rate k or large production rate f of cAMP; large chemotactic strength ν; large dimensions L of the medium; a small value of n (thus, low wave number, or long wavelength, perturbations are less stabilizing than short wavelength, or high wave number, perturbations); decreasing the diffusion constant of the cAMP. Figure 6.6 shows time snapshots of the amoeba density for the mode $n = 2$ when it is unstable. The regions where the amplitude is high corresponds to higher concentrations of amoeba, or regions of aggregation.

EXERCISES

1. Consider the problem for Fisher's equation with Dirichlet boundary conditions:

$$u_t = u_{xx} + u(1 - u), \quad -\frac{\pi}{2} < x < \frac{\pi}{2},$$

$$u = 3 \quad \text{at} \quad x = \pm\frac{\pi}{2}.$$

a) Show that $u_e(x) = \frac{3}{1+\cos x}$ is an equilibrium solution.

b) Define perturbations $U(x, t)$ by the equation $u = u_e(x) + U(x, t)$ and find the linearized perturbation equation and boundary conditions for $U(x, t)$.

c) Assume a solution to the linearized equation of the form $U = e^{\sigma t} g(x)$ and show that g must satisfy

$$g'' + \frac{\cos x - 5}{1 + \cos x} g = \sigma g, \quad g = 0 \text{ at } x = \pm \frac{\pi}{2}. \tag{6.58}$$

d) Show that if (6.58) has a nontrivial solution, then $\sigma < 0$, thereby showing local stability of the steady solution. Hint. Consider two cases, when g is positive and when g is negative on the interval and then examine at the signs of g'' and the other terms in (6.58) at a maximum or minimum point.

2. (**Turing system**). Consider the system of reaction-diffusion equations on the spatial domain $0 < x < L$ given by

$$u_t = \alpha u_{xx} + f(u, v), \quad v_t = \beta v_{xx} + g(u, v)$$

with no-flux boundary conditions $u_x = v_x = 0$ at $x = 0, L$. Let $u = \bar{u}$, $v = \bar{v}$ be an equilibrium solution and define small perturbations U and V from equilibrium given by

$$u = \bar{u} + U(x, t), v = \bar{v} + V(x, t).$$

a) Show that U and V satisfy no-flux boundary conditions and the linearized perturbation equations

$$
\begin{aligned}
U_t &= \alpha U_{xx} + f_u(\bar{u}, \bar{v})U + f_v(\bar{u}, \bar{v})V, & (6.59) \\
V_t &= \beta V_{xx} + g_u(\bar{u}, \bar{v})U + g_v(\bar{u}, \bar{v})V. & (6.60)
\end{aligned}
$$

b) Introduce matrix notation

$$\mathbf{w} = \begin{pmatrix} U \\ V \end{pmatrix}, \quad D = \begin{pmatrix} \alpha & 0 \\ 0 & \beta \end{pmatrix}, \quad J = \begin{pmatrix} f_u(\bar{u}, \bar{v}) & f_v(\bar{u}, \bar{v}) \\ g_u(\bar{u}, \bar{v}) & g_v(\bar{u}, \bar{v}) \end{pmatrix},$$

and show that (6.59)–(6.60) can be written as

$$\mathbf{w}_t = D\mathbf{w}_{xx} + J\mathbf{w}. \tag{6.61}$$

c) Assume modal solutions to (6.61) of the form

$$\mathbf{w} = \mathbf{c} e^{\sigma_n t} \cos \frac{n\pi x}{L}, \quad \mathbf{c} = \begin{pmatrix} c_{1n} \\ c_{2n} \end{pmatrix}, \quad n = 0, 1, 2, \ldots,$$

and show that for a nontrivial solution we must have

$$\det \left(\sigma_n I + \frac{n^2 \pi^2}{L^2} D - J \right) = 0. \tag{6.62}$$

(When expanded, this equation is a quadratic equation for the growth factor σ_n of the nth mode. The roots σ_n depend upon the diffusion constants α, β, the equilibrium solution $\overline{u}, \overline{v}$, the size of the medium L, and the wavelength $\pi L/n$ of the perturbation. If one can find values of the parameters that make one of the roots positive or have positive real part, then there is an unstable mode.)

3. Apply the method of Exercise 2 to examine the stability of the steady state of the Turing system

$$u_t = Du_{xx} + 1 - u + u^2 v, \quad v_t = v_{xx} + 2 - u^2 v, \quad 0 < x < \pi,$$

under no-flux boundary conditions. Specifically, find the condition (6.62) and determine values of D for which various modes (n) are unstable.

6.7 Distributions

6.7.1 Elliptic Problems

The notions of Green's functions and distributions carry over in a straightforward manner to several variables and to partial differential equations. We briefly state some of the main ideas. Let Ω be an open set in $\mathbb{R}^n$ (for definiteness the reader may take $n = 2$ or $n = 3$). We denote by $D(\Omega) \equiv C_0^\infty(\Omega)$ the set of functions defined on Ω that have continuous derivatives of all orders and that vanish outside a closed, bounded subset of Ω. The set $D(\Omega)$ is called the set of **test functions** on Ω. Then, a **distribution** u on $D(\Omega)$ is a continuous linear functional on $D(\Omega)$, and we denote the value, or action, of the distribution u by (u, ϕ), where $\phi \in D(\Omega)$. The set of all distributions defined on $D(\Omega)$ is denoted by $D'(\Omega)$. Every function u that is **locally integrable** on Ω, namely $\int_K |u(x)| \, dx < \infty$ for all closed bounded subsets K of Ω, generates a distribution via the formula

$$(u, \phi) = \int_\Omega u(x)\phi(x) \, dx, \quad \phi \in D(\Omega). \tag{6.63}$$

The **Dirac (delta) distribution** with pole at $\xi \in R^n$ is defined by

$$(\delta_\xi, \phi) = \phi(\xi), \quad \phi \in D(\Omega). \tag{6.64}$$

As in the one-dimensional case we often denote a locally integrable function and its associated distribution by the same symbol. Even if a distribution u is not generated by a locally integrable function as in (6.63), we often write it as

$u(x)$ even though it is not a pointwise function of x; and we write its action on ϕ symbolically in the integral form (6.63), even though the integral is not defined. Thus, for example, we write $\delta_\xi(x)$ or $\delta(x-\xi)$ for the delta distribution, and we define its action (6.64) symbolically as

$$\int_\Omega \delta(x - \xi)\phi(x)\, dx = \phi(\xi).$$

The multiplication of a distribution $u \in D'(\Omega)$ by a $C^\infty(\Omega)$ function a is defined by $(au, \phi) = (u, a\phi), \phi \in D(\Omega)$; moreover, the partial derivative of a distribution u is defined by

$$\left(\frac{\partial}{\partial x_k}u, \phi\right) = -\left(u, \frac{\partial}{\partial x_k}\phi\right), \quad \phi \in D(\Omega).$$

Second-order derivatives are defined by

$$\left(\frac{\partial^2}{\partial x_j \partial x_k}u, \phi\right) = \left(u, \frac{\partial^2}{\partial x_j x_k}\phi\right), \quad \phi \in D(\Omega).$$

All of these definitions for general distributions are motivated, as in the one-dimensional case, by the action of locally integrable functions. Higher-order distributional derivatives can be defined as well. Thus it makes sense to regard a linear, second-order partial differential operator L with C^∞-coefficients as a distribution in $D'(\Omega)$. If f is a distribution, then the differential equation

$$Lu = f \qquad (6.65)$$

can be regarded as an equation in distributions, and any distribution u that satisfies (6.65) is called a **distribution solution**. Thus, u is a distribution solution to (6.65) if, and only if,

$$(Lu, \phi) = (f, \phi), \quad \forall \phi \in D(\Omega). \qquad (6.66)$$

In particular, if $f \equiv \delta(x - \xi)$ is the delta distribution, then u is called a **fundamental solution** associated with the operator L. The formal adjoint operator associated with L is denoted by L^* and is defined by the relation

$$(Lu, \phi) = (u, L^*\phi), \quad \phi \in D(\Omega). \qquad (6.67)$$

Therefore, u is a distribution solution of (6.65) if, and only if,

$$(u, L^*\phi) = (f, \phi), \quad \forall \phi \in D(\Omega), \qquad (6.68)$$

and u is a fundamental solution (with pole at ξ) associated with the operator L if, and only if,

$$(u, L^*\phi) = \phi(\xi), \quad \forall \phi \in D(\Omega).$$

If u and f are locally integrable functions on Ω that satisfy (6.67), then we say u is a **weak solution** to (6.65). Thus, u is a weak solution to (6.65) if, and only if,

$$\int_\Omega u(x) L^* \phi(x) \, dx = \int_\Omega f(x) \phi(x) \, dx, \quad \forall \phi \in D(\Omega).$$

A solution to (6.65) that is twice continuously differentiable (we are restricting the discussion to second-order PDEs) on Ω is called a classical, or **genuine**, solution.

So, there are three levels of solutions: classical, weak, and distribution. Classical solutions are functions that have sufficiently many derivatives and satisfy the PDE (6.65) pointwise on the domain of interest. Weak solutions of (6.65), where f is locally integrable, are locally integrable functions that satisfy (6.67). Distribution solutions of (6.65) may not be functions or even generated by functions. Classical solutions are weak solutions, and weak solutions are distribution solutions, but the converses are not true.

Example 6.36

Show that the function

$$g(x, y) = \frac{1}{4\pi} \ln(x^2 + y^2) \tag{6.69}$$

is a distributional solution of the equation

$$\Delta u = \delta(x, y),$$

where $\Delta \equiv \partial^2/\partial x^2 + \partial^2/\partial y^2$ is the two-dimensional Laplacian, and $\delta(x, y)$ is the Dirac distribution with pole at $(0, 0)$. Thus we have to show that $(\Delta g, \phi) \equiv (g, \Delta \phi) = \phi(0, 0)$ for all $\phi \in D(\mathbb{R}^2)$. To this end, using polar coordinates, we have

$$(4\pi g, \Delta \phi) = \int_{R^2} \ln(x^2 + y^2)(\phi_{xx} + \phi_{yy}) \, dx \, dy$$

$$= \int_0^{2\pi} \int_0^\infty \ln r^2 \left(\phi_{rr} + \frac{1}{r} \phi_r + \frac{1}{r^2} \phi_{\theta\theta} \right) r \, dr \, d\theta$$

$$= \lim_{\epsilon \to 0} \int_\epsilon^\infty \int_0^{2\pi} \left(r \ln r^2 \phi_{rr} + \ln r^2 \phi_r + \frac{1}{r} \ln r^2 \phi_{\theta\theta} \right) d\theta \, dr.$$

The integral of the last term is zero by the θ periodicity of ϕ and its θ derivatives. So we need to calculate the other two terms. Integrating by parts gives

$$\int_\epsilon^\infty \ln r^2 \phi_r \, dr = \phi \ln r^2 |_\epsilon^\infty - \int_\epsilon^\infty \frac{2}{r} \phi \, dr$$

$$= -\phi(\epsilon, \theta) \ln \epsilon^2 - \int_\epsilon^\infty \frac{2}{r} \phi \, dr$$

and integrating by parts twice gives

$$\int_\epsilon^\infty r \ln r^2 \phi_{rr}\, dr = r\phi_r \ln r^2 - \int_\epsilon^\infty (2 + \ln r^2)\phi_r\, dr$$

$$= -\phi_r(\epsilon, \theta)\epsilon \ln \epsilon^2 + (2 + \ln \epsilon^2)\phi(\epsilon, \theta) + \int_\epsilon^\infty \frac{2}{r}\phi\, dr.$$

Putting these results together gives

$$(4\pi g, \Delta\phi) = \lim_{\epsilon\to 0} \int_0^{2\pi} \left(-\phi(\epsilon, \theta)\ln\epsilon^2 - \int_\epsilon^\infty \frac{2}{r}\phi\, dr\right) d\theta$$

$$+ \lim_{\epsilon\to 0} \int_0^{2\pi} \left((2 + \ln\epsilon^2)\phi(\epsilon, \theta) + \int_\epsilon^\infty \frac{2}{r}\phi\, dr\right) d\theta$$

$$= 2\lim_{\epsilon\to 0} \int_0^{2\pi} \phi(\epsilon, \theta)\, d\theta = 4\pi\phi(0, 0).$$

Therefore, the function defined by (6.69), which is called the **logarithmic potential**, is a fundamental solution associated with the two-dimensional Laplacian. Note that it is not correct to say (6.69) is a genuine solution or a weak solution to Laplace's equation $\Delta u = 0$ in $\mathbb{R}^2$. Observe that the Laplacian operator is invariant under a translation of coordinates, and so a fundamental solution associated with the Laplacian with pole at (ξ, η) is

$$g(x, y; \xi, \eta) = \frac{1}{4\pi} \ln((x - \xi)^2 + (y - \eta)^2). \tag{6.70}$$

Example 6.37

A fundamental solution associated with the Laplacian in $\mathbb{R}^3$ is the function $g(x, y, z) \equiv 1/4\pi r$, where $r = \sqrt{x^2 + y^2 + z^2}$. This means $\Delta g = \delta(x, y, z)$ in the sense of distributions. Again, the distribution g can be translated to any point (ξ, η, ζ) to obtain a fundamental solution with pole (ξ, η, ζ). In three dimensions g is called the **Newtonian potential**. We remark that the fundamental solutions associated with the Laplacian are radial solutions in that they depend only on the distance from the pole.

The Green's function associated with a partial differential operator, such as the Laplacian, is a fundamental solution that also satisfies the homogeneous boundary conditions. As such, the Green's function is the equilibrium response of the physical system under investigation caused by a unit point source. Mathematically, the Green's function is the kernel of the integral operator that represents the inverse of the partial differential operator. There is no single, universal method to construct Green's function. On finite domains one may use separation of variables, and on infinite domains one may try transform methods. Often geometric or physical arguments are helpful.

Example 6.38

Find the Green's function associated with the Laplacian operator on the two-dimensional domain $-\infty < x < \infty, y > 0$ with a Dirichlet boundary condition on $y = 0$. This means that we seek a distributional solution $G = G(x, y; \xi, \eta)$ to the problem

$$G_{xx} + G_{yy} = \delta(x, y; \xi, \eta), \qquad x, \xi \in \mathbb{R}; \ y, \eta > 0, \qquad (6.71)$$

$$G(x, 0; \xi, \eta) = 0, \quad x, \xi \in \mathbb{R}; \ \eta > 0. \qquad (6.72)$$

Physically, G could represent the potential in the half-plane $y > 0$ of an static electric field generated by a unit point positive charge at (ξ, η), with the condition that the potential vanish along $y = 0$. We shall construct the Green's function G by the so-called **method of images**, or reflection principle. We know that the fundamental solution g from equation (6.70) in Example 6.36 satisfies the equation $\Delta g = \delta(x, y; \xi, \eta)$ for a positive charge at (ξ, η), but it does not satisfy the boundary condition. To compensate we locate an "image" charge of opposite sign at $(\xi, -\eta)$, the reflection of the point (ξ, η) through the y axis. The potential due to that charge would be $\tilde{g} \equiv -g(x, y; \xi, -\eta)$. Now take $G \equiv g + \tilde{g}$, or

$$G(x, y, \xi, \eta) = \frac{1}{4\pi} \ln \frac{(x - \xi)^2 + (y - \eta)^2}{(x - \xi)^2 + (y + \eta)^2} \qquad (6.73)$$

Clearly G satisfies the homogeneous boundary condition (6.72). Moreover, $\Delta G = \Delta g + \Delta \tilde{g} = \delta(x, y, \xi, \eta) + 0 = \delta(x, y, \xi, \eta)$ in the upper half-plane $y > 0$. Here we used the fact that $\Delta \tilde{g} = 0$ in the *upper* half-plane; its pole is in the lower half-plane, and everywhere but at its pole it satisfies Laplace's equation.

Knowing Green's function allows us to solve the nonhomogeneous problem

$$u_{xx} + u_{yy} = \rho(x, y), \quad x \in \mathbb{R}, \ y > 0, \qquad (6.74)$$

$$u(x, 0) = 0, \quad x \in \mathbb{R}. \qquad (6.75)$$

The Green's function G given by (6.73) is the response of the system for a unit point source (say a unit charge) at (ξ, η). Therefore, to find the response to a problem with a continuum of sources (a charge density) of magnitude ρ we can superimpose all the point source responses over all ξ and η to obtain

$$u(x, y) = \int_0^\infty \int_{-\infty}^\infty G(x, y; \xi, \eta)\rho(\xi, \eta) \, dx \, dy. \qquad (6.76)$$

One can show that (6.76) is the solution to (6.74)–(6.75) under reasonable conditions on ρ (see, e.g., McOwen (2003)).

6.7.2 Tempered Distributions

In problems involving point sources we are faced with a Dirac distribution in the partial differential equation. If we want transform methods to apply, then we must make sense of the transform of a distribution. A formal calculation gives the correct answer. For example, consider taking the Fourier transform of the delta distribution $\delta(x - x_0)$ with pole at x_0; we have

$$(\mathcal{F}(\delta(x - x_0))(\xi) = \int_{-\infty}^{\infty} \delta(x - x_0)e^{i\xi x}\, dx = (\delta(x - x_0), e^{i\xi x}) = e^{i\xi x_0}.$$

Thus

$$\widehat{\delta(x - x_0)}(\xi) = e^{i\xi x_0}.$$

Many readers will be satisfied with this formal calculation. However, in order to take some of the mystery out of it we give a brief explanation of its underpinnings. Suppose u is a distribution in $D'(\mathbb{R})$ that is generated by a locally integrable function u, and assume that the Fourier transform $\hat{u}$ is also locally integrable (here we are using the convention of denoting a function and its associated distribution by the same symbol). Then, for any $\phi \in D(\mathbb{R})$ we have

$$(\hat{u}, \phi) = \int_{-\infty}^{\infty} \left(\int_{-\infty}^{\infty} u(x)e^{i\xi x}\, dx \right) \phi(\xi)\, d\xi$$

$$= \int_{-\infty}^{\infty} \left(\int_{-\infty}^{\infty} \phi(\xi)e^{i\xi x}\, d\xi \right) u(x)\, dx = (u, \hat{\phi})$$

This seems correct, but it would require that both ϕ and its Fourier transform $\hat{\phi}$ to be test functions in $D(\mathbb{R})$. If both are test functions, then one can prove that ϕ must be identically zero. Therefore we cannot define $\hat{u}$ on the set of test functions $D(\mathbb{R})$. The way out of this problem is to consider distributions on a larger class of test functions, namely the Schwartz class $\mathcal{S}$. A **tempered distribution** is a continuous linear functional on $\mathcal{S}$. Now it is possible to define the Fourier transform of a tempered distribution u to be the tempered distributor, $\hat{u}$ whose action on $\mathcal{S}$ is given by

$$(\hat{u}, \phi) = (u, \hat{\phi}), \quad \phi \in \mathcal{S}.$$

Thus, for example, the Fourier transform of the Dirac distribution is defined by

$$(\widehat{\delta(x - x_0)}, \phi) = (\delta(x - x_0), \hat{\phi}), \quad \phi \in \mathcal{S}.$$

Therefore,

$$(\widehat{\delta(x - x_0)}, \phi) = \hat{\phi}(x_0) = \int_{-\infty}^{\infty} \phi(\xi) e^{i x_0 \xi} \, d = (e^{i x_0 \xi}, \phi)$$

for all $\phi \in \mathcal{S}$. We conclude that, in the sense of tempered distributions,

$$\widehat{\delta(x - x_0)} = e^{i x_0 \xi}.$$

In particular, if $x_0 = 0$, then we have

$$\widehat{\delta(x)} = 1.$$

We will not discuss this topic further. Our purpose was only to show that it is possible to give mathematical meaning to the idea of taking a Fourier transform of a distribution. The reader interested in more detail should consult the references (see Strichartz (1994) for a particularly readable treatment).

6.7.3 Diffusion Problems

Now let us consider problems involving time. To fix the idea we examine the heat operator in one spatial dimension and time defined by

$$Lu \equiv u_t - k u_{xx}.$$

Based on our discussion of elliptic problems, we say that u is a fundamental solution associated with L if u is a distributional solution of

$$Lu = \delta(x, t; \xi, \tau), \tag{6.77}$$

where the right side is interpreted as a unit heat source applied at $x = \xi$ and at the instant of time $t = \tau$. Here, the domain Ω is a space–time domain consisting of coordinates $(x, t) \in \mathbb{R}^2$. To find the fundamental solution we take the source to be at $\xi = 0$ and $\tau = 0$. We can then examine the initial value problem

$$u_t - k u_{xx} = 0, \quad t > 0, \tag{6.78}$$

$$u(x, 0) = \delta(x). \tag{6.79}$$

(If we want to think about this carefully, we can regard (6.78) as a one-parameter family of equations for $u(\cdot, t)$ in $D'(\mathbb{R})$ for each value of the parameter $t > 0$, and (6.79) as an equation in $D'(\mathbb{R})$). Now let us solve (6.78)–(6.79) using Fourier transforms, assuming that the distributions are tempered. Taking transforms of (6.78) gives (we use ζ for the transform variable)

$$\hat{u}_t(\zeta, t) + k \zeta^2 \hat{u}(\zeta, t) = 0.$$

Solving this ordinary differential equation for $\hat{u}$ yields

$$\hat{u}(\zeta, t) = c(\zeta)e^{-k\zeta^2 t},$$

where $c(\zeta)$ is arbitrary. The Fourier transform of the initial condition (6.79) gives $\hat{u}(\zeta, 0) = 1$, and therefore $c(\zeta) = 1$. Consequently

$$\hat{u}(\zeta, t) = e^{-k\zeta^2 t}$$

So the Fourier transform of the solution is a Gaussian function. We know that the inverse transform is also a Gaussian and is in fact given by

$$u(x, t) = \frac{1}{\sqrt{4\pi kt}}e^{-x^2/4kt}.$$

This is the solution to the initial value problem (6.78)–(6.79). Finally, translating the source to $x = \xi, t = \tau$ and multiplying by a Heaviside function to turn on the solution at $t = \tau$, we obtain

$$K(x - \xi, t - \tau) \equiv \frac{H(t - \tau)}{\sqrt{4\pi k(t - \tau)}}e^{-(x-\xi)^2/4k(t-\tau)}, \qquad (6.80)$$

which is the fundamental solution associated with the heat operator. One can verify that (6.80) is a distributional solution to (6.77).

To understand the fundamental solution of the heat equation, let us fix $\tau = 0$ and consider

$$K(x - \xi, t) \equiv \frac{1}{\sqrt{4\pi kt}}e^{-(x-\xi)^2/4kt}, \qquad t > 0. \qquad (6.81)$$

We observe that

$$\lim_{t \to 0^+} K(x - \xi, t) = 0, \quad x \neq \xi,$$

$$\lim_{t \to \infty} K(x - \xi, t) = 0, \quad x \in \mathbb{R},$$

$$\lim_{t \to 0^+} K(x - \xi, t) = +\infty, \quad x = \xi.$$

Therefore, for large times the fundamental solution goes to zero for all x. As $t \to 0^+$, however, the solution develops an infinite spike at $x = \xi$ while tending to zero for $x \neq \xi$. So, physically, we interpret the fundamental solution $K(x - \xi, t)$ as the temperature distribution in an infinite bar initially at zero degrees with an instantaneous unit heat source applied at $x = \xi$ and $t = 0$. It is a unit heat source because, as one can check,

$$\int_{-\infty}^{\infty} K(x - \xi, t)\, dx = 1, \quad t > 0.$$

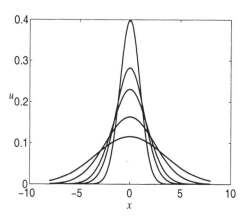

Figure 6.7 Temperature profiles for different times t caused by a unit energy source at $x = \xi = 0$ and $t = 0$.

That is, the area under the temperature profile curve, which is proportional to the energy, is unity. Stated differently, the function $K(x - \xi, t)$ is the temperature effect at (x, t) caused by a unit heat source applied initially at $x = \xi$. Figure 6.7 shows time snapshots of the function K for different values of t. The fundamental solution (6.81) associated with the heat operator is useful in constructing solutions to a variety of initial boundary value problems involving the heat equation.

Example 6.39

Consider the pure initial value problem for the heat equation

$$u_t - ku_{xx} = 0, \quad x \in \mathbb{R}, \quad t > 0; \quad u(x, 0) = f(x), \quad x \in \mathbb{R}. \tag{6.82}$$

By our previous discussion $K(x - \xi, t)$ is the heat response of the system when a unit point source is applied at $x = \xi$ and $\tau = 0$. Thus, to solve (6.82) we superimpose these solutions for point sources of magnitude $f(\xi)$ for all $\xi \in \mathbb{R}$. Therefore the solution to (6.82) is

$$u(x, t) = \int_{-\infty}^{\infty} K(x - \xi, t) f(\xi) \, d\xi$$

$$= \int_{-\infty}^{\infty} \frac{1}{\sqrt{4\pi kt}} e^{-(x-\xi)^2/4kt} f(\xi) \, d\xi.$$

Example 6.40

Consider the nonhomogeneous problem

$$u_t - ku_{xx} = F(x,t), \quad x \in \mathbb{R}, \ t > 0; \quad u(x,0) = 0, \quad x \in \mathbb{R}. \tag{6.83}$$

The solution for a point source at (ξ, τ) is $K(x - \xi, t - \tau)$. To find the solution for distributed sources $F(x,t)$ we superimpose to obtain

$$u(x,t) = \int_{-\infty}^{\infty} \int_{-\infty}^{\infty} \frac{H(t-\tau)}{\sqrt{4\pi k(t-\tau)}} e^{-(x-\xi)^2/4k(t-\tau)} F(\xi, \tau) \, d\xi \, d\tau$$

$$= \int_0^t \int_{-\infty}^{\infty} \frac{1}{\sqrt{4\pi k(t-\tau)}} e^{-(x-\xi)^2/4k(t-\tau)} F(\xi, \tau) \, d\xi \, d\tau,$$

which is the solution to (6.83).

EXERCISES

1. Is the function e^{x^2} locally integrable on $\mathbb{R}$? Does it generate a distribution in $D'(\mathbb{R})$? Does it generate a tempered distribution?

2. Show that for any locally integrable function f on $\mathbb{R}$ the function $u(x,y) = f(x-y)$ is a weak solution to the equation $u_x + u_y = 0$ on $\mathbb{R}^2$.

3. In the quarter plane in $\mathbb{R}^2$ find the Green's function associated with the boundary value problem

$$\Delta u = \delta(x,y;\xi,\eta), \quad x,y,\xi,\eta > 0,$$
$$u(x,0) = 0, \ x > 0; \quad u(0,y) = 0, \ y > 0.$$

(*Hint:* Put image charges in the other quadrants.)

4. In the upper half plane in $\mathbb{R}^2$, use an image charge find Green's function for the Neumann problem

$$\Delta u = \delta(x,y;\xi,\eta), \quad x,\xi \in \mathbb{R}, \ y,\eta > 0,$$
$$u_y(x,0) = 0, \quad x \in \mathbb{R}.$$

5. At $t = 0$ at a point in a homogeneous medium an amount of heat energy E is released and allowed to diffuse outward. Deduce that the temperature

$u(r,t)$ at time t and distance r from the release point satisfies the initial boundary value problem

$$u_t - \frac{k}{r^2}(r^2 u_r)_r = 0, \quad r > 0, \quad t > 0,$$

$$u(r,0) = 0, \quad r > 0,$$

$$\lim_{r \to \infty} u(r,t) = 0, \quad t > 0,$$

$$4\pi c \int_0^\infty r^2(r,t)\, dr = E, \quad t > 0,$$

where k is the diffusivity of the medium and c is its heat capacity. Use dimensional analysis and the Pi theorem to show that

$$u = \frac{E}{c}(kt)^{-3/2} U\left(\frac{r}{\sqrt{kt}}\right)$$

for some function U. Determine U to conclude that

$$u(r,t) = \frac{E/c}{(4\pi kt)^{3/2}} e^{-r^2/4kt}.$$

6. In $\mathbb{R}^2$ consider the operator

$$L = \frac{\partial^2}{\partial x^2} - \frac{\partial^2}{\partial x \partial y} - \frac{\partial}{\partial y}.$$

Find the adjoint operator L^*.

REFERENCES AND NOTES

There is a large number of excellent elementary textbooks introducing partial differential equations, to numerous to mention. The list below is a small selection that this author often finds useful. Strauss (1991), Farlow (1993), and Logan (2004) give entry-level introductions. Strauss' text, as well as Zauderer (1989), have thorough treatments of eigenvalue problems. At the next level, Evans (1998) and McOwen (2003) give excellent introductions to the theory. Texts that contain biological applications are Britton (2003), Edelstein-Keshet (1988), Grindrod (1991), and Kot (2001). Nonlinear equations are discussed in Logan (1994), which is an elementary treatment, and in the classic treatise by Whitham (1974), which emphasizes hyperbolic problems. Excellent, elementary treatments of Fourier transforms and distributions can be found in Strichartz (1994). Both Keener (2000) and Stakgold (1998) are advanced texts on applied mathematics that contain theoretical treatments of Green's functions and eigenvalue problems.

Britton, N. 2003. *Essential Mathematical Biology*, Springer-Verlag, New York.

Edelstein-Keshet, L. 1988. *Mathematical Models in Biology*, reprinted by Soc. Industr. Appl. Math., Philadelphia.

Evans, L. 1998. *Partial Differential Equations*, Amer. Math. Soc., Providence.

Farlow, S. 1993. *Partial Differential Equations for Scientists and Engineers*, Dover Publications, New York.

Grindrod, P. 1991. *Patterns and Waves*, Oxford University Press, Oxford.

Keener, J. 2000. *Principles of Applied Mathematics* (revised edition), Westview Press (Perseus Book Group), Cambridge, MA.

Kot, M. 2001. *Elements of Mathematical Ecology*, Cambridge University Press, Cambridge.

Logan, J. D. 1994. *An Introduction to Nonlinear Partial Differential Equations*, Wiley-Interscience, New York.

Logan, J. D. 2004. *Applied Partial Differential Equations*, 2nd ed., Springer-Verlag, New York.

McOwen, R. C. 2003. *Partial Differential Equations*, 2nd ed., Pearson Education, Upper Saddle River, NJ.

Stakgold, I. 1998. *Green's Functions and Boundary Value Problems*, 2nd ed., Wiley-Interscience, New York.

Strauss, W. 1991. *Partial Differential Equations*, John Wiley and Sons, New York.

Strichartz, R. 1994. *A Guide to Distribution Theory and Fourier Transforms*, CRC Press, Boca Raton, FL.

Whitham, G. B. 1974. *Linear and Nonlinear Waves*, Wiley-Interscience, New York.

Zauderer, E. 1989. *Partial Differential Equations of Applied Mathematics*, 2nd ed., Wiley-Interscience, New York.

7

Wave Phenomena

Two of the fundamental processes in nature are diffusion and wave propagation. In the last chapter we studied the equation of heat conduction, a parabolic partial differential equation that is the prototype of equations governing linear diffusion processes. In the present chapter we investigate wave phenomena and obtain equations that govern the propagation of waves first in simple model settings. The evolution equations governing such phenomena are hyperbolic and are fundamentally different from their parabolic counterparts in diffusion and from elliptic equations that govern equilibrium states.

7.1 Wave Propagation

7.1.1 Waves

A *wave* is defined as an identifiable signal or disturbance in a medium that is propagated in time, carrying energy with it. A few familiar examples are electromagnetic waves, waves on the surface of water, sound waves, and stress waves in solids, as occur in earthquakes. Material or matter is not necessarily convected with the wave; it is the disturbance, that carries energy, that is propagated. In this section we investigate several model equations that occur in wave phenomena and point out basic features that are encountered in the study of the propagation of waves.

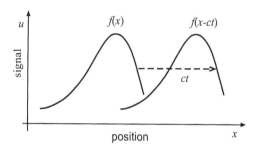

Figure 7.1 Right traveling wave.

One simple mathematical model of a wave is the function

$$u(x, t) = f(x - ct), \tag{7.1}$$

which represents an undistorted **right-traveling wave** moving at constant velocity c. The coordinate x represents position, t time, and u the strength of the disturbance. At $t = 0$ the wave profile is $u = f(x)$ and at $t > 0$ units of time the disturbance has moved to the right ct units of length (see Fig. 7.1). Of particular importance is that the wave profile described by (7.1) moves without distortion. Not all waves have this property; it is characteristic of linear waves, or wave profiles that are solutions to linear partial differential equations. On the other hand, waves that distort, and possibly break, are characteristic of nonlinear processes. To find a model that has (7.1) as the solution we compute u_t and u_x to get

$$u_t = -cf'(x - ct), \qquad u_x = f'(x - ct),$$

Hence

$$u_t + cu_x = 0. \tag{7.2}$$

Equation (7.2) is a first-order linear partial differential equation that, in the sense just described, is the simplest wave equation. It is called the **advection equation** and its general solution is (7.1), where f is an arbitrary differentiable function. As we noted in Section 6.2, the advection equation describes how a quantity is carried along with the bulk motion of a medium. Similarly, a traveling wave of the form $u = f(x + ct)$ is a left moving wave and is a solution of the partial differential equation $u_t - cu_x = 0$.

Other waves of interest in many physical problems are periodic, or sinusoidal waves. These traveling waves are represented by expressions of the form

$$u(x, t) = A \cos(kx - \omega t) \tag{7.3}$$

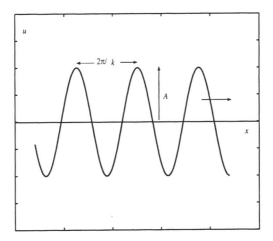

Figure 7.2 Plane wave.

(see Fig. 7.2). The positive number A is the **amplitude**, k is the **wave number** (the number of oscillations in 2π units of space, observed at a fixed time), and ω the **angular frequency** (the number of oscillations in 2π units of time, observed at a fixed location x). The number $\lambda = 2\pi/k$ is the **wavelength** and $P = 2\pi/\omega$ is the *time period*. The wavelength measures the distance between successive crests and the time period is the smallest time for an observer located at a fixed position x to see a repeat pattern. If we write (7.3) as

$$u = A \cos k \left(x - \frac{\omega}{k} t \right),$$

then we note that (7.3) represents a traveling wave moving to the right with velocity $c = \omega/k$. This number is called the **phase velocity**, and it is the speed one would have to move to remain at the same point on the traveling wave. For calculations, the complex exponential form

$$u(x, t) = e^{i(kx - \omega t)} \tag{7.4}$$

is often used rather than (7.3). These waves are called **plane waves**. Computations involving differentiations are easier with (7.4) and afterward, making use of Euler's formula $\exp(i\theta) = \cos\theta + i\sin\theta$, the real or imaginary part may be taken to recover real solutions. Again, waves of the type (7.3) [or (7.4)] are characteristic of linear processes and linear equations.

Not every wave propagates so that its profile remains unchanged or undistorted; a surface wave on the ocean is an obvious example. Less familiar perhaps, but just as common, are strong stress or pressure waves that propagate

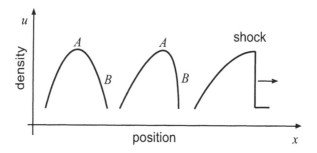

Figure 7.3 Distortion of a wave into a shock.

in solids or gases. To fix the idea and to indicate how nonlinearity affects the
shape of a wave, let us consider a strong pressure wave propagating in air,
caused, for example, by an explosion. (This is not the case of acoustics, where
sound waves propagate at relative small pressure changes.) The distortion of
a wave profile results from the property of most materials to transmit signals
at a speed that increases with increasing pressure. Therefore a strong pressure
wave that is propagating in a medium will gradually distort and steepen until
it propagates as an idealized discontinuous disturbance, or shock wave. This
is what also occurs in sonic booms. Figure 7.3 shows various snapshots of a
stress wave propagating into a material. The wave steepens as time increases
because signals or disturbances travel faster when the pressure is higher. Thus
the point A moves to the right faster than the point B. The shock that forms
in pressure is accompanied by discontinuous jumps in the other parameters,
such as density, particle velocity, temperature, energy, and entropy.

   Physically, a shock wave is not a strict discontinuity but rather an extremely
thin region where the change in the state is steep. The width of the shock is
small, usually of the order of a few mean free paths, or average distance to
a collision, of the molecules. In a shock there are two competing effects that
cause this thinness, the nonlinearity of the material that is causing the shock
to form, and the dissipative effects (e.g., viscosity) that are tending to smear
the wave out. Usually these two effects just cancel and the front assumes a
shape that does not change in time. The same mechanism that causes pressure
waves to steepen into shocks, that is, an increase in signal transmission speed
at higher pressures, causes release waves or **rarefaction waves** to form that
lower the pressure. Figure 7.4 shows a spreading rarefaction. Again, point A
at higher pressure moves to the right faster than point B, thereby flattening
the wave. So far we have mentioned two types of waves, those that propagate
undistorted at constant velocity and those that distort because the speed of
propagation depends on the amplitude of the wave. There is a third interesting

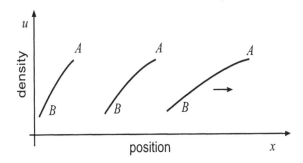

Figure 7.4 A rarefaction wave.

phenomenon that is also relevant, namely that of **dispersion**. In this case the speed of propagation depends on the wavelength of the particular wave. So, for example, longer waves can travel faster than shorter ones. Thus an observer of a wave at fixed location x_0 may see a different temporal wave pattern from another observer at fixed location x_1. Dispersive wave propagation arises from both linear and nonlinear models.

The behavior of linear partial differential equations with constant coefficients can be characterized by the plane wave solutions they admit. When we substitute a plane wave form (7.4) into such an equation, we obtain a relation between the frequency and the wave number,

$$\omega = \omega(k).$$

This equation is called the **dispersion relation**. We say the equation is **diffusive** if $\omega(k)$ is complex, and **dispersive** if $\omega(k)$ is real and $\omega(k)/k$ is independent of k; if $\omega(k)$ is real and $\omega(k)/k$ depends on k, then we say the equation is **hyperbolic**.

Example 7.1

The reader should check that the advection–diffusion equation $u_t + \gamma u_x = D u_{xx}$ has complex dispersion relation $\omega = \gamma k - iDk^2$. Therefore the advection–diffusion equation is diffusive. Plane wave solutions are given by

$$u(x,t) = e^{-Dk^2 t} e^{i(kx - \gamma t)}.$$

The factor $\exp(ik(x - \gamma t))$ represents a periodic, right-traveling wave with wave number k, and the factor $\exp(-Dk^2 t)$ represents a decaying amplitude. Qualitatively two conclusions can be drawn:

(i) For waves of constant wavelength (k constant) the attenuation of the wave increases with increasing D; hence D is a measurement of diffusion.

(ii) For constant γ the attenuation increases as k increases; hence smaller wavelengths attenuate faster than longer wavelengths.

From this example the reader should note that the diffusion equation ($\gamma = 0$) has plane wave solutions $u(x,t) = e^{-Dk^2t}e^{ikx}$, which are not traveling waves, but stationary waves that decay.

Example 7.2

Consider the partial differential equation

$$u_{tt} + \gamma u_{xxxx} = 0, \qquad \gamma > 0,$$

which arises in studying the transverse vibrations of a beam. Substituting (7.4) into this equation gives the dispersion relation

$$\omega = \pm\sqrt{\gamma}k^2,$$

which is real, and $\omega(k)/k$ is not constant. Hence the beam equation is dispersive and the phase velocity is $\pm\sqrt{\gamma}k$, which depends on the wave number. Therefore shorter wavelength vibrations travel faster.

Generally, in the dispersive case the plane wave solution takes the form of a right-traveling wave,

$$u(x,t) = e^{i(kx-\omega(k)t)} = e^{ik(x-\frac{\omega(k)}{k}t)}.$$

The wave speed $\omega(k)/k$, or phase velocity, is dependent upon the wave number k. Therefore, wave packets of different wavelength will propagate at different speeds, causing a signal to disperse. Common example of dispersive waves are water waves, vibrations in solids, and electromagnetic waves in dielectrics.

We may think of plane wave solutions of the form $e^{i(kx-\omega(k)t)}$ as being a one-parameter family of solutions, with $k \in \mathbb{R}$. If we superimpose them over all wave numbers, we obtain

$$u(x,t) = \int_{-\infty}^{\infty} a(k)e^{i(kx-\omega(k)t)}dk.$$

Under reasonable conditions we expect this to be a solution of the differential equation under investigation. If

$$u(x,0) = f(x), \qquad x \in \mathbb{R},$$

is an initial condition, then

$$u(x,0) = f(x) = \int_{-\infty}^{\infty} a(k)e^{ikx} dk,$$

which is the Fourier transform of the unknown coefficient function $a(k)$. Therefore

$$a(k) = \frac{1}{2\pi} \int_{-\infty}^{\infty} f(x)e^{-ikx} dx,$$

and we have obtained a formal solution to a Cauchy problem for the partial differential equation having the form

$$
\begin{aligned}
u(x,t) &= \frac{1}{2\pi} \int_{-\infty}^{\infty} \left(\int_{-\infty}^{\infty} f(y)e^{-iky} dy \right) e^{i(kx-\omega(k)t)} dk \\
&= \frac{1}{2\pi} \int_{-\infty}^{\infty} \left(\int_{-\infty}^{\infty} e^{-ik(y-x)} e^{-i\omega(k)t} dk \right) f(y)\, dy.
\end{aligned}
$$

So, again, Fourier analysis is at the heart of the problem. In principle, this integral could be calculated to find the solution.

7.1.2 The Advection Equation

We showed that the simplest wave equation, the **advection equation**,

$$u_t + cu_x = 0, \quad x \in \mathbb{R}, \ t > 0, \tag{7.5}$$

has general solution

$$u = f(x - ct), \tag{7.6}$$

which is a right-traveling wave propagating at constant velocity c, where f is an arbitrary function. If we impose the initial condition

$$u(x,0) = \phi(x), \quad x \in \mathbb{R}, \tag{7.7}$$

then it follows that $f(x) = \phi(x)$, and so the solution to the initial value problem (7.5) and (7.7) is

$$u(x,t) = \phi(x - ct). \tag{7.8}$$

The straight lines $x - ct = $ constant, which are called **characteristics**, play a special role in this problem. We interpret characteristics as lines in space-time along which signals are carried (see Fig. 7.5), here the signal being a constant value of the initial data. Furthermore, on these characteristics, the partial differential equation (7.5) reduces to the ordinary differential equation

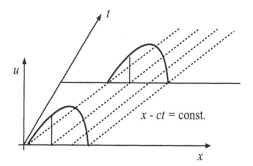

Figure 7.5 Solution to $u_t + cu_x = 0$, $c > 0$, and its characteristic curves.

$du/dt = 0$. That is to say, if C is a typical characteristic $x = ct + k$ for some constant k, then the directional derivative of u along that curve is

$$\frac{du}{dt}(x(t), t) = u_x(x(t), t)\frac{dx}{dt} + u_t(x(t), t)$$
$$= u_x(x(t), t)c + u_t(x(t), t),$$

which is the left side of (7.5) evaluated along C. Notice that the family of characteristic curves $x - ct = k$ propagates at speed c, which is the reciprocal of their slope in the xt coordinate system. Now let us complicate the partial differential equation (7.5) by replacing the constant c by a function of the independent variables t and x and consider the initial value problem

$$u_t + c(x, t)u_x = 0, \quad x \in \mathbb{R}, \ t > 0, \tag{7.9}$$
$$u(x, 0) = \phi(x), \quad x \in \mathbb{R},$$

where $c(x, t)$ is a given function. Let C be the family of curves defined by the differential equation

$$\frac{dx}{dt} = c(x, t). \tag{7.10}$$

Then along a member of C

$$\frac{du}{dt} = u_x\frac{dx}{dt} + u_t = u_xc(x, t) + u_t = 0.$$

Hence u is constant on each member of C. The curves C defined by (7.10) are the *characteristic curves*. Notice that they propagate at a speed $c(x, t)$, which depends on both time and the position in the medium.

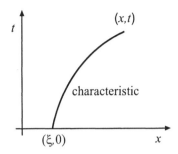

Figure 7.6 Characteristic $x = t^2 + \xi$.

Example 7.3

Consider the initial value problem

$$u_t + 2tu_x = 0, \quad x \in \mathbb{R}, \ t > 0,$$
$$u(x,0) = \exp(-x^2), \quad x \in \mathbb{R}, \ t > 0.$$

The characteristic curves are defined as solutions of the differential equation $dx/dt = 2t$. We obtain the family of parabolas

$$x = t^2 + k,$$

where k is a constant. Because u is constant on these characteristic curves, we can easily find a solution to the initial value problem. Let (x, t) be an arbitrary point where we want to obtain the solution. The characteristic curve through (x, t) passes through the x axis at $(\xi, 0)$, and it has equation $x = t^2 + \xi$ (see Fig. 7.6). Since u is constant on this curve

$$u(x,t) = u(\xi,0) = e^{-\xi^2} = e^{-(x-t^2)^2},$$

which is the unique solution to the initial value problem. The speed of the signal at (x, t) is $2t$, which is dependent upon t. The wave speeds up as time increases, but it retains its initial shape.

The preceding equation is defined on an infinite spatial domain $-\infty < x < \infty$. Now, suppose there are boundaries in the problem. To illustrate the basic points it is sufficient to study the simple advection equation

$$u_t + cu_x = 0, \quad 0 < x < 1, \ t > 0,$$

with $c > 0$. Assume an initial condition

$$u(x,0) = f(x), \quad 0 < x < 1.$$

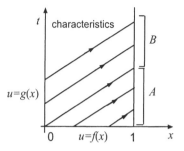

Figure 7.7 Characteristic diagram.

On the unbounded interval $(-\infty, \infty)$ the solution is a unidirectional right-traveling wave $u = f(x - ct)$. On a bounded interval what kinds of boundary conditions can be imposed at $x = 0$ and $x = 1$? See Fig. 7.7. Because u is given by the initial condition $f(x)$ along the initial line $t = 0$, $0 < x < 1$, data cannot be prescribed arbitrarily on the segment A along the boundary $x = 1$. This is because the characteristics $x - ct = $ constant are moving to the right and they carry the initial data to the segment A. Boundary data can be imposed along the line $x = 0$, since that data is carried along the forward-going characteristics to the segment B along $x = 1$. Then boundary conditions along B cannot be prescribed arbitrarily. Thus, it is clear that the problem

$$u_t + cu_x = 0, \quad 0 < x < 1, \quad t > 0,$$
$$u(x, 0) = f(x), \quad 0 < x < 1,$$
$$u(0, t) = g(t), \quad t > 0,$$

is properly posed. There are no backward-going characteristics in this problem, so there are no left traveling waves. Thus waves are not reflected from the boundary $x = 1$. To find the solution to this initial boundary value problem we must separate the domain into two portions, $0 < x < ct$ and $x > ct$. The solution in $x > ct$ is determined by the initial data, whereas the solution in $0 < x < ct$ is determined by the boundary data. These two regions are separated by the characteristic $x = ct$; we refer to the region $x > ct$ as the region of spacetime *ahead* of this limiting characteristic, and the region $0 < x < ct$ as the region *behind* the limiting characteristic. In summary, care must be taken to properly formulate and solve boundary value problems for one-dimensional, unidirectional wave equations. The situation is much different for second-order hyperbolic partial differential equations like $u_{tt} - c^2 u_{xx} = 0$. In this case both forward- and backward-going characteristics exist, and so left-traveling waves are also possible, as well as a mechanism for reflections from boundaries. This seems plausible because we can factor the differential operator into a product of two advection-like operators, $\frac{\partial^2}{\partial t^2} - c^2 \frac{\partial^2}{\partial x^2} = \left(\frac{\partial}{\partial t} - c\frac{\partial}{\partial x}\right)\left(\frac{\partial}{\partial t} + c\frac{\partial}{\partial x}\right)$, one right-moving and one left-moving.

EXERCISES

1. For the following equations find the dispersion relation and phase velocity. Classify as diffusive, dispersive, or hyperbolic.

   a) $u_t + cu_x + ku_{xxx} = 0$.

   b) $u_{tt} - c^2 u_{xx} = 0$.

   c) $u_{tt} + u_{xx} = 0$. (Note that this is an elliptic equation.)

2. Verify that the equation $u_t + c(x,t)u_x = 0$ has a solution $u(x,t) = f(a(x,t))$ for an arbitrary differentiable function f, where $a(x,t) = $ constant are the integral curves of $dx/dt = c(x,t)$.

3. Solve the following initial value problems on $t > 0$, $x \in \mathbb{R}$. Sketch the characteristic curves and several time snapshots of the wave in each case.

   a) $u_t + 3u_x = 0$, $\quad u(x,0) = \frac{1}{1+x^2}$.

   b) $u_t + xu_x = 0$, $\quad u(x,0) = \exp(-x)$.

   c) $u_t - x^2 t u_x = 0$, $\quad u(x,0) = x + 1$.

4. On the domain $0 < x < 1$, $t > 0$, consider the problem

$$u_t + cu_x = 0, \quad 0 < x < 1, \ t > 0,$$
$$u(x,0) = 0, \quad 0 < x < 1,$$
$$u(0,t) = te^{-t}, \quad t > 0,$$

   with $c > 0$. Find a formula for the solution, and sketch a graph of $u(1,t)$ for $t > 0$ when $c = 2$.

5. Find the solution and sketch a characteristic diagram

$$u_t + cu_x = 0, \qquad x \in \mathbb{R}, \ t > 0, \ c > 0$$
$$u(x,0) = \begin{cases} 1 - x^2, & |x| \le 1 \\ 0, & |x| > 1 \end{cases}.$$

6. Consider the the *linearized Korteweg-de Vries* equation

$$u_t + u_{xxx} = 0.$$

   a) Find the dispersion relation and determine how the phase velocity depends on the wave number.

b) By superimposing the plane wave solutions, find an integral representation for the solution to the Cauchy problem on $\mathbb{R}$. Write your answer in terms of the **Airy function**

$$\text{Ai}(x) = \frac{1}{2\pi} \int_{-\infty}^{\infty} \cos\left(\frac{z^3}{3} + xz\right) dz.$$

c) Take the initial condition to be $f(x) = e^{-x^2}$ and use a use a computer algebra system to plot solution profiles when $t = 0.5$ and $t = 1$.

7.2 Nonlinear Waves

7.2.1 Nonlinear Advection

In the last section we examined the two simple model wave equations, $u_t + cu_x = 0$ and $u_t + c(x,t)u_x = 0$, that are both first-order and linear. Now we study the same type of equation when a nonlinear advection term is introduced. In particular we consider

$$u_t + c(u)u_x = 0, \quad x \in \mathbb{R}, \ t > 0, \tag{7.11}$$
$$u(x,0) = \phi(x), \quad x \in \mathbb{R}, \tag{7.12}$$

where $c'(u) > 0$, with initial condition Using the guidance of the earlier examples we define the **characteristic curves** by the differential equation

$$\frac{dx}{dt} = c(u). \tag{7.13}$$

Then along a particular characteristic curve $x = x(t)$ we have

$$\frac{du}{dt}(x(t), t) = u_x(x(t), t)c(u(x(t), t)) + u_t(x(t), t) = 0.$$

Therefore u is constant along the characteristics, and the characteristics are straight lines since

$$\frac{d^2x}{dt^2} = \frac{d}{dt}\left(\frac{dx}{dt}\right) = \frac{d}{dt}c(u(x(t), t)) = c'(u)\frac{du}{dt} = 0.$$

In the nonlinear case, however, the speed of the characteristics as defined by (7.13) depends on the value u of the solution at a given point. To find the equation of the characteristic C through (x, t) we note that its speed is

$$\frac{dx}{dt} = c(u(\xi, 0)) = c(\phi(\xi))$$

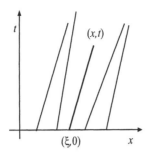

Figure 7.8 Characteristics moving at different speeds.

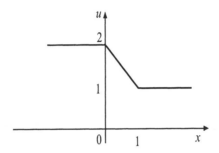

Figure 7.9 Initial wave profile.

(see Fig. 7.8). This results from applying (7.13) at $(\xi, 0)$. Hence, after integrating,

$$x = c(\phi(\xi))t + \xi \tag{7.14}$$

gives the equation of the desired characteristic C. Equation (7.14) defines $\xi = \xi(x, t)$ implicitly as a function of x and t, and the solution $u(x, t)$ of the initial value problem (7.11) and (7.12) is given by

$$u(x, t) = \phi(\xi) \tag{7.15}$$

where ξ is defined by (7.14). Consider the initial value problem

$$u_t + uu_x = 0, \quad x \in \mathbb{R}, \quad t > 0,$$

$$u(x, 0) = \phi(x) = \begin{cases} 2, & x < 0, \\ 2 - x, & 0 \leq x \leq 1, \\ 1, & x > 1. \end{cases}$$

The initial curve is sketched in Fig. 7.9. Since $c(u) = u$ the characteristics are straight lines emanating from $(\xi, 0)$ with speed $c(\phi(\xi)) = \phi(\xi)$. These are

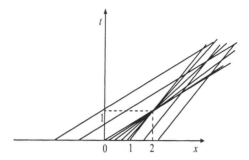

Figure 7.10 Characteristic diagram showing colliding characteristics.

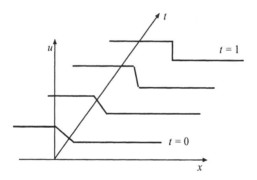

Figure 7.11 Solution surface with time profiles.

plotted in Fig. 7.10. For $x < 0$ the lines have speed 2; for $x > 1$ the lines have speed 1; for $0 \le x \le 1$ the lines have speed $2 - x$ and these all intersect at (2, 1). Immediately one observes that a solution cannot exist for $t > 1$, since the characteristics cross beyond that time and they carry different constant values of u. Figure 7.11 shows several wave profiles that indicate the steepening that is occurring. At $t = 1$ *breaking* of the wave occurs, which is the first instant when the solution becomes multiple valued. To find the solution for $t < 1$ we first note that $u(x,t) = 2$ for $x < 2t$ and $u(x,t) = 1$ for $x > t+1$. For $2t < x < t+1$ equation (7.14) becomes

$$x = (2 - \xi)t + \xi,$$

which gives

$$\xi = \frac{x - 2t}{1 - t}.$$

Equation (7.15) then yields

$$u(x,t) = \frac{2 - x}{1 - t}, \qquad 2t < x < t+1, \ t < 1.$$

This explicit form of the solution also indicates the difficulty at the breaking time $t = 1$.

In general the initial value problem (7.11)–(7.12) may have a solution only up to a finite time t_b, which is called the **breaking time**. Let us assume in addition to $c'(u) > 0$ that the initial wave profile satisfies the conditions

$$\phi(x) \geq 0, \qquad \phi'(x) < 0.$$

At the time when breaking occurs the gradient u_x will become infinite. To compute u_x we differentiate (7.14) implicitly with respect to x to obtain

$$\xi_x = \frac{1}{1 + c'(\phi(\xi))\phi'(\xi)t}.$$

Then from (7.15)

$$u_x = \frac{\phi'(\xi)}{1 + c'(\phi(\xi))\phi'(\xi)t}$$

The gradient catastrophe will occur at the minimum value of t, which makes the denominator zero. Hence

$$t_b = \min_{\xi} \frac{-1}{\phi'(\xi)c'(\phi(\xi))}, \qquad t_b \geq 0$$

In Example 2 we have $c(u) = u$ and $\phi(\xi) = 2 - \xi$. Hence $\phi'(\xi)c'(\phi(\xi)) = (-1)(1) = -1$ and $t_b = 1$ is the time when breaking occurs.

In summary we have observed that the nonlinear partial differential equation

$$u_t + c(u)u_x = 0, \quad c'(u) > 0$$

propagates the initial wave profile at a speed $c(u)$, which depends on the value of the solution u at a given point. Since $c'(u) > 0$, large values of u are propagated faster than small values and distortion of the wave profile occurs. This is consistent with our earlier remarks of a physical nature, namely that wave distortion and shocks develop in materials because of the property of the medium to transmit signals more rapidly at higher levels of stress or pressure. Mathematically, distortion and the development of shocks or discontinuous solutions are distinctively nonlinear phenomena caused by the term $c(u)u_x$ in the last equation.

EXERCISES

1. Consider the Cauchy problem

$$u_t = xuu_x, \quad x \in \mathbb{R}, \ t > 0,$$
$$u(x, 0) = x, \quad x \in \mathbb{R}.$$

Find the characteristics, and find a formula that determines the solution $u = u(x, t)$ implicitly as a function of x and t. Does a smooth solution exist for all $t > 0$?

2. Consider the initial value problem

$$u_t + uu_x = 0, \quad x \in \mathbb{R}, \ t > 0,$$

$$u(x,0) = \begin{cases} 1 - x^2, & |x| \leq 1, \\ 0, & |x| > 1. \end{cases}$$

Sketch the characteristic diagram. At what time t_b does the wave break? Find a formula for the solution.

3. Consider the initial value problem

$$u_t + uu_x = 0, \quad x \in \mathbb{R}, \ t > 0,$$

$$u(x,0) = \exp(-x^2), \quad x \in \mathbb{R}.$$

Sketch the characteristic diagram and find the point (x_b, t_b) in space–time where the wave breaks.

4. Consider the Cauchy problem

$$u_t + c(u)u_x = 0, \quad x \in \mathbb{R}, \ t > 0,$$

$$u(x,0) = f(x), \quad x \in \mathbb{R}.$$

Show that if the functions $c(u)$ and $f(x)$ are both nonincreasing or both nondecreasing, then no shocks develop for $t \geq 0$.

5. Consider the problem

$$u_t + u^2 u_x = 0, \quad x \in \mathbb{R}, \ t > 0,$$

$$u(x,0) = x, \quad x \in \mathbb{R}.$$

Derive the solution

$$u(x,t) = \begin{cases} x, & t = 0, \\ \frac{\sqrt{1+4xt}-1}{2t}, & t \neq 0, \ 1 + 4tx > 0. \end{cases}$$

When do shocks develop? Verify that $\lim_{t \to 0+} u(x,t) = x$.

6. Consider the **signaling problem**

$$u_t + c(u)u_x = 0, \quad t > 0, \ x > 0,$$

$$u(x,0) = u_0, \quad x > 0,$$

$$u(0,t) = g(t), \quad t > 0,$$

where c and g are given functions and u_0 is a positive constant. If $c'(u) > 0$, under what conditions on the signal g will no shocks form? Determine the solution in this case in the domain $x > 0$, $t > 0$.

7.2.2 Traveling Wave Solutions

One approach to analysis of partial differential equations is to examine model equations to gain a sense of the role played by various terms in the equation and how those terms relate to fundamental physical processes. In the preceding paragraphs we observed that the advection equation

$$u_t + cu_x = 0 \qquad\qquad (7.16)$$

propagates an initial disturbance or signal at velocity c, while maintaining the precise form of the signal. On the other hand, the nonlinear equation

$$u_t + uu_x = 0 \qquad\qquad (7.17)$$

propagates signals that distort the signal profile; that is, the nonlinear advection term uu_x causes a shocking up or rarefaction effect. We know from Chapter 6 that the diffusion equation

$$u_t - Du_{xx} = 0, \qquad\qquad (7.18)$$

which contains a diffusion term Du_{xx}, causes spreading of a density profile. Insights can be gained about nature of the various terms in an evolution equation by attempting to find either traveling wave solutions

$$u = U(x - ct), \qquad\qquad (7.19)$$

or plane wave solutions

$$u = A \ \exp[i(kx - \omega t)]. \qquad\qquad (7.20)$$

The latter is applicable to linear equations with constant coefficients, whereas traveling wave may be characteristic of linear or nonlinear processes.

In general, **traveling wave solutions** have no conditions at the boundaries (plus or minus infinity). However, there are often boundary conditions. Traveling wave solutions that approach constant states at $\pm\infty$ are called **wave front** solutions. They correspond to a traveling wave moving into and from a constant states. If these two states are equal, then we say the wave front is a **pulse**.

Example 7.4

(Burgers' equation) Consider the nonlinear equation

$$u_t + uu_x - Du_{xx} = 0, \quad D > 0, \qquad\qquad (7.21)$$

which is known as **Burgers' equation**. The term uu_x will have a shocking up effect that will cause waves to break and the term Du_{xx} is a diffusion term, like the one occurring in the diffusion equation (7.18). We attempt to find a traveling wave solution of (7.21) of the form

$$u = U(x - ct), \tag{7.22}$$

where the wave profile U and and the wave speed c are to be determined. Substituting (7.22) into (7.21) gives

$$-cU'(s) + U(s)U'(s) - DU''(s) = 0,$$

where

$$s = x - ct.$$

We interpret s as a moving coordinate. Noting that $UU' = \frac{1}{2}\frac{d}{ds}U^2$ and performing an integration gives

$$-cU + \frac{1}{2}U^2 - DU' = B,$$

where B is a constant of integration. Hence

$$\frac{dU}{ds} = \frac{1}{2D}(U - f_1)(U - f_2), \tag{7.23}$$

where

$$f_1 = c - \sqrt{c^2 + 2B}, \qquad f_2 = c + \sqrt{c^2 + 2B}.$$

We assume $c^2 > -2B$, and so $f_2 > f_1$. Separating variables in (7.23) and integrating again gives

$$\frac{s}{2D} = \int \frac{dU}{(U - f_1)(U - f_2)} = \frac{1}{f_2 - f_1}\ln\frac{f_2 - U}{U - f_1}, \qquad f_1 < U < f_2.$$

Solving for U yields

$$U(s) = \frac{f_2 + f_1 e^{Ks}}{1 + e^{Ks}}, \tag{7.24}$$

where

$$K = \frac{1}{2D}(f_2 - f_1) > 0.$$

For large positive s we have $U(s) \to f_1$, and for large negative s we have $U(s) \to f_2$. It is easy to see that $U'(s) < 0$ for all s and that $U(0) = \frac{1}{2}(f_1 + f_2)$. The graph of U is shown in Fig. 7.12 for different values of D. The weaker the diffusive effect (small D), the sharper the gradient in U. The traveling wave solution to (7.21) is

$$u(x, t) = \frac{f_2 + f_1 e^{K(x-ct)}}{1 + e^{K(x-ct)}},$$

where the speed is determined from the definition of f_1 and f_2 to be

$$c = \frac{1}{2}(f_1 + f_2).$$

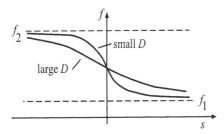

Figure 7.12 Traveling wave front solutions to Burgers' equation.

Graphically, the traveling wave solution is the profile $U(s)$ in Fig. 7.12 moving to the right at speed c. The solution of Burgers' equation, because it resembles the actual profile of a shock wave, is called the *shock structure* solution; it joins the asymptotic states f_1 and f_2. Without the Du_{xx} term the solutions of (7.21) would shock up and tend to break. The presence of the diffusion term prevents this breaking effect by countering the nonlinearity. The result is competition and balance between the nonlinear advection term uu_x and the diffusion term $-Du_{xx}$, much the same as occurs in a real shock wave in the narrow region where the gradient is steep. In this shock wave context the $-Du_{xx}$ term could also be interpreted as a viscosity term, as will become clear in the final chapter.

Example 7.5

(Korteweg–de Vries equation) A perturbation analysis of the equations that govern long waves in shallow water leads to another fundamental equation of applied mathematics, the **Korteweg–de Vries** *equation* (**KdV** *equation* for short)

$$u_t + uu_x + ku_{xxx} = 0, \quad k > 0. \tag{7.25}$$

This equation is a model equation for a nonlinear dispersive process. The uu_x term causes a shocking-up effect. In Burgers' equation this effect was balanced by a diffusion term $-Du_{xx}$, creating a shock structure solution. Now we replace this term by ku_{xxx}, which is interpreted as a dispersion term (plane wave solutions of the linear equation $u_t + ku_{xxx} = 0$ are dispersive). The KdV equation and modifications of it arise in many physical contexts (see C. S. Gardner et al., *SIAM Review*, Vol. 18, p. 412 (1976)). We follow the same reasoning as in Burgers' equation. That is, we assume a solution of the form

$$u = U(s), \quad s = x - ct, \tag{7.26}$$

where the waveform U and the wave speed c are to be determined. Substituting (7.26) into (7.25) gives

$$-cU' + \tfrac{1}{2}\frac{d}{ds}U^2 + kU''' = 0,$$

which when integrated gives

$$-cU + \tfrac{1}{2}U^2 + kU'' = A.$$

Multiplying by U' and integrating again yields

$$-\frac{c}{2}U^2 + \frac{1}{6}U^3 + \tfrac{1}{2}kU'^2 = AU + B.$$

One can then write

$$\frac{dU}{ds} = \pm\sqrt{\frac{1}{3k}}\ \phi(U)^{1/2}, \qquad \phi(U) = -U^3 + 3cU^2 + 6AU + 6B. \qquad (7.27)$$

Since $\phi(U)$ is a cubic polynomial, there are five possibilities:

(i) ϕ has one real root α

(ii) ϕ has three distinct real roots $\gamma < \beta < \alpha$.

(iii) ϕ has three real roots satisfying $\gamma = \beta < \alpha$.

(iv) ϕ has three real roots satisfying $\gamma < \alpha = \beta$.

(v) ϕ has a triple root γ.

Clearly, if α is a real root of ϕ, then $U = \alpha$ is a constant solution. We seek real, nonconstant, bounded solutions that will exist only when $\phi(U) \geq 0$. On examining the direction field it is easy to see that only unbounded solutions of (7.27) exist for cases (i) and (iv). We leave case (v) and case (ii) to the reader. Case (ii) leads to cnoidal waves. We examine case (iii) in which a class of solutions known as *solitons* arise. In this case (7.27) becomes

$$\frac{ds}{\sqrt{3k}} = \frac{dU}{(U - \gamma)\sqrt{\alpha - U}}. \qquad (7.28)$$

Letting $U = \gamma + (\alpha - \gamma)\mathrm{sech}^2 w$ and noting $dU = -2(\alpha - \gamma)\mathrm{sech}^2 w \tanh w\ dw$, it follows that

$$\frac{ds}{\sqrt{3k}} = \frac{-2(\alpha - \gamma)\mathrm{sech}^2 w \tanh w\ dw}{(\alpha - \gamma)\mathrm{sech}^2 w\sqrt{(\alpha - \gamma) - (\alpha - \gamma)\mathrm{sech}^2 w}} = \frac{-2dw}{\sqrt{\alpha - \gamma}}.$$

Integrating gives

$$w = -\sqrt{\frac{\alpha - \gamma}{12k}}\ s,$$

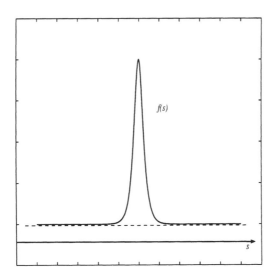

Figure 7.13 Soliton solution to the KdV equation.

and therefore a solution to (7.28) is given by

$$U(s) = \gamma + (\alpha - \gamma)\text{sech}^2\left(\sqrt{\frac{\alpha - \gamma}{12k}}\, s\right). \tag{7.29}$$

Clearly $U(s) \to \gamma$ as $s \to \pm\infty$. A graph of the waveform U is shown in Fig. 7.13. It is instructive to write the roots α and γ in terms of the original parameters. To this end

$$\begin{aligned}
\phi(U) &= -U^3 + 3cU^2 + 6AU + 6B \\
&= (U - \gamma)^2(\alpha - U) \\
&= -U^3 + (\alpha + 2\gamma)U^2 + (-2\alpha\gamma - \gamma^2)U + \gamma^2\alpha.
\end{aligned}$$

Thus the wave speed c is given by

$$c = \frac{\alpha + 2\gamma}{3},$$

and we can write the solution as

$$u(x, t) = \gamma + a\,\text{sech}^2\left(\sqrt{\frac{a}{12k}}\left(x - \left[\gamma + \frac{a}{3}\right]t\right)\right), \qquad a \equiv \alpha - \gamma.$$

Let us note several features of this wave front solution, or pulse. The velocity relative to the asymptotic state at $\pm\infty$ is proportional to the amplitude a. The width of the wave, defined by $\sqrt{12k/a}$, increases as k increases; that is, the wave disperses. Finally, the amplitude a is independent of the asymptotic state at $\pm\infty$. Such a waveform is known as a **soliton**, and many important equations of mathematical physics have soliton-like solutions (for example, the Boussinesq equation, the Sine–Gordon equation, the Born–Infeld equation, and nonlinear Schrödinger-type equations). In applications, the value of such solutions is that if a pulse or signal travels as a soliton, then the information contained in the pulse can be carried over long distances with no distortion or loss of intensity. Solitons, or solitary waves, have been observed in canals and waterways.

Reaction–advection–diffusion equations of the form

$$u_t = Du_{xx} + f(u, u_x)$$

form an important class of equations that often admit traveling wave solutions. If we assume $u = U(z)$, $z = x - ct$, then

$$-cU' = DU'' + F(U, U').$$

For wave front solutions we require $F(u_\pm, 0) = 0$, where $u_\pm$ are the constant states at $\pm\infty$. This equation is often examined in the UV phase plane, where $V = U'$. Then

$$U' = V, \quad V' = -\frac{1}{D}(F(U, V) + cV).$$

Showing the existence of traveling wave solutions then amounts to showing there is an orbit connecting the critical points $(u_-, 0)$ and $(u_+, 0)$.

7.2.3 Conservation Laws

The first order partial differential equation

$$u_t + J(u)_x = 0, \quad x \in \mathbb{R}, \ t > 0, \tag{7.30}$$

which represents a local conservation law, was derived under the conditions of smoothness of the density and flux. If there are discontinuities, we must take the integral form of the conservation law,

$$\frac{d}{dt} \int_a^b u(x, t) \, dx = J(u(a, t)) - J(u(b, t)), \tag{7.31}$$

which holds in all cases.

Example 7.6

The model nonlinear equation

$$u_t + u u_x = 0$$

can be written in conservation form as

$$u_t + (\tfrac{1}{2}u^2)_x = 0,$$

where the flux is $J(u) = \tfrac{1}{2}u^2$. The integral form of this law is

$$\frac{d}{dt} \int_a^b u(x,t)\ dx = \frac{1}{2}(u(a,t)^2 - u(b,t)^2).$$

In general we can write (7.30) in the form

$$u_t + c(u)u_x = 0, \tag{7.32}$$

where $c(u) = J'(u)$. We noted earlier that the initial value problem associated with (7.30) does not always have a solution for all $t > 0$. We showed that u is constant on the characteristic curves defined by $dx/dt = c(u)$, and the initial condition

$$u(x,0) = f(x)$$

is propagated along straight lines with speed $c(u)$. If $c'(u) > 0$ and $f'(x) < 0$, then the characteristics issuing from two points ξ_1 and ξ_2 on the x axis with $\xi_1 < \xi_2$ have speeds $c(f(\xi_1))$ and $c(f(\xi_2))$, respectively. It follows that $c(f(\xi_1)) > c(f(\xi_2))$, and therefore the characteristics must intersect (see Fig. 7.14). Because u is constant on characteristics, the solution is meaningless at the value of t where the intersection occurs. The first such t, denoted earlier by t_b, is the breaking time. A smooth solution exists only up to time t_b; beyond t_b something else must be tried. The key to continuing the solution for $t > t_b$ is to take a hint from the propagation of stress waves in a real physical continuum. There, under certain conditions, a smooth wave profile evolves into a shock wave. Hence, it is reasonable to expect that a discontinuous solution may exist and propagate for $t > t_b$. This discontinuous solution cannot satisfy the differential form (7.32) of the conservation law, but the integral form (7.31) remains valid. Therefore, we turn to a study of the simplest kinds of such solutions in an effort to obtain a condition that must hold at a discontinuity. Let $J = J(u)$ be continuously differentiable, and let $x = s(t)$ be a smooth curve in space-time across which u is discontinuous. Suppose u is smooth on each side of the curve (see Fig. 7.15). Next, select a and b such that the curve $x = s(t)$

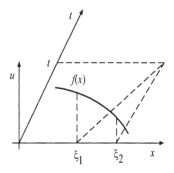

Figure 7.14 Characteristics determined by the initial condition.

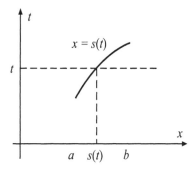

Figure 7.15 Shock path in space–time.

intersects $a \leq x \leq b$ at time t and let u_0 and u_1 denote the right and left limits of u at $s(t)$, respectively. That is,

$$u_0 = \lim_{x \to s(t)^+} u(x,t), \quad u_1 = \lim_{x \to s(t)^-} u(x,t).$$

From the conservation law (7.31) and Leibniz' rule for differentiating integrals,

$$J(u(a,t)) - J(u(b,t)) = \frac{d}{dt} \int_a^{s(t)} u(x,t)\ dx + \frac{d}{dt} \int_{s(t)}^b u(x,t)\ dx$$

$$= \int_a^{s(t)} u_t(x,t)\ dx + u_1 \frac{ds}{dt} + \int_{s(t)}^b u_t(x,t)\ dx - u_0 \frac{ds}{dt}.$$

Because u_t is bounded in each of the intervals $[a, s(t)]$ and $[s(t), b]$ separately, both integrals tend to zero in the limit as $a \to s(t)^-$ and $b \to s(t)^+$. Therefore

$$J(u_1) - J(u_0) = (u_1 - u_0)\frac{ds}{dt}. \tag{7.33}$$

Figure 7.16 Characteristic diagram.

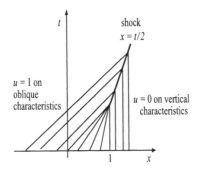

Figure 7.17 Characteristics and shock path.

Equation (7.33) is a **jump condition** relating the values of u and the flux in front of and behind the discontinuity to the speed ds/dt of the discontinuity. Thus the integral form of the conservation law provides a restriction on possible jumps across a simple discontinuity. A conventional notation for (7.33) is

$$[J(u)] = [u]\frac{ds}{dt},$$

where $[\cdot]$ indicates the jump in a quantity across the discontinuity.

Example 7.7

Consider the Cauchy problem

$$u_t + (\tfrac{1}{2}u^2)_x = 0, \quad x \in \mathbb{R}, \ t > 0,$$

$$u(x,0) = f(x) = \begin{cases} 1, & x \le 0, \\ 1 - x, & 0 < x < 1, \\ 0, & x \ge 1. \end{cases}$$

The characteristic diagram is shown in Fig. 7.16. Lines emanating from the x axis have speed $c(f(x)) = f(x)$. It is clear from the geometry that $t_b = 1$ and that a single valued solution exists for $t < 1$. For $t > 1$ we fit a shock or discontinuity beginning at $(1,1)$, separating the state $u_1 = 1$ on the left from the state $u_0 = 0$ on the right. Thus $[u] = 1 - 0 = 1$ and $[F] = \tfrac{1}{2}u_1^2 - \tfrac{1}{2}u_0^2 = \tfrac{1}{2}$. Therefore the discontinuity must have speed $s'(t) = [F]/[u] = \tfrac{1}{2}$. The resulting characteristic diagram is shown in Fig. 7.17. Therefore we have obtained a solution for all $t > 0$ satisfying the conservation law.

EXERCISES

1. Consider the problem

$$2u_t + u_x = 0, \quad x > 0, \ t > 0,$$
$$u(0,t) = 0, \quad t > 0,$$
$$u(x,0) = 1, \quad x \ge 0.$$

In what sense are $u_1(x,t) = 1 - H(t - 2x)$ and $u_2(x,t) = 1 - H(t - x)$ both solutions to the problem? Which one is the appropriate solution? (H is the Heaviside function.)

2. Consider the equation $u_t + c(u)u_x = 0$, $x \in \mathbb{R}$, $t > 0$, with initial data $u(x,0) = u_1$ for $x < 0$, and $u(x,0) = u_0$ for $x \ge 0$, with $c'(u) > 0$.

   a) Find a nonsmooth solution in the case $u_0 > u_1$

   b) Find a discontinuous (shock) solution in the case $u_0 < u_1$

   c) Discuss (a) and (b) in the context of shock waves and rarefaction waves.

3. In the traveling wave solution to Burgers' equation, the **shock thickness** is defined by $(f_2 - f_1)/\max |U'(s)|$. Give a geometric interpretation of this condition, and show that the shock thickness is given by $8D/(f_2 - f_1)$.

4. Find all wave front solutions of the reaction–advection equation

$$u_t + u_x = u(1 - u).$$

5. Consider the nonlinear advection–diffusion equation

$$\left(u - \frac{1}{2}u^2\right)_t = u_{xx} - u_x.$$

a) What is the speed c of right traveling wave front solutions $u = U(z)$, $z = x - ct$, that satisfy the conditions $U(-\infty) = 1$, $U(\infty) = 0$?

b) Determine explicitly the right-traveling wave front in (a) that satisfies the condition $U(0) = \frac{1}{2}$.

6. Show that Burgers' equation $u_t + uu_x = Du_{xx}$ can be reduced to the diffusion equation via the transformation $u = -2D\phi_x/\phi$ (this is the **Cole–Hopf** transformation). If the initial condition $u(x,0) = f(x)$ is imposed, derive the solution

$$u(x,t) = \int_{-\infty}^{\infty} \frac{x - \xi}{t} \exp\left(\frac{-G}{2D}\right) d\xi \Big/ \int_{-\infty}^{\infty} \exp\left(\frac{-G}{2D}\right) d\xi,$$

where

$$G = G(\xi, x, t) = \int_0^{\xi} f(\eta) \, d\eta + \frac{(x - \xi)^2}{2t}.$$

7.3 Quasi-linear Equations

A quasi-linear partial differential equation is an equation that is linear in its derivatives. Many of the equations that occur in applications (for example, the nonlinear equations of fluid mechanics that we will develop in a subsequent section) are quasi-linear. In this section we focus on the initial value problem for first-order equations and introduce the idea of the general solution. Finally, as an application, we introduce age-structured models in population dynamics.

In this section we study the Cauchy problem for the quasi-linear equation

$$u_t + c(x, t, u)u_x = f(x, t, u), \quad x \in \mathbb{R}, t > 0, \tag{7.34}$$

where the initial condition has the form

$$u(x, 0) = \phi(x), \quad x \in \mathbb{R}. \tag{7.35}$$

Motivated by our earlier approach, we define a family of curves by the differential equation

$$\frac{dx}{dt} = c(x, t, u). \tag{7.36}$$

On these curves u is no longer constant, but rather

$$\frac{du}{dt} = f(x, t, u). \tag{7.37}$$

The pair of differential equations (7.36)–(7.37) is called the **characteristic system** associated with (7.34). The initial data can be represented

$$x = \xi, \quad u = \phi(\xi) \quad \text{on} \quad t = 0, \tag{7.38}$$

where ξ is a parameter representing an arbitrary value on the x axis. We may interpret equations (7.36)–(7.37) as a system of two nonautonomous equations for $x = x(t)$ and $u = u(t)$, (the values of x and u along the characteristic curves) with initial values given in (7.38). The general solution of (7.36)–(7.37) involves two arbitrary constants and has the form

$$x = F(t, c_1, c_2), \quad u = G(t, c_1, c_2).$$

The initial data imply $\xi = F(0, c_1, c_2)$, $\phi(\xi) = G(0, c_1, c_2)$, which gives the constants in terms of ξ, or $c_1 = c_1(\xi)$, $c_2 = c_2(\xi)$. Therefore, x and u are given as functions of t and their initial values,

$$x = F(t, c_1(\xi), c_2(\xi)), \quad u = G(t, c_1(\xi), c_2(\xi)). \tag{7.39}$$

In principle, the first of these equations can be solved to obtain $\xi = \xi(x, t)$, which can then be substituted into the second equation to obtain an explicit representation $u = u(x, t)$ to the solution to (7.34)–(7.35).

Example 7.8

Solve the IVP

$$
\begin{aligned}
u_t + u u_x &= -u, & x \in \mathbb{R}, t > 0, \\
u(x, 0) &= -\frac{x}{2}, & x \in \mathbb{R}.
\end{aligned}
$$

The characteristic system is

$$\frac{dx}{dt} = u, \quad \frac{du}{dt} = -u,$$

and the initial condition gives

$$x = \xi, \quad u = -\frac{\xi}{2} \quad \text{on} \quad t = 0.$$

The general solution of the characteristic system is

$$x = -c_1 e^{-t} + c_2, \quad u = c_1 e^{-t}.$$

Applying the initial data gives $\xi = -c_1 + c_2$ and $-\frac{\xi}{2} = c_1$. Therefore

$$x = \frac{\xi}{2} e^{-t} + \frac{\xi}{2}, \quad u = -\frac{\xi}{2} e^{-t}.$$

Therefore $\xi = \xi(x, t) = 2x/(1 + e^{-t})$, and so

$$u(x, t) = \frac{-xe^{-t}}{1 + e^{-t}}.$$

We have carried out this solution process formally. In most problems it will not be possible to solve the characteristic system or the resulting algebraic equations that determine the constants $c_1 = c_1(\xi)$, $c_2 = c_2(\xi)$ as functions of ξ. As a consequence, we often have to be satisfied with the parametric representation (7.39). In practice, numerical methods are usually applied to (7.36)–(7.37) to advance the initial conditions forward in time.

In some contexts the general solution of the quasi-linear equation (7.34) is needed, in terms of a single, arbitrary function. To determine the general solution we proceed with some general definitions and observations. An expression $\psi(x, t, u)$ is called a **first integral** of the characteristic system (7.36)–(7.37) if $\psi(x, t, u) = k$ (constant), on solutions to (7.36)–(7.37). That is, $\psi(X(t), t, U(t)) = k$ for all t in some interval I, if $x = X(t)$, $u = U(t)$, $t \in I$, is a solution to (7.36)–(7.37). Taking the total derivative of this last equation with respect to t and using the chain rule gives

$$\psi_x c + \psi_t + \psi_u f = 0, \quad t \in I, \tag{7.40}$$

where each of the terms in this equation are evaluated at $(X(t), t, U(t))$.

Furthermore, if $\psi(x, t, u)$ is a first integral of the characteristic system, then the equation $\psi(x, t, u) = k$ defines, implicitly, a surface $u = u(x, t)$, on some domain D in the xt plane, provided $\psi_u \neq 0$. Thus

$$\psi(x, t, u(x, t)) = k, \quad (x, t) \in D.$$

Using the chain rule, we can calculate the partial derivatives of u as

$$\psi_t + \psi_u u_t = 0, \quad \psi_x + \psi_u u_x = 0,$$

or

$$u_t = -\frac{\psi_t}{\psi_u}, \quad u_x = \frac{\psi_x}{\psi_u}, \quad (x, t) \in D.$$

Each term on the right is evaluated at $(x, t, u(x, t))$. Observe also that the curve $(X(t), t, U(t))$ lies on this surface because $\psi(X(t), t, U(t)) = k$.

We claim that this surface $u = u(x, t)$ is a solution to the partial differential equation (7.34). To prove this, fix an arbitrary point (ξ, τ) in D. Then

$$u_t(\xi, \tau) + c u_x(\xi, \tau) = -\frac{\psi_t}{\psi_u} - c\frac{\psi_x}{\psi_u}, \tag{7.41}$$

where c and all the terms on the right side are evaluated at (ξ, τ, ω), where $\omega = u(\xi, \tau)$. But a solution curve $(X(t), t, U(t))$ lies on the surface and passes

through the point (ξ, τ, ω) at $t = \tau$. That is, $(\xi, \tau, \omega) = (X(\tau), \tau, U(\tau))$. Therefore, the right side of (7.41) may be evaluated at $(X(\tau), \tau, U(\tau))$.

In general, we expect to find two independent first integrals to the characteristic system, $\psi(t, x, u)$ and $\chi(t, x, u)$. The **general solution** of (7.34) is
then

$$H(\psi(t, x, u), \chi(t, x, u)) = 0, \tag{7.42}$$

where H is an arbitrary function. We can formally solve for one of the variables
and write this as $\psi(t, x, u) = G(\chi(t, x, u))$, where G is an arbitrary function. We
leave it as an exercise using the chain rule to show that if $\psi(t, x, u)$ and $\chi(t, x, u)$
are first integrals of the characteristic system, then (7.42) defines, implicitly, a
solution $u = u(x, t)$ to (7.34) for any function H, provided $H_\psi \psi_u + H_\chi \chi_u \neq 0$.

Example 7.9

Consider the equation

$$u_t + 2tu_x = \frac{x}{2u + 1}.$$

The characteristic system is

$$\frac{dx}{dt} = 2t, \quad \frac{du}{dt} = \frac{x}{2u + 1}.$$

The first equation gives first integral $\psi = x - t^2 = c_1$. Then the second equation
becomes

$$\frac{du}{dt} = \frac{t^2 + c_1}{2u + 1}.$$

Separating variables and integrating gives another first integral $\chi = u^2 + u +
\frac{2}{3}t^3 - xt = c_2$. Therefore, the general solution is

$$H\left(x - t^2, u^2 + u + \frac{2}{3}t^3 - xt\right) = 0,$$

where H is an arbitrary function. This expression defines implicit solutions. It
can also be written

$$u^2 + u + \frac{2}{3}t^3 - xt = G\left(x - t^2\right),$$

where G is an arbitrary function. This quadratic in u can be solved to determine
explicit solutions.

7.3.1 Age-Structured Populations

Elementary population models in biology lump the total population at time t into a single number $p(t)$, regardless of the age of the individuals. We have studied these models, e.g., the Malthus law and the logistics law, in earlier chapters. However, because individuals may differ in their vital statistics (birth- and deathrates, fertility), in some demographic models it is necessary to superimpose an age distribution at each time. These models are called **age-structured** models.

We consider a population of females whose age structure at time $t = 0$ is $f(a)$. That is, $f(a)\,da$ is approximately the initial number of females between age a and $a + da$. Females are usually used in demographic models because they have a definite beginning and end to their reproductive capacity. And, for simplicity, we impose the age range to be $0 \leq a < \infty$, even though the age at death is finite. Given the mortality and fecundity rates of females, the goal is to determine the age structure, or population density, $u(a, t)$ at time t, where $u(a, t)\,da$ is approximately the number of females between age a and $a + da$ at time t. The total number of females at time t is

$$N(t) = \int_0^\infty u(a, t)\,da.$$

We can derive a partial differential equation for the density by accounting for the change in population over a small time dt, where members leave the population only because of death. In the time interval from t to $t + dt$ all females of age a become females of age $a + da$. The number of females at time t between age a and $a + da$ is $u(a, t)\,da$, and the number between $a + da$ and $a + 2da$ at time $t + dt$ is $u(a + da, t + dt)\,da$. The difference between these two populations is the number that die during the time interval. If λ represents the mortality rate, or the fraction that die per unit time, then the number that die from time t to $t + dt$ is $\lambda[u(a, t)\,da]\,dt$. The mortality rate λ could depend upon age a, time t, age density u, or even the total population $N(t)$. Consequently, to leading order we have

$$u(a + da, t + dt)da - u(a, t)\,da = -\lambda u(a, t)\,da\,dt + \cdots.$$

By Taylor's expansion, $u(a + da, t + dt) = u(a, t) + u_t(a, t)\,dt + u_a(a, t)\,da + \cdots$. Therefore $u_t(a, t)\,dt + u_a(a, t)\,da = -\lambda u(a, t)\,dt + \cdots$. Dividing by dt, using $da = dt$, and taking the limit as $dt \to 0$ gives

$$u_t + u_a = -\lambda u, \tag{7.43}$$

which is the **McKendrick–Von Foerster equation**. The initial condition is

$$u(a, 0) = f(a). \tag{7.44}$$

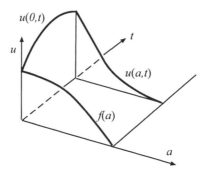

Figure 7.18 Graphical representation of an age-structured model showing age profiles at time $t = 0$ ($f(a)$) and at time t ($u(a,t)$). The boundary condition $u(0,t)$ represents the number of newborns.

The boundary condition at age $a = 0$ is not as simple as it may first appear because $u(0,t) = B(t)$, the number of newborns at time t, is not known *a priori*. Rather, it depends upon the fertility and ages of the reproducing females at time t. To formulate the boundary condition let $b(a,t)$ denote the average number of offspring per female of age a at time t. We call b the fecundity rate or maternity function. We expect b to be zero up until the age of maturity and zero after menopause. The total number of offspring produced by females in the age interval a to $a + da$ is $b(a,t)u(a,t)\, da$. Thus the total number of offspring $B(t)$ produced at time t is

$$B(t) = u(0,t) = \int_0^\infty b(a,t)u(a,t)\, da. \tag{7.45}$$

Therefore the boundary condition at $a = 0$ depends upon the solution to the differential equation. Such boundary conditions are called *nonlocal* because they involve an integration of the unknown in the problem. The initial boundary value problem (7.43)–(7.45) is an example of an age-structured model. These nonlocal, nonlinear models can be complex, and thus it is common to make assumptions that lead to simplification and the possibility of a tractable problem. For example, we assume the mortality rate λ is constant, and the fecundity rate b is dependent only upon age. Then we obtain the simplified McKendrick–Von Foerster model

$$
\begin{aligned}
u_t + u_a &= -\lambda u, & a > 0, t > 0, \\
u(a,0) &= f(a), & a > 0, \\
u(0,t) &= \int_0^\infty b(a)u(a,t)\, da, & t > 0.
\end{aligned}
$$

The characteristic system for the partial differential equation is

$$\frac{da}{dt} = 1, \quad \frac{du}{dt} = -\lambda u.$$

It is easy to find the first integrals $a - t = c_1$ and $ue^{\lambda t} = c_2$. Therefore the general solution to the pde is

$$u(a, t) = g(a - t)e^{-\lambda t},$$

where g is an arbitrary function. In the domain $a > t$ the solution is determined by the initial condition at $t = 0$. Therefore $g(a) = f(a)$ and we obtain

$$u(a, t) = f(a - t)e^{-\lambda t}, \quad a > t.$$

In the domain $0 < a < t$ the solution is determined by the boundary condition. From the general solution we have

$$u(0, t) = B(t) = g(-t)e^{-\lambda t},$$

which determines the arbitrary function, $g(t) = B(-t)e^{-\lambda t}$. Therefore

$$u(a, t) = B(t - a)e^{-\lambda a}, \quad 0 < a < t.$$

Hence, we will have determined the solution once the function B is found.

To determine B we use the nonlocal boundary condition to obtain

$$
\begin{aligned}
B(t) &= \int_0^\infty b(a)u(a, t)\, da \\
&= \int_0^t b(a)u(a, t)\, da + \int_t^\infty b(a)u(a, t)\, da \\
&= \int_0^t b(a)B(t - a)e^{-\lambda a}\, da + \int_t^\infty b(a)f(a - t)e^{-\lambda t}\, da.
\end{aligned}
$$

This equation is a nonhomogeneous Volterra integral equation for B and it can be written in the form

$$B(t) = \int_0^t b(a)B(t - a)e^{-\lambda a}\, da + F(t), \tag{7.46}$$

where $F(t) = \int_t^\infty b(a)f(a - t)e^{-\lambda t}\, da$. Equation (7.46) is called the **renewal equation**. In summary, the solution to McKendrick–Von Foerster model has been reduced to the solution of an integral equation.

We show how an asymptotic solution for large times can be obtained. For large t the nonhomogeneous term $F(t)$ vanishes because the fecundity b is zero at some finite age. Thus, the renewal equation can be approximated by

$$B(t) = \int_0^\infty b(a)B(t - a)e^{-\lambda a}\, da, \quad t \text{ large}.$$

Note that we can replace the upper limit by infinity since the fecundity rate vanishes at a finite age. Let us assume an exponential solution of the form $B(t) = U_0 e^{rt}$, where r.is an unknown growth rate. Substituting gives

$$1 = \int_0^\infty b(a)e^{-(r+\lambda)a}da. \tag{7.47}$$

This is the **Euler–Lotka equation**, and it determines the growth rate r. The population density in this asymptotic case is

$$u(a,t) = U_0 e^{rt} e^{-(r+\lambda)a}, \quad t > a, \quad t \text{ large}.$$

The decaying function $U(a) = e^{-(r+\lambda)a}$ is called the **stable age structure**.

More extensive material on age structured problems can be found in Brauer & Castillo-Chavez (2001) and Kot (2001). Age-structured problems are a special case of more general physiological structured problems where the age variable a can be replaced by length, size, weight, development, and so on.

EXERCISES

1. Insofar as possible, solve the following pure initial value problems:

   a) $u_t + u_x = -2tu, \quad u(x,0) = \phi(x)$.

   b) $u_t + u_x = g(x,t), \quad u(x,0) = \phi(x)$.

   c) $u_t + xu_x = e^t, \quad u(x,0) = \phi(x)$.

2. Use characteristics to find the general solution to the following equations.

   a) $u_t + u_x = -u^2$.

   b) $t^2 u_t + x^2 u_x = 2xt$.

   c) $uu_t = x - t$.

   d) $tu_t + xu_x = xt(1 + u^2)$.

   e) $tu_t + uu_x = x$.

   f) $u(u_t - u_x) = x - t$.

   g) $uu_t + xu_x = t$.

3. Consider the initial value problem

$$u_t + uu_x + u = 0, \quad x \in \mathbb{R}, \ t > 0,$$
$$u(x,0) = -2x, \quad x \in \mathbb{R}.$$

   Find the time t_b that a gradient catastrophe occurs and beyond which there is no continuously differentiable solution.

4. Consider the initial value problem

$$u_t + uu_x + \alpha u^2 = 0, \quad x \in \mathbb{R}, \ t > 0; \ \alpha > 0,$$
$$u(x,0) = 1, \quad x \in \mathbb{R}.$$

Derive the solution $u(x,t) = (1 + \alpha t)^{-1}$.

5. Consider the initial value problem

$$u_t + uu_x = -u, \quad x \in \mathbb{R}, \ t > 0,$$
$$u(x,0) = ke^{-x^2}, \quad x \in \mathbb{R},$$

where $k > 0$. Determine the values of k for which the solution blows up.

6. Consider the initial value problem for the damped advection equation,

$$u_t + uu_x = -\alpha u, \quad x \in \mathbb{R}, \ t > 0,$$
$$u(x,0) = \phi(x), \quad x \in \mathbb{R},$$

where $\alpha > 0$. Show that the solution is given implicitly by

$$x = \phi(\xi)e^{-\alpha t}, \quad u = \frac{\phi(\xi)}{\alpha}(1 - e^{-\alpha t}) + \xi.$$

Determine a condition on ϕ that guarantees no shocks will form. If a shock does form, show that the breaking time is greater than that for the undamped equation when $\alpha = 0$.

7. Consider

$$u_t + uu_x + u = 0, \quad x \in \mathbb{R}, \ t > 0,$$
$$u(x,0) = k\exp(-x^2), \quad x \in \mathbb{R}; \ k > 0.$$

For what values of k does breaking occur?

8. Solve the signaling problem

$$u_t + u_x = u^2, \quad x > 0, \ -\infty < t < \infty,$$
$$u(0,t) = g(t), \quad -\infty < t < \infty.$$

9. Consider an age-structured female population where $u = u(a,t)$ is the age density of females at age a at time t and where no individual survives beyond age δ. The model is

$$u_t + u_a = -\frac{\gamma}{\delta - a}u, \quad 0 < a < \delta, \ t > 0.$$

Find the solution on the given domain subject to the initial and boundary conditions

$$u(a,0) = f(a), \quad u(0,t) = B(t); \quad 0 < a < \delta, \ t > 0,$$

where f and B are given functions.

10. Consider a population of female organisms whose mortality rate is λu, with λ constant, and whose fecundity is given by $b(a) = ae^{-\gamma a}$. From the Euler–Lotka equation determine the growth rate r of the stable age-structured.population.

11. Consider the age-structured model

$$
\begin{aligned}
u_t + u_a &= -\lambda(N(t))u, \quad a > 0, t > 0, \\
u(a, 0) &= f(a), \quad a > 0, \\
u(0, t) &= \int_0^\infty b(N(t))u(a, t)da, \quad t > 0,
\end{aligned}
$$

where the mortality and fecundity rates depend upon the total population $N(t)$, with $b'(N) \leq 0$, $\lambda'(N) \geq 0$.

a) Show that $N'(t) = (b(N) - \lambda(N))N$.

b) Assuming the differential equation in part (a) has a nonzero equilibrium N^*, show that there is an equilibrium population density $u = U(a)$ given by

$$
U(a) = N^* \lambda(N^*) e^{-\lambda(N^*)a}.
$$

12. In the McKendrick–Von Foerster model assume the mortality rate λ and the fecundity rate b are constant.

a) Show that $B' = (b - \lambda)B$.

b) Find $B(t)$ and the population density $u(a, t)$.

c) What is the size $N(t)$ of the population at time t?

7.4 The Wave Equation

The science of acoustics is the study of sound propagation in solids, fluids, and gases. The fundamental equation is the wave equation that was obtained as an approximation for small-amplitude signals.

7.4.1 The Acoustic Approximation

As before we imagine a gas flowing in a tube of constant cross-sectional area A in such a way that the physical parameters are constant in any cross section. We fix two cross sections $x = a$ and $x = b$. The principle of mass conservation states simply that mass is neither created nor destroyed. Therefore, the rate of change of mass inside the interval $a \leq x \leq b$ equals the mass flux in at $x = a$ less the mass flux out at $x = b$. In symbols

$$\frac{d}{dt} \int_a^b A\rho(x,t)\, dx = A\rho(a,t)v(a,t) - A\rho(b,t)v(b,t), \qquad (7.48)$$

where $\rho(x,t)$ is the density and $v(x,t)$ is the velocity. The quantity $A\rho(x,t)v(x,t)$ gives the amount of fluid that crosses the location x per unit time and is therefore the mass flux. Equation (7.48) can be written

$$\int_a^b \rho_t(x,t)\, dx = - \int_a^b (\rho(x,t)v(x,t))_x\, dx.$$

Because the interval $a \leq x \leq b$ is arbitrary it follows that

$$\rho_t + (\rho v)_x = 0, \qquad (7.49)$$

which is the **continuity equation**, or the partial differential equation expressing conservation of mass .

Momentum balance requires that the rate of change of momentum inside $[a,b]$ must equal the transport of momentum carried into and out of the region at $x = a$ and $x = b$ plus the momentum created inside $[a,b]$ by the force (pressure times area) p acting at a and b. This is, of course, motivated by Newton's second law of motion, which states that the rate of change of momentum is the force, or $\frac{d}{dt}(mv) = F$. In mathematical terms,

$$\frac{d}{dt} \int_a^b A\rho(x,t)v(x,t)\, dx = A\rho(a,t)v^2(a,t) - A\rho(b,t)v^2(b,t) + Ap(a,t) - Ap(b,t).$$
$$(7.50)$$

The term $A\rho(x,t)v^2(x,t)$ the momentum times the velocity at x and is therefore the **momentum flux** at x. Bringing the time derivative into the integral and using the fundamental theorem of calculus turns (7.50) into

$$\int_a^b (\rho(x,t)v(x,t))_t\, dx = - \int_a^b (\rho(x,t)v^2(x,t) + p(x,t))_x\, dx.$$

The arbitrariness of the interval $a \leq x \leq b$ forces

$$(\rho v)_t + (\rho v^2 + p)_x = 0, \qquad (7.51)$$

which is a partial differential equation expressing **conservation of momentum**. Both (7.49) and (7.51) are of the form

$$\frac{\partial}{\partial t}[\cdots] + \frac{\partial}{\partial x}[\cdots] = 0,$$

and are in **conservation form**.

In the simplest gas dynamical calculations equations (7.49) and (7.51) may be supplemented by the barotropic equation of state

$$p = f(\rho), \quad f'(\rho) > 0, \tag{7.52}$$

which gives pressure as a function of density only. It is convenient to introduce the **local sound speed** c defined by

$$c^2 = f'(\rho). \tag{7.53}$$

Then (7.49) and (7.51) may be written

$$p_t + v p_x + c^2 \rho v_x = 0, \tag{7.54}$$

$$\rho v_t + \rho v v_x + p_x = 0, \tag{7.55}$$

where we have used $p_x = c^2 \rho_x$ and $p_t = c^2 \rho_t$ to rewrite (7.49), and we have used (7.49) to replace the quantity ρ_t in (7.51), thereby giving (7.55). Equations (7.54) and (7.55) along with (7.52) provide a starting point for the study of flow of a gas.

The equations governing the adiabatic flow of a gas are nonlinear and cannot be solved in general. Rather, only special cases can be resolved. The simplest case, that of acoustics, arises when the deviations from a constant equilibrium state $v = 0$, $\rho = \rho_0$, and $p = p_0$ are assumed to be small. Therefore we consider the isentropic equations

$$\rho_t + v \rho_x + \rho v_x = 0, \tag{7.56}$$

$$\rho v_t + \rho v v_x + p_x = 0, \tag{7.57}$$

with the barotropic equation of state

$$p = F(\rho).$$

It is common in acoustics to introduce the *condensation* δ defined by

$$\delta \equiv \frac{\rho - \rho_0}{\rho_0},$$

which is the relative change in density from the constant state ρ_0. We assume that v, δ, and all their derivatives are small compared to unity. Then the equation of state can be expanded in a Taylor series as

$$p = F(\rho_0) + F'(\rho_0)(\rho - \rho_0) + \tfrac{1}{2} F''(\rho_0)(\rho - \rho_0)^2 + O((\rho - \rho_0)^3)$$
$$= F(\rho_0) + \rho_0 F'(\rho_0)\delta + \tfrac{1}{2}\rho_0^2 F''(\rho_0)\delta^2 + O(\delta^3).$$

Whence

$$p_x = \rho_0 F'(\rho_0)\delta_x + O(\delta^2),$$

and therefore (7.56) and (7.57) become

$$\delta_t + v\delta_x + (1 + \delta)v_x = 0,$$
$$(1 + \delta)v_t + (1 + \delta)vv_x + F'(\rho_0)\delta_x + O(\delta^2) = 0.$$

Discarding products of small terms gives a set of linearized equations for the small deviations δ and v, namely

$$\delta_t + v_x = 0, \quad v_t + F'(\rho_0)\delta_x = 0 \tag{7.58}$$

These equations are known as the **acoustic approximation**. Eliminating v from (7.58) gives

$$\delta_{tt} - c_0^2\delta_{xx} = 0, \tag{7.59}$$

where

$$c^2 = F'(\rho_0). \tag{7.60}$$

Equation (7.59) is a second-order linear partial differential equation known as the **wave equation**. The quantity c in (7.60) is the propagation speed of acoustic waves. For an ideal gas $p = F(\rho) = k\rho^\gamma$, and therefore

$$c = \sqrt{\frac{\gamma p_0}{\rho_0}}$$

is the speed that small-amplitude signals are propagated in an ideal gas. By eliminating δ from equations (7.58) we note that the small velocity amplitude v satisfies the wave equation as well.

At this point it is worthwhile to test the acoustic approximation to determine its range of validity. Let U and Γ denote scales for the velocity v and condensation δ, respectively. From Chapter 1 we know that such scales can be defined as the maximum value of the quantity, for example, $U = \max |v|$ and $\Gamma = \max |\delta|$. If L and τ denote length and time scales for the problem, then the balance of the terms in (7.58) requires

$$\Gamma\tau^{-1} \approx UL^{-1}, \qquad U\tau^{-1} \approx c_0^2\Gamma L^{-1}.$$

Thus

$$c \approx L\tau^{-1}.$$

In the approximation we assumed that the convective term vv_x was small compared to v_t. Consequently, for acoustics

$$U^2L^{-1} \ll U\tau^{-1},$$

or

$$U \ll c. \tag{7.61}$$

That is, the acoustic approximation can be justified when the maximum particle velocity is small compared to the speed c that sound waves travel in the medium. The ratio U/c is called the **Mach number** M and therefore acoustics is valid when $M \ll 1$. For phenomena involving ordinary audible sounds the approximation is highly satisfactory, but for flows resulting from the presence of high-speed aircraft the full nonlinear equations (7.56) and (7.57) would be required for an accurate description. In the theory of *nonlinear acoustics* a few, but not all, some of the nonlinear terms are retained in the perturbation equations.

7.4.2 Solutions to the Wave Equation

The one-dimensional wave equation

$$u_{tt} - c^2 u_{xx} = 0 \tag{7.62}$$

arises naturally in acoustics and it describes how small amplitude waves (density and particle velocity) propagate. There are many other physical situations modeled by this important equation, for example, electromagnetism, wave on taut strings (guitar strings), and vibrations in elastic media. From earlier remarks we note that the wave equation is hyperbolic. It is easy to see (Exercise 2) that the general solution of the wave equation is given by

$$u(x,t) = f(x + ct) + g(x - ct), \tag{7.63}$$

where f and g are arbitrary functions; hence solutions are the superposition of right and left traveling waves moving at speed c. The functions f and g are determined by the initial and boundary data, although it is difficult to do this in specific cases.

The Cauchy problem for the wave equation is

$$u_{tt} - c^2 u_{xx} = 0, \quad x \in \mathbb{R}, \ t > 0, \tag{7.64}$$
$$u(x,0) = F(x), \quad u_t(x,0) = G(x), \quad x \in \mathbb{R},$$

where F and G are given functions, and it can be solved exactly by **D'Alembert's solution**.

Theorem 7.10

In (7.64) let $F \in C^2(\mathbb{R})$ and $G \in C^1(\mathbb{R})$. Then the solution of (7.64) is given by

$$u(x,t) = \frac{1}{2}(F(x + ct) + F(x - ct)) + \frac{1}{2c} \int_{x-ct}^{x+ct} G(y) \, dy. \tag{7.65}$$

Proof

The derivation of (7.65) follows directly by applying the initial conditions to determine the arbitrary functions f and g in (7.63). We have

$$u(x,0) = f(x) + g(x) = F(x), \qquad (7.66)$$

$$u_t(x,0) = cf'(x) - cg'(x) = G(x). \qquad (7.67)$$

Dividing (7.67) by c and integrating yields

$$f(x) - g(x) = \frac{1}{c} \int_0^x G(y)\, dy + A, \qquad (7.68)$$

where A is a constant of integration. Adding and subtracting (7.66) and (7.68) yield the two equations

$$f(x) = \frac{1}{2} F(x) + \frac{1}{2c} \int_0^x G(y)\, dy + \frac{1}{2} A,$$

$$g(x) = \frac{1}{2} F(x) - \frac{1}{2c} \int_0^x G(y)\, dy - \frac{1}{2} A,$$

from which (7.65) follows. Since $F \in C^2$ and $G \in C$, it follows that $u \in C^2$ and satisfies the wave equation. The solution can be verified using Leibniz' rule. $\square$

Much insight can be gained by examining a simple problem. Therefore let us consider the initial value problem

$$u_{tt} - c^2 u_{xx} = 0, \quad x \in \mathbb{R},\ t > 0,$$

$$u(x,0) = F(x), \quad u_t(x,0) = 0, \quad x \in \mathbb{R}.$$

By (7.65) the solution is

$$u(x,t) = \tfrac{1}{2}(F(x - ct) + F(x + ct)).$$

To fix the idea, suppose $F(x)$ is the initial signal shown in Fig. 7.19. At time $t > 0$ we note that u is the average of $F(x - ct)$ and $F(x + ct)$, which is just the average of the two signals resulting from shifting $F(x)$ to the right ct units and to the left ct units. A sequence of time snapshots at $t_1 < t_2 < t_3$ in Fig. 7.20 shows how the two signals $F(x - ct)$ and $F(x + ct)$ are averaged to produce u. The initial signal splits into two signals moving in opposite directions, each at speed c. The left-moving signal travels along the characteristics $x + ct =$ constant, and the right-moving signal travels along the characteristics $x - ct =$ constant. Therefore, there are two families of characteristic lines along which disturbances propagate. If the initial data F and G in (7.64) are supported in

Figure 7.19 Initial curve.

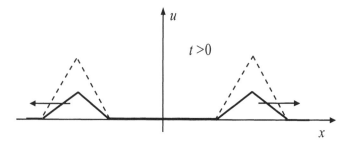

Figure 7.20 Time snapshots showing $u(x,t)$ (solid) as the average of two profiles $F(x+ct)$ and $F(x-ct)$ (dashed).

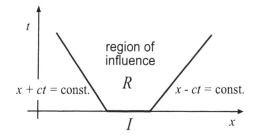

Figure 7.21 Region of influence R of an interval I.

an interval I on the x axis (i.e., F and G are zero outside I), then the region R of the xt plane that is affected by the disturbances within I is called the **region of influence** of I (see Fig. 7.21). This region has as its lateral boundaries the two characteristics $x + ct = $ constant and $x - ct = $ constant emanating from the left and right endpoints of I, respectively. A signal in I can never affect the solution outside R; or stated differently, if the initial data are supported in

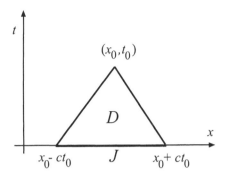

Figure 7.22 Domain D and its interval of dependence J.

I, then the solution is supported in R. These statements follow directly from (7.65).

We also pose the question of determining which initial values affect the value of u at a given point (x_0, t_0). From D'Alembert's solution,

$$u(x_0, t_0) = \frac{1}{2}(F(x_0 - ct_0) + F(x_0 + ct_0)) + \frac{1}{2c} \int_{x_0 - ct_0}^{x_0 + ct_0} G(y)\, dy.$$

Because of the integrated term, u will depend only on the values in the interval $I\colon [x_0 - ct_0, x_0 + ct_0]$. This interval, called the **interval of dependence**, is found by tracing the characteristics $x - ct = x_0 - ct_0$ and $x + ct = x_0 + ct_0$ back to the initial $t = 0$ line (Fig. 7.22).

When boundary conditions are present the situation is more complicated because waves are reflected from the boundaries. For example, consider the mixed initial boundary value problem

$$u_{tt} - c^2 u_{xx} = 0, \quad 0 < x < l,\ t > 0,$$
$$u(x, 0) = F(x), \quad u_t(x, 0) = G(x), \quad 0 \le x \le l,$$
$$u(0, t) = a(t), \quad u(l, t) = b(t), \quad t > 0,$$

where $F \in C^2[0, l]$, $G \in C^1[0, l]$, and $a, b \in C^2[0, \infty)$. If $u(x, t)$ is a class C^2 solution on $t \ge 0$ and $0 \le x \le l$, then necessarily the compatibility conditions $F(0) = a(0)$, $F(l) = b(0)$, $G(0) = a'(0)$, $g(l) = b(0)$, $a''(0) = c^2 F''(0)$, and $b''(0) = c^2 F(l)$ must hold at the corners $x = 0$, $x = l$, and $t = 0$. This problem can be transformed into one with homogeneous boundary conditions, which can be solved by eigenfunction expansions.

A geometric construction of the solution based on the following theorem gives insight into the nature of the solution to the wave equation.

Theorem 7.11

Let $u(x,t)$ be of class C^3 for $t > 0$ and $x \in \mathbb{R}$. Then u satisfies the wave equation (7.62) if, and only if, u satisfies the difference equation

$$u(x - ck, t - h) + u(x + ck, t + h) = u(x - ch, t - k) + u(x + ch, t + k) \quad (7.69)$$

for all $h, k > 0$.

Proof

Necessity follows easily from the fact that the general solution of the wave equation is given by (7.63) and both the forward- and backward-going waves $f(x + ct)$ and $g(x - ct)$ satisfy the difference equation. Conversely, if u satisfies the difference equation, then set $h = 0$, add $-2u(x, t)$ to both sides and then divide by $c^2 k^2$ to get

$$\frac{u(x - ck, t) + u(x + ck, t) - 2u(x, t)}{c^2 k^2} = \frac{u(x, t - k) + u(x, t + k) - 2u(x, t)}{c^2 k^2}.$$

$$(7.70)$$

By Taylor's theorem

$$u(x \pm ck, t) = u(x, t) \pm u_x(x, t)ck + \tfrac{1}{2}u_{xx}(x, t)c^2 k^2 + o(k^2),$$
$$u(x, t \pm k) = u(x, t) \pm u_t(x, t)k + \tfrac{1}{2}u_{tt}(x, t)k^2 + o(k^2).$$

Substituting these quantities into (7.70) gives

$$u_{xx}(x, t) = \frac{1}{c^2}u_{tt}(x, t) + \frac{o(k^2)}{k^2}.$$

Taking $k \to 0$ shows that u satisfies (7.62) and completes the proof. $\qquad\square$

Geometrically, the points $A\colon (x - ck, t - h)$, $B\colon (x + ch, t + k)$, $C\colon (x + ck, t + h)$, and $D\colon (x - ch, t - k)$ are the vertices of a **characteristic parallelogram** $ABCD$ formed by two pairs of characteristics, one forward-going pair and one backward-going pair. Then (7.69) may be written

$$u(A) + u(C) = u(B) + u(D)$$

This equation can be used to geometrically construct a solution to the initial boundary value problem as follows. We divide the region $0 < x < l$, $t > 0$ into regions I, II, III,..., as shown in Fig. 7.23, by drawing in the characteristics emanating from the lower corners $(0, 0)$ and $(l, 0)$ and continually reflecting them from the boundaries. The solution in region I is completely determined by D'Alembert's formula. To find the solution at a point R in region II we sketch the characteristic parallelogram $PQRS$ and use Theorem x to find $u(R) =$

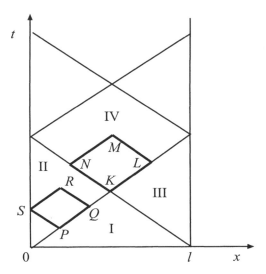

Figure 7.23 Geometrical construction of the solution to the wave equation.

$u(S)+u(Q)-u(P)$. The quantity $u(S)$ is determined by the boundary condition at $x = 0$, and $u(P)$ and $u(Q)$ are known from the solution in region I. A similar argument can be made for points in region III. To determine the solution at a point M in region IV we complete the characteristic parallelogram $KLMN$ as shown in the diagram and use Theorem 7.11 to conclude $u(M) = u(L) + u(N) - u(K)$. Again the three quantities on the right are known and we can proceed to the next step and obtain u in the entire region $t > 0$, $0 < x < l$.

7.4.3 Scattering and Inverse Problems

By a scattering problem we mean the following. Consider a situation where an object, called a **scatterer**, lies in some medium. An incoming *incident wave* impinges on the scatterer and produces a **reflected wave** and a **transmitted wave** (Fig. 7.24). The **direct scattering problem** is to determine the reflected and transmitted waves (amplitude, wave number, frequency, etc.) if the properties of the incident wave and the scatterer are known. The **inverse scattering problem** is to determine the properties of the scatterer, given the incident, reflected, and and possibly the transmitted waves. Inverse scattering problems arise in a number of physical situations. For example, in radar or sonar theory a known incident wave and observed reflected wave are used to detect the properties or presence of aircraft or submarine objects. In tomogra-

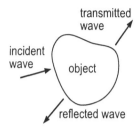

Figure 7.24 Object scattering an incident wave.

phy X-rays and sound waves are used to determine the presence or properties of tumors by detecting density variations; in this case the incident reflected, and transmitted waves are all known. In geologic exploration explosions may be set off on the earth's surface producing waves that, when reflected from underground layers, may indicate the presence of oil or other structures of geologic interest. In nondestructive testing, analysis of reflected and transmitted waves can indicate flaws in a material, for example, the nose cone of a shuttle craft. Inverse scattering theory is an active area of applied mathematics research. In the succeeding paragraphs we set up and solve a simple inverse problem involving the wave equation. We assume that an infinite bar $-\infty < x < \infty$ is composed of two homogeneous materials separated by an interface at $x = 0$. One material in region I ($x < 0$) has constant density ρ_1 and stiffness (Young's modulus) y_1 and the other in region II ($x > 0$) has constant density ρ_2 and stiffness y_2. The sound velocities in region I and region II are c_1 and c_2, respectively. Recall that $c^2 = y/\rho$. The inverse scattering problem can be stated as follows: Suppose an observer at $x = -\infty$ sends out an incident right traveling wave $\exp(i(x - c_1 t))$ of unit amplitude and unit wave number. When the wave impinges on the interface a wave is reflected back into region I and another wave is transmitted into region II (see Fig. 7.25). If the material properties ρ_1 and y_1 of region I are known, it is possible to determine the properties of ρ_2 and y_2 of region II by measuring the amplitude and wave number of the reflected wave? We answer this question in the context of small displacement theory where the displacements u_1 and u_2 in region I and region II, respectively, satisfy the one-dimensional wave equation

$$u_{tt} - c^2 u_{xx} = 0. \tag{7.71}$$

At the interface $x = 0$ the displacements must be continuous, since we assume that the regions do not separate. Therefore

$$u_1(0^-, t) = u_2(0^+, t), \quad t > 0 \tag{7.72}$$

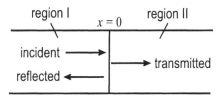

Figure 7.25 The interface between two materials scattering an incident wave.

Here $u_1(0^-, t)$ denotes the limit of $u_1(x, t)$ as $x \to 0^-$ and $u_2(0^+, t)$ denotes the limit of $u_2(x, t)$ as $x \to 0^+$. Further, we require that the force across the interface be continuous or

$$y_1 \frac{\partial u_1}{\partial x}(0^-, t) = y_2 \frac{\partial u_2}{\partial x}(0^+, t), \quad t > 0. \tag{7.73}$$

Equations (7.72) and (7.73) are called the *interface conditions*. We seek u_1 and u_2 satisfying (7.71) subject to (7.72) and (7.73). A reasonable attempt at a solution is to take u_1 to be the incident wave superimposed with the reflected wave and u_2 to be the transmitted wave. Hence we try a solution of the form

$$u_1 = \exp(i(x - c_1 t)) + A \exp(i\alpha(x + c_1 t)),$$
$$u_2 = B \exp(i\beta(x - c_2 t)).$$

The quantities A and α denote the amplitude and wave number of the reflected wave moving at speed c_1 in region I to the left, and B and β denote the amplitude and wave number of the transmitted wave moving at speed c_2 in region II to the right. Clearly u_1 and u_2 satisfy the wave equation with $c_1^2 = y_1/\rho_1$ and $c_2^2 = y_2/\rho_2$, respectively. The inverse scattering problem can now be stated in analytic terms as follows: Given A, α, ρ_1, and y_1, can ρ_2 and y_2 be determined?

The interface conditions (7.72) and (7.73) will yield relations among the constants. Condition (7.72) implies

$$\exp(-ic_1 t) + A \exp(i\alpha c_1 t) = B \exp(-i\beta c_2 t),$$

which, in turn, gives

$$\alpha = -1, \qquad \beta = \frac{c_1}{c_2}, \qquad B = 1 + A. \tag{7.74}$$

Therefore

$$u_1 = \exp[i(x - c_1 t)] + A \exp[-i(x + c_1 t)],$$
$$u_2 = (1 + A) \exp\left[i\frac{c_1}{c_2}(x - c_2 t)\right].$$

An application of (7.73) forces

$$y_1(1 - A) = y_2(1 + A)\frac{c_1}{c_2}. \tag{7.75}$$

A close examination of (7.74) and (7.75) yields the information we seek. Knowledge of A, y_1, and c_1 permits the calculation of the ratio y_2/c_2 or $\sqrt{y_2 \rho_2}$. Consequently both y_2 and ρ_2 cannot be calculated, only their product. If, however, there is means to measure the speed c_2 of the transmitted wave or determine its wave number β, then the material properties of region II can be found.

The direct scattering problem has a simple solution. If all of the material parameters ρ_1, y_1, ρ_2, and y_2 are known, relations (7.74) and (7.75) uniquely determine the reflected and transmitted waves.

EXERCISES

1. Maxwell's equations for a nonconducting medium with permeability μ and permittivity ε are

$$\frac{\partial \mathbf{B}}{\partial t} + \nabla \times \mathbf{E} = 0, \qquad \varepsilon \frac{\partial \mathbf{E}}{\partial t} = \frac{1}{\mu} \nabla \times \mathbf{B},$$

$$\nabla \cdot \mathbf{B} = 0, \qquad \nabla \cdot \mathbf{E} = 0,$$

   where $\mathbf{B}$ is the magnetic induction and $\mathbf{E}$ is the electric field. Show that the components of $\mathbf{B}$ and $\mathbf{E}$ satisfy the three-dimensional wave equation $u_{tt} - c^2(u_{xx} + u_{yy} + u_{zz}) = 0$ with propagation speed $c = (\varepsilon\mu)^{-1/2}$, where $u = u(t, x, y, z)$.

2. Show that the transformation

$$\xi = x - ct, \quad \eta = x + ct,$$

   transforms the wave equation (7.62) into the partial differential equation

$$\varphi_{\xi\eta} = 0, \quad \varphi(\xi, \eta) = u(x, y).$$

   As a consequence, show that the general solution to (7.62) is given by (7.63).

3. Solve

$$u_{tt} - u_{xx} = 0, \quad 0 < x < \pi, \ t > 0,$$
$$u(0, t) = u(\pi, t) = 0, \quad t > 0,$$
$$u(x, 0) = 0, \quad u_t(x, 0) = 4\sin x, \quad 0 < x < \pi.$$

4. Solve

$$u_{tt} - u_{xx} = 0, \quad x \in \mathbb{R}, \ t > 0,$$
$$u(x,0) = \sin x, \quad u_t(x,0) = 1, \quad x \in \mathbb{R}.$$

5. Let $v \in C^2$ satisfy the boundary value problem

$$v_{tt} - \frac{y_0}{\rho_0} v_{xx} = 0, \quad 0 < x < l, \ t > 0,$$
$$v(x,0) = v_t(x,0) = 0, \quad 0 < x < l,$$
$$v(0,t) = v(l,t) = 0, \quad t > 0.$$

a) Prove that $v = 0$ by introducing the **energy integral**

$$E(t) = \int_0^l \left(\tfrac{1}{2} \rho_0 v_t^2 + \tfrac{1}{2} y_0 v_x^2 \right) dx$$

and showing that $E(t) \equiv 0$ for all $t \geq 0$.

b) Prove that solutions to the boundary value problem

$$u_{tt} - c^2 u_{xx} = 0, \quad 0 < x < l, \ t > 0,$$
$$u(x,0) = F(x), \quad u_t(x,0) = G(x), \quad 0 < x < l,$$
$$u(0,t) = H(t), \quad u(l,t) = K(t), \quad t > 0,$$

are unique.

6. Let $u = u(x,y)$ be a solution to the boundary value problem in the unit square $\Omega = \{(x,y) \colon 0 < x, \ y < 1\}$:

$$u_{xy} = 0 \ \text{ on } \ \Omega, \qquad u = f \ \text{ on } \ \partial\Omega.$$

Show that f must satisfy the relation $f(O) + f(P) = f(Q) + f(R)$, where $O = (0,0)$, $P = (1,1)$, $Q = (1,0)$, and $R = (0,1)$. Discuss well-posedness for this hyperbolic equation.

7. Use Fourier transforms to find the solution to

$$u_{tt} - c^2 u_{xx} = 0, \quad x \in \mathbb{R}, \quad t > 0$$
$$u(x,0) = 0, \quad u_t(x,0) = G(x), \quad x \in \mathbb{R},$$

where G vanishes outside some closed, bounded interval.

8. Consider the boundary value problem

$$u_{tt} - c^2 u_{xx} = 0, \quad t > 0, \quad 0 < x < 1,$$
$$u(x,0) = F(x), \quad u_t(x,0) = G(x), \quad 0 < x < 1,$$
$$u(0,t) = a(t), \quad u(1,t) = b(t), \quad t > 0.$$

Transform this problem to one with a nonhomogeneous partial differential equation and homogeneous boundary conditions.

9. Consider the nonhomogeneous problem

$$u_{tt} - c^2 u_{xx} = f(x,t), \quad x \in \mathbb{R}, \ t > 0,$$
$$u(x,0) = 0, \quad u_t(x,0) = 0, \quad x \in \mathbb{R}.$$

By integrating over the characteristic triangle T bounded by characteristics emanating backward from the point (x_0, t_0) to the x axis, show that

$$u(x_0, t_0) = \frac{1}{2c} \int \int_T f(x,t) \, dx \, dt.$$

10. Consider a bar with constant stiffness y_0 and density ρ_0 occupying the region $x \leq 0$. An incident small displacement wave $\exp(i(x - c_0 t))$ impinges on $x = 0$ from $x = -\infty$.

   a) Find the reflected wave if the end at $x = 0$ is fixed.

   b) Find the reflected wave if the end at $x = 0$ is free.

   c) Find the reflected wave if a mass M is attached to the end at $x = 0$. (Hint: Show that the boundary condition is $M u_{tt} + y_0 u_x = 0$.)

11. Find the solution to the *outgoing signaling problem*

$$u_{tt} - c^2 u_{xx} = 0, \quad x > 0, \ -\infty < t < \infty.$$
$$u_x(0, t) = Q(t), \quad -\infty < t < \infty.$$

(Hint: There are no boundaries to the right, so attempt a right-traveling wave solution.)

12. A bar of constant cross-sectional area is composed of two materials. For $-\infty < x < 0$ and $1 < x < \infty$ the material parameters are y_0 and ρ_0 and for $0 < x < 1$ the parameters are y_1 and ρ_1. An incident displacement wave $\exp(i(x - c_0 t))$ impinges on the system from $-\infty$. Assuming the linearized theory and the fact that displacements are continuous across the discontinuities what are the frequency and wave number of the transmitted wave in the region $x > 1$?

13. In three dimensions the wave equation is

$$u_{tt} - c^2 \Delta u = 0,$$

where Δ is the Laplacian. For waves with spherical symmetry $u = u(r, t)$ and $\Delta u = u_{rr} + (2/r)u_r$. By introducing the variable $U = ru$, show that the general solution for the spherically symmetric wave equation is

$$u = \frac{1}{r}f(r - ct) + \frac{1}{r}g(r + ct).$$

14. Show that $u(x, t) = f(x - ct)$ is a weak solution to the wave equation $u_{tt} - c^2 u_{xx} = 0$, $(x, t) \in \mathbb{R}^2$, for any locally integrable function f on $\mathbb{R}$.

15. Use D'Alembert's formula to formally derive the solution to

$$u_{tt} - c^2 u_{xx} = 0, \quad x \in \mathbb{R}, t > 0,$$
$$u(x, 0) = 0, \ u_t(x, 0) = \delta(x - \xi), \quad x, \xi \in \mathbb{R}.$$

in the form

$$u(x, t) = \frac{1}{2c}(H(x + ct) - H(x - ct)),$$

where H is the Heaviside function. Comment on the physical meaning.

16. Find an integral representation for the solution to the initial boundary value problem for $u = u(x, y, t)$:

$$u_t = u_{yy}, \quad x, \ y > 0, \ t > 0,$$
$$u(x, y, 0) = 0, \quad x, \ y > 0,$$
$$u(0, 0, t) = u_0(t), \quad t > 0,$$
$$u_t(x, 0, t) = -u_x(x, 0, t) + u_y(x, 0, t), \quad x > 0, \ t > 0$$
$$\lim_{y \to \infty} u(x, y, t) = 0.$$

17. Show that the nonlinear wave equation

$$u_{tt} - u_{xx} + 4u^3 = 0$$

has solutions of the form $u(x, t) = F(\theta)$, where $\theta = kx - \omega t$, $\omega^2 > k^2$, and where F is a periodic function. (Hint: Transform to the FF'-phase plane.) Show that the period is given by

$$P = \sqrt{2(\omega^2 - k^2)} \int_{-a^{1/4}}^{a^{1/4}} \frac{d\vartheta}{\sqrt{a - \vartheta^4}},$$

where a is a constant defining characterizing an orbit.

REFERENCES AND NOTES

Most all books in partial differential equations discuss wave phenomena. The texts listed below are only a small, specialized sample. Whitham (1974) and Logan (1994) contain a large amount of material on linear and nonlinear waves. Brauer & Castillo-Chavez (2001), Murray (2001), and Kot (2001) discuss age structure and partial differential equations.

Brauer F. & Castillo-Chavez, C. 2001. *Mathematical Models in Population Biology and Epidemiology*, Springer-Verlag, New York.

Kot, M. 2001. *Elements of Mathematical Ecology*, Cambridge University Press, Cambridge.

Logan, J. D. 1994. *An Introduction to Nonlinear Partial Differential Equations*, Wiley-Interscience, New York.

Murray, J. D. 2001. *Mathematical Biology*, Vol 1, Springer–Verlag, New York.

Whitham, G. B. 1974. *Linear and Nonlinear Waves*, Wiley-Interscience, New York.

8

Mathematical Models of Continua

Continuum mechanics (fluid dynamics, elasticity, gasdynamics, etc.) is the study of motion, kinematics, and dynamics of continuous systems of particles such as fluids, solids, and gases. Out of this subject evolve some of the most important partial differential equations of mathematical physics and engineering and some of the most important techniques for solving applied problems. For example, the origins of singular perturbation theory lie in the study of the flow of air around a wing or airfoil, and many of the problems of bifurcation theory have their beginnings in fluid mechanics. More than any other area of engineering or physics, continuum mechanics remains a paradigm for the development of techniques and examples in applied mathematics.

The field equations, or governing equations, of continuum mechanics are a set nonlinear partial differential equations for unknown quantities such as density, pressure, displacement, particle velocity, and so on, which describe the state of the continuum at any instant. They arise, as in the case of the heat conduction equation from (a) conservation laws such as conservation of mass, momentum, and energy and (b) assumptions regarding the makeup of the continuum, called constitutive relations or equations of state. Under certain cases, say, when the amplitude of waves is small, the equations reduce to some of the simpler equations of applied mathematics, like the one-dimensional wave equation.

The equations are developed from the concept that the material under investigation is a continuum; that is, it exhibits no structure however finely it is divided. This development gives rise to a model in which the fluid parameters are defined at all points as continuous functions of space and time. The molecular aggregation of matter is completely extraneous to the continuum model. The model may be expected to be invalid when the size of the continuum region of interest is the same order as the characteristic dimension of the molecular structure. For gases this dimension is on the order of 10^{-7} meters, which is the mean free path, and for liquids it is on the order of 10^{-10} meters, which is a few intermolecular spacings. Thus the continuum model is violated only in extreme cases.

8.1 Kinematics

The underlying physical model to which we apply the general ideas is a one-dimensional continuum that can be thought of as a cylindrical tube through which a fluid is flowing. The fluid may be a liquid or gas and the lateral wall of the pipe is assumed to have no effect on the flow parameters. Throughout the motion we assume that each cross-section remains planar and moves longitudinally down the cylinder and that there is no variation of any of the flow parameters in any cross section. It is this assumption, that the only variation is longitudinal, that gives the description a one-dimensional character. The situation is the same as in the one-dimensional models studied in Chapters 6 and 7.

To understand how physical quantities are defined in a one-dimensional continuum we consider the following thought experiment to observe how the density of a fluid is related to its molecular structure. At time t we consider a disk D of fluid of width α centered at x_0 and having cross-sectional area A. See Fig. 8.1. The average density of the fluid in D is $\rho_\alpha = M_\alpha/\alpha A$, where M_α is the mass of fluid in D. To define the density $\rho(x_0, t)$ we examine what happens as α approaches zero. The graph in Fig. 8.2 records the results of the experiment. In region II, because there are many particles (molecules) inside D, one would expect that the average density ρ_α would vary very little. If, however, α were on the order of molecular distances, say $\alpha = 10^{-9}$ meters, there may be only a few molecules in D and we might expect large fluctuations in ρ_α even for small changes in α. Such rapid fluctuations are depicted in region I in Fig. 8.2. It seems unreasonable, therefore, to define $\rho(x_0, t)$ as the limiting value of ρ_α as $\alpha \to 0$. Rather, $\rho(x_0, t)$ should be defined as

$$\rho(x_0, t) = \lim_{\alpha \downarrow \alpha^*} \rho_\alpha,$$

where α^* is the value of α where density nonuniformities begin to occur; here, for example, $\alpha^* = 10^{-9}$ meters. In a similar fashion other physical quantities can be considered as point functions in continuously distributed matter without

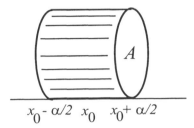

Figure 8.1 A fluid disk D with width α, centered at x_0.

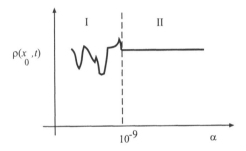

Figure 8.2 Plot of the average density ρ_α vs. the disk of width α (meters).

regard to its molecular or atomistic structure. This continuum assumption is often phrased as follows—the fluid is composed of small regions or *fluid elements*
that can be idealized as points at which the flow variables become continuous functions of position and time, but they are not so small that discernable
fluctuations of these quantities exist within the element. We now discuss the
kinematics of fluid motion. In a one-dimensional moving fluid there are two
basic coordinate systems used to keep track of the motion. For example, suppose water is flowing in a stream and one wishes to measure the temperature.
One can stand on the bank at a fixed location x, measured from some reference
position $x = 0$, and insert a thermometer, thereby measuring the temperature
$\theta(x,t)$ as a function of time and the fixed position x. The coordinate x is called
a laboratory or **Eulerian coordinate**. On the other hand, one can measure
the temperature from a boat riding with the flow. In this case, at time $t = 0$
each particle (section) is labeled with a particle label h, and each particle always retains its label as it moves downstream. The result of the measurement
is the temperature $\Theta(h,t)$ as a function of time t and the **Lagrangian** or
material coordinate h. The variable x is a fixed spatial coordinate, and the
representation of the field functions or physical variables (temperature, density,
pressure, etc.) as functions of t and x gives a *Eulerian* description of the flow.

A representation in terms of t and the material variable h gives a *Lagrangian* description of the flow. For physical variables we reserve lowercase letters for Eulerian quantities and capital letters for Lagrangian quantities. Thus, $f(x,t)$ denotes the measurement of a physical quantity in Eulerian coordinates, and $F(h,t)$ denotes the corresponding description in terms of Lagrangian coordinates.

Now let I be an interval representing the one-dimensional region along the axis of the cylinder where the fluid is in motion. At time $t = 0$ we label all of the particles (in a one-dimensional continuum the word *particle* is synonymous with *cross section,* and we shall use them interchangeably) with a Lagrangian coordinate h such that

$$x = h \quad \text{at} \quad t = 0,$$

where x is a fixed Eulerian coordinate in I. By a **fluid motion**, or flow, we mean a twice continuously differentiable mapping $\phi \colon I \times [0, t_1] \to \mathbb{R}$ defined by

$$x = \phi(h, t), \tag{8.1}$$

which for each $t \in [0, t_1]$ is invertible on I. Rather than write ϕ in (8.1), we use the symbol x and write

$$x = x(h, t). \tag{8.2}$$

The convention of using x to denote both a spatial coordinate and a function is common and often preferred over (8.1). In words, (8.2) gives the position x at time t of the particle, or cross section, labeled h. For a fixed $h = h_0$ the curve

$$x = x(h_0, t), \quad 0 \le t \le t_1$$

defines the **particle path** of the particle labeled h_0.

The invertibility assumption on ϕ guarantees that (8.2) can be solved for h to obtain

$$h = h(x, t). \tag{8.3}$$

This equation determines the particle, or cross section, at position x at time t. Since (8.2) and (8.3) are inverses,

$$x = x(h(x, t), t), \tag{8.4}$$

and

$$h = h(x(h, t), t). \tag{8.5}$$

Consequently, if $f(x, t)$ is a physical quantity in Eulerian form, then

$$F(h, t) = f(x(h, t), t) \tag{8.6}$$

gives the Lagrangian description. Conversely, if $F(h, t)$ is a Lagrangian quantity, then

$$f(x, t) = F(h(x, t), t) \tag{8.7}$$

gives its description in Eulerian form. The duality expressed by (8.6) and (8.7) is physically meaningful; for example, (8.7) states that the Eulerian measurement f made by an individual at x at time t coincides with the measurement F made by one moving on the particle h, at the instant t, when h is at x. The chain rule shows that the derivatives are related by

$$F_h(h, t) = f_x(x(h, t), t)x_h(h, t) \tag{8.8}$$

and

$$F_t(h, t) = f_t(x(h, t), t) + f_x(x(h, t), t)x_t(h, t), \tag{8.9}$$

where $f_x(x(h, t), t)$ means $f_x(x, t)$ evaluated at $x = x(h, t)$, for example. Good notation in fluid dynamics is essential. Here we have indicated explicitly the points at which the derivatives are evaluated. Shortcuts can sometimes lead to confusing and little understood expressions and equations, and writing out the general formulas in detail at least once can dismiss much of the confusion.

As in classical mechanics we define the **velocity** of a particle or cross section h as the time rate of change of its position, or

$$V(h, t) = x_t(h, t).$$

Therefore the Eulerian form of the velocity is defined by

$$v(x, t) = V(h(x, t), t),$$

which gives the velocity of the cross section now at location x. The **acceleration** of the cross section labeled h is

$$A(h, t) = V_t(h, t) = x_{tt}(h, t).$$

Therefore, the acceleration of the cross section now at location x is given in Eulerian form by

$$a(x, t) = V_t(h(x, t), t) = V_t(h, t)|_{h(x,t)}. \tag{8.10}$$

From (8.9) it follows that

$$a(x, t) = v_t(x, t) + v(x, t)v_x(x, t).$$

The time rate of change of a physical quantity following a cross section or particle h that is now located at x, such as occurs on the right side of (8.10), is called the material derivative of that quantity. Precisely, the **material derivative** of $f(x, t)$ is defined by

$$\frac{Df}{Dt}(x, t) \equiv F_t(h(x, t), t) = F_t(h, t)|_{h=h(x,t)}, \tag{8.11}$$

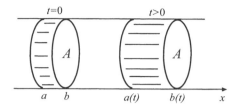

Figure 8.3 A material quantity of fluid at $t = 0$ and at $t > 0$.

where on the right we have explicitly noted that the derivative is taken before the evaluation at $h(x, t)$ is made. The quantity Df/Dt may be interpreted as the time derivative of F following the cross section h, frozen at the instant h is located at x. From (8.9) it easily follows that

$$\frac{Df}{Dt} = f_t + v f_x. \tag{8.12}$$

Therefore, in Eulerian coordinates, the acceleration can be written

$$a = \frac{Dv}{Dt}.$$

Another kinematic relation that plays an important role in the derivation of the governing equations is the time rate of change of the Jacobian of the mapping ϕ defining the flow. The **Jacobian** is defined by

$$J(h, t) \equiv x_h(h, t).$$

Hence

$$J_t(h, t) = \frac{\partial}{\partial h} x_t(h, t) = V_h(h, t) = v_x(x(h, t), t) x_h(h, t),$$

or

$$J_t(h, t) = v_x(x(h, t), t) J(h, t). \tag{8.13}$$

Equation (8.13) is the **Euler expansion formula**. To obtain its Eulerian form, we let $h = h(x, t)$, so

$$\frac{Dj}{Dt}(x, t) = v_x(x, t) j(x, t),$$

where $j(x, t) = J(h(x, t), t)$.

8.1.1 Mass Conservation

We now derive the *field equations* that govern the motion of a one-dimensional continuous medium. These equations express conservation of mass, momentum, and energy, and are universal in that they are valid for any medium. The derivation of the field equations is based on a Lagrangian approach. First, we find the analytic implications of the fact that the mass in an arbitrary material portion of the cylinder at $t = 0$ does not change as that portion of material moves in time. At time $t = 0$ we consider an arbitrary portion of fluid between $x = a$ and $x = b$ (see Fig. 8.3), and after time t we suppose that this portion of fluid has moved to the region between $x = a(t) \equiv x(a, t)$ and $x = b(t) \equiv x(b, t)$. If $\rho(x, t)$ denotes the density of the fluid, then the amount of fluid between $a(t)$ and $b(t)$ is

$$\int_{a(t)}^{b(t)} \rho(x, t) A \, dx.$$

Mass conservation requires

$$\frac{d}{dt} \int_{a(t)}^{b(t)} \rho(x, t) \, dx = 0. \tag{8.14}$$

Using Leibniz's formula for differentiating integrals with variable limits we could proceed directly and compute the left side of (8.14). We prefer, however, a method that easily generalizes to higher dimensions and is more in the spirit of techniques in fluid dynamics. We change variables in (8.14) according to $x = x(h, t)$. Then $dx = J(h, t) \, dh$, and (8.14) becomes

$$\frac{d}{dt} \int_a^b R(h, t) J(h, t) \, dh = 0, \tag{8.15}$$

where $R(h, t)$ (read as an uppercase ρ) is the Lagrangian density defined by $R(h, t) = \rho(x(h, t), t)$. Note that (8.14) has been changed to an integral with constant limits so that the derivative may be brought directly under the integral because R and J are continuously differentiable. Consequently,

$$\int_a^b (RJ_t + JR_t) \, dh = \int_a^b [R(h, t) v_x(x(h, t), t) + R_t(h, t)] J \, dh = 0,$$

where we used the Euler expansion formula (8.13). Because a and b are arbitrary and J is nonzero, we conclude that

$$R_t(h, t) + R(h, t) v_x(x(h, t), t) = 0, \tag{8.16}$$

from which we get

$$R_t + \frac{R}{J} V_h = 0, \tag{8.17}$$

which is the **conservation of mass** law in Lagrangian form. The Eulerian form can be obtained directly from (8.16) by substituting $h = h(x, t)$. We get

$$\frac{D\rho}{Dt} + \rho v_x = 0, \tag{8.18}$$

or

$$\rho_t + v\rho_x + \rho v_x = 0. \tag{8.19}$$

Equation (8.18), or (8.19), is known as the **continuity equation** and it is one of the fundamental equations of one-dimensional flow. It is a first-order nonlinear partial differential equation in terms of the density ρ and velocity v. The fluid motion is said to be **incompressible** if

$$\frac{D\rho}{Dt} = 0 \quad \text{or} \quad R_t = 0,$$

and *steady* if ρ and v are independent of t.

8.1.2 Momentum Conservation

In classical mechanics, a particle of mass m having velocity $\mathbf{v}$ has linear momentum $m\mathbf{v}$. Newton's second law asserts that the time rate of change of momentum of the particle is equal to the net external force acting upon it. To generalize this law to one-dimensional continuous media we assume the *balance of linear momentum principle*, which states that the time rate of change of linear momentum of any portion of the fluid equals the sum of the external forces acting upon it. The **linear momentum** at time t of material in the material region $a(t) \leq x \leq b(t)$ is defined by

$$\mathbf{i} A \int_{a(t)}^{b(t)} \rho(x, t) v(x, t) \, dx,$$

where $\mathbf{i}$ is the unit vector in the positive x direction, $v(x, t)$ is the velocity, and A is the cross-sectional area.

The precise characterization of the forces acting on a material region in a continuous medium was the result of ideas evolving from works of Newton, Euler, and Cauchy. Basically, there are two types of forces that act on the material region, body forces and surface forces. **Body forces** are forces such as gravity, or an electric or magnetic field. Such a force is assumed to act on each cross section of the region and it is represented by

$$\mathbf{f}(x, t) = f(x, t)\mathbf{i}.$$

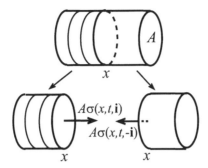

Figure 8.4 Stress vectors.

The units of f are force per unit mass, and the total body force acting on the region $a(t) \leq x \leq b(t)$ is therefore

$$\mathbf{i}A \int_{a(t)}^{b(t)} \rho(x,t)f(x,t) \; dx.$$

Surface forces are forces like pressure that act across sections in the fluid medium. More specifically, consider a cross section at time t located at x. By $\sigma(x,t,\mathbf{i})$ we denote the force per unit area on the material *on* the negative (left) side of the cross section *due to* the material on the positive (right) side of the section. Similarly $\sigma(x,t,-\mathbf{i})$ denotes the force per unit area exerted on the material *on* the right side of the section *by* the material on the left side of the section. By convention, the third argument in σ, here either $\mathbf{i}$ or $-\mathbf{i}$, is a vector that points outward and normal from the surface on which the force is acting. The vectors $\sigma(x,t,\mathbf{i})$ and $\sigma(x,t,-\mathbf{i})$ are called surface tractions or **stress vectors**. Figure 8.4 depicts a geometric description of these notions. Before a cross section is indicated, no stress is defined; however, once a cross section x is indicated, the forces $A\sigma(x,t,\mathbf{i})$ and $A\sigma(x,t,-\mathbf{i})$, which are exerted on the shaded and unshaded portions, respectively, are defined. In anticipation of what we prove later, we have drawn these forces opposite and equal in Fig. 8.4. At present it is not known in which direction these point. And we emphasize that the third argument in σ, either $\mathbf{i}$ or $-\mathbf{i}$, does not define the direction of the stress but serves only to indicate the orientation, or normal direction, of the surface of the section. We can now write a quantitative description of the balance of

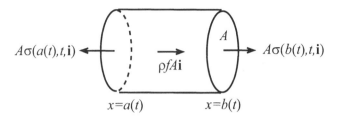

$$A\sigma(a(t),t,\mathbf{i}) \longleftarrow \qquad\qquad \xrightarrow{} \qquad\qquad \xrightarrow{A} A\sigma(b(t),t,\mathbf{i})$$

$$\rho f A\mathbf{i}$$

$$x=a(t) \qquad\qquad x=b(t)$$

Figure 8.5 Forces on a fluid element.

the linear momentum principle for a material region $a(t) \leq x \leq b(t)$,

$$\mathbf{i}\frac{d}{dt}A\int_{a(t)}^{b(t)} \rho(x,t)v(x,t)\,dx$$

$$= \mathbf{i}A\int_{a(t)}^{b(t)} \rho(x,t)f(x,t)\,dx + A\sigma(b(t),t,\mathbf{i}) + A\sigma(a(t),t,-\mathbf{i}). \qquad (8.20)$$

In words, the time rate of change of momentum of the material region $[a(t),b(t)]$ equals the total body force plus the surface forces on $[a(t),b(t)]$. A schematic of the forces is shown in Fig. 8.5. To put (8.20) in a more workable form we calculate the left side in a similar manner as was done for the integral form of the conservation of mass equation in the preceding section. Changing variables to the Lagrangian coordinate h and using the Euler expansion formula gives

$$\frac{d}{dt}\int_{a(t)}^{b(t)} \rho v\,dx = \frac{d}{dt}\int_{a}^{b} RVJ\,dh = \int_{a}^{b}\left(J\frac{\partial}{\partial t}(RV) + RVJ_t\right)dh$$

$$= \int_{a}^{b}\left[\frac{\partial}{\partial t}(RV) + RVv_x\right]J\,dh = \int_{a(t)}^{b(t)}\left[\frac{D}{Dt}(\rho v) + \rho vv_x\right]dx.$$

From the mass conservation equation (8.18) it follows that

$$\frac{D}{Dt}(\rho v) = \rho\frac{Dv}{Dt} + v\frac{D\rho}{Dt} = \rho\frac{Dv}{Dt} - v\rho v_x.$$

Hence,

$$\frac{d}{dt}\int_{a(t)}^{b(t)} \rho v\,dx = \int_{a(t)}^{b(t)} \rho\frac{Dv}{Dt}\,dx, \qquad (8.21)$$

and (8.20) can be written

$$\mathbf{i}\int_{a(t)}^{b(t)}\left(\rho\frac{Dv}{Dt} - \rho f\right)dx = \sigma(b(t),t,\mathbf{i}) + \sigma(a(t),t,-\mathbf{i}). \qquad (8.22)$$

This equation can be further simplified using the following result, which is the continuum mechanical version of the action–reaction principle expressed in Newton's third law.

Theorem 8.1

The balance of linear momentum principle expressed by (8.22) implies

$$\sigma(x, t, -\mathbf{i}) = -\sigma(x, t, \mathbf{i}) \tag{8.23}$$

for any cross section x.

Proof

Let x_0 be arbitrary with $a(t) < x_0 < b(t)$. Applying (8.22) to the region between $a(t)$ and x_0 and to the region between x_0 and $b(t)$, we get

$$\mathbf{i} \int_{a(t)}^{x_0} \left[\rho \frac{Dv}{Dt} - \rho f \right] dx = \sigma(x_0, t, \mathbf{i}) + \sigma(a(t), t, -\mathbf{i}),$$

and

$$\mathbf{i} \int_{x_0}^{b(t)} \left[\rho \frac{Dv}{Dt} - \rho f \right] dx = \sigma(b(t), t, \mathbf{i}) + \sigma(x_0, t, -\mathbf{i}).$$

Adding these two equations and then subtracting (8.22) from the result gives

$$\sigma(x_0, t, \mathbf{i}) + \sigma(x_0, t, -\mathbf{i}) = 0,$$

which, because of the arbitrariness of x_0, proves (8.23). $\qquad \square$

This result permits us to define the scalar **stress** component $\sigma(x, t)$ by the equation

$$\sigma(x, t, \mathbf{i}) = \sigma(x, t)\mathbf{i}.$$

Therefore (8.22) becomes

$$\int_{a(t)}^{b(t)} \left(\rho \frac{Dv}{Dt} - \rho f \right) dx = \sigma(b(t), t) - \sigma(a(t), t)$$

$$= \int_{a(t)}^{b(t)} \sigma_x(x, t) \, dx.$$

Because the interval $[a(t), b(t)]$ is arbitrary, we have

$$\rho \frac{Dv}{Dt} = \rho f + \sigma_x, \tag{8.24}$$

which is the Eulerian form expressing balance of linear momentum. A familiar form of the equation can be written if we define the **pressure** p by

$$p(x, t) = -\sigma(x, t).$$

Then (8.24) becomes

$$v_t + v v_x + \frac{1}{\rho} p_x = f. \tag{8.25}$$

8.1.3 Thermodynamics and Energy Conservation

In summary, the Eulerian equations of motion for a one-dimensional continuum with no body forces are

$$\rho_t + v\rho_x + \rho v_x = 0, \tag{8.26}$$

$$v_t + vv_x + \frac{1}{\rho}p_x = 0. \tag{8.27}$$

This is a pair of first-order nonlinear partial differential equations for the three unknowns ρ, v, and p. Intuition tells us that a third equation is needed. Further reflection dictates that particular physical properties of the medium must play a role, and such properties are not included in (8.26)–(8.27), which are completely general and hold for any continuum. Equations that specify properties of the medium are known as equations of state or **constitutive relations**. Such equations give relations between observable effects and the internal constitution of the material, and they are generally expressible in terms of thermodynamic variables such as density, pressure, energy, entropy, temperature, and so on.

Example 8.2

A simple equation of state is one in which the pressure depends only upon the density, or

$$p = F(\rho), \tag{8.28}$$

where F is a given differentiable function and $F'(\rho) > 0$. Equation (8.28) is known as a **barotropic equation of state**, and it allows the pressure to be eliminated from (8.27) to obtain

$$v_t + vv_x + \frac{F'(\rho)}{\rho}\rho_x = 0. \tag{8.29}$$

Therefore equations (8.26) and (8.29) govern *barotropic flow* and represent a complete set of equations for the two unknowns ρ and v. It is not evident at this point, but implicit in (8.28) are assumptions regarding the other thermodynamic variables that may, or may not, be valid for a general fluid motion. We recall from Chapter 7 that linearization of these equations about some constant, ambient state gave rise to the equations of acoustics, and ultimately the wave equation.

In contrast to (8.28), it is more often the case that the equation of state introduces yet another unknown variable in the problem, for example, the temperature. If that were the case, then still another equation is required. Such an equation comes from energy conservation. For general fluid motions a complete set of field equations usually consists of conservation equations for mass, momentum, and energy, as well as one or more constitutive relations. A discussion of energy conservation naturally entails the development of some basic concepts in equilibrium thermodynamics. Classic thermodynamics deals with relations between equilibrium states of uniform matter and the laws that govern changes in those states. It is implicit that the changes occur through a sequence of uniform equilibrium states, and the usual results of equilibrium thermodynamics apply. (It is possible to imagine processes that take place so rapidly that local fluid elements are incapable of establishing instantaneous equilibrium. Such processes belong to the study of nonequilibrium fluid dynamics.) Each of the thermodynamic quantities, such as temperature, pressure, and so on, are assumed to be functions of position and time and thermodynamical relations are assumed to hold locally. To illustrate this principle, consider a hypothetical gas in a container and let P, R, and Θ denote its pressure, density, and temperature, respectively. It is observed experimentally that the ratio of pressure to density is proportional to the temperature, or

$$\frac{P}{R} = \mathcal{R}\Theta, \tag{8.30}$$

where $\mathcal{R}$ is the constant of proportionality, characteristic of the gas. Equation (8.30) is a Lagrangian statement because it holds for a material volume of gas. The assumption of local thermodynamic equilibrium allows us to assume that

$$\frac{p}{\rho} = \mathcal{R}\theta,$$

where p, ρ, and the temperature θ are the local Eulerian quantities that are functions of x and t.

Example 8.3

(Ideal gas) Under normal conditions most gases obey the *ideal gas law*

$$p = \mathcal{R}\rho\theta, \tag{8.31}$$
$$e = c_v\theta + \text{constant}, \tag{8.32}$$

where p is pressure, ρ is density, θ is temperature, and e is the internal energy per unit mass. The constant c_v is the specific heat at constant volume and $\mathcal{R}$ is the *gas constant* for the particular gas. A gas satisfying (8.31) and (8.32) is called an **ideal gas** and many compressible fluids of practical interest can be approximately treated as ideal gases. Hence (8.31) and (8.32) describe a wide range of phenomena in fluid dynamics. These equations coupled with (8.26) and (8.27) give only four equations for the five unknowns p, v, ρ, θ, and e, so it is again evident that another equation is required.

A common statement of the first law of thermodynamics is that energy is conserved if heat is taken into account. Thus, the first law of thermodynamics provides a concise statement of energy conservation. It comes from considering the consequences of adding a small amount of heat to a unit mass of a material substance in a way that equilibrium conditions are established at each step. Some of the energy will go into the work $pd(1/\rho)$ done in expanding the specific volume $1/\rho$ by $d(1/\rho)$, and the remainder will go into increasing the internal energy e by de. The precise relationship is

$$q = de + pd\rho^{-1}, \tag{8.33}$$

where the differential form q is the heat added. For an equilibrium process equation (8.33) is the **first law of thermodynamics**. In general, q is not exact, that is, there does not exist a state function Q for $q = dQ$. If $q = 0$, then the process is called *adiabatic*.

Example 8.4

(Equations of state) For an ideal gas described by equations (8.31) and (8.32) we have

$$q = de + pd\rho^{-1} = c_v \, d\theta + \mathcal{R}\theta \, d(\ln \rho^{-1}). \tag{8.34}$$

From (8.34) we notice that θ^{-1} is an integrating factor for the differential form q. Therefore

$$\theta^{-1}q = d(c_v \ln \theta + \mathcal{R} \ln \rho^{-1}) = ds,$$

where

$$s \equiv c_v \ln \theta + \mathcal{R} \ln \rho^{-1} + \text{constant} \tag{8.35}$$

is the *entropy*. Consequently, for an ideal gas the first law (8.33) takes the form

$$de = \theta \, ds - pd\rho^{-1}. \tag{8.36}$$

Equation (8.35) may also be written in terms of the pressure as

$$p = kp^\gamma \exp\left(\frac{s}{c_v}\right), \tag{8.37}$$

where

$$\gamma \equiv 1 + \frac{\mathcal{R}}{c_v}$$

and k is a constant. Combining (8.31) and (8.32) gives yet another form of the equation of state, namely

$$e = \frac{p}{(\gamma - 1)\rho} + \text{constant} \tag{8.38}$$

Finally, in some contexts it is useful to introduce the enthalpy h defined by $h = e + p/\rho$. For an ideal gas $h = c_p\theta$, where $c_p = \mathcal{R} + c_v$ is the specific heat at constant pressure. Hence $\gamma = c_p/c_v$ is the ratio of specific heats. For air, $\gamma = 1.4$, and for a monatomic gas, $\gamma = \frac{5}{3}$. Generally, $\gamma > 1$.

Other equations of state have been proposed to include various effects. The *Abel* or *Clausius equation of state*

$$p\left(\frac{1}{\rho} - \alpha\right) = \mathcal{R}\theta, \qquad \alpha \text{ constant}$$

introduces a constant covolume α to account for the size of the molecules. The *van der Waals equation of state*

$$(p + \beta\rho^2)\left(\frac{1}{\rho} - \alpha\right) = \mathcal{R}\theta, \qquad \alpha, \beta \text{ constant}$$

contains the term $\beta\rho^2$ to further account for intermolecular forces. The *Tait equation*

$$\frac{p + B}{p} = \left(\frac{\rho}{\rho_0}\right)^{\bar{\gamma}},$$

where $\bar{\gamma}$ and B are constants, has been used to model the behavior of liquids at high pressures.

We now formulate a partial differential equation governing the flow of energy in a system. The approach is consistent with the earlier development of balance laws in integral form. Let $[a(t), b(t)]$ be a one-dimensional material region with cross-sectional area A. We define the kinetic energy of the fluid in the region by $A\int_{a(t)}^{b(t)} \frac{1}{2}\rho v^2 \, dx$ and the internal energy in the region by $A\int_{a(t)}^{b(t)} \rho e \, dx$. By the general conservation of energy principle the time rate of change of the total energy equals the rate that the forces do work on the region plus the rate that heat flows into the region. There are two forces, the body force $f(x,t)$ acting at each cross section of the region, and the stress $\sigma(x,t)$ acting at the two ends. Since force times velocity equals the rate work is done, the total rate that work is done on the region is $A\int_{a(t)}^{b(t)} \rho f v \, dx - A\sigma(a(t), t)v(a(t), t) + A\sigma(b(t), t)v(b(t), t)$. The rate that heat flows into the region is $AJ(a(t), t) - AJ(b(t), t)$, where $J(x,t)$ is the heat flux in energy units per unit area per time. Therefore, we *postulate* the balance of energy law

$$\frac{d}{dt}\int_{a(t)}^{b(t)} \left(\tfrac{1}{2}\rho v^2 + \rho e\right) \, dx$$

$$= \int_{a(t)}^{b(t)} \rho f v \, dx - \sigma(a(t), t)v(a(t), t) + \sigma(b(t), t)v(b(t), t)$$

$$+ J(a(t), t) - J(b(t), t),$$

or

$$\frac{d}{dt} \int_{a(t)}^{b(t)} \left(\tfrac{1}{2}\rho v^2 + \rho e\right) dx = \int_{a(t)}^{b(t)} (\rho f v + (\sigma v)_x - J_x) \, dx.$$

Assuming sufficient smoothness of the state variables and using the arbitrariness of the interval $[a(t), b(t)]$, we have the following Eulerian differential form of the conservation of energy law:

$$\frac{\rho}{2}\frac{Dv^2}{Dt} + \rho\frac{De}{Dt} = \rho f v + (\sigma v)_x - J_x. \tag{8.39}$$

Equation (8.39) can be cast into various forms. First, multiplication of the momentum balance equation (8.24) by v gives

$$\frac{\rho}{2}\frac{Dv^2}{Dt} = \rho f v + v\sigma_x.$$

Subtracting this from (8.39) gives an equation for the rate of change of internal energy

$$\rho\frac{De}{Dt} = \sigma v_x - J_x. \tag{8.40}$$

An alternate approach to energy conservation comes from an examination of the first law of thermodynamics (8.33). The second law of thermodynamics states that the differential form q in general has integrating factor θ^{-1} and $\theta^{-1}q = ds$, where s is the **entropy** and θ is the absolute temperature. Thus

$$\theta \, ds = de + p d\rho^{-1}.$$

This combined form of the first and second laws of thermodynamics can be reformulated as a partial differential equation. Because it refers to a given material region, we *postulate*

$$\theta\frac{Ds}{Dt} = \frac{De}{Dt} + p\frac{D(1/\rho)}{Dt}, \tag{8.41}$$

which is another local form of the conservation of energy principle. Noting that $D\rho^{-1}/Dt = -(1/\rho^2)D\rho/Dt = -(1/\rho)v_x$ and putting $\sigma = -p$, equation (8.41) may be subtracted from (8.40) to obtain

$$\theta\frac{Ds}{Dt} = -\frac{1}{\rho}\phi_x,$$

which relates the entropy change to the heat flux. If we assume the *constitutive relation* $J = -K\theta_x$, which is Fourier's law, then the energy equation (8.41) can be expressed as

$$\frac{De}{Dt} + p\frac{D(1/\rho)}{Dt} = \frac{1}{\rho}(K\theta_x)_x. \tag{8.42}$$

Example 8.5

(Adiabatic flow) In adiabatic flow $\theta Ds/Dt = 0$, and the energy equation (8.41) becomes

$$\frac{De}{Dt} + p\frac{D(1/\rho)}{Dt} = 0. \tag{8.43}$$

Equations (8.26), (8.27), and (8.43), along with a *thermal equation of state* $p = p(\rho, \theta)$ and a *caloric equation of state* $e = e(\rho, \theta)$, give a set of five equations for the unknowns ρ, v, p, e, and θ.

Example 8.6

(Ideal gas) For adiabatic flow of an ideal gas, expression (8.38) for the energy can be substituted into (8.43) to obtain

$$\frac{Dp}{Dt} - \frac{\gamma p}{\rho}\frac{D\rho}{Dt} = 0. \tag{8.44}$$

Equations (8.26), (8.27), and (8.44) give three equations for ρ, v, and p. Because $Ds/Dt = 0$, the entropy is constant for a given fluid particle (cross section); in Lagrangian form $S = S(h)$. In this case the equations governing **adiabatic flow** are often written

$$\frac{D\rho}{Dt} + \rho v_x = 0, \quad \rho\frac{Dv}{Dt} + p_x = 0, \quad \frac{Ds}{Dt} = 0,$$

along with the equation of state (8.37), giving $p = p(s, \rho)$. If the entropy is constant initially, that is, $S(h, 0) = s_0$ for all h, then $s(x, t) = s_0$ for all x and t, and the resulting adiabatic flow is called **isentropic**. This case reduces to barotropic flow as discussed earlier.

In general, for nonadiabatic flow involving heat conduction, equations (8.26), (8.27), and (8.42), along with thermal and caloric equations of state give five equations for ρ, v, p, e, and θ. If there is no motion ($v = 0$) and $s = c_v \ln\theta+$ const., then (8.42) reduces to the classic diffusion equation

$$\rho c_v \theta_t = (K\theta_x)_x.$$

8.1.4 Stress Waves in Solids

The general conservation laws for continuous media also hold for solid mechanics. We write the governing equations as (see (8.19) and (8.24))

$$\rho_t + v\rho_x + \rho v_x = 0, \tag{8.45}$$

$$v_t + vv_x = f + \frac{1}{\rho}\sigma_x, \tag{8.46}$$

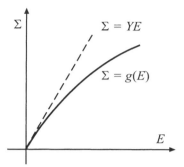

Figure 8.6 Stress vs. strain.

Our physical model is a cylindrical bar in which we seek to describe longitudinal vibrations, that is, the motion of the planar cross sections of the bar. In this case the equation of state or constitutive relation is generally a relation between the distortion (compression or elongation) the bar undergoes subject to an applied stress. To define the distortion we consider, at time $t = 0$, a small portion of the bar between h and $h + \Delta h$. At time $t > 0$ this material is located between $x(h, t)$ and $x(h + \Delta h, t)$. The *distortion* is the fractional change given by

$$\text{distortion} = \frac{\text{new length} - \text{original length}}{\text{original length}}$$
$$= \frac{x(h + \Delta h, t) - x(h, t) - \Delta h}{\Delta h}$$
$$= \frac{\partial x}{\partial h}(h, t) - 1 + O(\Delta h).$$

We define the **strain** E to be the lowest-order approximation of the distortion, that is,

$$E \equiv \frac{\partial x}{\partial h}(h, t) - 1.$$

If $U(h, t)$ denotes the *displacement* of a cross section h at time t, then $x(h, t) = h + U(h, t)$ and

$$E = \frac{\partial U}{\partial h}(h, t). \tag{8.47}$$

The constitutive relation is an equation that gives the Lagrangian stress $\Sigma(h, t)$ as a definite function of the strain $E(h, t)$. Quite generally, the graph may look like the stress–strain curve shown in Fig. 8.6. If, however, only small strains are of interest, then the stress–strain curve can be approximated by a straight line by a linear equation

$$\Sigma(h, t) = Y(h)E(h, t). \tag{8.48}$$

The proportionality factor $Y(h)$ is called *Young's modulus* or the *stiffness*, and the linear **stress–strain relation** (8.48) is called **Hooke's law**.[1] In Eulerian coordinates, (8.48) becomes

$$\sigma(x,t) = y(x,t)\varepsilon(x,t), \tag{8.49}$$

where $\sigma(x,t) = \Sigma(h(x,t),t)$, $y(x,t) = Y(h(x,t))$, and $\varepsilon(x,t) = E(h(x,t),t)$. Using the relation

$$\left.\frac{\partial U}{\partial h}\right|_{h=h(x,t)} = \frac{u_x}{1-u_x},$$

where u is the Eulerian displacement, Hooke's law (8.49) can be written

$$\sigma(x,t) = y(x,t)\frac{u_x(x,t)}{1-u_x(x,t)}. \tag{8.50}$$

Then, σ can be eliminated from the momentum equation (8.45) to obtain

$$\rho v_t + \rho v v_x = \rho f + \frac{\partial}{\partial x}\left(\frac{y u_x}{1-u_x}\right). \tag{8.51}$$

Therefore, with the assumption of a constitutive relation (8.48), we have obtained two equations, (8.50) and (8.51), for the three unknowns v, u, and ρ. The third equation is the defining relationship between u and v, or

$$v = \frac{Du}{Dt} = u_t + v u_x. \tag{8.52}$$

As they stand, equations (8.45), (8.51), and (8.52) are nonlinear and cannot be resolved analytically. As in the case of the acoustic approximation, however, we can obtain a linearized theory for small displacements u. Therefore consider a bar with initial density $\rho_0(x)$ and stiffness $y_0(x)$. For a small displacement theory assume u and its derivatives are small compared to unity. From (8.52)

$$v = u_t(1-u_x)^{-1} = u_t(1 + u_x + u_x^2 - \cdots) = u_t + O(u^2),$$

and

$$v_t = u_{tt} + O(u^2).$$

Let $\overline{\rho}(x,t)$ denote the small deviation from the initial density $\rho_0(x)$; that is,

$$\rho(x,t) = \rho_0(x) + \overline{\rho}(x,t).$$

---

[1] We assume here that Y does not depend on time t. Materials for which Y is time-dependent are said to have *memory*.

By introducing the Lagrangian density R and using the mean value theorem,

$$\begin{aligned}
|\overline{\rho}(x,t)| &= |\rho(x,t) - \rho_0(x)| \\
&= |R(x-u,t) - R(x,0)| \\
&\leq |R_h(\tilde{x},\tilde{t})u| + |R_t(\tilde{x},\tilde{t})|t,
\end{aligned}$$

where $0 < \tilde{t} < t$ and $x < \tilde{x} < x - u$. Hence the deviation $\overline{\rho}$ is small provided that u *and* t are small and that the derivatives R_h and R_t are bounded. A similar calculation shows that if $y(x,t) = y_0(x) + \overline{y}(x,t)$, then the deviation $\overline{y}$ satisfies the equality $|\overline{y}(x,t)| = |Y_h(\tilde{x})u|$, and therefore $\overline{y}$ is small if u is small and Y_h is bounded.

With all of this information, the momentum equation (8.51) can be written (taking $f = 0$)

$$(\rho_0 + \overline{\rho})(u_{tt} + O(u^2)) + (\rho_0 + \overline{\rho})(vv_x) = \frac{\partial}{\partial x}[(y_0 + \overline{y})(u_x)(1 + u_x + O(u^2))].$$

Retaining the lowest-order terms gives the linearized, small displacement equation

$$\rho_0(x)u_{tt} = \frac{\partial}{\partial x}(y_0(x)u_x). \tag{8.53}$$

Subject to the assumptions in the last paragraph we may expect (8.53) to govern the small longitudinal vibrations of a bar of density $\rho_0(x)$ and stiffness $y_0(x)$.

We notice that the unknown density dropped out of (8.51) and could be replaced by the initial density $\rho_0(x)$. Therefore (8.53) is one equation in the single unknown u; the conservation of mass equation is then a consistency relation. If $\rho_0(x) = \rho_0$ and $y_0(x) = y_0$, where ρ_0 and y_0 are constants, then (8.53) reduces to the wave equation

$$u_{tt} - c^2 u_{xx} = 0, \quad c^2 = \frac{y_0}{\rho_0}.$$

Generally, a rod of finite extent will be subject to boundary conditions at its ends, say at $x = 0$ and at $x = l$. For definiteness we focus on the end $x = l$. Clearly if the end is held fixed, then

$$u(l,t) = 0, \quad t > 0 \qquad \text{(fixed end)}.$$

If the end is free, or no force acts on the face at $x = l$, then $\Sigma(l,t) = 0$ or $Y(l)E(l,t) = 0$. Hence

$$Y(l)\frac{\partial U}{\partial h}(l,t) = 0, \quad t > 0,$$

or in Eulerian variables,

$$y(l,t)\frac{u_x(l,t)}{1 - u_x(l,t)} = 0, \quad t > 0 \qquad \text{(free end)}.$$

A properly posed problem for determining the small displacements $u(x,t)$ of a bar of length l with stiffness $y_0(x)$ and density $\rho_0(x)$ consists of the partial differential equation (8.53) along with boundary conditions at $x = 0$ and $x = l$, and with initial conditions of the form

$$u(x,0) = u_0(x), \quad u_t(x,0) = v_0(x), \quad 0 < x < l,$$

where $u_0(x)$ and $v_0(x)$ are the given initial displacement and velocity, respectively.

EXERCISES

1. Verify (8.8) and (8.9).

2. The reciprocal W of the Lagrangian density R is called the *specific volume*. If at time $t = 0$ the density is a constant ρ_0, prove that conservation of mass and momentum can be expressed as

$$W_t - \rho_0^{-1}V_h = 0,$$
$$V_t + \rho_0^{-1}P_h = 0.$$

3. Show that the Jacobian $J(h,t)$ is given by

$$J(h,t) = \frac{R(h,0)}{R(h,t)},$$

and therefore measures the ratio of the initial density to the density at time t.

4. Prove that

$$\frac{d}{dt}\int_{a(t)}^{b(t)} \rho(x,t)g(x,t)\,dx = \int_{a(t)}^{b(t)} \rho(x,t)\frac{Dg}{Dt}(x,t)\,dx$$

for any sufficiently smooth function g, where $a(t) \le x \le b(t)$ is a material region.

5. Write the equation $F_t + F^2 F_h = 0$, where $F = F(h,t)$, in Eulerian form.

6. A one-dimensional flow is defined by

$$x = (h - t)(1 + ht)^{-1}, \quad 0 < t < t_1.$$

a) Sketch the particle paths on xt and ht diagrams.

b) Find $V(h, t)$ and $v(x, t)$.

c) Verify the Eulerian expansion formula.

d) Verify that $Dv/Dt = v_t + vv_x$.

e) If the density is $R(h, t) = h^2$, find the density that an observer would measure at $x = 1$.

f) Repeat (a) through (e) for a fluid motion defined by $x = \frac{1}{2}h(1 + e^t)$.

7. Derive the conservation of mass and momentum equations for one-dimensional flow in a cylinder of variable (Lagrangian) cross-sectional area $A(h, t)$. Assume $A(h, t) = A(h, 0)$ for all $t > 0$. Specifically, show

$$\frac{D(a\rho)}{Dt} + a\rho v_x = 0, \qquad \rho a \frac{Dv}{Dt} = \rho a f + \frac{\partial}{\partial x}(a\sigma),$$

where $a(x, t) = A(h(x, t), 0)$.

8. By introducing the enthalpy $h = e + p/\rho$ into the energy equation $De/Dt + pD(\rho^{-1})/Dt = 0$, show that for steady flow

$$h_x - \frac{1}{\rho} p_x = 0.$$

Prove that

$$h + \tfrac{1}{2}v^2 = \text{constant}.$$

9. A solution of the wave equation

$$\rho_0(x)u_{tt} = \frac{\partial}{\partial x}(y_0(x)u_x), \quad 0 < x < l,$$

of the form

$$u(x, t) = w(x)\exp(i\omega t),$$

subject to boundary conditions at $x = 0$ and $x = l$, is called a time-harmonic standing wave. The set of values ω and the corresponding spatial distributions $w(x)$ are called the **fundamental frequencies** and **normal modes** of vibration, respectively. Determine the fundamental frequencies and normal modes for the following wave equations.

a) $u_{tt} - c^2 u_{xx} = 0, \quad 0 < x < \pi,$

   $u(0, t) = 0, \quad u_x(\pi, t) = 0.$

b) $u_{tt} - c^2 u_{xx} = 0, \quad 0 < x < \frac{\pi}{2},$

   $u(0, t) = 0, \quad u_x\left(\frac{\pi}{2}, t\right) + \frac{3}{2}u\left(\frac{\pi}{2}, t\right) = 0.$

c) $u_{tt} = \frac{\partial}{\partial x}(x^2 u_x)$, $1 < x < e$,

   $u(1,t) = u(e,t) = 0$.

d) $x u_{tt} = (x u_x)$, $0 < x < 2$,

   $\lim_{x \to 0^+} x u_x(x,t) = 0$, $u(2,t) = 0$.

10. Derive the continuity equation (8.19) directly by calculating

$$\frac{d}{dt} \int_{a(t)}^{b(t)} \rho(x,t) \, dx$$

using Leibniz' formula.

11. Derive (8.37) and (8.41).

12. Show that the equations governing the one-dimensional, time-dependent flow of a compressible fluid under constant pressure are

$$v_t + v v_x = 0,$$
$$\rho_t + (\rho v)_x = 0,$$
$$e_t + (ev)_x + p v_x = 0.$$

Subject to initial conditions $v(x,0) = f(x)$, $\rho(x,0) = g(x)$, and $e(x,0) = h(x)$, where f, g, and h are given smooth functions with nonnegative derivatives, derive the solution $v = f(x - vt)$ and

$$\rho = \frac{g(x - vt)}{1 + t f'(x - vt)}, \qquad e = \frac{h(x - vt) + p}{1 + t f'(x - vt)} - p.$$

Here, e is measured in energy per unit volume.

8.2 Gas Dynamics

8.2.1 Riemann's Method

Heretofore there has been little discussion of how fluid motion begins. It is common in gas dynamics, or compressible fluid dynamics, to imagine the flow being induced by the motion of a piston inside the cylindrical pipe. Such a device is not as unrealistic as it may first appear. The piston may represent the fluid on one side of the valve after it is opened or it may represent a detonator in an explosion process; while in aerodynamics it may represent a blunt object moving into a gas.

Figure 8.7 Piston problem.

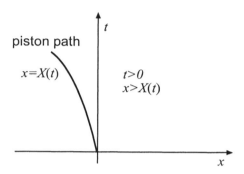

Figure 8.8 Space-time diagram of the piston problem.

Accordingly we set up and solve a simple problem whose method of solution, called **Riemann's method**, is typical of more general problems in gas dynamics. We consider a gas in a tube initially in the constant state

$$v = 0, \quad \rho = \rho_0, \quad p = p_0, \quad c = c_0, \tag{8.54}$$

with equation of state

$$p = k\rho^\gamma, \quad k, \gamma > 0. \tag{8.55}$$

To initiate the motion a piston located initially at $x = 0$ is withdrawn slowly according to

$$x = X(t), \quad X'(t) < 0, \quad X''(t) < 0, \tag{8.56}$$

where X is given function (see Fig. 8.7). The problem is to determine v, p, c, and ρ for all $t > 0$ and $X(t) < x < \infty$. We may regard this problem as a boundary value problem where initial conditions are given along the positive x axis and v is given by the piston velocity along the piston path (see Fig. 8.8). The method of solution is motivated by the study of the simple nonlinear model equation $u_t + c(u)u_x = 0$ from Chapter 7. There we were able to define a family of curves called characteristics along which signals propagated and along which the partial differential equation reduced to an ordinary differential equation.

In the case of this model equation $du/dt = 0$ along $dx/dt = c(u)$. We follow a similar strategy for (7.54) and (7.55) and attempt to find characteristic curves along which the partial differential equations reduce to simpler equations. To this end we multiply (7.55) by c and then add and subtract it from (7.54) to obtain

$$p_t + (v+c)p_x + \rho c(v_t + (v+c)v_x) = 0, \tag{8.57}$$

$$p_t + (v-c)p_x - \rho c(v_t + (v-c)v_x) = 0. \tag{8.58}$$

Hence, along the families of curves C^+ and C^- defined by

$$C^+ : \frac{dx}{dt} = v + c, \tag{8.59}$$

$$C^- : \frac{dx}{dt} = v - c, \tag{8.60}$$

we have

$$\frac{dp}{dt} + \rho c \frac{dv}{dt} = 0, \qquad \text{on } C^+, \tag{8.61}$$

$$\frac{dp}{dt} - \rho c \frac{dv}{dt} = 0, \qquad \text{on } C^-. \tag{8.62}$$

Equations (8.61) and (8.62) may be rewritten as

$$\frac{c}{\rho} \frac{dp}{dt} \pm \frac{dv}{dt} = 0, \qquad \text{on } C^+, C^-.$$

Integrating gives

$$\int \frac{c(\rho)}{\rho} d\rho + v = \text{constant} \qquad \text{on } C^+, \tag{8.63}$$

$$\int \frac{c(\rho)}{\rho} d\rho - v = \text{constant} \qquad \text{on } C^-. \tag{8.64}$$

The left sides of (8.63) and (8.64) are called the **Riemann invariants**; they are quantities that are constant along the characteristic curves C^+ and C^- defined by (8.59) and (8.60). If the equation of state is defined by (8.55), then

$$c^2 = k\gamma \rho^{\gamma - 1},$$

and

$$\int \frac{c(\rho)}{\rho} d\rho = \frac{2c}{\gamma - 1}.$$

Therefore the Riemann invariants are

$$r_+ \equiv \frac{2c}{\gamma - 1} + v = \text{constant} \qquad \text{on } C^+, \tag{8.65}$$

$$r_- \equiv \frac{2c}{\gamma - 1} - v = \text{constant} \qquad \text{on } C^-, \tag{8.66}$$

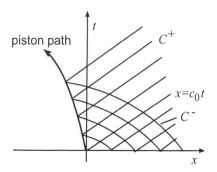

Figure 8.9 C^+ and C^- characteristics.

We have enough information from the Riemann invariants to determine the solution of the piston withdrawal problem. First consider the C^- characteristics. Since $c = c_0$ when $v = 0$ the characteristics that begin on the x axis leave with speed $-c_0$ and the constant in (8.66) has valued $2c_0/(\gamma - 1)$. Hence on the C^- characteristics

$$\frac{2c}{\gamma - 1} - v = \frac{2c_0}{\gamma - 1}. \tag{8.67}$$

Since the last equation must hold along every C^- characteristic it must hold everywhere and therefore r_- is constant in the entire region $t > 0$, $X(t) < x < \infty$. The C^- characteristics must end on the piston path, since their speed $v_p - c_p$ at the piston is more negative than the speed v_p of the piston (see Fig. 8.9). It is easy to see that the C^+ characteristics are straight lines. Adding and subtracting (8.67) and (8.65) shows

$$c = \text{constant} \quad \text{on a } C^+ \text{ characteristic,}$$
$$v = \text{constant} \quad \text{on a } C^+ \text{ characteristic.}$$

We have used the fact that (8.67) holds everywhere. The speed $v + c$ of a C^+ characteristic is therefore constant and thus the characteristic is a straight line. The C^+ characteristics emanating from the x axis have speed c_0 (since $v = 0$ on the x axis) and carry the constant state $v = 0$, $c = c_0$, $p = p_0$, $\rho = \rho_0$ into the region $x > c_0 t$. This is just the uniform state ahead of the signal $x = c_0 t$ beginning at the origin traveling at speed c_0 into the constant state. This signal indicates the initial motion of the piston.

A C^+ characteristic beginning on the piston at $(X(\tau), \tau)$ and passing through (x, t) has equation

$$x - X(\tau) = (c + v)(t - \tau). \tag{8.68}$$

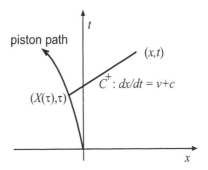

Figure 8.10 Positive characteristic carrying information from the piston path.

The speed $v + c$ can be calculated as follows. Clearly $v = X'(t)$, which is the velocity of the piston. From (8.67)

$$\frac{2c}{\gamma - 1} - X'(\tau) = \frac{2c_0}{\gamma - 1},$$

and hence

$$c = c_0 + \frac{\gamma - 1}{2}X'(\tau).$$

Thus

$$x - X(\tau) = \left(c_0 + \frac{\gamma + 1}{2}X'(\tau)\right)(t - \tau) \tag{8.69}$$

is the equation of the desired C^+ characteristic (see Fig. 8.10). The fact that v, c, p, and ρ are constant on the C^+ characteristics gives the solution

$$v(x,t) = X'(\tau),$$

where τ is given implicitly by (8.69). Obviously

$$c(x,t) = c_0 + \frac{\gamma - 1}{2}v(x,t).$$

The quantities p and ρ may be calculated from the equation of state and the definition of c^2. Qualitatively we think of C^- characteristics as carrying the values

$$r_- = \frac{2c}{\gamma - 1} - v$$

from the constant state back into the flow. The C^+ characteristics carry information from the piston forward into the flow. Whenever one of the Riemann invariants is constant throughout the flow, we say that the solution in the nonuniform region is a **simple wave**. It can be shown that a simple wave

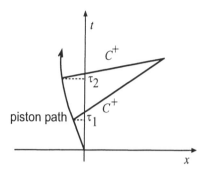

Figure 8.11 Colliding positive characteristics.

solution always exists adjacent to a uniform state provided that the solution remains smooth. A complete analysis is given in Courant and Friedrichs (1948).

In the piston withdrawal problem previously discussed we assumed that the piston path $x = X(t)$ satisfied the conditions $X'(t) < 0$ and $X''(t) < 0$, which means the piston is always accelerating backward. If there is ever an instant of time when the piston slows down, that is, $X''(t) > 0$, then a smooth solution cannot exist for all $t > 0$, for two distinct C^+ characteristics emanating from the piston will cross. Suppose, for example, that τ_1 and τ_2 are two values of t in a time interval where $X''(t) > 0$. The speed of characteristics leaving the piston is

$$v + c = c_0 + \frac{\gamma + 1}{2} X'(\tau).$$

If $\tau_1 < \tau_2$, then $X'(\tau_1) < X'(\tau_2)$ and it follows that $v + c$ at τ_1 is smaller than $v + c$ at τ_2. Therefore the characteristic emanating from $(X(\tau_2), \tau_2)$ is faster than the characteristic emanating from $(X(\tau_1), \tau_1)$. Thus the two characteristics must cross (see Fig. 8.11).

Example 8.7

Beginning at time $t = 0$ a piston located at $x = 0$ moves forward into a gas under uniform conditions with equation of state given by (8.55). Its path is given by $X(t) = at^2$, $a > 0$. We determine the first instant of time that two characteristics cross and the wave breaks. For this problem the preceding analysis remains valid and the C^+ characteristics emanating from the piston have equation (8.69) or

$$x - a\tau^2 = (c_0 + (\gamma + 1)a\tau)(t - \tau),$$

or

$$F(x, t, \tau) \equiv \gamma a\tau^2 + (c_0 - (\gamma + 1)at)\tau + x - c_0 t = 0. \tag{8.70}$$

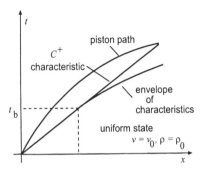

Figure 8.12 The positive characteristics emanating from the piston path and forming an envelope. The characteristics intersect in the region between the envelop and the curve $x = c_0 t$. The shock begins at time t_b

Along such a characteristic v is constant and hence

$$v(x, t) = X'(\tau) = 2a\tau,$$

where $\tau = \tau(x, t)$ is to be determined from (8.70). To solve (8.70) for τ requires by the implicit function theorem that $F_\tau(x, t, \tau) \neq 0$. Since $F_\tau = 2\gamma a\tau + (c_0 - (\gamma + 1)at)$, the first instant of time t that (8.70) cannot be solved for τ is

$$t_b = \min_{t \geq 0} \frac{c_0 + 2\gamma a\tau}{(\gamma + 1)a} = \frac{c_0}{(\gamma + 1)a}.$$

That is, the breaking time t_b will occur along the first characteristic (indexed by τ) where $F_\tau = 0$. The characteristic diagram is shown in Fig. 8.12 in the special case $a = c_0 = 1$ and $\gamma = 3$.

8.2.2 The Rankine–Hugoniot Conditions

As we observed earlier smooth solutions break down when characteristics intersect, since constant values are carried along the characteristics. The solution that develops is a discontinuous one and it propagates as a shock wave. We now determine what conditions hold across such a discontinuity. We proceed as in Section 8.1 where conservation laws are discussed.

The integral form of the conservation of mass law, namely

$$\frac{d}{dt} \int_a^b \rho(x, t) \, dx = -\rho(x, t) v(x, t)\big|_a^b \tag{8.71}$$

holds in all cases, even when the functions ρ and v are not smooth. In Chapter 7 we derived the jump condition across a shock, which we briefly review. Let $x = s(t)$ be a smooth curve in space–time that intersects the interval $a \leq x \leq b$ at time t, and suppose v, ρ, and p suffer simple discontinuities along the curve. Otherwise v, ρ, and p are assumed to be C^1 functions with finite limits on each side of $x = s(t)$. Then, by Leibniz' rule,

$$
\begin{aligned}
\frac{d}{dt}\int_a^b \rho(x,t)\,dx &= \frac{d}{dt}\int_a^{s(t)} \rho(x,t)\,dx + \frac{d}{dt}\int_{s(t)}^b \rho(x,t)\,dx \\
&= \int_a^{s(t)} \rho_t(x,t)\,dx + \rho(s(t)^-,t)s'(t) \\
&\quad + \int_{s(t)}^b \rho_t(x,t)\,dx - \rho(s(t)^+,t)s'(t).
\end{aligned}
$$

Both integrals on the right side approach zero as $a \to s(t)^-$ and $b \to s(t)^+$ and therefore from (8.71) it follows that

$$
(\rho(s(t)^-,t) - \rho(s(t)^+,t))s'(t) = -\rho(x,t)v(x,t)\big|_{s(t)-}^{s(t)+}. \tag{8.72}
$$

If the values (one-sided limits) of ρ and v on the right and left sides of $x = s(t)$ are denoted by the subscripts zero and one, respectively, for example,

$$
\rho_0 = \lim_{x\to s(t)^+} \rho(x,t), \qquad \rho_1 = \lim_{x\to s(t)^-} \rho(x,t),
$$

then (8.72) can be written

$$
(\rho_1 - \rho_0)s'(t) = \rho_1 v_1 - \rho_0 v_0. \tag{8.73}
$$

In a similar fashion the condition

$$
(\rho_1 v_1 - \rho_0 v_0)s'(t) = \rho_1 v_1^2 + p_1 - \rho_0 v_0^2 - p_0 \tag{8.74}
$$

can be obtained from the integral form (7.50) of the conservation of momentum law. The two conditions (8.73) and (8.74) are known as the **Rankine–Hugoniot jump conditions**; they relate the values ρ_0, v_0, p_0 ahead of the discontinuity to the speed s' of the discontinuity and to the values ρ_1, v_1, and p_1 behind it. If the state ahead is at rest, that is, $v_0 = 0$, then the conditions become

$$
\begin{aligned}
(\rho_1 - \rho_0)s' &= \rho_1 v_1, & (8.75) \\
\rho_1 v_1 s' &= \rho_1 v_1^2 + p_1 - p_0. & (8.76)
\end{aligned}
$$

Supplemented with the equation of state

$$
p_1 = f(\rho_1), \tag{8.77}
$$

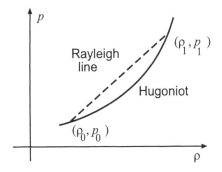

Figure 8.13 The Hugoniot diagram.

the equations can be regarded as three equations in the four unknowns ρ_1, v_1, p_1, and s'. If any one of these quantities is known, the remaining three may be determined.

It is helpful to picture the information contained in (8.75)–(8.77) on a $p\rho$ diagram as shown in Fig. 8.13. The solid curve is the graph of the equation of state $p = f(\rho)$. It is known as the **Hugoniot curve**; and all states, both initial and final, must lie on this curve. The straight line connecting the state ahead (ρ_0, p_0) to the state behind (ρ_1, p_1) is the **Rayleigh line**. If (8.76) is rewritten using equation (8.75) as

$$p_1 - p_0 = \rho_0 v_1 s',$$

then it is clear that the slope of the Rayleigh line is given by

$$\frac{p_1 - p_0}{\rho_1 - \rho_0} = \frac{\rho_0}{\rho_1} s'^2.$$

These ideas extend beyond this simple case to more general problems in gas-dynamics (see Courant and Friedrichs (1948)).

EXERCISES

1. Derive the jump condition (8.74).

2. Discuss the piston withdrawal problem when the piston path is given by $x = -V_0 t$, where V_0 is a positive constant. Assume an equation of state $p = k\rho^\gamma$ with uniform conditions $v = 0$, $p = p_0$, $\rho = \rho_0$, $c = c_0$ ahead of the piston at $t = 0$.

3. At $t = 0$ the gas in a tube $x \geq 0$ is at rest. For $t > 0$ a piston initially located at $x = 0$ moves according to the law $x = X(t)$, where $X(t)$ is small. Show that in the acoustic approximation the motion induced in the gas is

$$v = X'\left(t - \frac{x}{c_0}\right),$$

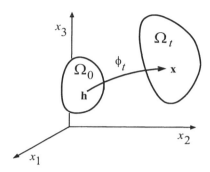

Figure 8.14 Fluid motion.

and find the corresponding density variation.

4. A sphere of initial radius r_0 pulsates according to $r(t) = r_0 + A\sin\omega t$. In the linearized theory find the motion of the gas outside the sphere.

8.3 Fluid Motions in $\mathbb{R}^3$

8.3.1 Kinematics

In the early sections of this chapter the equations governing fluid motion were developed in one spatial dimension. This limitation is severe, however, because most interesting fluid phenomena occur in higher dimensions. In this section we derive the field equations for fluid dynamics in three dimensions, with the two-dimensional case being an obvious corollary. Let Ω_0 be an arbitrary closed bounded set in $\mathbb{R}^3$. We imagine that Ω_0 is a region occupied by a fluid at time $t = 0$. By a **fluid motion** we mean a mapping $\phi_t\colon \Omega_0 \to \Omega_t$ defined for all t in an interval I containing the origin, which maps the region Ω_0 into $\Omega_t = \phi_t(\Omega_0)$, the latter being the region occupied by the same fluid at time t (see Fig. 8.14). We assume that ϕ_t is represented by the formula

$$\mathbf{x} = \mathbf{x}(\mathbf{h}, t), \tag{8.78}$$

where $\mathbf{h} = (h_1, h_2, h_3)$ is in Ω_0 and $\mathbf{x} = (x_1, x_2, x_3) = (x, y, z)$, with

$$\mathbf{x}(\mathbf{h}, 0) = \mathbf{h}.$$

Thus $\mathbf{h}$ is a Lagrangian coordinate or particle label given to each fluid particle at $t = 0$, and $\mathbf{x}$ is the Eulerian coordinate representing the laboratory position of the particle $\mathbf{h}$ at time t. We further assume that the function $\mathbf{x}(\mathbf{h}, t)$ is twice continuously differentiable on its domain $\Omega_0 \times I$ and that for each $t \in I$ the mapping ϕ_t has a unique inverse defined by

$$\mathbf{h} = \mathbf{h}(\mathbf{x}, t).$$

Hence

$$\mathbf{x}(\mathbf{h}(\mathbf{x}, t), t) = \mathbf{x}, \qquad \mathbf{h}(\mathbf{x}(\mathbf{h}, t), t) = \mathbf{h}.$$

By convention, functions of $\mathbf{h}$ and t that represent the results of Lagrangian measurements will be denoted by capital letters, and functions of $\mathbf{x}$ and t that represent the results of a fixed laboratory measurement will be denoted by lowercase letters. The measurements are connected by the formulas

$$f(\mathbf{x}, t) = F(\mathbf{h}(\mathbf{x}, t), t), \quad F(\mathbf{h}, t) = f(\mathbf{x}(\mathbf{h}, t), t).$$

In the sequel we often use the notation $F|_\mathbf{x}$ and $f|_\mathbf{h}$ to denote the right sides of these equations, respectively.

For a given fluid motion (8.78) the is defined by

$$\mathbf{V}(\mathbf{h}, t) = \frac{\partial \mathbf{x}}{\partial t}(\mathbf{h}, t).$$

Then

$$\mathbf{v}(\mathbf{x}, t) = \mathbf{V}(\mathbf{h}(\mathbf{x}, t), t).$$

$\mathbf{V}(\mathbf{h}, t) = \langle V_1(\mathbf{h}, t), V_2(\mathbf{h}, t), V_3(\mathbf{h}, t) \rangle$ is the actual velocity vector of the fluid particle $\mathbf{h}$ at time t, whereas $\mathbf{v}(\mathbf{x}, t) = \langle v_1(\mathbf{x}, t), v_2(\mathbf{x}, t), v_3(\mathbf{x}, t) \rangle$ is the velocity vector measured by a fixed observer at location $\mathbf{x}$. If $\mathbf{h}_0$ is a fixed point in Ω_0, then the curve $\mathbf{x} = \mathbf{x}(\mathbf{h}_0, t)$, $t \in I$, is called the **particle path** of the particle $\mathbf{h}_0$. The following theorem states that knowledge of the Eulerian velocity field $\mathbf{v}(\mathbf{x}, t)$ is equivalent to knowledge of all the particle paths.

Theorem 8.8

If the velocity vector field $\mathbf{v}(\mathbf{x}, t)$ is known, then all of the particle paths can be determined, and conversely.

Proof

First we prove the converse. If each particle path is given, then the function $\mathbf{x} = \mathbf{x}(\mathbf{h}, t)$ is known for all $\mathbf{h} \in \Omega_0$. Consequently, $\mathbf{h} = \mathbf{h}(\mathbf{x}, t)$ can be determined and $\mathbf{v}$ can be calculated from

$$\mathbf{v}(\mathbf{x}, t) = \mathbf{V}(\mathbf{h}(\mathbf{x}, t), t) = \mathbf{x}_t(\mathbf{h}(\mathbf{x}, t), t).$$

Now assume $\mathbf{v}(\mathbf{x}, t)$ is given. Then

$$\frac{\partial \mathbf{x}}{\partial t}(\mathbf{h}, t) = \mathbf{v}(\mathbf{x}(\mathbf{h}, t), t),$$

or

$$\frac{\partial \mathbf{x}}{\partial t} = \mathbf{v}(\mathbf{x}, t), \tag{8.79}$$

where explicit dependence on $\mathbf{h}$ has been dropped. Equation (8.79) is a first-order system of differential equations for $\mathbf{x}$. Its solution subject to the initial condition

$$\mathbf{x} = \mathbf{h} \quad \text{at} \quad t = 0$$

gives the particle paths $\mathbf{x} = \mathbf{x}(\mathbf{h}, t)$. $\qquad\qquad\qquad\qquad\qquad\qquad\square$

It is interesting to contrast the particle paths with the so-called **stream-lines** of the flow. The latter are integral curves of the vector field $\mathbf{v}(\mathbf{x}, t_0)$ frozen at some fixed but arbitrary instant t_0 of time. Thus the streamlines are found by solving the system

$$\frac{d\mathbf{x}}{ds} = \mathbf{v}(\mathbf{x}, t_0), \tag{8.80}$$

where $\mathbf{x} = \mathbf{x}(s)$ and s is a parameter along the curves. If $\mathbf{v}$ is independent of t, then the flow is called **steady** and the streamlines coincide with the particle paths. In a time-dependent flow they need not coincide.

Example 8.9

Consider the fluid motion

$$x_1 = t + h_1, \quad x_2 = h_2 e^t, \quad x_3 = th_1 + h_3, \quad t > 0.$$

Then

$$\mathbf{V} = \left\langle \frac{\partial x_1}{\partial t}, \frac{\partial x_2}{\partial t}, \frac{\partial x_3}{\partial t} \right\rangle = \langle 1, h_2 e^t, h_1 \rangle.$$

Inverting the motion,

$$h_1 = x_1 - t, \quad h_2 = x_2 e^{-t}, \quad h_3 = x_3 - tx_1 + t^2.$$

Therefore

$$\mathbf{v} = \langle 1, x_2, x_1 - t \rangle,$$

and thus the motion is not steady. The streamlines at time t_0 are given by solutions to (8.80), or

$$\frac{dx_1}{ds} = 1, \quad \frac{dx_2}{ds} = x_2, \quad \frac{dx_3}{ds} = x_1 - t_0.$$

Hence

$$x_1 = s + c_1, \quad x_2 = c_2 e^s, \quad x_3 = \tfrac{1}{2}s^2 + (c_1 - t_0)s + c_3,$$

where c_1, c_2, and c_3 are constants.

As in the one-dimensional case we define the **material derivative** by

$$\frac{Df}{Dt}(\mathbf{x}, t) \equiv F_t(\mathbf{h}, t)|_{\mathbf{h}=\mathbf{h}(\mathbf{x},t)},$$

where f and F are the Eulerian and Lagrangian representations of a given measurement, respectively. A straightforward calculation using the chain rule shows that

$$F_t(\mathbf{h}, t) = \frac{\partial}{\partial t} f(\mathbf{x}(\mathbf{h}, t), t)$$

$$= f_t(\mathbf{x}(\mathbf{h}, t), t) + \sum_{j=1}^{3} f_{x_j}(\mathbf{x}(\mathbf{h}, t), t) V_j(\mathbf{h}, t)$$

$$= f_t(\mathbf{x}(\mathbf{h}, t), t) + \mathbf{V} \cdot \nabla f|_h.$$

Evaluating at $\mathbf{h} = \mathbf{h}(\mathbf{x}, t)$ gives

$$\frac{Df}{Dt}(\mathbf{x}, t) = f_t(\mathbf{x}, t) + \mathbf{v}(\mathbf{x}, t) \cdot \nabla f(\mathbf{x}, t). \tag{8.81}$$

If $\mathbf{f} = \langle f_1, f_2, f_3 \rangle$ is a vector function, then componentwise,

$$\frac{Df_i}{Dt} = \frac{\partial f_i}{\partial t} + \sum_{j=1}^{3} v_j \frac{\partial f_i}{\partial x_j}, \qquad i = 1, 2, 3,$$

and we write

$$\frac{D\mathbf{f}}{Dt} = \mathbf{f}_t + \sum_{j=1}^{3} v_j \frac{\partial \mathbf{f}}{\partial x_j} \equiv \mathbf{f}_t + (\mathbf{v} \cdot \nabla)\mathbf{f}.$$

Here $\mathbf{v} \cdot \nabla$ is a special notation for the operator $v_1 \partial/\partial x_1 + v_2 \partial/\partial x_2 + v_3 \partial/\partial x_3$. The material derivative Df/Dt at $(\mathbf{x}, t)$ is a measure of the time rate of change of the quantity F moving with the particle $\mathbf{h}$ frozen at the instant the particle is at $\mathbf{x}$.

Another kinematical result is the three-dimensional analog of the **Euler expansion theorem**. It gives the time rate of change of the Jacobian

$$J(\mathbf{h}, t) = \det \left(\frac{\partial x_i}{\partial h_j} \right)$$

of the transformation ϕ_t. We record this as a theorem.

Theorem 8.10

In Lagrangian and Eulerian form, respectively,

$$\begin{aligned} J_t(\mathbf{h}, t) &= J(\mathbf{h}, t) \nabla \cdot \mathbf{v}(\mathbf{x}, t)|_\mathbf{h}, \\ \frac{Dj}{Dt}(\mathbf{x}, t) &= j(\mathbf{x}, t) \nabla \cdot v(\mathbf{x}, t), \end{aligned}$$

where $j(\mathbf{x}, t) = J(\mathbf{h}(\mathbf{x}, t), t)$.

Details of the proof are left as an exercise.

In the one-dimensional case the derivation of many of the results depended on knowing the time derivative of an integral over the moving material region. For three dimensions we state this as a fundamental theorem.

Theorem 8.11

(**The convection theorem**) Let $g = g(\mathbf{x}, t)$ be a continuously differentiable function. Then

$$\frac{d}{dt} \int_{\Omega_t} g \, d\mathbf{x} = \int_{\Omega_t} \left(\frac{Dg}{Dt} + g \nabla \cdot \mathbf{v} \right) d\mathbf{x}. \qquad (8.82)$$

The region of integration Ω_t of the integral on the left side of (8.82) depends on t, and therefore the time derivative cannot be brought under the integral sign directly. A change of variables in the integral from Eulerian coordinates $\mathbf{x}$ to Lagrangian coordinates $\mathbf{h}$ will transform the integral to one over the fixed, time-independent volume Ω_0. The derivative may then be brought under the integral sign and the calculation can proceed. Letting $\mathbf{x} = \mathbf{x}(\mathbf{h}, t)$, we have

$$\int_{\Omega_t} g(\mathbf{x}, t) \, d\mathbf{x} = \int_{\Omega_0} G(\mathbf{h}, t) J(\mathbf{h}, t) \, d\mathbf{h}.$$

Here we have used the familiar formula for changing variables in a multiple integral where the Jacobian enters as a factor in the new volume element. Then, also using Theorem 8.10,

$$\begin{aligned}
\frac{d}{dt} \int_{\Omega_t} g \, d\mathbf{x} &= \int_{\Omega_0} \frac{\partial}{\partial t} (GJ) \, d\mathbf{h} \\
&= \int_{\Omega_0} (G_t + G \nabla \cdot \mathbf{v}|_{\mathbf{x} = \mathbf{x}(\mathbf{h}, t)}) J \, d\mathbf{h} \\
&= \int_{\Omega_t} \left(\frac{Dg}{Dt} + g \nabla \cdot \mathbf{v} \right) d\mathbf{x},
\end{aligned}$$

where in the last step we transformed back to Eulerian coordinates and used (8.81).

The **Reynold's transport theorem** is an immediate corollary.

Corollary 8.12

We have

$$\frac{d}{dt} \int_{\Omega_t} g \, d\mathbf{x} = \int_{\Omega_t} g_t \, d\mathbf{x} + \int_{\partial \Omega_t} g \mathbf{v} \cdot \mathbf{n} \, d\tau, \qquad (8.83)$$

where $\mathbf{n}$ is the outer unit normal to the surface $\partial \Omega_t$.

Proof

The integrand on the right side of (8.82) is

$$\frac{Dg}{Dt} + g\nabla \cdot \mathbf{v} = g_t + \nabla \cdot g\,\mathbf{v}.$$

Thus

$$\int_{\Omega_t} \left(\frac{Dg}{Dt} + g\nabla \cdot \mathbf{v}\right) dx = \int_{\Omega_t} g_t\,dx + \int_{\Omega_t} \nabla \cdot g\mathbf{v}\,dx.$$

An application of the divergence theorem yields

$$\int_{\Omega_t} \nabla \cdot g\mathbf{v}\,dx = \int_{\partial\Omega_t} g\mathbf{v} \cdot \mathbf{n}\,dA,$$

and the corollary follows. $\qquad\qquad\square$

From calculus we recall that the integral $\int_{\partial\Omega_t} g\mathbf{v} \cdot \mathbf{n}\,dA$ represents the flux of the vector field $g\mathbf{v}$ through the surface $\partial\Omega_t$. Hence (8.83) states that the time rate of change of the quantity $\int_{\Omega_t} g$, where both the integrand and the region of integration depend on t, frozen in time, of the change in g plus the flux or convection of g through the boundary $\partial\Omega_t$. Equation (8.83) is the three-dimensional version of Leibniz' formula.

With all this machinery, it is a simple matter to obtain a mathematical expression of mass conservation. We take it as a physical axiom that a given material volume of fluid has the same mass as it evolves in time. Symbolically

$$\int_{\Omega_t} \rho(\mathbf{x}, t)\,dx = \int_{\Omega_0} \rho(\mathbf{x}, 0)\,dx,$$

where $\rho(\mathbf{x}, t)$ is the Eulerian density. Thus

$$\frac{d}{dt}\int_{\Omega_t} \rho(\mathbf{x}, t)\,dx = 0,$$

and Theorem 8.11 gives, upon taking $g = \rho$,

$$\frac{D\rho}{Dt} + \rho\nabla \cdot \mathbf{v} = 0, \qquad\qquad (8.84)$$

because the volume Ω_t is arbitrary. The first-order nonlinear partial differential equation (8.84) is the **continuity equation**, and it is a mathematical expression for conservation of mass. An important class of fluid motions are those in which a material region maintains the same volume. Thus we say a fluid motion $\phi_t \colon \Omega_0 \to \Omega_t$ is **incompressible** if

$$\frac{d}{dt}\int_{\Omega_t} dx = 0, \quad \text{for all } \Omega_0.$$

Theorem 8.13

The following are equivalent:

(i) ϕ_t *is incompressible,*

(ii) $\nabla \cdot \mathbf{v} = 0$,

(iii) $\Delta_t = 0$,

(iv) $\frac{D\rho}{Dt} = 0$,

where $\Delta(\mathbf{h}, t) = \rho(\mathbf{x}(\mathbf{h}, t), t)$ is the Lagrangian density.

The proof is straightforward. By definition, (iii) and (iv) are equivalent, while (ii) and (iv) are equivalent by conservation of mass (8.84). Finally (i) and (ii) are equivalent by setting $g = 1$ in the convection theorem.

8.3.2 Dynamics

The nature of the forces on a fluid element was presented in Section 8.1.2 for one-dimensional flows. The idea in three dimensions is again to generalize Newton's second law, which states for a particle of mass m that $d\mathbf{p}/dt = \mathbf{F}$, or the time rate of change of momentum is the total force. For a material region Ω_t of fluid we define the **momentum** by

$$\int_{\Omega_t} \rho\mathbf{v}\,d\mathbf{x}.$$

The forces on the fluid region are of two types, body forces, which act at each point of the region, and surface forces or tractions, which act on the boundary of the region. Body forces will be denoted by $\mathbf{f}(\mathbf{x}, t)$, which represents the force per unit mass acting at the point $\mathbf{x}$ at time t. Thus the total body force on the region Ω_t is

$$\int_{\Omega_t} \rho\mathbf{f}\,d\mathbf{x}.$$

At each point $\mathbf{x}$ on the boundary $\partial\Omega_t$ of the given region Ω_t we assume that there exists a vector $\sigma(\mathbf{x}, t; \mathbf{n})$ called the **stress vector**, which represents the force per unit area acting at $\mathbf{x}$ at time t on the surface $\partial\Omega_t$ by the material exterior to Ω_t. We note that the stress at $\mathbf{x}$ depends on the orientation of the surface through $\mathbf{x}$ Simply put, different surfaces will have different stresses (see Fig. 8.15). The dependence of σ upon the orientation of the surface is denoted by the argument $\mathbf{n}$ in $\sigma(\mathbf{x}, t; \mathbf{n})$, where $\mathbf{n}$ is the outward unit normal to the

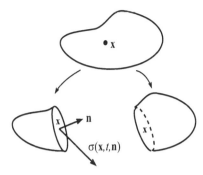

Figure 8.15 Diagram showing the dependence of the stress vector σ on the unit normal $\mathbf{n}$.

surface. We think of $\mathbf{n}$ as pointing toward the material that is causing the stress. Therefore the total surface force or traction on the region Ω_t is given by

$$\int_{\partial \Omega_t} \sigma(x, t; \mathbf{n}) \, dA.$$

We now postulate the **balance of momentum** principle, or Cauchy's stress principle, which states that for all material regions Ω_t

$$\frac{d}{dt} \int_{\Omega_t} \rho \mathbf{v} \, d\mathbf{x} = \int_{\Omega_t} \rho \mathbf{f} \, d\mathbf{x} + \int_{\partial \Omega_t} \sigma(\mathbf{x}, t; \mathbf{n}) \, dA. \tag{8.85}$$

In words, the time rate of change of momentum equals the total force. Because

$$\frac{d}{dt} \int_{\Omega_t} \rho \mathbf{v} \, d\mathbf{x} = \int_{\Omega_t} \rho \frac{D\mathbf{v}}{Dt} \, d\mathbf{x},$$

we may write (8.85) as

$$\int_{\Omega_t} \rho \left(\frac{D\mathbf{v}}{Dt} - \mathbf{f} \right) d\mathbf{x} = \int_{\partial \Omega_t} \sigma(\mathbf{x}, t; \mathbf{n}) \, dA. \tag{8.86}$$

At a given point $\mathbf{x}$ in a fluid on a surface defined by the orientation $\mathbf{n}$ the stress on the fluid exterior to Ω_t caused by the fluid interior to Ω_t is equal and opposite to the stress on the interior caused by the fluid exterior to Ω_t. More precisely, see the following theorem.

Theorem 8.14

(Action–reaction) We have

$$\sigma(\mathbf{x}, t; -\mathbf{n}) = -\sigma(\mathbf{x}, t; \mathbf{n}). \tag{8.87}$$

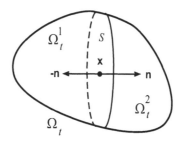

Figure 8.16 Action-reaction.

Proof

Let Ω_t be a material region and divide it into two regions Ω_t^1 and Ω_t^2 by a surface S that passes through an arbitrary point $\mathbf{x}$ in Ω_t (refer to Fig. 8.16). Applying (8.86) to Ω_t^1 and Ω_t^2, we obtain

$$\int_{\Omega_t^1} \rho \left(\frac{D\mathbf{v}}{Dt} - \mathbf{f} \right) dx \; = \; \int_{\partial \Omega_t^1} \sigma \, dA, \tag{8.88}$$

$$\int_{\Omega_t^2} \rho \left(\frac{D\mathbf{v}}{Dt} - \mathbf{f} \right) dx \; = \; \int_{\partial \Omega_t^2} \sigma \, dA. \tag{8.89}$$

Subtracting (8.86) from the sum of (8.88) and (8.89) gives

$$0 = \int_S \left(\sigma(\mathbf{x}, t; \mathbf{n}) + \sigma(\mathbf{x}, t; -\mathbf{n}) \right) dA,$$

where $\mathbf{n}$ is the outward unit normal to Ω_t^1. The integral mean value theorem implies

$$[\sigma(\mathbf{z}, t; \mathbf{n}_1) + \sigma(\mathbf{z}, t; -\mathbf{n}_1)] \cdot \text{Area}\,(S) = \mathbf{0}, \tag{8.90}$$

where $\mathbf{z}$ is some point on S and $\mathbf{n}_1$ is the outward unit normal to Ω_t^1 at $\mathbf{z}$. Taking the limit as the volume of Ω_t goes to zero in such a way that the area of S goes to zero and $\mathbf{x}$ remains on S we get $\mathbf{z} \to \mathbf{x}$ and $\mathbf{n}_1 \to \mathbf{n}$, and so (8.90) implies the result.

As it turns out the stress vector $\sigma(\mathbf{x}, t; \mathbf{n})$ depends in a very special way on the unit normal $\mathbf{n}$. The next theorem is one of the fundamental results in fluid mechanics. $\qquad\qquad\square$

Theorem 8.15

(Cauchy theorem) Let $\sigma_i(\mathbf{x}, t; \mathbf{n})$, $i = 1, 2, 3$, denote the three components of the stress vector $\sigma(\mathbf{x}, t; \mathbf{n})$, and let $\mathbf{n} = (n_1, n_2, n_3)$. Then there exists a matrix function $\sigma_{ji}(\mathbf{x}, t)$ such that

$$\sigma_i = \sum_{j=1}^{3} \sigma_{ji} n_j. \tag{8.91}$$

In fact

$$\sigma_{ji}(\mathbf{x}, t) = \sigma_i(\mathbf{x}, t; \mathbf{e}_j), \tag{8.92}$$

where $\mathbf{e}_j$, $j = 1, 2, 3$, denote the unit vectors in the direction of the coordinate axes x_1, x_2, and x_3, respectively.

Proof

Let t be a fixed instant of time and let Ω_t be a tetrahedron at $\mathbf{x}$ as shown in Fig. 8.17 with the three faces S_1, S_2, and S_3 parallel to the coordinate planes and the face S is the oblique face. Using the integral mean value theorem, we have

$$
\begin{aligned}
\int_{\partial \Omega_t} \sigma \, dA &= \sum_{j=1}^{3} \int_{S_j} \sigma \, dA + \int_S \sigma \, dA \\
&= A(S_1)\sigma(\mathbf{z}_1, t; -\mathbf{e}_1) + A(S_2)\sigma(\mathbf{z}_2, t; -\mathbf{e}_2) \\
&\quad + A(S_3)\sigma(\mathbf{z}_3, t; -\mathbf{e}_3) + A(S)\sigma(\mathbf{z}, t; \mathbf{n}),
\end{aligned}
$$

where $\mathbf{z}_i$ is a point on S_i and $\mathbf{z}$ is a point on S. Here A denotes the area function and $\mathbf{n}$ is the outer normal to S. Because

$$A(S_j) = A(S)n_j \equiv l^2 n_j$$

(we have defined l to be $\sqrt{A(S)}$), it follows that

$$\frac{1}{l^2} \int_{\partial \Omega_t} \sigma \, dA = -\sum_{j=1}^{3} \sigma(\mathbf{z}_j, t; \mathbf{e}_j) n_j + \sigma(\mathbf{z}, t; \mathbf{n}),$$

where Theorem 8.14 has been applied. Taking the limit as $l \to 0$ gives

$$\lim_{l \to 0} \frac{1}{l^2} \int_{\partial \Omega_t} \sigma \, dA = -\sum_{j=1}^{3} \sigma(\mathbf{x}, t; \mathbf{e}_j) n_j + \sigma(\mathbf{x}, t; \mathbf{n}).$$

We now show that

$$\lim_{l \to 0} \frac{1}{l^2} \int_{\partial \Omega_2} \sigma(\mathbf{x}, t; \mathbf{n}) \, dA = 0. \tag{8.93}$$

To this end, it follows from (8.86) that

$$\left| \int_{\partial \Omega_t} \sigma(\mathbf{x}, t; \mathbf{n}) \, dA \right| \leq \int_{\Omega_t} \left| \rho \left(\frac{D\mathbf{v}}{Dt} - \mathbf{f} \right) \right| d\mathbf{x} \leq M l^3,$$

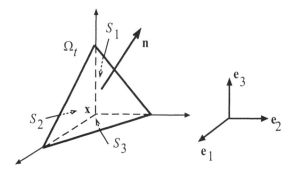

Figure 8.17 Proof of Cauchy's theorem.

where the last inequality results from the boundedness of the integrand. Dividing by l^2 and taking the limit as $l \to 0$ proves (8.93). Consequently

$$\sigma(\mathbf{x}, t; \mathbf{n}) = \sum_{j=1}^{3} \sigma(\mathbf{x}, t; \mathbf{e}_j) n_j, \tag{8.94}$$

or in component form

$$\sigma_i(\mathbf{x}, t; \mathbf{n}) = \sum_{j=1}^{3} \sigma_i(\mathbf{x}, t; \mathbf{e}_j) n_j, \qquad i = 1, 2, 3.$$

Defining the σ_{ji} by

$$\sigma_{ji}(\mathbf{x}, t) \equiv \sigma_i(\mathbf{x}, t; \mathbf{e}_j)$$

gives the final result.   □

The nine quantities σ_{ji} are the components of the **stress tensor**. By definition, σ_{ji} is the ith component of the stress vector on the face whose normal is $\mathbf{e}_j$. For example, σ_{21}, σ_{22}, and σ_{23} are shown in Fig. 8.18. Equation (8.94) shows that $\sigma(\mathbf{x}, t; \mathbf{n})$ can be resolved into a linear combination of $\sigma(\mathbf{x}, t; \mathbf{e}_1), \sigma(\mathbf{x}, t; \mathbf{e}_2)$, and $\sigma(\mathbf{x}, t, \mathbf{e}_3)$, that is, into three stresses that are the stresses on the coordinate planes at $\mathbf{x}$. These stresses are shown in Fig. 8.19. The components of the three vectors shown are the σ_{ji}.

With the aid of Theorem 8.15 a vector partial differential equation can be formulated that expresses the momentum balance law (8.86). This calculation is facilitated by adopting a notational convention that saves in writing summation signs in complicated expressions. This practice is called the *summation convention* and it assumes that a sum is taken over any repeated index in a given term. Thus

$$a_i b_{ij} \quad \text{means} \quad \sum_i a_i b_{ij},$$

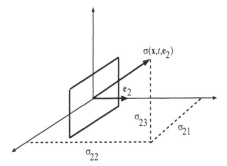

Figure 8.18 The stress vector $\sigma(\mathbf{x}, t, \mathbf{e}_2)$ and its three stress components.

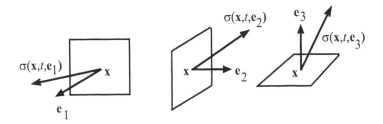

Figure 8.19 Stress tensor components.

and

$$a_i b_{ij} c_j \quad \text{means} \quad \sum_i \sum_j a_i b_{ij} c_j.$$

The range of the summation index (or indices) is determined from context. For example (8.91) may be written $\sigma_i = \sigma_{ji} n_j$ with a sum over $j = 1$ to $j = 3$ assumed on the right side. Any index not summed in a given expression is called a free index. Free indices vary over their appropriate ranges; for example, in (8.91) the index i is free and ranges over $i = 1, 2, 3$. In (8.92) both i and j are free with $i, j = 1, 2, 3$. In terms of this convention the divergence theorem may be expressed

$$\int_\Omega \frac{\partial}{\partial x_i} f_i \, dx = \int_{\partial\Omega} f_i n_i \, dA, \tag{8.95}$$

where $\mathbf{f} = (f_1, f_2, f_3)$ and $\mathbf{n} = (n_1, n_2, n_3)$. On both sides of the equation i is assumed to be summed from $i = 1$ to $i = 3$.

In component form (8.86) is

$$\int_{\Omega_t} \rho \left(\frac{Dv_i}{Dt} - f_i \right) d\mathbf{x} = \int_{\partial\Omega_t} \sigma_i \, dA. \tag{8.96}$$

But the right side is

$$\int_{\partial\Omega_t} \sigma_i \, dA = \int_{\partial\Omega_t} \sigma_{ji} n_j \, dA = \int_{\Omega_t} \frac{\partial}{\partial x_j} \sigma_{ji} \, d\mathbf{x},$$

where we have used (8.91) and (8.95). Then (8.96) becomes

$$\int_{\Omega_t} \left(\rho \frac{Dv_i}{Dt} - \rho f_i - \frac{\partial}{\partial x_j} \sigma_{ji} \right) d\mathbf{x} = 0, \qquad i = 1, 2, 3.$$

Because of the arbitrariness of the region of integration

$$\rho \frac{Dv_i}{Dt} = \rho f_i + \frac{\partial}{\partial x_j} \sigma_{ji}, \qquad i = 1, 2, 3. \tag{8.97}$$

Equations (8.97) are the **Cauchy equations** or *equations of motion*, and they represent balance of linear momentum. Along with the continuity equation (8.84) we have four equations for ρ, the three components v_i of the velocity, and the nine components σ_{ji} of the stress tensor.

There is a common way to write the Cauchy stress principle in terms of a fixed laboratory volume Ω_1. It is expressed by the following theorem.

Theorem 8.16

Cauchy's stress principle (8.85) is equivalent to

$$\frac{d}{dt} \int_{\Omega_1} \rho v_i \, d\mathbf{x} + \int_{\partial\Omega_1} \rho v_i v_j n_j \, dA = \int_{\Omega_1} \rho f_i \, d\mathbf{x} + \int_{\partial\Omega_1} \sigma_i \, dA, \tag{8.98}$$

where Ω_1 is any fixed region in $\mathbb{R}^3$.

Proof

Using Corollary 8.12 with $g \equiv \rho v_i$ gives

$$\frac{d}{dt} \int_{\Omega_t} \rho v_i \, d\mathbf{x} = \int_{\Omega_t} (\rho v_i)_t \, d\mathbf{x} + \int_{\partial\Omega_t} \rho v_i v_j n_j \, dA.$$

Therefore

$$\int_{\Omega_t} \rho \frac{Dv_i}{Dt} \, d\mathbf{x} = \int_{\Omega_t} (\rho v_i)_t \, d\mathbf{x} + \int_{\partial\Omega_t} \rho v_i v_j n_j \, dA.$$

Now let Ω_1 be the volume that coincides with Ω_t at $t = t_1$. Then

$$\int_{\Omega_1} \rho \frac{Dv_i}{Dt} \, d\mathbf{x} = \frac{d}{dt} \int_{\Omega_1} \rho v_i \, d\mathbf{x} + \int_{\partial\Omega_1} \rho v_i v_j n_j \, dA.$$

Applying (8.86) gives (8.98).   □

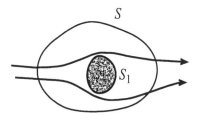

Figure 8.20 Flow around an obstacle.

Example 8.17

As an application of (8.98) we compute the force on a stationary object in a steady fluid flow. Suppose S_1 is the boundary of the object and let S be an imaginary surface that contains the object (see Fig. 8.20). Assume the flow is steady and that there is no body force ($\mathbf{f} = \mathbf{0}$). Letting Ω_1 be the region in between S and S_1 and applying the vector form of (8.98) gives

$$\int_{\partial \Omega_1} (\rho \mathbf{v}(\mathbf{v} \cdot \mathbf{n}) - \sigma)\, dA = \mathbf{0}, \qquad \partial \Omega_1 = S_1 \cup S,$$

or

$$\int_S (\rho \mathbf{v}(\mathbf{v} \cdot \mathbf{n}) - \sigma)\, dA = \int_{S_1} (\sigma - \rho \mathbf{v}(\mathbf{v} \cdot \mathbf{n}))\, dA = \int_{S_1} \sigma\, dA,$$

where the fact that $\mathbf{v} \cdot \mathbf{n} = 0$ on S_1 has been used. But the force $\mathbf{F}$ on the obstacle is

$$\begin{aligned} \mathbf{F} &= -\int_{S_1} \sigma\, dA \\ &= \int_S (\sigma - \rho \mathbf{v}(\mathbf{v} \cdot \mathbf{n}))\, dA. \end{aligned}$$

This is a useful result since it permits the calculation of $\mathbf{F}$ at a control surface at a distance from the obstacle; it may be virtually impossible to obtain the stresses on the surface of the obstacle itself.

8.3.3 Energy

The four equations exhibited in (8.84) and (8.97) clearly do not represent a complete determined system for all of the unknowns. At this point in the one-dimensional case we introduced an equation of state or constitutive relation. In the present case specification of the stress tensor σ_{ij} is required before further progress can be made.

For a fluid (gas or liquid) in a motionless state it is clear that the stress vector σ at a point on a surface in the fluid is always normal to the surface, that is, it is in the direction of the unit normal $\mathbf{n}$ (or $-\mathbf{n}$). Such is not true for solids, since they can sustain complicated shear or tangential stresses and not undergo motion. This normality property extends to fluids in uniform motion where the velocity field is constant over some time interval. One of the simplest classes of general nonuniform fluid motions is that class of motions termed **inviscid** where the same normality assumption holds, namely

$$\sigma(\mathbf{x}, t; \mathbf{n}) = -p(\mathbf{x}, t)\mathbf{n}. \tag{8.99}$$

Hence, for inviscid flow the stress across a surface is proportional to the normal vector where the proportionality factor p is the **pressure**. The concept of an inviscid fluid is highly useful in technology, since many real fluids are actually modeled by (8.99) and the calculations are far simpler using (8.99) than for more complicated constitutive relations. An **ideal fluid** is an inviscid fluid undergoing an incompressible motion, and much of the literature in fluid mechanics deals with ideal fluids.

From (8.99) it is easy to calculate the components σ_{ij} of the stress tensor. We have

$$\sigma_{ij} = -p(\mathbf{x}, t)\delta_{ij}, \tag{8.100}$$

where δ_{ij} is the Kronecker delta symbol defined by $\delta_{ij} = 1$ if $i = j$ and $\delta_{ij} = 0$ if $i \neq j$. Because

$$\frac{\partial}{\partial x_j}\sigma_{ji} = -\frac{\partial}{\partial x_j}\left(p(\mathbf{x}, t)\delta_{ij}\right) = -\frac{\partial}{\partial x_i}p(\mathbf{x}, t),$$

we may write the Cauchy equations (8.97) as

$$\rho\frac{Dv_i}{Dt} = \rho f_i - \frac{\partial p}{\partial x_i}, \qquad i = 1, 2, 3,$$

or in vector form as

$$\rho\frac{D\mathbf{v}}{Dt} = \rho\mathbf{f} - \nabla p. \tag{8.101}$$

Equation (8.101) is called the **Euler equation**. Equations (8.84) and (8.101) represent four equations for ρ, $\mathbf{v}$, and p. Supplemented by equations of state and an energy conservation equation, a complete set of equations can be found. For inviscid fluids the discussion now follows the one-dimensional analysis presented in Section 8.2.

Equation (8.100) suggests that in the most general case the stress tensor takes the form

$$\sigma_{ij} = -p\delta_{ij} + \tau_{ij}, \tag{8.102}$$

where p is identified with the thermodynamic pressure and τ_{ij} define the components of the **viscous stress tensor**. These components represent effects caused by viscosity, for example, where adjacent fluid elements undergo shear stresses across surfaces separating them.

As in the one-dimensional case, we postulate a **balance of energy law** of the form

$$\frac{d}{dt} \int_{\Omega_t} \left(\frac{1}{2} \rho \mathbf{v} \cdot \mathbf{v} + \rho e \right) dx = \int_{\Omega_t} \rho \mathbf{f} \cdot \mathbf{v} \, dx + \int_{\partial \Omega_t} \sigma \cdot \mathbf{v} \, dA - \int_{\partial \Omega_t} \mathbf{q} \cdot \mathbf{n} \, dA, \quad (8.103)$$

where Ω_t is an arbitrary material region with exterior unit normal $\mathbf{n}$ and $\mathbf{q}$ represents the heat flux density. In words, the time rate of change of the kinetic plus internal energy in Ω_t equals the rate work is done by the body forces $\mathbf{f}$, plus the rate work is done by the surface stresses σ, plus the rate heat flows into the region. A differential form of the energy balance law can be easily obtained by appealing to the arbitrariness of Ω_t after an application of the divergence theorem and Cauchy's theorem. We record as follows.

Theorem 8.18

If the functions are sufficiently smooth, equation (8.103) implies

$$\rho \frac{D}{Dt} \left(\frac{1}{2} \mathbf{v} \cdot \mathbf{v} + e \right) = \rho \mathbf{f} \cdot \mathbf{v} + \frac{\partial}{\partial x_j} (\sigma_{ji} v_i) - \nabla \cdot \mathbf{q}. \quad (8.104)$$

The proof is left as an exercise. We also have the following theorem.

Theorem 8.19

The change in internal energy is given by

$$\rho \frac{De}{Dt} = -\nabla \cdot \mathbf{q} + \sigma_{ji} \frac{\partial v_i}{\partial x_j}. \quad (8.105)$$

Proof

Multiplying the momentum balance law (8.97) by v_i and summing over $i = 1, 2, 3$ gives

$$\frac{\rho}{2} \frac{D}{Dt} (\mathbf{v} \cdot \mathbf{v}) = \rho f_i v_i + \frac{\partial}{\partial x_j} (\sigma_{ji} v_i) - \sigma_{ji} \frac{\partial v_i}{\partial x_j}.$$

$\square$

Subtracting from (8.104) yields (8.105) and hence the result.

From the combined form of the first and second laws of thermodynamics,

$$\theta \frac{Ds}{Dt} = \frac{De}{Dt} + p \frac{D(1/\rho)}{Dt}, \tag{8.106}$$

we can obtain an expression for how the entropy of a fluid particle changes. Combining (8.105) and (8.106) and using $D\rho^{-1}/Dt = -\rho^{-2}D\rho/Dt = \rho^{-1}\nabla \cdot \mathbf{v}$, we get the following theorem.

Theorem 8.20

Energy balance may be written

$$\rho\theta \frac{Ds}{Dt} = p\nabla \cdot \mathbf{v} - \nabla \cdot \mathbf{q} + \sigma_{ji} \frac{\partial v_i}{\partial x_j}. \tag{8.107}$$

The term $-\nabla \cdot \mathbf{q}$ represents the heat flow, and the sum $\Psi \equiv p\nabla \cdot \mathbf{v} + \sigma_{ji}\partial v_i/\partial x_j = \tau_{ji}\partial v_i/\partial x_j$ represents the heat generation due to deformation. Thus we may write

$$\rho \frac{Ds}{Dt} = -\frac{1}{\theta}\nabla \cdot \mathbf{q} + \frac{1}{\theta}\Psi. \tag{8.108}$$

The function Ψ is called the **dissipation function,** and it is the rate per unit volume that mechanical energy is dissipated into heat. Thermodynamics also requires that the entropy increase equal or exceed the heat added divided by the absolute temperature. Thus we *postulate* the inequality

$$\frac{d}{dt} \int_{\Omega_t} \rho s \, d\mathbf{x} \geq - \int_{\partial \Omega_t} \frac{1}{\theta} \mathbf{q} \cdot \mathbf{n} \, dA.$$

The differential form of this axiom is called the **Clausius–Duhem inequality:**

$$\rho \frac{Ds}{Dt} \geq -\nabla \cdot (\theta^{-1}\mathbf{q}). \tag{8.109}$$

Equations (8.108) and (8.109) may be combined to obtain

$$\theta^{-1}\Psi - \theta^{-2}\mathbf{q} \cdot \nabla \theta \geq 0. \tag{8.110}$$

Sufficient conditions for (8.110) are

$$\Psi \geq 0, \qquad -\mathbf{q} \cdot \nabla \theta \geq 0.$$

That is, deformation does not convert heat into mechanical energy, and heat flows against the temperature gradient.

To proceed further, some assumption is required regarding the form of the viscous stress tensor τ_{ij}. A **Newtonian fluid** is one in which there is a linear dependence of τ_{ij} on the **rate of deformation** $D_{ij} \equiv \frac{1}{2}(\partial v_i/\partial x_j + \partial v_j/\partial x_i)$, that is,

$$\tau_{ij} = C_{ijrs}D_{rs},$$

where the coefficients C_{ijrs} may depend on the local thermodynamic states θ and ρ. Arguments of symmetry and invariance of the stress tensor under translations and rotations lead to (see, e.g., Segel and Handelman (1977))

$$\tau_{ij} = 2\mu D_{ij} + \lambda \delta_{ij} \nabla \cdot \mathbf{v},$$

where μ and λ are coefficients of viscosity. Generally μ and λ may depend on temperature or density, but here we assume they are constant. The previous assumption forces the stress tensor to have the form

$$\sigma_{ij} = (-p + \lambda \nabla \cdot \mathbf{v})\delta_{ij} + 2\mu D_{ij},$$

and therefore the momentum law (8.97) becomes

$$\rho \frac{Dv_i}{Dt} = \rho f_i - \frac{\partial}{\partial x_i}(p - (\lambda + \mu)\nabla \cdot \mathbf{v}) + \mu \nabla^2 v_i \qquad (8.111)$$

for $i = 1, 2, 3$. In the incompressible case where $\nabla \cdot \mathbf{v} = 0$, (8.111) reduces in vector form to

$$\rho \frac{D\mathbf{v}}{Dt} = \rho \mathbf{f} - \nabla p + \mu \nabla^2 \mathbf{v}, \qquad (8.112)$$

which are the **Navier–Stokes equations**. Exact solutions of (8.112) in special cases are given in the exercises.

For viscous flows it is generally assumed that the fluid adheres to a rigid boundary. Thus we assume the *adherence boundary condition*

$$\lim_{\mathbf{x} \to \mathbf{x}_b} \mathbf{v}(\mathbf{x}) = \mathbf{v}(\mathbf{x}_b),$$

where $\mathbf{x}_b$ is a point on a rigid boundary and $\mathbf{v}(\mathbf{x}_b)$ is the known velocity of the boundary.

EXERCISES

1. Prove the Euler expansion when $n = 2$. Hint: Recall the rule for differentiating a determinant:

$$\frac{d}{dt} \begin{vmatrix} A & B \\ C & D \end{vmatrix} = \begin{vmatrix} \frac{dA}{dt} & B \\ \frac{dC}{dt} & D \end{vmatrix} + \begin{vmatrix} A & \frac{dB}{dt} \\ C & \frac{dD}{dt} \end{vmatrix}.$$

2. Let $\Delta(\mathbf{h}, t)$ be the Lagrangian density defined by $\Delta(\mathbf{h}, t) = \rho(\mathbf{x}(\mathbf{h}, t), t)$. Show that

$$J(\mathbf{h}, t) = \frac{\Delta(\mathbf{h}, 0)}{\Delta(\mathbf{h}, t)}.$$

3. Derive the Lagrangian form of the mass conservation law

$$\Delta_t + \Delta(\nabla \cdot \mathbf{v}|_\mathbf{h}) = 0.$$

4. Prove that if mass conservation holds, then

$$\frac{d}{dt} \int_{\Omega_t} \rho g \, d\mathbf{x} = \int_{\Omega_t} \rho \frac{Dg}{Dt} \, d\mathbf{x}$$

for any continuously differentiable function g. Show that the scalar function g can be replaced by a vector function $\mathbf{g}$.

5. If $\sigma(\mathbf{x}, t; \mathbf{n}) = -p(\mathbf{x}, t)\mathbf{n}$, prove that $\sigma_{ij} = -p(\mathbf{x}, t)\delta_{ij}$.

6. A two-dimensional fluid motion is given by

$$x_1 = h_1 e^t, \quad x_2 = h_2 e^{-t}, \quad t > 0.$$

a) Find $\mathbf{V}$ and $\mathbf{v}$.

b) Find the streamlines and show they coincide with the particle paths.

c) If $\rho(\mathbf{x}, t) = x_1 x_2$, show that the motion is incompressible.

7. The Eulerian velocity of a two-dimensional fluid flow is

$$\mathbf{v} = \langle t, x_1 + 1 \rangle, \quad t > 0.$$

Find the streamlines passing through the fixed point (a_0, b_0) and find the particle path of the particle $(h_1, h_2) = (a_0, b_0)$. Does the particle path ever coincide with a streamline at any time t_0?

8. A steady flow in two dimensions is defined by

$$\mathbf{v} = \langle 2x_1 + 3x_2, x_1 - x_2 \rangle, \quad t > 0.$$

Find the particle paths. Compute the acceleration measured by an observer located at the fixed point $(3, 4)$.

9. A flow is called **potential** on a region D if there exists a continuously differentiable function $\phi(t, \mathbf{x})$ on D for which $\mathbf{v} = \nabla \phi$, and a flow is called **irrotational** on D if $\nabla \times \mathbf{v} = \mathbf{0}$.

a) Prove that a potential flow is irrotational.

b) Show that an irrotational flow is not always potential by considering

$$\mathbf{v} = \left\langle \frac{-y}{x^2 + y^2}, \frac{x}{x^2 + y^2} \right\rangle \quad \text{on } \mathbb{R}^2 - \{0\}.$$

c) Show that if D is simply connected then irrotational flows are potential.

10. a) Show that
$$(\mathbf{v} \cdot \nabla)\mathbf{v} = \tfrac{1}{2}\nabla|\mathbf{v}|^2 + \nabla \times (\mathbf{v} \times \mathbf{v}).$$

b) Prove *Bernoulli's theorem*: For a potential flow of an inviscid fluid, suppose there exists a function $\psi(t,\mathbf{x})$ for which $\mathbf{f} = -\nabla\psi$. Then for any path C connecting two points $\mathbf{a}$ in $\mathbf{b}$ in a region where $\mathbf{v} = \nabla\phi$,
$$\left(\phi_t + \frac{1}{2}|\mathbf{v}|^2 + \psi\right)\bigg|_a^b + \int_C \frac{dp}{\rho} = 0.$$

(Hint: Integrate the equation of motion and use (a).)

11. In a region D where a flow is steady, irrotational, and incompressible with constant density ρ, prove that
$$\frac{1}{2}|\mathbf{v}|^2 + \frac{1}{\rho}p + \psi = \text{constant},$$

where $\mathbf{f} = -\nabla\psi$.

12. For a fluid motion with $\sigma_{ij} = \sigma_{ji}$ prove that
$$\frac{dK}{dt} = \int_{\Omega_t} \rho\mathbf{f} \cdot \mathbf{v}\, d\mathbf{x} + \int_{\partial\Omega_t} \sigma \cdot \mathbf{v}\, dA - \int_{\Omega_t} \sigma_{ij} D_{ij}\, d\mathbf{x}$$

where $K(\Omega_t) \equiv \tfrac{1}{2}\int_{\Omega_t} \rho\mathbf{v} \cdot \mathbf{v}\, dx$ is the kinetic energy of Ω_t. This result is the *energy transport theorem*.

13. For an ideal fluid prove that $K(\Omega) = \text{constant}$, where Ω is a fixed region in $\mathbb{R}^3$ and $\mathbf{v}$ is parallel to $\partial\Omega$ (see Exercise 12).

14. The alternating symbol ε_{ijk} is defined by
$$\varepsilon_{ijk} = \begin{cases} 0, & \text{if any two indices are equal,} \\ 1, & \text{if } (ijk) \text{ is an even permutation of } 123, \\ -1, & \text{if } (ijk) \text{ is an odd permutation of } 123. \end{cases}$$

A permutation (ijk) is an even (odd) permutation of (123) if it takes an even (odd) number of switches of adjacent elements to get it to the form (123). For example,
$$321 \rightarrow 231 \rightarrow 213 \rightarrow 123,$$

so (321) is an odd permutation of (123). Show that if $\mathbf{a} = \langle a_1, a_2, a_3 \rangle$ and $\mathbf{b} = \langle b_1, b_2, b_3 \rangle$, then the kth component of the cross product $\mathbf{a} \times \mathbf{b}$ is given by
$$(\mathbf{a} \times \mathbf{b})_k = \varepsilon_{ijk} a_i b_j, \qquad \text{sum on } i, j = 1, 2, 3.$$

15. (*Conservation of Angular Momentum*) A particle of mass m moving in $\mathbb{R}$ with velocity vector $\mathbf{v}$ has linear momentum $\mathbf{p} = m\mathbf{v}$ and angular momentum about the origin $\mathbf{0}$ given by $\mathbf{L} = \mathbf{x} \times \mathbf{p}$, where $\mathbf{x}$ is its position vector. If $\mathbf{F}$ is a force on the particle, then $\mathbf{F} = d\mathbf{p}/dt$ by Newton's second law. Show that $\mathbf{N} = d\mathbf{L}/dt$, where $\mathbf{N} = \mathbf{x} \times \mathbf{F}$ is the torque (or moment of force) about $\mathbf{0}$. For a continuum argue that one should postulate

$$\frac{d}{dt} \int_{\Omega_t} (\mathbf{x} \times \rho \mathbf{v}) \, d\mathbf{x} = \int_{\Omega_t} (\mathbf{x} \times \rho \mathbf{f}) \, d\mathbf{x} + \int_{\partial \Omega_t} (\mathbf{x} \times \sigma) \, dA, \qquad (8.113)$$

where Ω_t is any material region. Write (8.113) in component form using the alternating symbol ε_{ijk}. Note that (8.113) does not account for any internal angular momentum of the fluid particles that would occur in a fluid where the particles are rotating rods; such fluids are called *polar*.

16. Prove that (8.113) of Exercise 15 is equivalent to the symmetry of the stress tensor σ_{ij} (i.e., $\sigma_{ij} = \sigma_{ji}$). *Hint:*

$$\frac{d}{dt} \int_{\Omega_t} \varepsilon_{ijk} x_i \rho v_j \, d\mathbf{x} = \int_{\Omega_t} \varepsilon^{ijk} x_i \rho \, \frac{Dv_j}{Dt} \, d\mathbf{x}.$$

Use the equation of motion and

$$\frac{\partial}{\partial x_l} (x_i \sigma_{lj}) = x_i \, \frac{\partial}{\partial x_l} \sigma_{lj} + \sigma_{ij},$$

followed by an application of the divergence theorem. This theorem shows that balance of angular momentum is equivalent to symmetry of the stress tensor. Thus, of the nine components of σ_{ij} only six are independent.

17. Let $\Gamma \subseteq \Omega_0$ be a simple closed curve and let $\Gamma_t = \phi_t(\Gamma)$, where ϕ_t is an isentropic fluid motion. The *circulation* about Γ_t is defined by the line integral $C_{\Gamma_t} \equiv \int_{\Gamma_t} \mathbf{v} \cdot d\mathbf{l}$. Prove *Kelvin's theorem*, which asserts that C_{Γ_t} is a constant in time.

18. The *vorticity* ω of a flow is defined by $\omega = \nabla \times \mathbf{v}$. Determine the vorticity of the following flows and sketch the streamlines in each case.

    a) $\mathbf{v} = v_0 \exp(-x_2^2)\mathbf{j}$.

    b) $\mathbf{v} = v_0 \exp(-x_2^2)\mathbf{j}$.

    c) $\mathbf{v} = \omega_0(-x_2\mathbf{i} + x_1\mathbf{j})$.

19. If ρ is constant and $\mathbf{f} = \nabla \psi$ for some ψ, prove that (8.112) implies the *vorticity equation*

$$\frac{D\omega}{Dt} = (\omega \cdot \mathbf{grad})\mathbf{v} + \frac{\mu}{\rho} \, \Delta\omega.$$

20. Consider an incompressible viscous flow of constant density ρ_0 under the influence of no body forces governed by the Navier–Stokes equation (8.112).

   a) If l, ν/l^2, U, and $\rho_0 \nu U/l$, where $v \equiv \mu/\rho_0$ are length, time, velocity, and pressure scales, show that in dimensionless form the governing equations can be written

$$\nabla \cdot \mathbf{v} = 0, \qquad \mathbf{v}_t + \mathrm{Re}(\mathbf{v} \cdot \nabla)\mathbf{v} = -\nabla p + \Delta \mathbf{v},$$

   where $\mathrm{Re} \equiv Ul/\nu$ is a constant called the *Reynolds number*. The constant ν is the kinematic viscosity.

   b) For flows with small Reynolds number, that is, $\mathrm{Re} \ll 1$, show that

$$\Delta p = 0, \qquad \frac{\partial}{\partial t}(\Delta \mathbf{v}) = \Delta(\Delta \mathbf{v}).$$

   (Hint: $\nabla \times (\nabla \times \mathbf{v}) = \nabla(\nabla \cdot \mathbf{v}) - \Delta \mathbf{v}$.)

21. (*Plane Couette flow*) An incompressible viscous fluid of constant density under no body forces is confined to lie between two infinite flat plates at $x_3 = 0$ and $x_3 = d$. The lower plate is stationary and the upper plate is moved at constant velocity U in the x_1 direction. Show that the fluid motion is given by

$$v_1 = \frac{U}{d}\,x_3, \quad v_2 = v_3 = 0, \quad p = p_0 = \text{constant}.$$

Show that the stress on the lower plate is $\mu U/d$.

22. (*Plane Poiseuille flow*) An incompressible viscous fluid of constant density under no body forces is confined to lie between two stationary infinite planes $x_3 = -d$ and $x_3 = d$. A flow is forced by a constant pressure gradient $\nabla p = \langle -C, 0, 0 \rangle$, $C > 0$. Show that the fluid velocity is given by

$$v_1 = \frac{C}{2\mu}\,(d^2 - x_3^2), \quad v_2 = v_3 = 0.$$

Show that the mass flow per unit width in the x_2 direction is given by $2Cd^3/3\mu$.

REFERENCES AND NOTES

Continuum mechanics has long been a central part of applied mathematics, and it is an entity within itself. As such, it is not possible to reference a large selection of texts or articles. Here we mention a few elementary introductions that relate to our material and to various extensions. A clean, readable, elementary introduction is Chorin & Marsden (1993). A thorough presentation is given in Segel & Handelman (1987). Classic texts for gas dynamics and shock waves include Courant & Friedrichs (1948) and Whitham (1974). Applications to detonation and combustion, and to porous media, are given in the author's texts (Logan 1994, 2001).

Chorin, A. J. & Marsden, J. 1993. *A Mathematical Introduction to Fluid Mechanics*, 3rd ed., Springer-Verlag, New York.

Courant, R. & K. O. Friedrichs, K. O. 1948. *Supersonic Flow and Shock Waves*, Wiley-Interscience, New York (reprinted by Springer-Verlag, New York, 1976).

Logan, J. D. 1994. An Introduction to Nonlinear Partial Differential Equations, Wiley-Interscience, New York.

Logan, J. D. 2001. *Transport Modeling in Hydrogeochemical Systems*, Springer-Verlag, New York.

Segel, L. A. & Handelman, G. H. 1977. *Mathematics Applied to Continuum Mechanics*, Macmillan, New York (reprinted by Dover Publications, New York 1987).

Whitham, G. B., 1974. *Linear and Nonlinear Waves*, Wiley-Interscience, New York.

Index